Textbook of Bacteriology

Textbook of Bacteriology

Edited by Ricky Parks

SYRAWOOD
PUBLISHING HOUSE

New York

Published by Syrawood Publishing House,
750 Third Avenue, 9th Floor,
New York, NY 10017, USA
www.syrawoodpublishinghouse.com

Textbook of Bacteriology
Edited by Ricky Parks

International Standard Book Number: 978-1-68286-517-0 (Hardback)

Cataloging-in-Publication Data

Textbook of bacteriology / edited by Ricky Parks.
    p. cm.
Includes bibliographical references and index.
ISBN 978-1-68286-517-0
1. Bacteriology. 2. Molecular biology. 3. Microbiology.
I. Parks, Ricky.
QR46 .T49 2018
576.3--dc23

TABLE OF CONTENTS

PREFACE

Bacteriology studies bacteria and bacteriologic etiology. This book on bacteriology discusses topics related to bacterial cultures as well as microbial resistance and evolution. The study of bacteria is important for the medical and industrial processes, ranging from parasitology to fermentation techniques in the food processing industry. Discovering potent bacteriostatic agents are another important function of this field. This book provides significant information of this discipline to help develop a good understanding of bacteriology and its related fields. With its detailed analyses and data, the book will prove immensely beneficial to professionals and students involved in this area at various levels.

This book is a comprehensive compilation of works of different researchers from varied parts of the world. It includes valuable experiences of the researchers with the sole objective of providing the readers (learners) with a proper knowledge of the concerned field. This book will be beneficial in evoking inspiration and enhancing the knowledge of the interested readers.

In the end, I would like to extend my heartiest thanks to the authors who worked with great determination on their chapters. I also appreciate the publisher's support in the course of the book. I would also like to deeply acknowledge my family who stood by me as a source of inspiration during the project.

Editor

Whole Genome Sequencing and Analysis of Plant Growth Promoting Bacteria Isolated from the Rhizosphere of Plantation Crops Coconut, Cocoa and Arecanut

Alka Gupta[1,9], Murali Gopal[1*,9], George V. Thomas[1], Vinu Manikandan[2], John Gajewski[3], George Thomas[4], Somasekar Seshagiri[5], Stephan C. Schuster[3,6], Preeti Rajesh[2*,9], Ravi Gupta[2*,9]

1 Central Plantation Crops Research Institute, Kasaragod, Kerala, India, 2 SciGenom Labs Pvt. Ltd., Plot 43A, SDF 3rd Floor CSEZ, Kakkanad, Cochin, Kerala, India, 3 Center for Comparative Genomics and Bioinformatics, Pennsylvania State University, 310 Wartik Lab, University Park, Pennsylvania, United States of America, 4 SciGenom Research Foundation, Cochin, Kerala, India, 5 Department of Molecular Biology, Genentech Inc., South San Francisco, California, United States of America, 6 Singapore Centre on Environmental Life Sciences Engineering, Nanyang Technical University, Singapore, Singapore

Abstract

Coconut, cocoa and arecanut are commercial plantation crops that play a vital role in the Indian economy while sustaining the livelihood of more than 10 million Indians. According to 2012 Food and Agricultural organization's report, India is the third largest producer of coconut and it dominates the production of arecanut worldwide. In this study, three Plant Growth Promoting Rhizobacteria (PGPR) from coconut (CPCRI-1), cocoa (CPCRI-2) and arecanut (CPCRI-3) characterized for the PGP activities have been sequenced. The draft genome sizes were 4.7 Mb (56% GC), 5.9 Mb (63.6% GC) and 5.1 Mb (54.8% GB) for CPCRI-1, CPCRI-2, CPCRI-3, respectively. These genomes encoded 4056 (CPCRI-1), 4637 (CPCRI-2) and 4286 (CPCRI-3) protein-coding genes. Phylogenetic analysis revealed that both CPCRI-1 and CPCRI-3 belonged to *Enterobacteriaceae* family, while, CPCRI-2 was a *Pseudomonadaceae* family member. Functional annotation of the genes predicted that all three bacteria encoded genes needed for mineral phosphate solubilization, siderophores, acetoin, butanediol, 1-aminocyclopropane-1-carboxylate (ACC) deaminase, chitinase, phenazine, 4-hydroxybenzoate, trehalose and quorum sensing molecules supportive of the plant growth promoting traits observed in the course of their isolation and characterization. Additionally, in all the three CPCRI PGPRs, we identified genes involved in synthesis of hydrogen sulfide (H_2S), which recently has been proposed to aid plant growth. The PGPRs also carried genes for central carbohydrate metabolism indicating that the bacteria can efficiently utilize the root exudates and other organic materials as energy source. Genes for production of peroxidases, catalases and superoxide dismutases that confer resistance to oxidative stresses in plants were identified. Besides these, genes for heat shock tolerance, cold shock tolerance and glycine-betaine production that enable bacteria to survive abiotic stress were also identified.

Editor: Matteo Pellegrini, UCLA-DOE Institute for Genomics and Proteomics, United States of America

Funding: The study was internally funded by SciGenom Labs Pvt Ltd. The funder had no role in study design, data collection and analysis, decision to publish, or preparation of the manuscript.

Competing Interests: Some of the authors as noted are employees of SciGenom Pvt Ltd. S.S. is an employee of Genentech and holds shares in Roche. The study was internally funded by SciGenom Labs Pvt Ltd.

* Email: mgcpcri@yahoo.co.in (MG); preeti@scigenom.com (PR); ravig@scigenom.com (RG)

9 These authors contributed equally to this work.

Introduction

Plant rhizosphere harbors numerous bacteria capable of stimulating and aiding plant growth and are termed plant growth promoting rhizobacteria (PGPR) [1]. They exert their beneficial effects through direct or indirect mechanisms. The direct mechanisms include biofertilization, stimulation of root growth, rhizo-remediation and plant stress control [2]. Indirect mechanisms primarily involve biological control comprised of antibiosis, induction of systemic resistance and competition for nutrition and niches [2]. Owing to their diverse plant growth promoting capabilities, PGPRs have become the new inoculants for biofertilizer technology [3]. To improve the biofertilizer technology, understanding the molecular mechanisms of plant growth

promotion and biocontrol by rhizobacteria is important [4]. Identification of genes that contribute to the beneficial activity of rhizobacteria, besides adding to our understanding of the molecular mechanisms, will aid in developing better biofertilizers.

Next generation sequencing technologies (NGS) have enabled whole genome sequencing of bacteria and other organisms [5]. Systematic analysis of whole genome data has aided the understanding of the molecular genetics of many bacterial species [6]. Recently, NGS has been employed to study genomes of several PGPRs, mainly isolated from crop species such as wheat [7], *Miscanthus* [8], pepper [9]. PGPRs from soil have also been sequenced directly [10]. However, thus far, genome sequences of

PGPRs isolated from plantation crops, particularly from coconut, cocoa and arecanut, have not been reported.

Coconut (*Cocos nucifera* L.), cocoa (*Theobroma cacao* L.) and arecanut (*Areca catechu* L.) are three important plantation crops grown in 2.2 million hectares in India. These plantation crops harbor plant-beneficial microorganisms in their rhizospheres [11–15] and some of these have been utilized for growth promotion [16–20] as they offer an opportunity for ecologically safe nutrient management [21]. In this study, we have performed deep sequencing analysis of three PGPRs, CPCRI-1, CPCRI-2 and CPCRI-3, isolated from coconut [16], cocoa [20] and arecanut [22] rhizosphere, respectively.

Results

PGPR strains

Soil samples collected from rhizospheres of coconut, cocoa and arecanut grown in different agro-ecological zones of India were used to isolate 1512 morphologically distinct heterotrophic bacteria [13,15,18,22]. The details of places from which the soil samples were collected, soil types and their pH along with isolation media are given in Table S1. The isolates were screened *in vitro* for several important plant growth promoting functions (Table 1). The isolates that gave best results in the *in vitro* testing were then studied for plant growth promotion using rice and cowpea seeds in environmental growth chamber and green house conditions [18,19]. They were also then tested on coconut [16], cocoa [20] and arecanut [22] seedlings grown in polybags.

Three PGPRs designated CPCRI-1 (RNF-267 from coconut) [16], CPCRI-2 (KGSF-20 from cocoa) [15,20] and CPCRI-3 (KtRA5-88 from arecanut) [22] were selected for further studies. All the three isolates had rod shape morphology and were negative for Gram's staining. CPCRI-1 showed good phosphate solubilizing capacity and promoted growth of coconut seedlings [16]. CPCRI-2 was capable of promoting growth of cocoa seedlings [20]. CPCRI-3, isolated from arecanut rhizosphere, was able to tolerate low pH and possessed plant growth promoting attributes [22]. The plant growth promotion traits of the three isolates are summarized in Table 1. The morphological, biochemical and physiological attributes of the three PGPRs are summarized in

Table S2. Given the beneficial attributes of the three PGPRs, we chose to characterize them further at the genomic level.

Whole genome shotgun sequencing and assembly

We performed shotgun multiplexed sequencing of the genomes of CPCRI-1, CPCRI-2 and CPCRI-3 using the 454-sequencing platform. We obtained >300,000 quality-filtered reads each for CPCRI-1 and CPCRI-3 with an average read length of 465 bp and 421 bp, respectively. For CPCRI-2, we obtained >150,000 quality filtered reads with an average read length of 408 bp (Table 2). We assembled the sequencing reads for each of the three genomes using GS *de novo* assembler version 2.6 [23]. Of the total reads obtained ~90% were assembled into contigs corresponding to each of the genomes. For CPCRI-1, 350,636 reads were assembled into 39 contigs (N50 of 242,562 bp; longest contig length of 730,806 bp; mean contig length of 114,755 bp) for a total of 4,475,442 nucleotides at ~30× coverage. The 144,293 reads obtained for CPCRI-2, were assembled into 101 contigs (N50 of 89,849 bp; longest contig length of 282,342 bp; mean contig length of 52,329 bp) for a total of 5,285,206 nucleotides at ~12× coverage. In the case of CPCRI-3, the 313,271 reads obtained were assembled into 47 contigs (N50 of 161,752 bp; longest contig length of 529,776 bp; mean contig length of 99,348 bp) for a total of 4,669,355 nucleotides at ~30× coverage (Table 2, Fig. S1).

The estimated genome size based on the sequence data was 4.7 Mb for CPCRI-1, 5.9 Mb for CPCRI-2 and 5.1 Mb for CPCRI-3. Phylogenetic analysis derived from comparison of 31 conserved housekeeping protein-coding genes [24] indicated that while CPCRI-1 and CPCRI-3 were members of the *Enterobacteriaceae* family, CPCRI-2 was a *Pseudomonadaceae* family member. Their estimated genome sizes are consistent with the sizes observed for other family members (Table S3 & S4). The GC content of the bacterial isolates was 56.0%, 63.6% and 54.8% for CPCRI-1, CPCRI-2 and CPCRI-3, respectively (Table 2).

Gene prediction and annotation

Glimmer-MG [25] predicted 4056, 4637 and 4286 protein-coding genes in CPCRI-1, CPCRI-2 and CPCRI-3, respectively (Table 3 & Table S5). Consistent with this the average predicted

Table 1. Biological and plant growth promotional properties of PGPR isolates.

S. No.	Attributes	CPCRI-1 (RNF 267) [16]	CPCRI-2 (KGSF20) [20]	CPCRI-3 (KtRA5-88) [22]
1.	pH tolerance levels	4.0 to 9.0	4.0 to 8.0	2.5 to 7.0
2.	Optimum pH for growth	5.0–6.0	7.0	4.5
3.	NaCl tolerance	upto 8%	upto 4%	upto 2%
4.	Temperature tolerance	15 to 40°C	15 to 40°C	30 to 40°C
5.	ACC deaminase activity	562 µmol α-ketobutyrate h^{-1} mg protein^{-1}	199 µmol α-ketobutyrate h^{-1} mg protein^{-1}	474 µmol α-ketobutyrate h^{-1} mg protein^{-1}
6.	Phosphate solubilization	217 µg ml^{-1}	99.8 µg ml^{-1}	82 µg ml^{-1}
7.	IAA production	2.4 µg ml^{-1}	1.5 µg ml^{-1}	1.2 µg ml^{-1}
8.	Growth on N-free agar medium	–	Growth observed	–
9.	Chitinase activity	–	–	–
10.	β-1,3-glucanase activity	20 µg glucose min^{-1} mg protein^{-1}	7.8 µg glucose min^{-1} mg protein^{-1}	2.4 µg glucose min^{-1} mg protein^{-1}
11.	Salicylic acid production	–	6.1 µg ml^{-1}	–
12.	Siderophore production	–	6 mm	11 mm

– indicates no growth/production/activity.

Table 2. Genome assembly statistics.

	CPCRI-1	CPCRI-2	CPCRI-3
# of reads	361,881	158,071	326,453
Average read length (bp)	465	408	421
# of bases (bp)	168,202,148	64,520,437	137,575,981
# of reads assembled	350,636 (96.9%)	144,293 (91.3%)	313,271 (96%)
Number of contigs (> = 500 bp)	39	101	47
N50 (bp)	242,562	89,849	161,752
Average contig length (bp)	114,755	52,329	99,348
Total length of contigs (bp)	4,475,442	5,285,206	4,669,355
GC Content (%)	56.0	63.6	54.8
Average contig size (bp)	114,755	52,329	99,348
Longest contig size (bp)	730,806	282,342	529,776
Q40 plus bases (%)	99.97	99.77	99.96

protein coding genes size in CPCRI-1, CPCRI-2 and CPCRI-3 was found to be 972 bp, 981 bp and 951 bp, respectively. In bacteria, a robust correlation exists between the genome size and the numbers of genes it encodes [26]. A comparison of 26 published complete genomes in the *Enterobacteriaceae* family revealed an average genome size of 4.8 Mb and the coded for an average of 4655 proteins (Table S3). Our estimate of 4056 genes in CPCRI-1 and 4286 genes in CPCRI-3 is consistent with this observation. The *Pseudomonas* genus had an average genome size of 6.0 Mb and encoded an average of 5366 protein coding genes (Table S4). Though CPCRI-2 had a genome size of 5.9 Mb, and coded for about 4637 proteins, this number is similar to those observed in *Pseudomonas putida* BIRD-1, a PGPR [10].

The average GC content of the protein coding genes in CPCRI-1 (56.55%), CPCRI-2 (64.12%) and CPCRI-3 (55.4%) and their relation with the genomic GC content were estimated and are presented in Table 3, Table S3, S4. The GC distribution for all three positions in the codon for protein-coding genes (Fig. S2) showed that the average GC content was the highest for the third base position and lowest for second position within the codon. Though the overall trend was similar between the bacteria, the GC content for CPCRI-2 at the third base was ~85% compared to only 68% for CPCRI-1 and CPCRI-3. The codon usage

analysis (Fig. S3) showed that CTG that codes for Leu (L) is the most used codon in all three bacteria.

In addition to the protein coding genes, we predicted 74, 61, and 74 tRNA genes in CPCRI-1, CPCRI-1, and CPCRI-3, respectively (Table S6A, B, C). The predicted genes represented 20–21 different tRNAs corresponding to the universal codons. Our analysis additionally identified 5–6 different rRNA genes (Table 3 & Table S7).

To further understand the bacterial strains, the predicted protein-coding genes identified using Glimmer-MG were compared against the non-redundant (nr) NCBI protein database using BLASTX [27]. We found that a majority of the predicted protein-coding genes (98%) had a homologous protein sequence in the NCBI non-redundant (nr) protein database (Fig. 1). Among the genes with homologs, >90% of genes had a high confidence match (E-value$< = 1.0$ e^{-50}) and >93% had identity of at least 80% with a putative homolog (Fig. 1A, B). Interestingly, 31, 59 and 91 genes from CPCRI-1, 2 and 3, respectively, showed no significant identity to sequences in the NCBI database. Of the protein with no significant identity, we annotated protein domains for 15, 21 and 29 genes from CPCRI-1, 2 and 3, respectively, using InterPro [28] and CDD [29]. We further annotated all the

Table 3. Gene prediction and annotation summary.

	CPCRI-1	CPCRI-2	CPCRI-3
# of predicted protein-coding gene	4,056	4,637	4,286
Average GC content of Protein-coding genes (%)	56.55	64.12	55.4
Average gene length (bp)	972	981	951
# of tRNA genes	74	61	74
Average length of tRNA gene (bp)	76	75	76
# of rRNA genes	5	6	6
# of protein-coding genes with at least one significant BLASTX result	4,053	4,596	4,216
# of BLASTX protein hit present in UniProt	3,747	4,551	3,674
# of protein-coding genes assigned GO term	2,906	3,379	2,515
# of protein-coding genes assigned an InterPro id	3,466	4,097	3,063

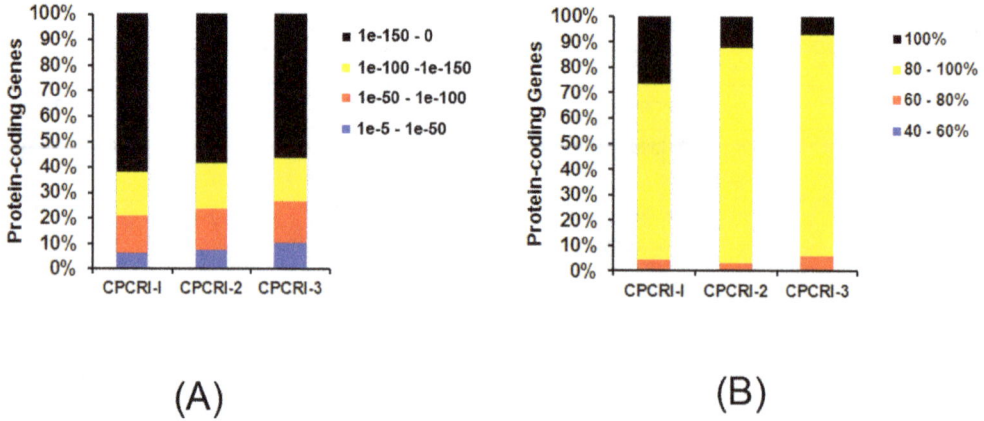

(A)

(B)

Figure 1. Protein coding genes. Stacked bar graph representing the percentage of predicted protein coding genes with significant matches (E-value$< = 1e10^{-5}$) (A) in the NCBI nr-protein database identified using BLASTX and (B) the proportion of proteins binned by percent identity measured by BLASTX alignment.

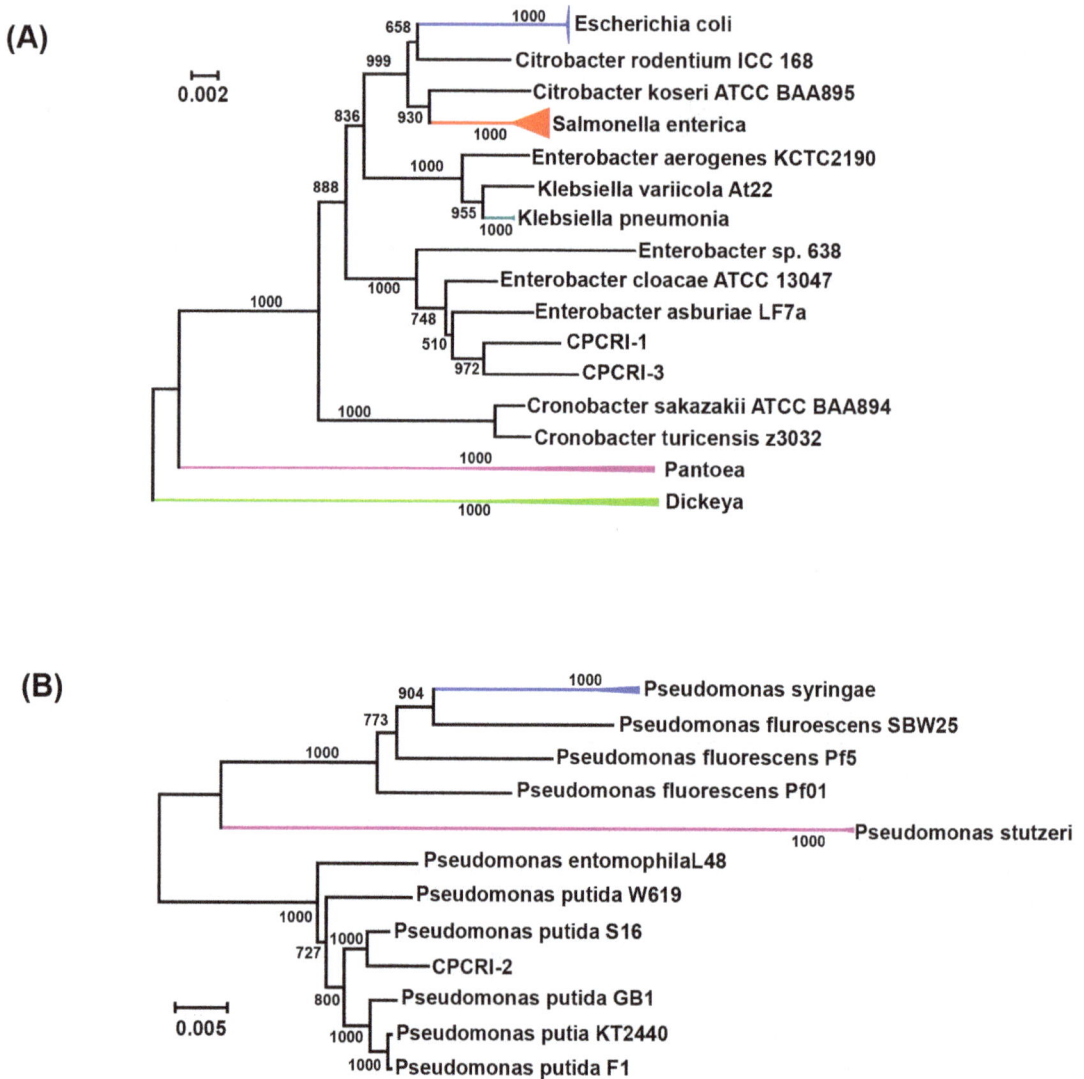

Figure 2. Phylogenetic tree. Using 31 conserved housekeeping protein-coding genes from (A) CPCRI-1 and CPCRI-3, (B) CPCRI-2, a phylogenetic tree was generated using AMPHORA2 [24,94] and ClustalW [95]. The colored branch/node represents node where multiple strains of the same species are collapsed into a single species for representation.

(A)

(B)

(C)

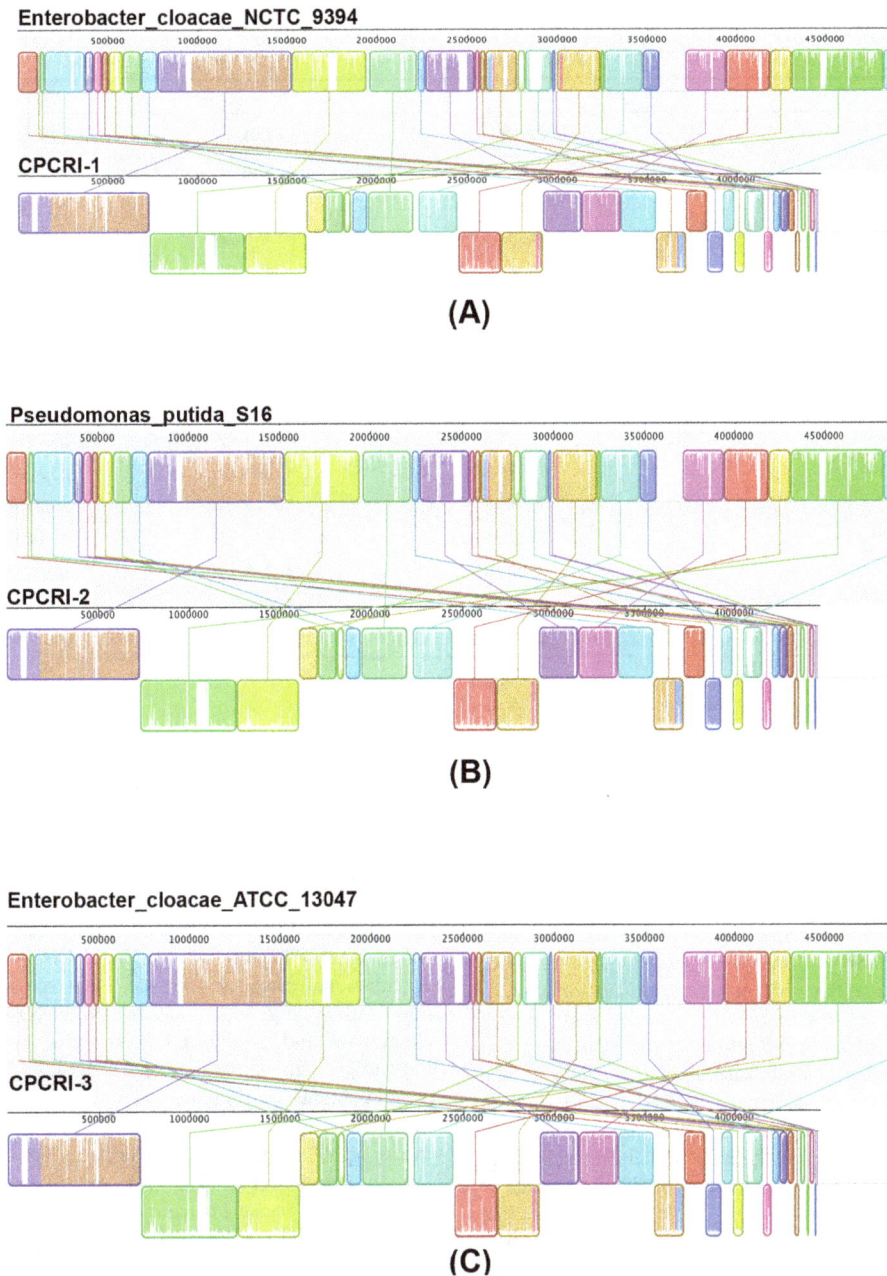

Figure 3. Genome comparison. Pairwise alignment of CPCRI-1, CPCRI-2, CPCRI-3 genome with *Enterobacter cloacae NCTC 9394*, *Pseudomonas putida S16* and *Enterobacter cloacae ATCC 13047*, respectively using the progressive Mauve aligner [34]. The colored blocks represent the homologous region between the genomes that are internally free from genomic rearrangement.

protein-coding genes with known protein domains using UniProt database (Table S8A, B, and C) [30].

Phylogeny analysis

The phylogenetic analysis performed using a set of 31 conserved housekeeping protein-coding genes [24] revealed that the CPCRI-1 and CPCRI-3 genomes were closely related to the *Enterobacter cloacae* group (Fig. 2A). The CPCRI-2 genome was found to be closely related to the *Pseudomonas putida* group. CPCRI-1 grouped with *Enterobacter asburiae* strain LF7a and *Enterobacter cloacae* ATCC 13047, both of which are members of the *Enterobacter cloacae* complex. CPCRI-3 clustered closely with

Enterobacter asburiae strain LF7a and *Enterobacter* sp. 638 [31], an endophyte of poplar trees. The closest relative to CPCRI-2 was *Pseudomonas putida* strain S16, a gram-negative soil bacterium with an ability to degrade aromatic and heterocyclic compounds, such as nicotine, benzoate, and phenylalanine [32]. Taxonomy based study using MEGAN4 [33] showed that CPCRI-1 and CPCRI-3 belong to *Enterobacteriacea* family and CPCRI-2 belong to *Pseudomonas* genus (Fig. S4, Result S1). Consistent with this, Biolog analysis indicated CPCRI-2 to be *Pseudomonas putida* [20]. Although Biolog analysis at low confidence level, indicated CPCRI-3 to be *Pantoea agglomerans* [22], a member of

Table 4. Pairwise comparison of CPCRI-1, CPCRI-2, and CPCRI-3 genomes against bacteria genomes using progressive Mauve aligner [34].

Bacteria	CPCRI-1	CPCRI-2	CPCRI-3
Enterobacter asburiae LF7a	84.94/83.15	65.25/14.5	81.26/78.61
Enterobacter cloacae ATCC 13047	86.20/87.75	65.88/14.67	**81.58/78.56***
Enterobacter cloacae EcWSU1	85.86/86.70	65.38/14.88	81.4/77.95
Enterobacter cloacae NCTC 9394	**92.69/89.02***	65.98/14.47	81.4/77.38
Enterobacter cloacae SCF1	79.12/72.00	66.85/12.69	78.3/66.83
Enterobacter sp. 638	82.0/78.73	64.79/13.36	80.57/75.66
Pseudomonas putida BIRD-1	65.60/18.09	87.84/81.05	65.19/17.06
Pseudomonas putida F1	65.50/18.94	88.05/83.70	65.2/17.04
Pseudomonas putida GB-1	65.47/18.7	87.7/82.97	65.16/17.57
Pseudomonas putida KT2440	65.57/18.14	87.69/83.0	65.21/17.0
Pseudomonas putida S16	65.91/18.01	**89.88/85.0***	65.31/16.91
Pseudomonas putida W619	65.50/18.46	84.67/78.64	65.14/16.84
Pseudomonas aeruginosa LESB58	65.87/15.85	75.35/59.15	65.56/15.53
Pseudomonas fluorescens F113	65.43/18.89	76.3/66.19	64.93/17.05
Pseudomonas syringae pv tomato str DC3000	65.26/17.91	74.91/55.95	64.6/16.93

Percentage sequence similarity/genome coverage in the conserved block is shown.
*Bacteria showing highest similarity/genome coverage with sequenced bacteria CPCRI-1, CPCRI-2 and CPCRI-3.

the *Enterobacteriaceae*, the genome sequence of CPCRI-1 did not reveal similarity to any sequenced bacterium at the species level.

Pairwise genome comparison with existing bacterial genomes

We performed a pairwise genome comparison of our assembled bacterial genomes against 40 different bacteria using progressive Mauve aligner [34]. The bacterial groups identified for analysis included *Enterobacter*, *Escherichia coli*, *Pseudomonas putida*, *Citrobacter*, *Dickeya*, *Klebsiella*, *Pantoea*, *Salmonella*, *Shigella*, *Azotobacter*, *Bradyrhizobium*, *Mesorhizobium* and *Rhizobium*. The genome level comparison showed that CPCRI-1 had the highest similarity to *Enterobacter cloacae NCTC 9394i* (similarity score of 92.69%, coverage of 89.02%). In addition, CPCRI-2 was closest to *Pseudomonas putida S16* (similarity score of 89.88%, coverage of 85.0%), and CPCRI-3 to be most similar to *Enterobacter cloacae ATCC 13047* (similarity score of 81.58%, coverage of 78.56%; Table 4, Table S9 and Fig. 3). These results are consistent with the phylogenetic analysis findings that showed CPCRI-1 and CPCRI-3 belong to *Enterobacter cloacae* group and CPCRI-2 to the *Pseudomonas putida* group.

Functional analysis of the bacterial genome

We performed functional analysis of the annotated genomes using gene onotology (GO), SEED classification and KEGG pathways. The GO based classification of the genes revealed 2,200 to 3,000 (2,562 for CPCRI-1, 3,020 for CPCRI-2, and 2,226 for CPCRI-3) genes associated with at least one molecular function, 1,500–1,900 (1,836 for CPCRI-1, 1,918 for CPCRI-2, and 1,555 for CPCRI-3) genes associated with at least one biological process, and 1,200–1,500 (1,487 for CPCRI-1, 1,530 for CPCRI-2, and 1,226 for CPCRI-3) genes associated with at least one cellular component (Table S10). The Carbohydrate metabolism, chemotaxis, cell adhesion, cilium or flagellum-dependent related motility, response to stress, iron ion binding, oxidoreductase activity are among the top 20 GO biological processes and molecular

functions found in the bacteria (Fig. 4A, B). These terms are related to the functional class of genes that aid the plant growth.

The SEED based classification [35] analysis of the proteins, performed using MEGAN4 [33] assigned functional roles to the annotated genes that were then grouped into one or more subsystems. MEGAN4 classified 1998, 2202 and 2030 annotated genes from CPCRI-1, CPCRI-2 and CPCRI-3, respectively into 25 functional categories (Fig. 4C). A large number of genes fall into carbohydrate metabolism, stress response, motility and chemotaxis and metabolism of aromatic compounds that helps in plant growth. Annotation against KEGG pathway classified the protein-coding genes into six different pathway categories: metabolism, genetic information processing, environmental information processing, cellular processes, organismal systems and human diseases (Fig. 4D). The metabolism and environment information processing category pathways were highly represented in all three bacteria. Overall we found that many genes fall into the functional classes that support plant growth.

Plant growth promoting properties

In the genomic sequence of three PGPRs sequenced we identified genes that can be attributed to their ability to improve nutrient availability, suppress pathogenic fungi, resist oxidative stress, quorum sensing and ability to break down aromatic and toxic compounds and other abiotic stress (Table 5). The genomes of CPCRI-1, 2 and 3 possessed genes encoding glucose dehydrogenase activity while, CPCRI-1 and 2 carried the co-factor pyrrolo-quinolone quinine (*pqq*) gene cluster which is involved in solubilization of mineral phosphates fixed in soil particles [36]. Indole acetic acid (IAA) is an important hormone that helps in plant growth [37]. Although IAA production was observed in culture from the three PGPRs was low (Table 1), CPCRI-1 and CPCRI-3 genomes contained *ipd*C that codes for indolepyruvate decarboxylase, an enzyme that produces indole acetic acid from tryptophan [38]. In these genomes we also found some of the trp cluster (*trp*A, B, D, C, R) genes involved in

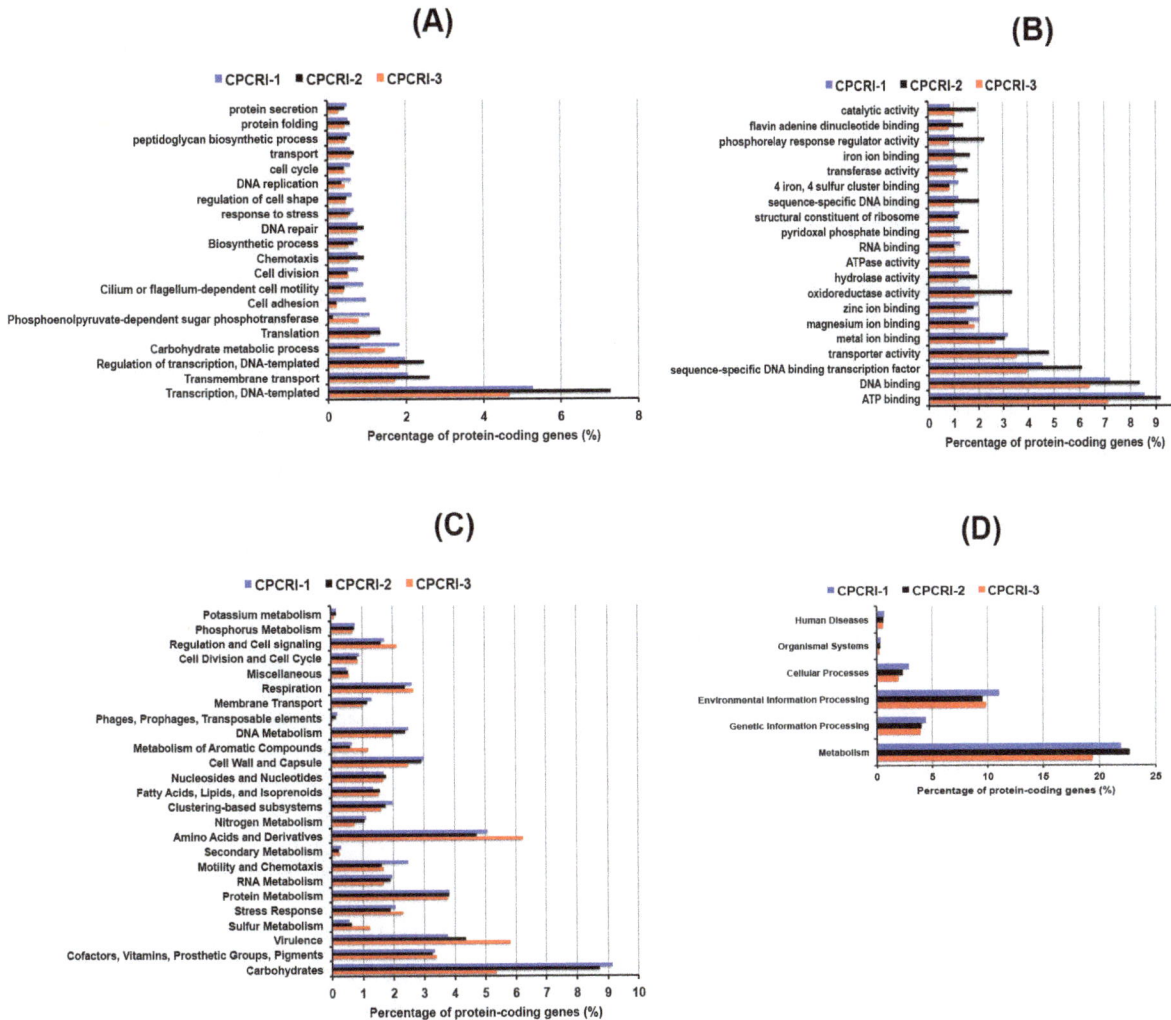

Figure 4. Functional annotation. The percentage of predicted coding genes of CPCRI PGPR genomes into (A) GO biological classes, (B) GO molecular classes, (C) SEED classification, (D) KEGG pathway classification.

tryptophan biosynthesis. These genes may play a role in synthesis of tryptophan used in multiple biological processes, including IAA biosynthesis. The 1-aminocyclopropane-1-carboxylate (ACC) de-aminase has been shown in symbiotic bacteria to function in lowering the plant ethylene known to inhibit the nodulation process [39]. We identified in CPCRI-2 acdS gene homologue that codes for ACC deaminase enzyme. In CPCRI-1 and 3, we found, rimM [37] and dcyD [40], both of which also code for ACC deaminase. In addition, we found genes involved in hydrogen sulfide (H_2S) biosynthesis in all the three PGPR genomes (CPCRI-1, Gene#2608; CPCRI-2, Gene#3152; CPCRI-3, Gene#3465-67, 3470-72). Recently, H_2S has been reported to increase plant growth and seed germination [41] and the H_2S production by the PGPRs may play an analogous role in plant roots they colonize.

Among recently described volatile molecules known to directly influence the plant growth promotion are acetoin and 2,3-butanediol [42]. The CPCRI genomes encode budC, budA [43] and als [44], all of which are involved in the production of acetoin and 2,3-butanediol.

Many PGPRs are known to have biocontrol activities. Production and secretion of siderophores widely used by bacteria for iron acquisition is one of the modes of biocontrol activity.

While CPCRI-2 encoded 41 genes involved in the production and utilization of pyoverdine, siderophore, CPCRI-1 and 3, have additional genes such as fpv and mbt [45] that are linked to pyoverdine production. Besides the genes for pyoverdine, the gene cluster responsible for synthesis of temperature regulated achro-mobactin siderophore, acrA and acrB, were also identified in all the three PGPR genomes. In addition to siderophores, chemicals such as phenazine and 4-hydroxybenzoate produced by the PGPRs act as antibiotics and suppress plant pathogenic microbes. In all the three genomes we were able to identify the phzF involved in phenazine synthesis and ubiC that codes for chorismatelyase involved in 4-hydroxybenzoate synthesis. Also, in CPCRI-2 we identified a homologue of gene associated with the synthesis of anti-microbial compound pyocin [46]. In addition to these, in CPCRI-1 and CPCRI-2 genomes we identified gabD and gabT involved in production of pest/disease suppressing γ-aminobutyric acid (GABA) [46]. The whole genomes of the three bacteria coded for several genes that encode peroxidases, catalases, superoxide dismutase, and glutathione transferases, all of which alleviate oxidative stress in plants (Fig. S5).

Table 5. List of genes attributable to plant growth promotion traits in the CPCRI PGPR genomes.

Plant growth promotion traits	CPCRI-1	CPCRI-2	CPCRI-3	Genes with potential for conferring PGP traits
Phosphate solubilization	+	+	+	pqq, glucose dehydrogenase gene homolog, pstA, B, C
IAA production	+	−	+	ipdC
Siderophore production	+	+	+	pvd, pyoverdine homologous genes, fpvA, mbtH, acrA,B, fhu
ACC deaminase activity	+	+	+	acdS, rimM, dcyD
Acetoin & butanediol synthesis	+	+	+	als, budA, C, poxB
Phenazine production	+	+	+	phzF
Chitinase production	+	+	+	Chitinase gene homolog
4-hydroxybenzoate production	+	+	+	ubiC
Pyocin	−	+	−	Pyocin gene homolog
Trehalose metabolism	+	+	+	Trehalose synthase gene homolog
H$_2$S production	+	+	+	cysC, J, I, N
Quorum sensing	+	−	+	luxS, lsr
Heat shock proteins	+	+	+	dnaJ, K, groE
Cold shock proteins	+	+	+	cspA, C, D, E
Glycine-betaine production	+	+	+	soxB, opu, proX, glycine betaine gene homolog
Peroxidases	+	+	+	osmC, glutathione peroxidase genes similar to Enterobacterasburiae, oxyR
Catalases	+	+	+	Catalase gene homolog
Superoxide dismutase	+	+	+	sodB,C, super oxide dismutase gene homologs
GABA production	+	+	−	gabD, T

+ indicates presence of genes.
− indicates absence of genes.

Carbohydrate metabolism

Analyses of CPCRI-1, CPCRI-2 and CPCRI-3 genomes showed that they carried genes consistent with their ability to survive in soil environment and plant rhizospheres. The genomes of all 3 bacteria encode genes for central carbohydrate metabolism, including the tricarboxylic acid cycle, the Entner-Doudoroff pathway, glycolysis, gluconeogenesis, pyruvate metabolism and the pentose-phosphate pathways. However, the methyl citrate cycle for propionate metabolism (Table S11) was identified only in CPCRI-2. All the three bacteria carried genes for galactose, fructose, mannose, gluconate and glycogen metabolism, however CPCRI-1 and CPCRI-3 genomes showed the presence of a larger number of metabolic pathways for monosaccharides, disaccharides, oligosaccharides, and polysaccharides, than the CPCRI-2 genome. This indicated that CPCRI-1 and CPCRI-3 could use a large variety of plant-derived carbohydrates as carbon source. Additionally, CPCRI-1 and CPCRI-3 encode genes that can support the use of L-rhamnose, L-arabinose, xylose, trehalose, maltose, lactose and β-glucosides as a carbon source, even though utilization of lactose as a sole carbon source is a characteristic of the *Enterobacteriaceae* family [31]. These findings are consistent with the Biolog studies that demonstrated the ability of CPCRI-1 and CPCRI-3 to use L-rhamnose, trehalose, maltose and lactose.

Trehalose, a disaccharide, is accumulated by many microorganisms growing under high salt or osmotic stress and has been shown to play an important role in *Rhizobium*-legume symbiosis [47]. Accumulation of trehalose in *Bradyrhizobium japonicum* enhances its survival under conditions of salinity stress and plays an important role in the development of symbiotic nitrogen-fixing root nodules on soybean plants [48]. We observed that while all the three PGPRs encoded genes that support trehalose biosyn-

thesis, CPCRI-1 and CPCRI-3 also encoded genes for exogenous trehalose uptake that can potential allow them to use exogenous trehalose.

Malonate metabolism has been characterized in various symbiotic bacteria, such as *A. calcoaceticus*, *K. pneumoniae*, *P. fluorescens* and *P. putida* [49,50]. All three bacterial genomes reported in this study contain genes involved in malonate metabolism. In CPCRI-1 genome, we observed a cluster of nine genes mdcABCDEFGHR (Fig. S6A; CPCRI-1, Gene#3919-3927; CPCRI-3 Gene#2395-2403) involved in malonate decarboxylation (Fig. S6C). Also, CPCRI-2 encodes a nine gene malonate cluster mdcMLHGEDCBA (CPCRI-2, Gene#3117-3125; Fig. S6B). This suggests that all the three bacteria are capable of malonate utilization.

Degradation of aromatic compounds

In addition to the carbohydrate metabolism pathway genes, the CPCRI-2 genome coded for genes involved in the degradation of various aromatic compounds such as benzoate, 2,4-dichlorobenzoate, 1,2-dichloroethane, tetrachloroethane and bisphenolA. The β-ketoadipate pathway, an important bacterial energy source, has been identified in the *Pseudomonas* species and many members of *Rhizobiaceae* family of soil microorganisms [51,52]. The CPCRI-2 genome contains β-ketoadipate pathway genes involved in degradation of lignin derived aromatic compounds. Further in CPCRI-2, we also found genes involved in the metabolism of polyhydroxybutyrate (PHB), an aliphatic polyester synthesized by several bacteria as a means of carbon storage and a source of reducing equivalents in starving conditions [53]. PHB is stored intracellularly as granules and improves bacterial tolerance to high temperatures, H$_2$O$_2$ exposure, UV-irradiation, desiccation, and osmotic stress [53,54]. Interestingly, the three genomes also encoded *ars*C gene [45] which may play a role in detoxifying arsenic.

Quorum sensing

Both CPCRI-1 and CPCRI-3 encoded the autoinducer-2 (AI-2) gene [*lux*S; Gene #3755 (CPCRI-1); Gene #3423 (CPCRI-3)]. AI-2 is a small molecule produced by a number of bacterial species, implicated in the regulation of biofilm formation, motility and production of virulence factors [55]. AI-2 has been suggested to act directly through quorum sensing or indirectly through modulation of cellular metabolism. AI-2 dependent quorum sensing system has been demonstrated to be crucial for symbiosis between *Sinorhizobium meliloti* and legumes. *S. meliloti* can respond to the AI-2 signal by up-regulating transcription of its *lsr*-like operon [56]. Genomes of CPCRI-1 and CPCRI-3 also code for *lsr* operon, which contains genes encoding the transport apparatus responsible for internalizing, phosphorylating and processing of the AI-2 signal. The *lsr* operon comprises of six genes, *lsr*ACDBFG [57] is present in CPCRI-1 (Gene#2142-2147) and CPCRI-3 (Gene#4093-4098) genomes. The *lsr*B encodes the ligand binding protein, *lsr*C and *lsr*D each encode a transmembrane protein, and *lsr*A encodes a cytoplasmic protein responsible for ATP hydrolysis during transport. In addition to AI-2 quorum systems, non-lux based quorum sensing proteins controlled by *rib*B gene [58] was found in the genomes of all the three CPCRI bacteria (CPCRI-1: Gene#2104, CPCRI-2: Gene#1422, Gene#1433 and for CPCRI-3: Gene#3641). Gene *rib*B is a homolog of the *Escherichia coli* gene for 3,4-dihydroxy-2-butanone 4-phosphate synthase, a key enzyme for riboflavin synthesis, which along with *qsr*P, *acf*A, *qsr*V, and *qsr*7 have been proved as non-lux based QSR protein producing genes [58]. In CPCRI-2 (*Pseudomonas putida*) genome, Lux-R system (Gene#3078, 3458, 3799, 3918, 4032, 4115, 4123, 4282, 4600 and 4637) that is involved in acyl-homoserine lactone (ACL) controlled quorum sensing system was found [59]. A collection of genes such as *gac*A (Gene#4600), *rsm*A (Gene#3492) and *rpo*S (Gene#3005) that are known to regulate and network LasRI and RhlRI quorum sensing systems in *Pseudomonas aeruginosa* [60] was also found in CPCRI-2. These observations suggest that CPCRI-1, 2 and 3 may have quorum sensing ability that can contribute to their symbiotic relationship with the host plant.

PGPR fitness conferring genes

Production of heat-shock proteins, cold-shock proteins and osmoregulants in the bacteria regulate survival under harsh conditions. The genomes of all the three CPCRI isolates carried heat-shock protein genes like *dna*J, K and *gro*E, cold-shock proteins genes such as *csp*A, C, D, and E, and several copies of osmoprotectant glycine betaine synthesis genes. Other genes, *gac*S [61], *sox*S, R, *oxy*R [62] involved in protecting plants against oxidative stress were also found in the CPCRI genomes. We found *xer*C gene [63] in all the three CPCRI genomes. The *xer*C gene product, a site recombinase, is critical for the PGPRs to be an effective rhizosphere colonizer [63].

Comparison with non-PGPR bacteria

We compared the genes present in CPCRI bacteria with non-PGPR bacteria of similar strain. The CPCRI-1, 3 genomes were compared with *Enterobacter cloacae EcWSU1* [64] and *Enterobacter cloacae subsp. cloacae ATCC 13047* [65], whereas CPCRI-2 genome was compared with *Pseudomonas putida strain S16* [32]. Comparison was performed at functional classification level using GO, SEED and KEGG annotation (Table S12A–I).

Comparison of CPCRI-1, 3 with non-PGPR showed that pyrroloquinoline quinone (*pqq*) biosynthetic gene which is involved in solubilization of mineral phosphates was only present in CPCRI-1, 3 genomes. The acetoin-production gene, which is associated with butanediol dehydrogenase activity, was absent in non-PGPRs. The iron-scavenging group of genes involved in siderophore synthesis and their uptake was more enriched in CPCRI-1, 3 as compared to non-PGPRs. Also, CPCRI-1 genome showed adhesion group of genes to be highly enriched as compared to non-PGPRs.

Comparison of CPCRI-2 with non-PGPR revealed many functional groups that included some key plant growth traits. The widespread colonization island, siderophore enterobactin, pyrroloquinoline quinone (*pqq*) biosynthetic and phenazine (*phz*) biosynthesis genes present in CPCRI-2 were completely absent in *Pseudomonas putida strain S16*. Genes related to adhesion, iron scavenging and sulfur metabolism were more enriched in CPCRI-2 as compared to *Pseudomonas putida strain S16*.

Discussion

In this study we reported the whole genome sequencing and analysis of three PGPRs, CPCRI-1, CPCRI-2 and CPCRI-3 isolated from coconut [16], cocoa [20] and arecanut [22], respectively. The genomic level characterization reported here of PGPRs, to our knowledge is the first for rhizobacteria isolated from coconut, cocoa and arecanut. Usually the bacterial genomes are compact and tightly packed with genes and other functional elements and range in size from 0.5 to 10 Mb, with coding regions averaging ∼1 Kb [66]. Following assembly we estimated the genome sizes to be 4.7 Mb for CPCRI-1, 5.9 Mb for CPCRI-2 and 5.1 Mb for CPCRI-3 and there was a good correlation observed between the genome size and genome numbers of the three PGPRs as earlier reported in other studies [26]. The genome size of our *Enterobacter* spp. (CPCRI-1 and 3) was comparable to those of the others isolated from plantation crops such as sugar cane [67] and poplar [31], which had 4.9 and 4.6 Mb sizes respectively. Similarly, the genome size of *Pseudomonas* from cocoa (CPCRI-2) matched with that of the *Pseudomonas aurantiaca* obtained from sugar cane [68]. The GC contents recorded for CPCRI-1 and 3 matched well within the range reported for *Enterobacteriaceae* (38–60%) family and the range reported for the genus *Enterobacter* (52–60%) [69]. Similarly, GC content of CPCRI-2 was observed to fit well in the range expected for *Pseudomonas* genus (58–69%) [70]. Earlier studies have revealed that the GC content of the total genome usually matched with GC content of protein coding genes, spacer genes and stable RNA genes [71]. We could also observe a strong positive correlation between the GC content of the protein coding genes with GC content of total genome of CPCRI-1, 2 and 3 (Table 3, Table S3 and S4). Another interesting observation was about the codon usage pattern: CTG that codes for Leucine (leu) (Fig. S2) was found to be the most preferred codon in the CPCRI isolates as reported earlier in *Escherichia coli* and *Drosophila melanogaster* [72–74].

We identified between 4000 and 4600 protein coding genes in each of the three genomes. While a majority of the genes had homologs in the published sequence database, for 31, 59 and 91 proteins in CPCRI-1, CPCRI-2 and CPCRI-3, respectively, no homologs were found, suggesting that these may have novel functions.

Phylogenetic analysis indicated that both the bacteria isolated from coconut and arecanut belonged to *Enterobacteriaceae* and may reflect the fact that both plantation crops belong to the

Arecaceae family and have similar root niche/environment. The cocoa isolate belonged to the *Pseudomonadaceae* family.

Consistent with the PGP properties we found several genes that function in mineral phosphate solubilization, ACC deaminase function, IAA, acetoin and butanediol production. Previously, genes with similar functions in other PGPRs have been reported [36,40,42,75–77]. The genome sequence of *Enterobacters* spp. of coconut and arecanut and *Pseudomonas* from cocoa possessed many genes that have been reported in PGPR isolated from the plantation crops such as poplar and sugarcane. For example, *sod*B, C controlling the superoxide dismutase activity in CPCRI-1 and 3, *oxy*R gene known to regulate production of anti-microbial compound 4-hydroxybenzoate in CPCRI-1, mobility genes *flg*, *flh*, *fim*, and *fli*, in CPCRI-3 had orthologs in *Enterobacter* sp. 638 PGPR isolated from poplar [31]. Similarly, phosphate transporter genes *pst*A, B and C found in CPCRI-1 and 3 had orthologs in the *Enterobacter* spp. SP1 PGPR isolated from sugarcane [67]. Additionally, comparison of CPCRI-1, 2 and 3 genomes against non-PGPR genomes of the same genus showed several plant growth related group of genes that were either absent, like the pyrroloquinoline quinone (*pqq*) biosynthetic process gene, or less enriched in non-PGPR genomes.

In addition to growth promoting functions, PGPRs also indirectly support plant growth by suppressing pathogens [2]. In the PGPR genomes reported in this study, we identified several genes that are known to support the production of antimicrobial compounds such as siderophores, phenazine, 4-hydroxybenzoate and GABA [46]. They also contained genes for chitinase enzyme that can dissolve cell walls of pathogenic fungi, nematodes and insect pests [46]. In addition, CPCRI-2 genome encoded a gene for production of pyocin, a compound that suppresses growth of other related species. The three PGPR genomes also encoded enzymes such as peroxidases, catalases, super oxide dismutases and glutathione transferases all of which are involved in the management of oxidative stresses in plants.

Sulfur is an essential nutrient for plant growth and development and is associated with stress tolerance in plants [78]. Crop plants generally rely on the soil for their sulfur requirement and the mobilization of this sulfur for assimilation by plants is mediated by the microbial community in the soil and rhizosphere [79]. Sulfur-deficient conditions can cause severe losses in crop yield [80]. Sulfur nutrition is demonstrated to be critical in cocoa somatic embryogenesis [81]. In cocoa, elemental sulfur was identified in the xylem of resistant genotypes after infection by the vascular fungal pathogen *Verticillium dahlia* [82]. We found genes involved in H_2S biosynthesis in all the three PGPRs sequenced and they may, in particular in cocoa PGPR (CPCRI-2), be an important source of sulfur. We have also identified protein coding genes in the three bacteria known to be involved in resistance to copper, cobalt, zinc, arsenic, mercury and cadmium, suggesting that they function in detoxification of these metals.

Sequence analysis also showed that all the three CPCRI isolates have complete gene clusters corresponding to Type II, VI, Sec and Twin arginine targeting gene complexes (Table S13). Some of the past studies have shown that the Type I–VI and Sec secretion systems in rhizobacteria *Pseudomonas fluorescens* and *Variovorax paradoxus* function in promoting plant growth [45,46,83,84]. The presence of these secretion systems in PGPRs may play a role in their plant growth promoting functions and also provide support for their rhizosphere colonization ability [85,86].

Among the many biological properties of CPCRI isolates, their ability to utilize different carbohydrate sources and survive and grow under a wide range of pH, NaCl concentrations, and temperature would able to help them establish well under changing soil conditions. Accumulation of disaccharide trehalose has been implicated in survival of some of the plant-beneficial symbiotic microorganisms under salt or osmotic stress conditions [47,48]. We observed that while all the three bacteria are capable of trehalose biosynthesis, CPCRI-1 and CPCRI-3 also have genes (*tre*Y, Z) that will support exogenous trehalose uptake, further indicating that they are capable of tolerating high salinity or osmotic stress. Presence of genes that regulate the production of heat-shock, cold-shock proteins and osmoregulants in CPCRI PGPRs indicate that they have the capabilities to adapt to harsh conditions for their survival.

The genomic information obtained support the observed traits making them ideal candidates for further development as biofertilizers. The genes identified in our draft genome can now be studied for specific functions using knockout strategies. Experiments can be designed to identify the genes involved in the plant colonization and plant growth promotion process. Genetic engineering can be used to further improve the plant growth promoting properties of these bacteria. These findings will help in designing comprehensive strategies for development and use of such PGPRs to support sustainable plantation crop cultivation.

Materials and Methods

PGPR strains

About 1512 morphologically distinct heterotrophic bacteria were isolated from coconut, cocoa and arecanut rhizosphere soil samples [13,15,18,22] collected from privately owned farms with the permission of the owner. The different agro-ecological zones in India from which the samples were collected are listed in Table S1. The bacteria were first screened *in vitro* for a dozen important plant growth promoting properties and then tested for growth promotion in rice (for coconut and arecanut isolates) and cowpea (for cocoa isolates). Also, they were tested for growth promotion activity in coconut, cocoa and arecanut seedlings [13,16,18,19]. Based on their plant growth promotion characteristics, three PGPRs, designated here as CPCRI-1 (from coconut), CPCRI-2 (from cocoa) and CPCRI-3 (from arecanut) were given bio labels as RNF267 [16], KGSF20 [20], and KtRA5-88 [22], respectively, based on place/source of isolation and were selected for whole genome sequencing studies.

Bacterial cell morphology was assessed microscopically. Gram's staining was also performed. PGPR identification was done by conventional biochemical assays and Biolog analysis [18,19]. Cultures grown for 24 h on Biolog universal growth (BUG) agar were collected and processed according to the manufacturer's instructions (Hayward, CA). Briefly, cultures were transferred to inoculating fluid A (IF-A) and inoculum density was adjusted to 98% T using Biolog turbidimeter (Hayward, CA). Using multi-channel pipette, cell suspension was inoculated into Biolog Gen III Microplates (100 µl/well) containing 96 wells that provides 94 phenotypic tests. Plates were incubated at 33°C for 24 h. The optical density at 590 nm produced from the reduction of tetrazolium violet in each well was read after 24 h using a Biolog Microplate reader (version 5.1.1). Identification was performed by comparing the pattern formed in culture wells with possible patterns in the Microstation/MicrologVersion 5.1.1 database. A species identification of the PGPRs isolated coconut, cocoa and arecanut was acknowledged when the similarity index (SIM) and distance (DIS) values were >0.5 and <5.0, respectively [13,15,16,19,20].

Genomic DNA isolation

The three PGPRs, CPCRI-1, CPCRI-2 and CPCRI-3, chosen based on their plant beneficial attributes towards coconut, cocoa and arecanut were grown in Tryptic Soy Broth (TSB) medium at 30°C for 24–48 h. Genomic DNA was extracted using Gen Elute bacterial genomic DNA kit (Sigma, USA) as per the manufacturer's instructions. The extracted DNA was resolved on 0.8% agarose gel to check its integrity. The quality of the genomic DNA samples was assessed using Bioanalyzer DNA 7500 chip (Agilent, CA). The DNA yield was estimated on a TBS-380 Mini-Fluorometer (Turner BioSystems, CA) using PicoGreen dsDNA Quantitation Reagent (Molecular Probes, OR)

Library preparation and multiplexed whole genome shotgun sequencing

Whole genome shotgun libraries were generated from 1 μg genomic DNA using the GS FLX Titanium Rapid Library Preparation Kit (Roche Applied Science, CA) according to the manufacturer's protocol. Rapid library MID Adaptors MID10, MID11 and MID12 (Roche Applied Science, CA) were ligated to the CPCRI-1, CPCRI-2 and CPCRI-3 libraries, respectively. The quality of the libraries (library size ~1,600 bp) was assessed using Bioanalyzer High sensitivity DNA chip (Agilent, CA). The libraries were titrated by emulsion titrations and based on the percent enrichment, appropriate amount of the libraries were used to set up the large volume emulsion PCRs for each of the individual libraries. The beads containing clonally amplified DNA were enriched and the sequencing primer was annealed. Finally, half of the beads containing the CPCRI-2 libraries were mixed with CPCRI-1 library beads and the other half with the CPCRI-3 library beads. Each set of bead mix was then loaded on a picoTiter plate (half the plate) and sequenced using the GS FLX Titanium Sequencing Kit XL+ (Roche Applied Science, CA). Upon sequencing and processing of the raw data, demultiplexed data were assembled using GS de novo assembler version 2.6 (Roche Applied Science, CA).

Gene prediction

The protein-coding genes prediction was performed using Glimmer-MG [25], a metagenomics gene prediction program that uses interpolated Markov models (IMMs) to identify the protein-coding regions in the genome. The default setting for Glimmer-MG was used for gene prediction. The tRNA genes in the genome were identified using tRNA-SE program [87]. The BLASTN program (E-value$< = 1.0$ e^{-10}) at WebMGA was used for predicting ribosomal RNA genes [88].

Gene annotation

For gene annotation, we first compared predicted protein-coding genes against the non-redundant (nr) NCBI protein database using BLASTX (E-value$< = 1.0 e^{-5}$) program [27,89]. BLASTX result was parsed and the top hit database accession numbers were extracted. The accession numbers were then compared against UniProt knowledgebase for annotating genes [90]. The BLASTX result was imported into MEGAN4 [33,91] to perform KEGG pathway analysis [92] and SEED classification [93] of the proteins. The annotated genes were inspected for identifying those involved in PGP functions, pathogen suppression, abiotic stress tolerance, rhizosphere competence, carbohydrate metabolism and other important relevant functions.

Phylogenetic analysis

Phylogenetic analysis was performed using AMPHORA2 [24,94], a phylogenomic inference tool used for genomic phylotyping of bacteria and archaeal genomes. It scans the genome for 31 marker genes, which are universally distributed in both phyla. The 31 marker genes identified in CPCRI-1, CPCRI-2, and CPCRI-3 genome were then aligned using ClustalW [95]. Phylogenetic tree was inferred using bootstrap method available in ClustalW package.

Genome comparison

We compared our assembled bacterial genomes with the available complete bacterial genomes using progressive Mauve aligner [34] using default settings. The published genomes used in the alignment were obtained from PATRIC database (http://patricbrc.vbi.vt.edu) [96]. The sequence alignment file generated by the aligner was parsed to calculate pairwise similarity. Briefly, we first extracted the conserved blocks from the alignment file and then regions with <50 continuous gaps were considered for computing similarity score based on a pairwise sequence similarity percentage and coverage score which represents the percentage of genome that could be aligned pairwise.

Supporting Information

Figure S1 CPCRI genomes. Circos plot representing the CPCRI-1 (A), CPCRI-2 (B) and CPCRI-3 (C) genomes. The innermost circle represents the GC content, the second circle from the innermost circle represent non-coding genes, the third circle from inside represents coding genes on negative strand, the fourth circle represents coding genes on positive strand, and the outermost circle represent contigs.

Figure S2 GC-content based on codon position. GC-content distribution at each of the three codon position dervied from proportion of genes with a given GC-content at that position is shown for CPCRI-1 (A) CPCRI-2 (B) and CPCRI-3 (C).

Figure S3 Codon usage. The proportion of each codon (%) used in the CPCRI PGPR genomes computed from the protein-coding genes.

Figure S4 Protein taxonomy tree. Proteins encoded by the CPCRI PGPR genomes analyzed using MEGAN4 [33]. The numbers in bracket represent total number of gene assigned based on MEGAN4 annotation. The number in the bracket correspond to CPCRI-1, CPCRI-2 and CPCRI-3 in that order.

Figure S5 Number of genes coding for oxidative stress response enzymes in each of the indicated CPCRI PGPR strains.

Figure S6 Malonate gene cluster in (A) CPCRI-1, (B) CPCRI-2, (C) CPCRI-3 genome.

Table S1 Isolation details of PGPR isolates.

Table S2 Biological properties of PGPRs (CPCRI-1, CPCRI-2, CPCRI-3).

Table S3 Genomic properties of *Enterobacteriaceae* family.

Table S4 Genomic properties of *Pseudomonadeceae* family.

Table S5 A. Predicted protein-coding genes in CPCRI-1. **B.** Predicted protein-coding genes in CPCRI-2. **C.** Predicted protein-coding genes in CPCRI-3.

Table S6 A. Predicted tRNA genes in CPCRI-1 genome. **B.** Predicted tRNA genes in CPCRI-2 genome. **C.** Predicted tRNA genes in CPCRI-3 genome.

Table S7 Predicted rRNA genes in CPCRI-1, CPCRI-2, CPCRI-3 genome.

Table S8 A. CPCRI-1 protein coding genes annotation. **B.** CPCRI-2 protein coding genes annotation. **C.** CPCRI-3 protein coding genes annotation.

Table S9 Pairwise comparison of CPCRI-1, CPCRI-2, and CPCRI-3 PGPR genomes against bacteria genomes using progressive Mauve aligner. Percentage sequence similarity/ genome coverage in the conserved block is shown.

Table S10 List of GO terms identified in the bacterial genomes.

Table S11 Propionate metabolism pathway genes in CPCRI-1, CPCRI-2 and CPCRI-3 PGPR genomes.

Table S12 A. SEED comparison summary for CPCRI-1, 3 vs non-PGPRs (*Enterobacter cloacae EcWSU1, Enterobacter cloacae subsp. cloacae ATCC 13047*). **B.** KEGG comparison summary for CPCRI-1, 3 vs non-PGPRs (*Enterobacter cloacae EcWSU1,*

Enterobacter cloacae subsp. cloacae ATCC 13047). **C.** SEED comparison summary for CPCRI-2 vs non-PGPRs (*Pseudomonas putida S16*). **D.** KEGG comparison summary for CPCRI-2 vs non-PGPRs (*Pseudomonas putida S16*) **E.** GO comparison summary of CPCRI-1 vs *Enterobacter cloacae EcWSU1* (a non-PGPR). **F.** GO comparison summary of CPCRI-1 vs *Enterobacter cloacae subsp. cloacae ATCC 13047* (a non-PGPR). **G.** GO comparison summary of CPCRI-2 vs *Pseudomonas putida strain S16* (a non-PGPR). **H.** GO comparison summary of CPCRI-3 vs *Enterobacter cloacae EcWSU1* (a non-PGPR). **I.** GO comparison summary of CPCRI-3 vs *Enterobacter cloacae subsp. cloacae ATCC 13047* (a non-PGPR).

Table S13 Bacteria secretion system KEGG pathway gene in CPCRI-1, CPCRI-2 and CPCRI-3 genomes.

Result S1 Protein taxonomy results using MEGAN4 program.

Acknowledgments

We thank Sam Santhosh for the support and being the driving force behind the project. We are grateful to Devi Santhosh and Sneha Somasekar for helping edit the manuscript.

Author Contributions

Conceived and designed the experiments: AG MG PR RG. Wrote the paper: GT AG MG GVT SS PR RG. Isolated the genomic DNA from selected PGPR strains: AG MG VM. Performed the sequencing: JG PR. Performed the analysis: MG VM RG. Provided technical advice, analysis support and oversight: SS GVT SCS.

References

1. Kloepper JW, Schroth MN (1978) Plant growth-promoting rhizobacteria on radishes. 4th Internat Conf on Plant Pathogenic Bacter Station de Pathologie Vegetale et Phytobacteriologie, INRA, Angers, France. pp. 879–882

2. Lugtenberg B, Kamilova F (2009) Plant-growth-promoting rhizobacteria. Annu Rev Microbiol 63: 541–556.

3. Nelson LM (2004) Plant growth promoting rhizobacteria (PGPR): Prospects for new inoculants. Crop Management 3: 301–305.

4. Bloemberg GV, Lugtenberg BJ (2001) Molecular basis of plant growth promotion and biocontrol by rhizobacteria. Curr Opin Plant Biol 4: 343–350.

5. Schuster SC (2008) Next-generation sequencing transforms today's biology. Nat Methods 5: 16–18.

6. MacLean D, Jones JD, Studholme DJ (2009) Application of 'next-generation' sequencing technologies to microbial genetics. Nat Rev Microbiol 7: 287–296.

7. Mathimaran N, Srivastava R, Wiemken A, Sharma AK, Boller T (2012) Genome sequences of two plant growth-promoting fluorescent Pseudomonas strains, R62 and R81. J Bacteriol 194: 3272–3273.

8. Song JY, Kim HA, Kim JS, Kim SY, Jeong H, et al. (2012) Genome Sequence of the Plant Growth-Promoting Rhizobacterium Bacillus sp. Strain JS. J Bacteriol 194: 3760–3761.

9. Ma M, Wang C, Ding Y, Li L, Shen D, et al. (2011) Complete genome sequence of Paenibacillus polymyxa SC2, a strain of plant growth-promoting Rhizobacterium with broad-spectrum antimicrobial activity. J Bacteriol 193: 311–312.

10. Matilla MA, Pizarro-Tobias P, Roca A, Fernandez M, Duque E, et al. (2011) Complete genome of the plant growth-promoting rhizobacterium Pseudomonas putida BIRD-1. J Bacteriol 193: 1290.

11. Bopaiah BM (1985) Occurrence of phosphate solubilising microorganisms in the root region of arecanut palms. J Plantation Crops 13: 60–62.

12. Bopaiah BM, Shetty HS (1991) Soil microflora and biological activities in the rhizospheres and root regions of coconut-based multistoreyed cropping and coconut monocropping systems. Soil Biol Biochem 23: 89–94.

13. George P, Gupta A, Gopal M, Thomas L, Thomas GV (2012) Screening and in vitro evaluation of phosphate solubilizing bacteria from rhizosphere and roots of coconut palms (Cocos nucifera L.) growing in different states of India. Journal of Plantation Crops 40: 61–65.

14. Ghai SK, Thomas GV (1989) Occurrence of Azospirillum in coconut based farming systems. Plant and Soil 114: 235–241.

15. Thomas L, Gupta A, Gopal M, ChandraMohanan R, George P, et al. (2011) Evaluation of rhizospheric and endophytic Bacillus spp. and fluorescent

Pseudomonas spp. isolated from Theobroma cacao L. for antagonistic reaction to Phytophthora palmivora, the causal organism of black pod disease of cocoa. J Plantation Crops 39: 370–376.

16. George P, Gupta A, Gopal M, Thomas L, Thomas GV (2013) Multifarious beneficial traits and plant growth promoting potential of Serratia marcescens KiSII and Enterobacter sp. RNF 267 isolated from the rhizosphere of coconut palms (Cocos nucifera L.). World J Microbiol Biotechnol 29: 109–117.

17. Gupta A, Gopal M, Thomas GV (2006) Bioaugmentation of Cocos nucifera L. seedlings with the plant growth promoting rhizobacteria, Bacillus coagulans and Brevibacillus brevis for growth promotion. XXXVI Conference and Annual Meeting of European Society for New Methods in Agricultural Research. Iasi, Romania.

18. Gupta A, Gopal M, Thomas GV (2013) Consolidated report (2009–2013) of the NAIP Project on Diversity analysis of Bacillus and other predominant genera in extreme environments and their utilization in agriculture. Central Plantation Crops Research Institute, Kasaragod, Kerala, India.

19. Thomas GV, Gupta A, Gopal M (2014) Consolidated report (2006–2014) of the ICAR Network Project on Application of Microorganisms in Agriculture and Allied Sectors: Development and application of PGPR formulations for growth improvement and disease suppression in coconut and cocoa. Central Plantation Crops Research Institute, Kasaragod, Kerala, India.

20. Thomas L (2013) Identification and evaluation of plant growth promoting rhizobacteria from the diverse bacilli and fluorescent pseudomonad population in rhizosphere and roots of cocoa (Theobroma cacao L.). PhD Thesis, Mangalore University, Karnataka, India.

21. Thomas GV, Iyer R, Bopaiah BM (1991) Beneficial microbes in the nutrition of coconut. J Plantation Crops 19: 127–138.

22. Anusree GK, Gupta A, Gopal M, Manjusha A, Thomas GV. Isolation and characterization of acid tolerant bacteria from the rhizosphere of arecanut growing in extremely acidic soils of Kerala and Karnataka. In preparation.

23. Margulies M, Egholm M, Altman WE, Attiya S, Bader JS, et al. (2005) Genome sequencing in microfabricated high-density picolitre reactors. Nature 437: 376–380.

24. Wu M, Scott AJ (2012) Phylogenomic analysis of bacterial and archaeal sequences with AMPHORA2. Bioinformatics 28: 1033–1034.

25. Kelley DR, Liu B, Delcher AL, Pop M, Salzberg SL (2012) Gene prediction with Glimmer for metagenomic sequences augmented by classification and clustering. Nucleic Acids Res 40: e9.

26. Konstantinidis KT, Tiedje JM (2004) Trends between gene content and genome size in prokaryotic species with larger genomes. Proc Natl Acad Sci U S A 101: 3160–3165.

27. Camacho C, Coulouris G, Avagyan V, Ma N, Papadopoulos J, et al. (2009) BLAST+: architecture and applications. BMC Bioinformatics 10: 421.

28. Jones P, Binns D, Chang HY, Fraser M, Li W, et al. (2014) InterProScan 5: genome-scale protein function classification. Bioinformatics.

29. Marchler-Bauer A, Lu S, Anderson JB, Chitsaz F, Derbyshire MK, et al. (2011) CDD: a Conserved Domain Database for the functional annotation of proteins. Nucleic Acids Res 39: D225–229.

30. UniProt C (2014) Activities at the Universal Protein Resource (UniProt). Nucleic Acids Res 42: D191–198.

31. Taghavi S, van der Lelie D, Hoffman A, Zhang YB, Walla MD, et al. (2010) Genome sequence of the plant growth promoting endophytic bacterium Enterobacter sp. 638. PLoS Genet 6: e1000943.

32. Yu H, Tang H, Wang L, Yao Y, Wu G, et al. (2011) Complete genome sequence of the nicotine-degrading Pseudomonas putida strain S16. J Bacteriol 193: 5541–5542.

33. Huson DH, Mitra S, Ruscheweyh HJ, Weber N, Schuster SC (2011) Integrative analysis of environmental sequences using MEGAN4. Genome Res 21: 1552–1560.

34. Darling AE, Mau B, Perna NT (2010) progressiveMauve: multiple genome alignment with gene gain, loss and rearrangement. PLoS One 5: e11147.

35. Overbeek R, Begley T, Butler RM, Choudhuri JV, Chuang HY, et al. (2005) The subsystems approach to genome annotation and its use in the project to annotate 1000 genomes. Nucleic Acids Res 33: 5691–5702.

36. Rodriguez H, Fraga R, Gozalez T, Bashan Y (2006) Genetics of phosphate solubilization and its potential applications for improving plant growth-promoting bacteria. Plant Soil pp. 15–21.

37. Duca D, Lorv J, Patten CL, Rose D, Glick BR (2014) Indole-3-acetic acid in plant-microbe interactions. Antonie Van Leeuwenhoek 106: 85–125.

38. Straub D, Yang H, Liu Y, Tsap T, Ludewig U (2013) Root ethylene signalling is involved in Miscanthus sinensis growth promotion by the bacterial endophyte Herbaspirillum frisingense GSF30(T). J Exp Bot 64: 4603–4615.

39. Shah S, Li J, Moffatt BA, Glick BR (1998) Isolation and characterization of ACC deaminase genes from two different plant growth-promoting rhizobacteria. Can J Microbiol 44: 833–843.

40. Medigue C, Krin E, Pascal G, Barbe V, Bernsel A, et al. (2005) Coping with cold: the genome of the versatile marine Antarctica bacterium Pseudoalteromonas haloplanktis TAC125. Genome Res 15: 1325–1335.

41. Dooley FD, Nair SP, Ward PD (2013) Increased growth and germination success in plants following hydrogen sulfide administration. PLoS One 8: e62048.

42. Ryu CM, Farag MA, Hu CH, Reddy MS, Wei HX, et al. (2003) Bacterial volatiles promote growth in Arabidopsis. Proc Natl Acad Sci U S A 100: 4927–4932.

43. Blomqvist K, Nikkola M, Lehtovaara P, Suihko ML, Airaksinen U, et al. (1993) Characterization of the genes of the 2,3-butanediol operons from Klebsiella terrigena and Enterobacter aerogenes. J Bacteriol 175: 1392–1404.

44. Renna MC, Najimudin N, Winik LR, Zahler SA (1993) Regulation of the Bacillus subtilis alsS, alsD, and alsR genes involved in post-exponential-phase production of acetoin. J Bacteriol 175: 3863–3875.

45. Duan J, Jiang W, Cheng Z, Heikkila JJ, Glick BR (2013) The complete genome sequence of the plant-growth-promoting bacterium Pseudomonas sp. UW4. PLoS One 8: e58640.

46. Loper JE, Hassan KA, Mavrodi DV, Davis EW 2nd, Lim CK, et al. (2012) Comparative genomics of plant-associated Pseudomonas spp.: insights into diversity and inheritance of traits involved in multitrophic interactions. PLoS Genet 8: e1002784.

47. Zahran HH (1999) Rhizobium-legume symbiosis and nitrogen fixation under severe conditions and in an arid climate. Microbiol Mol Biol Rev 63: 968–989, table of contents.

48. Sugawara M, Cytryn EJ, Sadowsky MJ (2010) Functional role of Bradyrhizobium japonicum trehalose biosynthesis and metabolism genes during physiological stress and nodulation. Appl Environ Microbiol 76: 1071–1081.

49. Kim YS (2002) Malonate metabolism: biochemistry, molecular biology, physiology, and industrial application. J Biochem Mol Biol 35: 443–451.

50. Koo JH, Kim YS (1999) Functional evaluation of the genes involved in malonate decarboxylation by Acinetobacter calcoaceticus. Eur J Biochem 266: 683–690.

51. Li D, Yan Y, Ping S, Chen M, Zhang W, et al. (2010) Genome-wide investigation and functional characterization of the beta-ketoadipate pathway in the nitrogen-fixing and root-associated bacterium Pseudomonas stutzeri A1501. BMC Microbiol 10: 36.

52. MacLean AM, MacPherson G, Aneja P, Finan TM (2006) Characterization of the beta-ketoadipate pathway in Sinorhizobium meliloti. Appl Environ Microbiol 72: 5403–5413.

53. Kadowaki MA, Muller-Santos M, Rego FG, Souza EM, Yates MG, et al. (2011) Identification and characterization of PhbF: a DNA binding protein with regulatory role in the PHB metabolism of Herbaspirillum seropedicae SmR1. BMC Microbiol 11: 230.

54. Ratcliff WC, Kadam SV, Denison RF (2008) Poly-3-hydroxybutyrate (PHB) supports survival and reproduction in starving rhizobia. FEMS Microbiol Ecol 65: 391–399.

55. Reading NC, Sperandio V (2006) Quorum sensing: the many languages of bacteria. FEMS Microbiol Lett 254: 1–11.

56. Pereira CS, McAuley JR, Taga ME, Xavier KB, Miller ST (2008) Sinorhizobium meliloti, a bacterium lacking the autoinducer-2 (AI-2) synthase, responds to AI-2 supplied by other bacteria. Mol Microbiol 70: 1223–1235.

57. Pereira CS, de Regt AK, Brito PH, Miller ST, Xavier KB (2009) Identification of functional LsrB-like autoinducer-2 receptors. J Bacteriol 191: 6975–6987.

58. Callahan SM, Dunlap PV (2000) LuxR- and acyl-homoserine-lactone-controlled non-lux genes define a quorum-sensing regulon in Vibrio fischeri. J Bacteriol 182: 2811–2822.

59. Fuqua C, Winans SC, Greenberg EP (1996) Census and consensus in bacterial ecosystems: the LuxR-LuxI family of quorum-sensing transcriptional regulators. Annu Rev Microbiol 50: 727–751.

60. Schuster M, Greenberg EP (2006) A network of networks: quorum-sensing gene regulation in Pseudomonas aeruginosa. Int J Med Microbiol 296: 73–81.

61. Whistler CA, Corbell NA, Sarniguet A, Ream W, Loper JE (1998) The two-component regulators GacS and GacA influence accumulation of the stationary-phase sigma factor sigmaS and the stress response in Pseudomonas fluorescens Pf-5. J Bacteriol 180: 6635–6641.

62. Ochsner UA, Vasil ML, Alsabbagh E, Parvatiyar K, Hassett DJ (2000) Role of the Pseudomonas aeruginosa oxyR-recG operon in oxidative stress defense and DNA repair: OxyR-dependent regulation of katB-ankB, ahpB, and ahpC-ahpF. J Bacteriol 182: 4533–4544.

63. Shen X, Hu H, Peng H, Wang W, Zhang X (2013) Comparative genomic analysis of four representative plant growth-promoting rhizobacteria in Pseudomonas. BMC Genomics 14: 271.

64. Humann JL, Wildung M, Cheng CH, Lee T, Stewart JE, et al. (2011) Complete genome of the onion pathogen Enterobacter cloacae EcWSU1. Stand Genomic Sci 5: 279–286.

65. Ren Y, Ren Y, Zhou Z, Guo X, Li Y, et al. (2010) Complete genome sequence of Enterobacter cloacae subsp. cloacae type strain ATCC 13047. J Bacteriol 192: 2463–2464.

66. Ochman H, Davalos LM (2006) The nature and dynamics of bacterial genomes. Science 311: 1730–1733.

67. Zhu B, Chen M, Lin L, Yang L, Li Y, et al. (2012) Genome sequence of Enterobacter sp. strain SP1, an endophytic nitrogen-fixing bacterium isolated from sugarcane. J Bacteriol 194: 6963–6964.

68. Mehnaz S, Bauer JS, Gross H (2014) Complete Genome Sequence of the Sugar Cane Endophyte Pseudomonas aurantiaca PB-St2, a Disease-Suppressive Bacterium with Antifungal Activity toward the Plant Pathogen Colletotrichum falcatum. Genome Announc 2.

69. Rameshkumar N, Lang E, Nair S (2010) Mangrovibacter plantisponsor gen. nov., sp. nov., a nitrogen-fixing bacterium isolated from a mangrove-associated wild rice (Porteresia coarctata Tateoka). Int J Syst Evol Microbiol 60: 179–186.

70. Couillerot O, Prigent-Combaret C, Caballero-Mellado J, Moenne-Loccoz Y (2009) Pseudomonas fluorescens and closely-related fluorescent pseudomonads as biocontrol agents of soil-borne phytopathogens. Lett Appl Microbiol 48: 505–512.

71. Muto A, Osawa S (1987) The guanine and cytosine content of genomic DNA and bacterial evolution. Proc Natl Acad Sci U S A 84: 166–169.

72. Hershberg R, Petrov DA (2009) General rules for optimal codon choice. PLoS Genet 5: e1000556.

73. Vicario S, Moriyama EN, Powell JR (2007) Codon usage in twelve species of Drosophila. BMC Evol Biol 7: 226.

74. Sharp PM, Cowe E, Higgins DG, Shields DC, Wolfe KH, et al. (1988) Codon usage patterns in Escherichia coli, Bacillus subtilis, Saccharomyces cerevisiae, Schizosaccharomyces pombe, Drosophila melanogaster and Homo sapiens; a review of the considerable within-species diversity. Nucleic Acids Res 16: 8207–8211.

75. Shankar M, Ponraj P, Ilakiam D, Rajendhran J, Gunasekaran P (2012) Genome sequence of the plant growth-promoting bacterium Enterobacter cloacae GS1. J Bacteriol 194: 4479.

76. Sharma V, Kumar V, Archana G, Kumar GN (2005) Substrate specificity of glucose dehydrogenase (GDH) of Enterobacter asburiae PSI3 and rock phosphate solubilization with GDH substrates as C sources. Can J Microbiol 51: 477–482.

77. Goldstein AH (1996) Involvement of the quinoprotein glucose dehydrogenase in the solubilization of exogenous phosphates by Gram-negative bacteria; Washington, DC. ASM Press. pp. 197–203.

78. Gill SS, Tuteja N (2011) Cadmium stress tolerance in crop plants: probing the role of sulfur. Plant Signal Behav 6: 215–222.

79. Kertesz MA, Fellows E, Schmalenberger A (2007) Rhizobacteria and plant sulfur supply. Adv Appl Microbiol 62: 235–268.

80. Nemat MA, El-Kader AAA, Attia M, Alva AK (2011) Effects of Nitrogen Fertilization and Soil Inoculation of Sulfur-Oxidizing or Nitrogen-Fixing Bacteria on Onion Plant Growth and Yield. International Journal of Agronomy 2011.

81. Emile M, Nicolas N, Auguste IE, Abdourahamane S, Omokolo DN (2010) Sulphur depletion altered somatic embryogenesis in Theobroma cacao L. Biochemical difference related to sulphur metabolism between embryogenic and non embryogenic calli. African Journal of Biotechnology 9: 5665–5675.

82. Cooper RM, Resende MLV, Flood J, Rowan MG, Beale MH, et al. (1996) Detection and cellular localization of elemental sulphur in disease-resistant genotypes of Theobroma cacao. Nature 379: 159–162.

83. Preston GM, Bertrand N, Rainey PB (2001) Type III secretion in plant growth-promoting Pseudomonas fluorescens SBW25. Mol Microbiol 41: 999–1014.

84. Han JI, Choi HK, Lee SW, Orwin PM, Kim J, et al. (2011) Complete genome sequence of the metabolically versatile plant growth-promoting endophyte Variovorax paradoxus S110. J Bacteriol 193: 1183–1190.

85. Viollet A, Corberand T, Mougel C, Robin A, Lemanceau P, et al. (2011) Fluorescent pseudomonads harboring type III secretion genes are enriched in the mycorrhizosphere of Medicago truncatula. FEMS Microbiol Ecol 75: 457–467.

86. Barret M, Egan F, Moynihan J, Morrissey JP, Lesouhaitier O, et al. (2013) Characterization of the SPI-1 and Rsp type three secretion systems in Pseudomonas fluorescens F113. Environ Microbiol Rep 5: 377–386.

87. Schattner P, Brooks AN, Lowe TM (2005) The tRNAscan-SE, snoscan and snoGPS web servers for the detection of tRNAs and snoRNAs. Nucleic Acids Res 33: W686–689.

88. Wu S, Zhu Z, Fu L, Niu B, Li W (2011) WebMGA: a customizable web server for fast metagenomic sequence analysis. BMC Genomics 12: 444.

89. Gish W, States DJ (1993) Identification of protein coding regions by database similarity search. Nat Genet 3: 266–272.

90. Magrane M, Consortium U (2011) UniProt Knowledgebase: a hub of integrated protein data. Database (Oxford) 2011: bar009.

91. Huson DH, Auch AF, Qi J, Schuster SC (2007) MEGAN analysis of metagenomic data. Genome Res 17: 377–386.

92. Kanehisa M, Goto S, Sato Y, Furumichi M, Tanabe M (2012) KEGG for integration and interpretation of large-scale molecular data sets. Nucleic Acids Res 40: D109–114.

93. Aziz RK, Bartels D, Best AA, DeJongh M, Disz T, et al. (2008) The RAST Server: rapid annotations using subsystems technology. BMC Genomics 9: 75.

94. Wu M, Eisen JA (2008) A simple, fast, and accurate method of phylogenomic inference. Genome Biol 9: R151.

95. Larkin MA, Blackshields G, Brown NP, Chenna R, McGettigan PA, et al. (2007) Clustal W and Clustal X version 2.0. Bioinformatics 23: 2947–2948.

96. Snyder EE, Kampanya N, Lu J, Nordberg EK, Karur HR, et al. (2007) PATRIC: the VBI PathoSystems Resource Integration Center. Nucleic Acids Res 35: D401–406.

## 2

Identification and Biochemical Characterization of an Acid Sphingomyelinase-Like Protein from the Bacterial Plant Pathogen *Ralstonia solanacearum* that Hydrolyzes ATP to AMP but Not Sphingomyelin to Ceramide

Michael V. Airola[1], Jessica M. Tumolo[2], Justin Snider[1], Yusuf A. Hannun[1]*

1 Department of Medicine and the Stony Brook University Cancer Center, Stony Brook University, Stony Brook, New York, United States of America, **2** Department of Biochemistry and Molecular Biology, Medical University of South Carolina, Charleston, South Carolina, United States of America

Abstract

Acid sphingomyelinase (aSMase) is a human enzyme that catalyzes the hydrolysis of sphingomyelin to generate the bioactive lipid ceramide and phosphocholine. ASMase deficiency is the underlying cause of the genetic diseases Niemann-Pick Type A and B and has been implicated in the onset and progression of a number of other human diseases including cancer, depression, liver, and cardiovascular disease. ASMase is the founding member of the aSMase protein superfamily, which is a subset of the metallophosphatase (MPP) superfamily. To date, MPPs that share sequence homology with aSMase, termed aSMase-like proteins, have been annotated and presumed to function as aSMases. However, none of these aSMase-like proteins have been biochemically characterized to verify this. Here we identify RsASML, previously annotated as RSp1609: acid sphingomyelinase-like phosphodiesterase, as the first bacterial aSMase-like protein from the deadly plant pathogen *Ralstonia solanacearum* based on sequence homology with the catalytic and C-terminal domains of human aSMase. A biochemical characterization of RsASML does not support a role in sphingomyelin hydrolysis but rather finds RsASML capable of acting as an ATP diphosphohydrolase, catalyzing the hydrolysis of ATP and ADP to AMP. In addition, RsASML displays a neutral, not acidic, pH optimum and prefers Ni^{2+} or Mn^{2+}, not Zn^{2+}, for catalysis. This alters the expectation that all aSMase-like proteins function as acid SMases and expands the substrate possibilities of this protein superfamily to include nucleotides. Overall, we conclude that sequence homology with human aSMase is not sufficient to predict substrate specificity, pH optimum for catalysis, or metal dependence. This may have implications to the biochemically uncharacterized human aSMase paralogs, aSMase-like 3a (aSML3a) and aSML3b, which have been implicated in cancer and kidney disease, respectively, and assumed to function as aSMases.

Editor: Vladimir N. Uversky, University of South Florida College of Medicine, United States of America

Funding: This work was supported by grants from the National Institutes of Health NIGMS R37 GM043825 (to Y.A.H.) and NIGMS F32 GM100679 (to M.V.A.). The MUSC Summer Undergraduate Research Program (SURP) provided additional support (to J.M.T.). The funders had no role in study design, data collection and analysis, decision to publish, or preparation of the manuscript.

Competing Interests: The authors have declared that no competing interests exist.

* Email: yusuf.hannun@sbumed.org

Introduction

Sphingomyelinases (SMases) are enzymes that catalyze the hydrolysis of sphingomyelin (SM) to generate ceramide (Cer) and phosphocholine [1–4]. Three families of SMases have been identified (acid, neutral, and alkaline) that are distinguished by their pH optima, protein fold, subcellular localization, primary structure, and metal dependence [1,3,5,6]. Acid SMase (aSMase) was the first identified human SMase and is encoded by the SMPD1 gene [2]. ASMase is required for SM turnover in the lysosome and aSMase deficiency is the underlying cause of the genetic diseases Niemann-Pick Type A and B [2]. ASMase has also been shown to play important roles in atherosclerosis [7], cystic fibrosis [8], Wilson's disease [9], bacterial infection [8], and apoptosis [2]. Two additional uncharacterized human proteins, aSMase-like 3a (aSML3a) and aSML3b, belong to the aSMase-

protein superfamily. ASML3a and aSML3b have been implicated in cancer [10–12] and kidney disease [13–16], respectively.

The genetic and biochemical properties of aSMase have been well characterized [2]. SM hydrolysis by aSMase displays an acidic pH optima and requires Zinc for activity [2,17]. The aSMase protein is comprised of three parts: a sphingolipid-activator protein (SAP)-like domain that aids in SM extraction from the membrane and exposure to the catalytic domain [18], a catalytic domain belonging to the metallo-phosphatase (MPP) superfamily, and a C-terminal domain of unknown structure and function (Fig. 1) [2]. A number of different inactivating mutations have been identified in Niemann-Pick Type A and B patients, which defines all three-protein domains as necessary for proper enzymatic function [2].

Ralstonia solanacearum is a deadly plant pathogen that causes southern bacterial wilt, infects agriculturally important crops

a

aSMase

| signal peptide | SAP | Pro rich | MPP (catalytic) | C-ter Dom | — 629 |

human aSML3a

| sp | MPP (catalytic) | C-ter Dom | — 445 |

32% identity
49% homology
with aSMase

R. solanacearum aSMase-like (RsASML)

| sp | MPP (catalytic) | C-ter Dom | — 476 |

24% identity
37% homology
with aSMase

human aSML3b

| sp | MPP (catalytic) | C-ter Dom | — 456 |

30% identity
46% homology
with aSMase

b

```
                    ••  •           • • •      •      ••   • •                                  •  •         •  •  ••
ASMase  182  PKPPPKPPSPPAPGAPVSRILFLTDLHWDHDYLEGTDPDCADPLCCRRGSGLPPASRPG.AGYWGEYSKCDLPLRTLESLLSGLGPAGPF.DMVYWTGDI
RsASM    30  QAAPPLDPPATGDTQAPGFFGALSDIHFNPFY....DPALVDRLAAAEPSAWDGIFKTSSITEPNDTD.AGPGYPLLKTTLDAIAPQARRLDYVILPGDF
                                                              *                                              *

              ••  •                                        •        ••                          •••  ••                • ••     •
ASMase  280  PAHD...........VWHQTRQDQLRALTTVTALVRKFLGPVPVYPAVGNHESIPVNSFPPPFIEGNHSSRWLYEAMAKAWEPWLPAEALRTLRIGGFY
RsASM   125  LTHDFRENYMLYASDKSDAAYRSFVLKTIRYVAMGLKARFPDVPVVATLGNNDSFCGDYQI...EPSSEFLYDLTATMAEAAG...NPAGFSAYPELGAY
                                            *                    *                               *                •

                                                                              H280
                 •                •         • ••   •                        •••• •                                   • •
ASMase  368  ALSPYP...GLRLISLNMNFCSRENFWLLIN.STDPAGQLQWLVGELQAAEDRGDKVHIIGHIPPGHCL.................KSWSWNYYRIVAR
RsASM   219  VIPHPRTARHYFVVLENTFLSAKYRNTCGLTYTNPSQALLLWLESTLYRMKRENATATLVMHIPSGIDAYSSTRACGFSSSPVPYFAAGSGDALANILQR
                                                                              *
                                                                 MPP domain (catalytic) | C-terminal domain

              •      • •     •     •     •           •          ••••            •           •       •    •       •  •
ASMase  446  YENTLAAQFFGHTHVDEFEVFYDEETLSRPLAVAFLAPSATTYIGLNPGYRVYQIDGNYSRSSHVVLDHETYILNLTQANIPGAIPHWQLLYRARETYGL
RsASM   319  YPDQVRAIFTGHSHMDDFRVLSDSGGK..PFAYERVIPSVTPFFRNNPGYQIYSYER.......ATGAPLDYWARIYAASGQSNTRAWRWEYGFQQAYNV
                *          *      *      *            *                   *                                 *    *

                    •           •           •                      ••
ASMase  546  PNTLPTAWHNLVYRMRGDMQLFQTFWFLYHKGHPPSEPCGTPCRLATLCAQLSARADSPALCRHLMPDGSLPEAQSLWPRPLFC
RsASM   401  GELSPDNLNTLAAAIAKDPATRAKYIAFYTGS.ANSGTITNRNWPAFACALTNLSAKAFSTCFCGGAQ
                  *                                               |
                                                   conserved disulfide
                                                    Cysteine pair
```

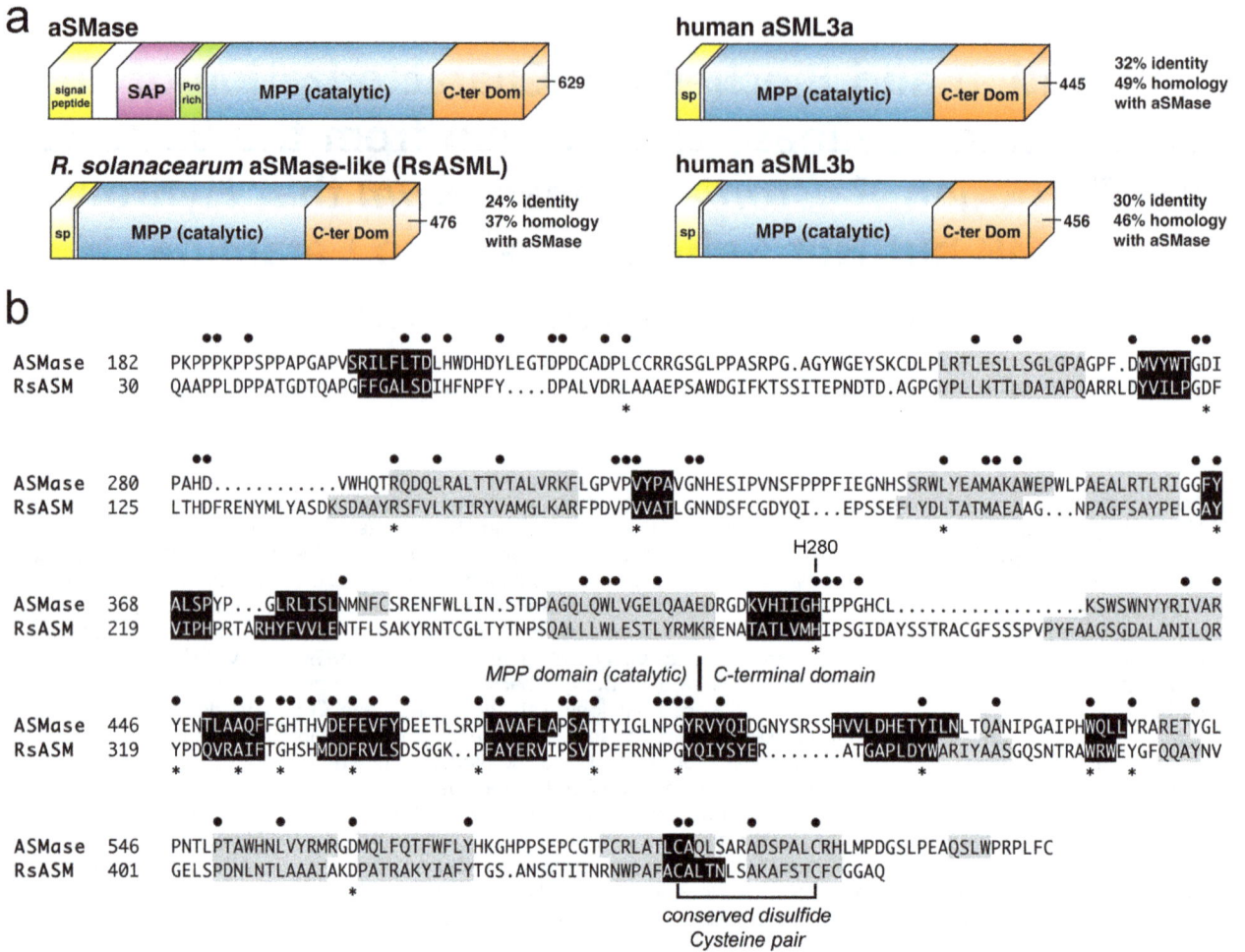

Figure 1. Domain architecture and sequence alignment of RsASML and human aSMase. (a) Domain architecture of acid SMase and acid-SMase-like proteins. Human aSMase contains three domains: a SAP (Sphingolipid-Activating Protein) involved in lipid binding, a MetalloPhosPhatase (MPP) catalytic domain, and a C-terminal domain required for activity but of unknown function. RsASML, from the bacteria *R. solanacearum*, and two human proteins, acid SMase-like 3a (aSML3a) and aSML3b of unknown function, share homology with the catalytic and C-terminal domains of human aSMase. (b) Sequence alignment of human aSMase and RsASML highlighting predicted secondary structure elements using Jpred3 (black = beta strands, grey = alpha helices). Black circles above denote identical residues. Asterisks below sequence indicate conserved residues identified in Niemann-Pick Type A or B patients. A known disulfide cysteine pair in human aSMase, conserved in RsASML, is noted.

(tomato, potato, pepper, eggplant, tobacco, banana, etc.), and has a broad geographic distribution [19,20]. *R. solanacearum* is extremely lethal and expresses over a hundred different pathogenicity factors upon plant infection [19–22]. Currently, there is no method to control this pathogen and infected fields can rarely be reused, even after crop rotation with nonhost plants [23]. As such, specific strains are under quarantine status in the United States, Europe, and around the world.

Table 1. List of bacterial aSMase homologues identified by a BLAST search.

NCBI#	Bacterial organism	Max Identity	Positive Homology	Gaps
YP_002951829	*Desulfovibrio magneticus RS-1*	27% (113/411)	40% (165/411)	21% (87/411)
YP_002138293	*Geobacter bemidjiensis Bem*	24% (107/449)	35% (160/449)	22% (101/449)
ZP_06188574	*Legionella longbeachae D-4968*	26% (78/304)	40% (123/304)	19% (58/304)
ZP_01909010	*Plesiocystis pacifica SIR-1*	25% (87/346)	42% (147/346)	21% (73/346)
YP_005055725	*Granulicella mallensis MP5ACTX8*	25% (74/298)	38% (115/298)	16% (48/298)
ZP_10138266	*Fluoribacter dumoffii Tex-KL*	25% (56/228)	45% (104/228)	9% (22/228)
YP_002257058	***Ralstonia solanacearum GMI1000***	**24% (107/447)**	**37% (168/447)**	**23% (103/447)**

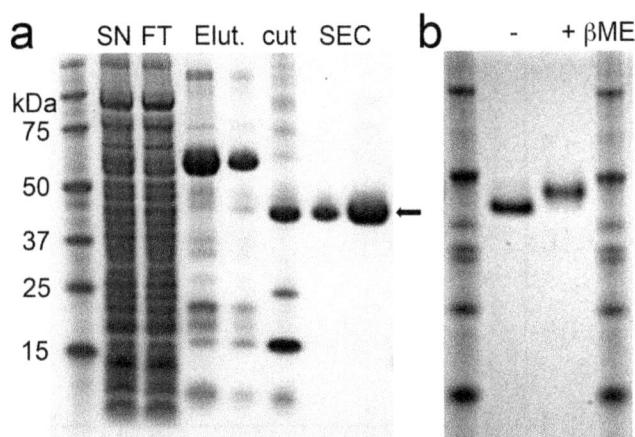

Figure 2. SDS-PAGE of purified RsASML protein. (a) SDS-PAGE of RsASML protein purification stained with coomassie blue. The arrow indicates the final protein purity used in biochemical assays. Abbreviations: SN: supernatant, FT: flow-through from Ni-Column, Elut: Elution from Ni-Column, Cut: Elution fractions incubated with ULP-1, a SUMO specific-protease, SEC: Size-exclusion chromatography. (b) SDS-PAGE of RsASML protein ran under non-reducing (−βME) and reducing (+βME) conditions induces a band shift, consistent with RsASML containing intra-molecular disulfide bonds.

Bacterial homologues of neutral SMases have been identified in many pathogens including *Bacillus cereus, Staphylococcus aureus, Clostridium perfringens, Listeria ivanovii* and *Streptomyces griseocarneus* [1,3,4,24,25]. In these pathogenic bacteria, the secreted neutral SMases are toxins that catalyze the hydrolysis of SM on the outer plasma membrane leaflet of erythrocytes and lymphocytes, causing hemolysis, lymphotoxicity, and septicemia [24–26]. Bacterial neutral SMases have served as important models for understanding the structure and biochemistry of the mammalian neutral SMase family [3]. Although bacterial neutral SMases are common, there have been no bacterial homologues identified to date for aSMase.

Here we identify and biochemically characterize the gene RSp1609, herein referred to as RsASML, as the first bacterial aSMase-like protein from the deadly plant pathogen *R. solanacearum*. We find that the RsASML protein, unlike aSMase, cannot hydrolyze SM to Cer but appears to be an ATP diphosphohydrolase, which catalyzes the hydrolysis of ATP and ADP to AMP. In addition, the biochemical properties of RsASML differ from aSMase, displaying a neutral pH optima and a Ni^{2+} metal dependence. Overall, this work broadens the substrates and pH optima of aSMase-like proteins and presents the possibility that *R. solanacearum* uses ATP hydrolysis to aid in plant pathogenicity. In addition, the identification of a bacterial aSMase-like protein suitable for X-ray crystallography studies may aid in future studies defining the structural features of the aSMase protein superfamily.

Experimental Procedures

Protein overexpression and purification

The RsASML gene, encompassing amino acids 30–476, which lacks a putative N-terminal secretory signal peptide (residues 1–29), from *R. solanacearum GMI1000* was PCR amplified from a cosmid (a kind gift from Stephane Genin, INRA, Toulouse, FRA). The PCR product was cloned into the *E. coli* overexpression vector ppSUMO using BamHI and NotI restriction sites. The resulting translated protein contained a cleavable N-terminal HisTag/SUMO-RsASML peptide. The RsASML ppSUMO plasmid was transformed into Origami 2 (DE3) cells for protein expression.

Cells harboring the RsASML ppSUMO plasmid were grown at 37°C in Terrific broth to an OD ~2.0. The temperature was reduced to 15°C, and after one hour at 15°C protein production was induced with 30 mg IPTG per liter. Twenty hours after IPTG induction, cells were centrifuged and cell pellets were stored at −80°C.

Cells were resuspended in Buffer A (25 mM HEPES, pH 7.5, 500 mM NaCl, 5% glycerol, and 60 mM imidazole) and lysed by sonication. After centrifugation the supernatant was applied to a 5 mL HisTrap FF column (GE Healthcare), washed extensively with Buffer A, and eluted with Buffer B (25 mM HEPES, pH 7.5, 500 mM NaCl, 5% glycerol, and 300 mM imidazole). Fractions containing the His/SUMO-RsASML protein were collected, and the His/SUMO-tag was removed by overnight incubation with the purified SUMO-protease, ULP-1. Digested protein was applied to a Hi-Load 26–60 Superdex 200 size-exclusion column equilibrated with 10 mM HEPES, pH 8.0, 50 mM NaCl.

Table 2. K_{metal} and V_{max} values for $NiCl_2$ and $MnCl_2$.

	$NiCl_2$	$MnCl_2$
V_{max}(μM/min)	9.47+/−0.38	0.104+/−0.004
K_{metal} (μM)	336.2+/−62.64	35.35+/−5.29
V_{max}/K_{metal}	0.0282	0.0029

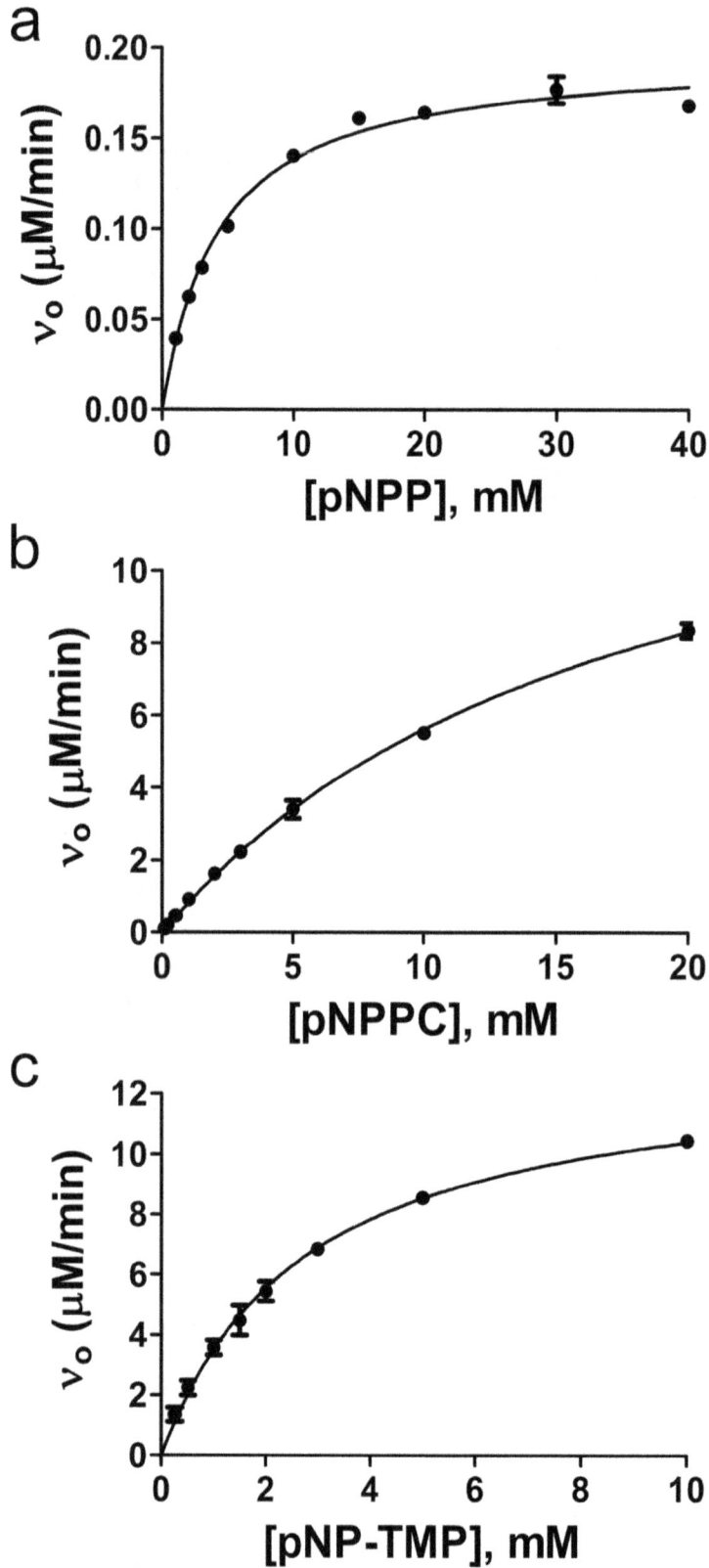

Figure 3. Michaelis-Menten kinetics of RsASML vs. different pNP-based substrates. Concentration dependence of RsASML activity towards (a) pNPP, (b) pNPPC, (c) pNP-TMP. All reactions were carried out in 10 mM HEPES, pH 8.0, 1 mM NiCl$_2$, with 100 nM RsASML protein.

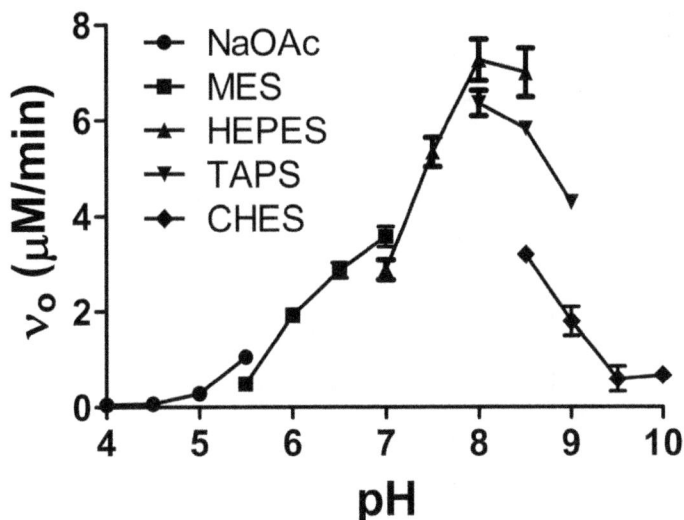

Figure 4. pH dependence of RsASML activity. RsASML activity towards pNPPC at different pH's. The buffers for each pH are as indicated. Reaction conditions were 10 mM HEPES, pH 8.0, 1 mM NiCl$_2$, 10 mM pNPPC with 100 nM RsASML protein.

Fractions containing the RsASML protein were pooled and concentrated to 5–10 mg/mL, aliquoted, and flash frozen.

Para-nitrophenol (pNP)-based assays

Purified RsASML protein was incubated with the pNP-based substrates: para-nitrophenol phosphate (pNPP), para-nitrophenol phosphocholine (pNPPC, synonym: O-(4-Nitrophenylphosphoryl)choline), and pNP-thymidine 5′-monophosphate (pNP-TMP, synonym: Thymidine 5′-monophosphate p-nitrophenyl ester) in a standard 96-well plate at 25°C. A temperature of 25°C was used for activity assays as incubation at 37°C resulted in a decrease of RsASML enzyme activity over time, presumably due to protein instability. pNP formation was followed by monitoring the change in absorbance at 405 nm over time using a BioTek Synergy HT microplate reader. The extinction coefficient for pNP at 405 nm is 18,000 M^{-1}cm^{-1}. K_m and V_{max} values were determined by plotting the initial velocity (μM/min) of the reaction versus the concentration of substrate and using a non-linear regression analysis. k_{cat} values (min^{-1}) were determined by dividing the V_{max} values (μM/min) by the concentration of the enzyme RsASML (μM). For pH screening, multiple endpoint assays were used to calculate the initial velocity, where the reaction was quenched by addition of 1 N NaOH and the absorbance was immediately measured.

SMase assays

Sphingomyelin, labeled with ^{14}C in the phosphocholine headgroup, was incorporated in Triton X-100 micelles by sonication and incubated with RsASML or Bc-nSMase (*Bacillus cereus* nSMase) proteins at room temperature in their respective optimal reaction buffers. Reactions were quenched by addition of 1.5 mL chloroform and 400 μL water. ^{14}C labeled phosphocholine was extracted using a standard Folch extraction by collecting the aqueous layer and the radioactive products were quantified using a Beckman LS6500 Scintillation Counter as described previously [27].

NBD-lyso-SM and NBD-lyso-PC assays

0.4 mM Nitrobenzoxadiazol-lyso-sphingomyelin (NBD-lyso-SM) and Nitrobenzoxadiazol-lyso-phosphatidylcholine (NBD-

lyso-PC) (Avanti Polar Lipids) stock solutions were prepared by first solubilizing in Methanol (final stock concentration = 8% v/v) and after addition of BSA (4 mg/mL). Stock solutions were stored at −20°C until use. Reactions were initiated by addition of RsASML or Bc-nSMase proteins to the NBD-lyso lipids (20 μM final) in their respective reaction buffers (RsASML: 1 mM NiCl$_2$, 10 mM HEPES, pH 8.0 or Bc-nSMase: 10 mM MgCl$_2$, 25 mM Tris, pH 7.5). The reaction was quenched by addition of 400 μL Chloroform/Methanol/HCl (100:200:1). An additional 120 μL Chloroform and 120 μL 2 M KCl were added to this solution, vortexed, and centrifuged at 3000 rpm for 10 min. 200 μL of the organic layer was removed, placed in a clean glass tube, and dried. The extracted lipids were resuspended in 50 μL of Chloroform/Methanol (1:1) and 5 μL was spotted onto a clean TLC plate. The plate was placed in a TLC chamber equilibrated with Chloroform/Methanol/Water (65:35:8) to separate the reaction products. Plates were visualized using a Typhoon FLA 7000 with excitation and emission.

ATP hydrolysis assays

RsASML was incubated with adenosine-based nucleotides (1 mM) in reaction buffer (10 mM HEPES, pH 8.0, 1 mM NiCl$_2$) for 1 hr at RT. Reactions were quenched by addition of 80 μL 1 M stock of Acetic Acid/Sodium Acetate (14.7/5.3), 560 μL water, and 30 μL of chloroacetaldehyde (50% solution). The reaction products were derivatized by incubation overnight at 60°C to form the fluorescent etheno-derivatived products.

High-pressure liquid chromatography (HPLC) was performed using a Waters 1525 Binary HPLC Pump, Waters 717 plus autosampler, and Shimandzu RF-551 Spectrofluorometric detector. The derivatized reaction products were diluted 100 fold in Buffer A (0.1 M Potassium phosphate buffer, pH 6.0) and applied to a Luna 5u C18 (2) 100A column. Products were separated using a stepwise gradient against 50/50 mixture of Buffer A and Methanol. Emission and excitation values of 275 nm and 415 nm were used to detect the fluorescent products.

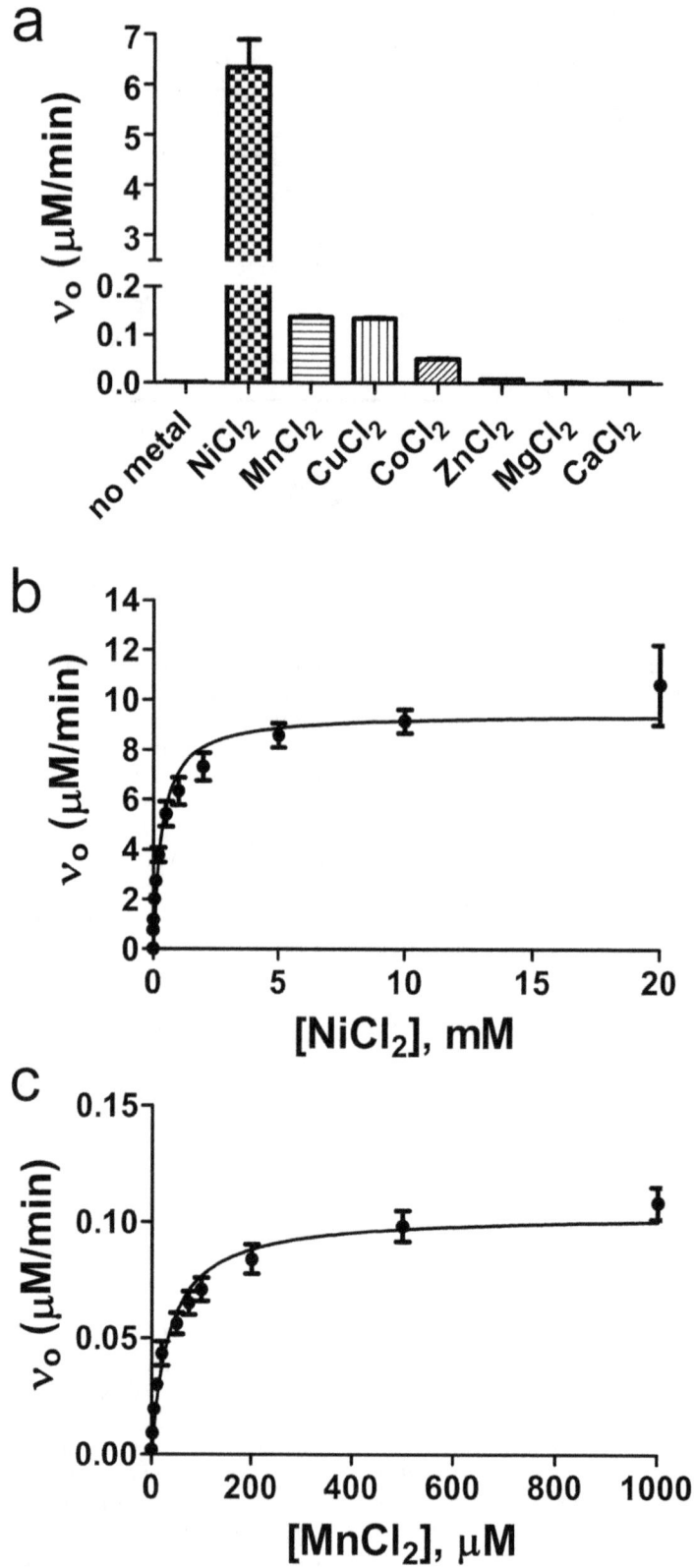

Figure 5. Metal dependence of RsASML activity. (a) RsASML activity towards pNPPC with different divalent metals. Reaction conditions were 10 mM HEPES, pH 8.0, 10 mM pNPPC with 100 nM RsASML protein at a concentration of 1 mM for each metal. (b) Dependence of RsASML activity on Ni^{2+} concentration. (c) Dependence on Mn^{2+} concentration.

Table 3. Michaelis-Menten values for pNP-based substrates.

	pNPP	pNPPC	pNP-TMP
k_{cat} (min^{-1})	2.01+/−0.04	163.67+/−7.46	135.52+/−4.65
K_m (mM)	4.29+/−0.27	18.55+/−1.41	2.79+/−0.22
k_{cat}/K_m	0.47	8.82	48.64

Results

Identification of a bacterial aSMase-like protein

Human aSMase is heavily glycosylated making it a difficult crystallization target [2,28,29]. To aid in structurally defining the aSMase protein fold, we searched for bacterial aSMase-like proteins that shared sequence homology with human aSMase and could be used for structural studies. A BLAST search [30] revealed a number of bacterial proteins that shared significant homology with the catalytic domain of human aSMase, and a few that shared homology with both the catalytic and C-terminal domains of aSMase (Table 1). None of these proteins shared additional homology with the SAP domain of aSMase.

Given that the well-characterized bacterial homologues to human neutral SMase (nSMase) derive from pathogenic organisms [1,3,31,32]; we reasoned that an aSMase-like protein from a bacterial pathogen was more likely to exhibit SMase activity. Given none of the bacterial proteins that we identified were from human pathogens, we selected the protein RsASML from the plant pathogen *R. solanacearum* for further biochemical analysis (Table 1). The primary sequence of RsASML shared positive homology and predicted secondary structure with human aSMase over the entire catalytic and C-terminal domains (Fig. 1b). This included a C-terminal cysteine pair known to form a disulfide in human aSMase [33], as well as 18 amino acids mutated in Niemann-Pick Type A or B patients [34–43]. A reverse BLAST search with the RsASML protein sequence identified RsASML as belonging to the Metallophosphatase (MPP)_aSMase protein superfamily with an E-value of $5.22e^{-48}$. The list of close sequence homologues to RsASML found 30 bacterial proteins sharing the highest degree of sequence homology, followed by a long list of eukaryotic homologues to aSML3a, aSML3b, and aSMase.

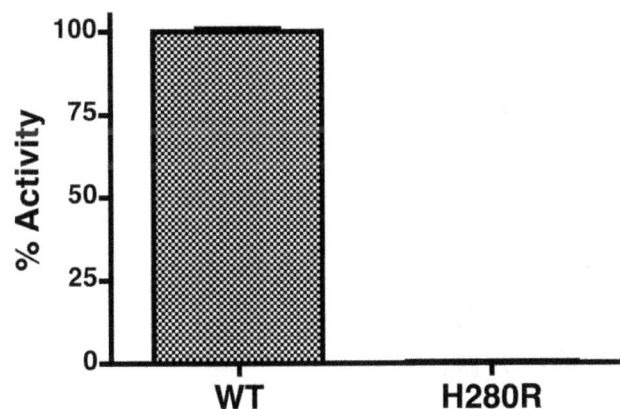

Figure 6. A corresponding Niemann-Pick mutation inactivates RsASML. RsASML activity of WT and H280R proteins towards pNPPC. The H280R substitution in RsASML corresponds to the H427R mutation in aSMase found in a subset of Niemann-Pick Type A patients.

Overall, this suggests that RsASML shares similar protein architecture to aSMase and is one of a few bacterial proteins belonging to the MPP_aSMase protein superfamily.

Expression of RsASML requires intra-molecular disulfide formation

Initial attempts to overexpress and purify RsASML from *E. coli* BL21 (DE3) cells failed. Human aSMase contains a number of intra-molecular disulfide bonds that are required for activity [33]. We hypothesized that RsASML may also require intra-molecular disulfide formation for proper folding and activity. Indeed, overexpression of a SUMO-RsASML fusion in Origami2 (DE3) cells, which promote disulfide formation in the *E. coli* cytoplasm, did result in the generation of soluble protein that could be purified to >95% homogeneity using Nickel and size-exclusion columns (Fig. 2a). SDS-PAGE samples of RsASML run under reducing and non-reducing conditions resulted in a shift in the protein band, which confirmed the presence of intra-molecular disulfides in RsASML (Fig. 2b).

Biochemical characterization of RsASML

To characterize the biochemical properties of RsASML we started by testing the ability of RsASML to hydrolyze generic phosphatase substrates belonging to the para-nitrophenol (pNP) family. RsASML was able to catalyze the hydrolysis of phosphate from para-nitrophenol phosphate (pNPP) (Fig. 3a), as well as a phosphocholine headgroup from para-nitrophenol phosphocholine (pNPPC) (Fig. 3b). Significantly higher activity was detected towards the soluble SM-mimic pNPPC, which shares the phosphocholine headgroup with SM (Fig. 3b). After confirming linearity of the reaction versus time and protein concentration, we determined that RsASML, unlike human aSMase, displayed activity over a broad range between pH 6–9, with maximal activity at the neutral pH of 8 (Fig. 4). RsASML activity was metal dependent with Ni^{2+} stimulating activity far above all other metals (Fig. 5). Significant activity over baseline was also detected for the metals Mn^{2+}>Cu^{2+}>Co^{2+}. Ni^{2+} displayed a relatively high K_{metal} of 336 μM and a V_{max} of 9.5 μM/min (Fig. 5b, Table 2). In comparison, Mn^{2+} had a lower K_{metal} of 35.4 μM but an ~100 fold lower V_{max} of 0.1 μM/min (Fig. 5c, Table 2). Notably, Zn^{2+}, which human aSMase utilizes for SM hydrolysis, provided the weakest activity of the transition metals tested (Fig. 5a).

In addition to pNPP and pNPPC, RsASML could also hydrolyze the substrate para-nitrophenol-thymidine 5′-monophosphate (pNP-TMP) (Fig. 3c), which requires nucleotide phosphodiesterase activity for hydrolysis. Interestingly, RsASML displayed higher activity towards pNP-TMP versus the SM-mimic pNPPC under identical conditions (Fig. 3). A comparison of the k_{cat} and K_m values found that RsASML had similar k_{cat} values for pNPPC and pNP-TMP (Fig. 3, Table 3). The increase in activity was due to a higher affinity of RsASML towards pNP-TMP versus pNPPC, with a K_m approximately 10 fold lower (Table 3).

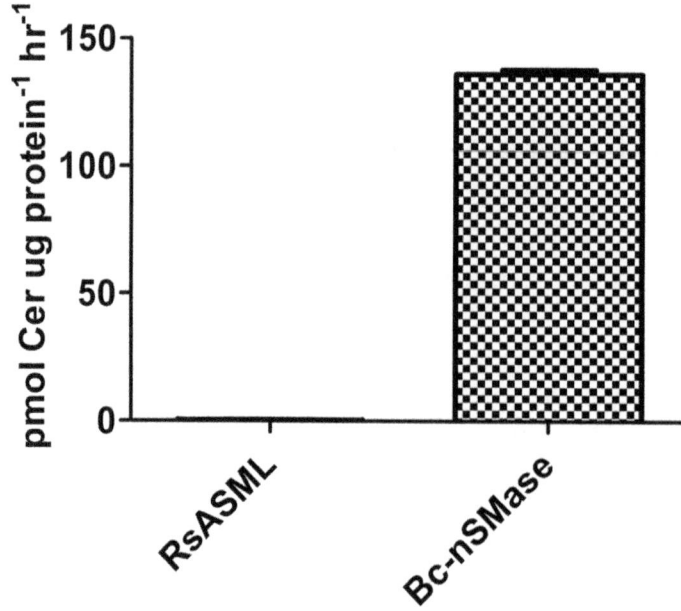

Figure 7. Sphingomyelinase activity of RsASML and Bc-nSMase. RsASML reaction conditions: 10 mM HEPES, pH 8.0, 1 mM NiCl₂. Bc-nSMase (*Bacillus cereus* nSMase) reactions conditions: 10 mM HEPES, pH 7.5, 10 mM MgCl₂.

To ensure that the activity we were observing was due to RsASML and not any minor contaminating *E. coli* proteins, we generated RsASML H280R by substituting Histidine 280 to Arginine using site-directed mutagenesis. This mutation in RsASML corresponds to the H427R mutation in human aSMase found in a subset of Niemann-Pick Type A patients [37]. RsASML H280R did not display any activity towards pNPPC (Fig. 6) confirming the hydrolysis was due to the presence of RsASML.

RsASML does not hydrolyze sphingomyelin or other phosphocholine-based lipids

Having established optimal reaction conditions; we assessed the ability of RsASML to hydrolyze SM and other lipid substrates containing a phosphocholine headgroup. Using a standard SMase

assay where SM is incorporated into Triton-X 100 mixed micelles, we found that RsASML did not catalyze SM hydrolysis (Fig. 7). For comparison, we used bacterial nSMase from *B. cereus* (Bc-nSMase) as a positive control. RsASML lacks the SAP domain found in human aSMase that is thought to aid in SM extraction from the membrane and facilitate SM availability to the catalytic domain. To ensure the lack of SMase activity of RsASML was not due to the absence of a SAP domain and lack of SM access to the catalytic domain, we assessed the activity of RsASML in more soluble lipid systems. These systems utilized the fluorescently labeled lipids NBD-lyso-SM and NBD-lyso-phosphatidylcholine (NBD-lyso-PC) complexed with fatty acid-free BSA. RsASML did not hydrolyze NBD-lyso-SM and NBD-lyso-PC, while Bc-nSMase hydrolyzed both (Fig. 8). From these experiments, we concluded

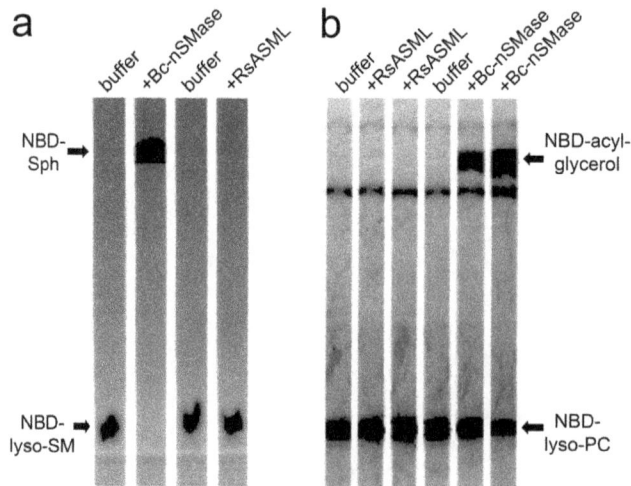

Figure 8. Phospholipase activity of RsASML and Bc-nSMase towards NBD-lyso-SM and NBD-lyso-PC. (a) Fluorescently imaged TLC plate of NBD-lyso-SM reactions. (b) Fluorescently imaged TLC plate of NBD-lyso-PC reactions. RsASML reaction conditions: 10 mM HEPES, pH 8.0, 1 mM NiCl₂. Bc-nSMase (*Bacillus cereus* nSMase) reactions conditions: 10 mM HEPES, pH 7.5, 10 mM MgCl₂.

Table 4. IC$_{50}$ values for inhibition of RsASML pNPPC activity.

Molecule	IC$_{50}$ (μM)
Adenosine	243.2
ADP	0.07049
AMP	0.07499
ATP	0.05864
β-NADPH	52.15
cAMP	63.72
cGMP	32028
CTP	1502
GTP	44.21
L-α-glycerolphosphate	745.9
L-α-glycerolphosphocholine	64.61
NAD$^+$	0.06632
Phosphate	265
Phospho-Threonine	6671
Phosphocholine	4430
Pyro-Phosphate	638
TMP	35.85
TTP	200.8
UTP	1362

IC$_{50}$ values were determined using a nonlinear regression (log [inhibitor] vs. normalized response) in the program PRISM. All reactions were carried out in 10 mM HEPES, pH 8.0, 1 mM NiCl$_2$ with 61.3 nM RsASML protein and 5 mM pNPPC.

that RsASML is unlikely to be a SMase and may catalyze the hydrolysis of a different substrate.

Competitive Inhibition Assay to Identify High Affinity Substrates

To aid in the identification of high affinity substrates, we used a competitive inhibition assay. The phosphatase activity of RsASML towards pNPPC was assessed in the presence of varying concentrations of phospho-containing small molecules. Lower IC$_{50}$ values indicate more potent inhibition. As an initial verification, we used the corresponding phospho-moieties from the set of pNP-based substrates. Consistent with the measured affinities of RsASML towards the pNP-based substrates, TMP displayed the lowest IC$_{50}$ compared to free phosphate and phosphocholine (Fig. 9, Table 4). Other small molecules, representing protein-phosphatase, poly-phosphatase, and phospholipase activities, yielded relatively high IC$_{50}$ values (Table 4).

Adenosine-based nucleotides are potent inhibitors of RsASML

Given that RsASML displayed nucleotide phosphodiesterase activity towards pNP-TMP, we assessed the IC$_{50}$'s of different nucleotides. ATP sharply inhibited RsASML activity towards pNPPC, while other tri-phosphate-nucleotides did not (Fig. 9, Table 4). A comparison of the IC$_{50}$'s of adenosine-based nucleotides found that ATP, ADP, and AMP all potently inhibited RsASML activity in the nanomolar range. Notably, the concentration of RsASML protein was nearly equivalent to the IC$_{50}$ concentrations of ATP, ADP, and AMP. Adenosine, which has no phosphate group, inhibited weakly with a significantly higher IC$_{50}$. NAD$^+$, which contains a di-phosphate adenosine moiety, had a comparable IC$_{50}$ to ATP/ADP/AMP. In contrast, additions or

modifications of the AMP backbone resulted in diminished inhibition as seen for β-NADPH and cyclic-AMP (cAMP) (Fig. 9, Table 4).

RsASML can hydrolyze ATP and ADP to AMP

The potent inhibition of ATP, ADP, and AMP suggested that RsASML might be a nucleotide phosphodiesterase specific for adenosine-based nucleotides. The ability of RsASML to hydrolyze adenosine-based nucleotides was directly assessed. RsASML catalyzed the hydrolysis of both ATP and ADP to form AMP (Fig. 10). AMP was not further hydrolyzed to adenosine. Overall, this suggests that RsASML is an ATP diphosphohydrolase.

Discussion

Here we identify and biochemically characterize RsASML, the first bacterial aSMase-like protein, as well as the first aSMase-like protein to be biochemically characterized. Although RsASML shares significant sequence homology with the catalytic and C-terminal domains of human aSMase, the protein does not display SMase activity. Rather, our in vitro characterization suggests RsASML may function as an ATP diphosphohydrolase. In addition to broadening the potential substrates of the MPP_aSMase protein superfamily, we find that RsASML displays a different pH optima and metal dependence to human aSMase. We note that the Ni^{2+} metal dependence is not likely to be biologically relevant as free Ni^{2+} concentrations are very low. It is more likely that Mn^{2+} is the physiological relevant metal for RsASML activity.

This work has a number of implications to the MPP_aSMase protein superfamily. To date, all aSMase-like proteins have been assumed to exhibit aSMase activity based simply on the substrate specificity of aSMase. This includes the uncharacterized human proteins aSML3a and aSML3b, which similar to RsASML also

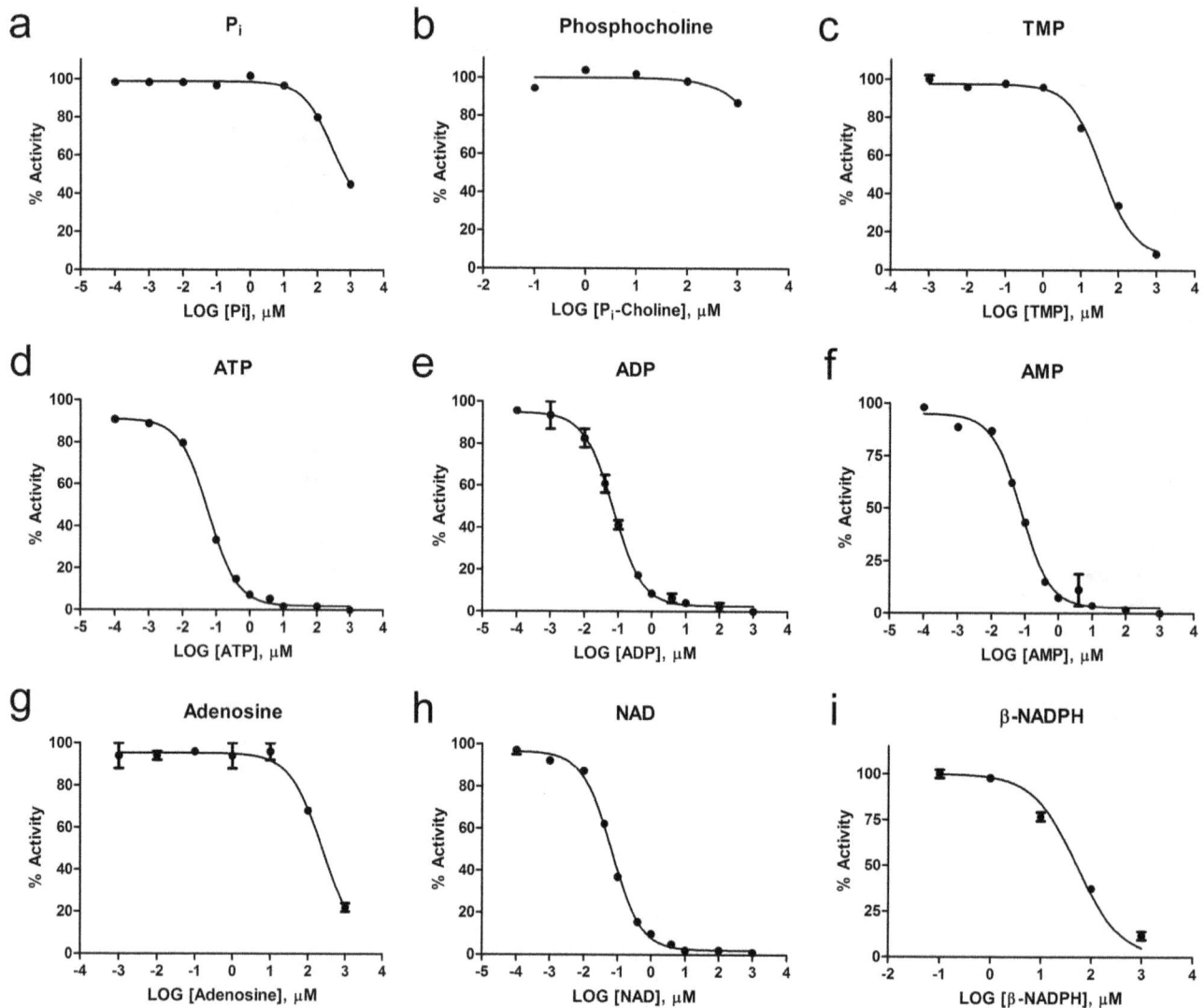

Figure 9. Competitive inhibition of RsASML activity by small molecules. RsASML activity towards pNPPC was assessed in the presence of varying concentrations of phospho-containing small molecules. Data points were fit using a nonlinear regression (log [inhibitor] vs. normalized response) in the program PRISM. All reactions were carried out in 10 mM HEPES, pH 8.0, 1 mM NiCl$_2$ with 61.3 nM RsASML protein and 5 mM pNPPC.

lack the SAP domain found in aSMase. This is important as aSML3a (gene name: SMPDL3A) and aSML3b (gene name: SMPDL3B) have recently been identified as necessary for cell division [44] and a potential target for treatment of the common kidney disease Focal Segmental Glomerulosclerosis (FSGS) [45], respectively. Notably, ceramide has been assumed to be the resultant bioactive product of aSML3a and aSML3b activity in these cases and others [10,11,14,44–47]. Based on the characterization of RsASML, the MPP_aSMase protein superfamily may catalyze the hydrolysis of a variety of substrates, similar to the ectonucleotide pyrophosphatase/phosphodiesterase (ENPP) protein superfamily, whose well-characterized members ENPP1, ENPP2 (more commonly referred to as Autotaxin), and ENPP7 (alkaline SMase) catalyze the hydrolysis of ATP to AMP, lyso-PC to lyso-PA, and SM to Cer, respectively [5,48–51]. That aSMase-like proteins may hydrolyze a variety of phosphate moieties is consistent with aSMase belonging to the general metallophosphatase protein superfamily, whose members include protein phosphatases and exonucleases. Therefore it is of utmost importance to define the substrate specificities of both aSML3a

and aSML3b, which may include SM, ATP, or other phospho-containing molecules, to help guide future studies.

These results also raise the possibility that the substrates of S-SMase, the secreted form of aSMase, which is located in the neutral pH environment of the extracellular space and where Zinc concentrations are very low, may not be limited to SM. How S-SMase functions outside its normal acidic pH optimum is a major question that still needs to be addressed. An interesting study found that S-SMase was capable of hydrolyzing artherogenic, modified LDL-bound SM at neutral pH but not normal LDL-bound SM [52]. The work presented here, raises the distinct possibility that S-SMase may be capable of hydrolyzing additional substrates, beyond artherogenic LDL-SM [52], in the neutral pH environment of the extracellular matrix.

Furthermore, it would be interesting to evaluate the role of RsASML as a putative ATP diphosphohydrolase in the pathogenesis of *R. solanacearum* in vivo. Recently, DORN1 was identified as the long sought after plant extracellular ATP receptor [53]. DORN1 is unique in that it shares no homology with the human extracellular ATP receptors, P2X and P2Y. Extracellular

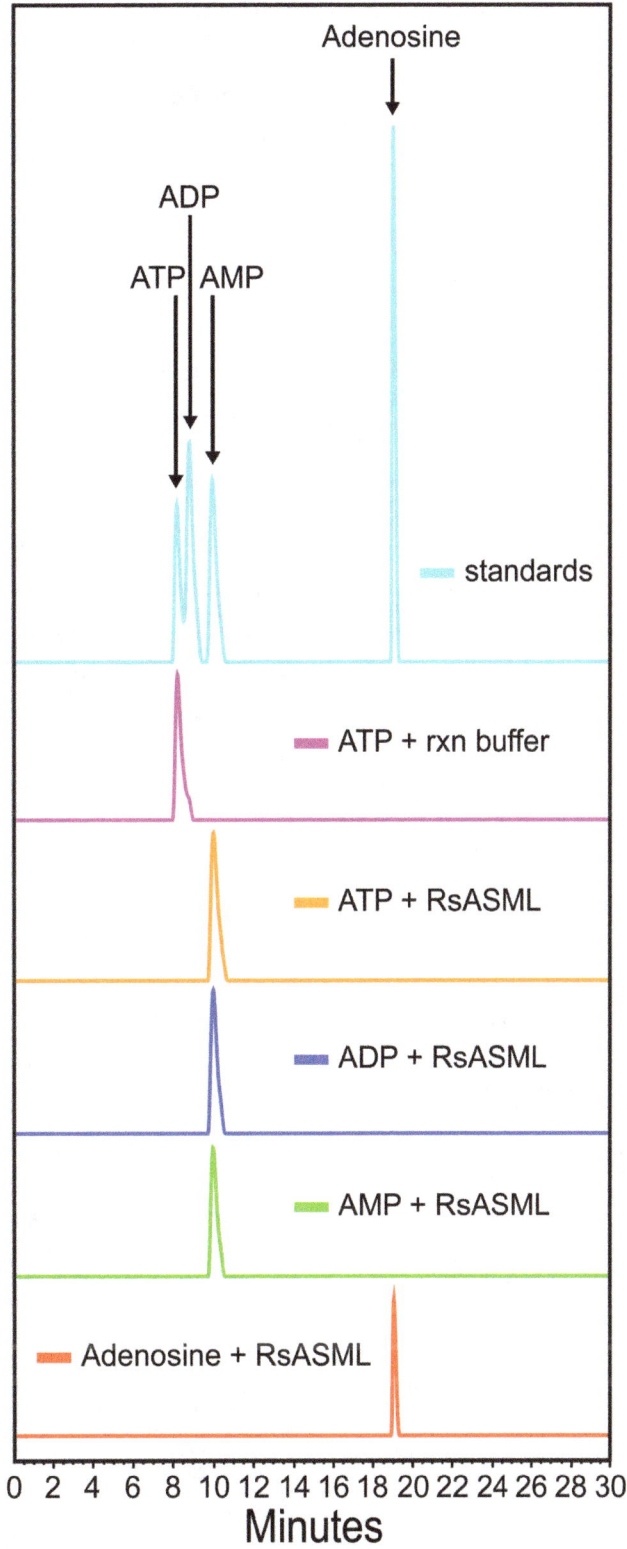

Figure 10. RsASML catalyzes the hydrolysis of ATP and ADP to AMP. HPLC chromatograms of adenosine-based nucleotides incubated with reaction buffer or RsASML protein. All reactions were carried out in 10 mM HEPES, pH 8.0, 1 mM NiCl$_2$ with 1 μM RsASML protein.

ATP was found to up-regulate a highly similar set of genes that correlate with early plant wounding [53]. Concentrations of extracellular ATP, as high as 40 μM, have been measured at sites of plant physical wounding [54]. Overall, this suggests that extracellular ATP may serve as a danger signal in plants in response to physical wounding [53]. Although future work is needed to verify a role for RsASML in extracellular ATP hydrolysis during plant infection, RsASML does contain a putative secretory signal peptide. Based on this, we hypothesize that RsASML may function to lower the concentration of an extracellular ATP danger signal during plant infection and thereby aid in the pathogenesis of *R. solanacearum*.

Acknowledgments

We thank Stephane Genin (INRA, Toulouse, FRA) for kindly providing the cosmid containing the RSp1609 gene (RsASML) for cloning. We also thank Nabil Matmati, David Montefusco, Michael J Pulkoski-Gross, Achraf Shamseddine, and Jaekyo Yi for their assistance.

Author Contributions

Conceived and designed the experiments: MVA YAH. Performed the experiments: MVA JMT. Analyzed the data: MVA YAH. Contributed reagents/materials/analysis tools: JS. Contributed to the writing of the manuscript: MVA YAH.

References

1. Airola MV, Hannun YA (2013) Sphingolipid Metabolism and Neutral Sphingomyelinases. Sphingolipids: Basic Science and Drug Development: Springer. 57–76.
2. Jenkins RW, Canals D, Hannun YA (2009) Roles and regulation of secretory and lysosomal acid sphingomyelinase. Cellular signalling 21: 836–846.
3. Clarke CJ, Snook CF, Tani M, Matmati N, Marchesini N, et al. (2006) The extended family of neutral sphingomyelinases. Biochemistry 45: 11247–11256.
4. Wu BX, Clarke CJ, Hannun YA (2010) Mammalian neutral sphingomyelinases: regulation and roles in cell signaling responses. Neuromolecular medicine 12: 320–330.
5. Duan R-D, Bergman T, Xu N, Wu J, Cheng Y, et al. (2003) Identification of human intestinal alkaline sphingomyelinase as a novel ecto-enzyme related to the nucleotide phosphodiesterase family. Journal of Biological Chemistry 278: 38528–38536.
6. Duan R-D (2006) Alkaline sphingomyelinase: an old enzyme with novel implications. Biochimica et Biophysica Acta (BBA)-Molecular and Cell Biology of Lipids 1761: 281–291.
7. Devlin CM, Leventhal AR, Kuriakose G, Schuchman EH, Williams KJ, et al. (2008) Acid sphingomyelinase promotes lipoprotein retention within early atheromata and accelerates lesion progression. Arteriosclerosis, thrombosis, and vascular biology 28: 1723–1730.
8. Yu H, Zeidan YH, Wu BX, Jenkins RW, Flotte TR, et al. (2009) Defective acid sphingomyelinase pathway with Pseudomonas aeruginosa infection in cystic fibrosis. American journal of respiratory cell and molecular biology 41: 367.
9. Lang PA, Schenck M, Nicolay JP, Becker JU, Kempe DS, et al. (2007) Liver cell death and anemia in Wilson disease involve acid sphingomyelinase and ceramide. Nature medicine 13: 164–170.
10. Buckhaults P, Rago C, Croix BS, Romans KE, Saha S, et al. (2001) Secreted and cell surface genes expressed in benign and malignant colorectal tumors. Cancer research 61: 6996–7001.
11. Wright KO, Messing EM, Reeder JE (2002) Increased expression of the acid sphingomyelinase-like protein ASML3a in bladder tumors. The Journal of urology 168: 2645–2649.
12. Nambiar PR, Nakanishi M, Gupta R, Cheung E, Firouzi A, et al. (2004) Genetic signatures of high-and low-risk aberrant crypt foci in a mouse model of sporadic colon cancer. Cancer research 64: 6394–6401.
13. Fornoni A, Sageshima J, Wei C, Merscher-Gomez S, Aguillon-Prada R, et al. (2011) Rituximab targets podocytes in recurrent focal segmental glomerulosclerosis. Sci Transl Med 3: 85ra46.
14. Sinha A, Bagga A (2013) Rituximab therapy in nephrotic syndrome: implications for patients' management. Nature Reviews Nephrology.
15. Fornoni A, Merscher S, Kopp JB (2014) Lipid biology of the podocyte [mdash] new perspectives offer new opportunities. Nature Reviews Nephrology.
16. Yoo T-H, Pedigo CE, Guzman J, Correa-Medina M, Wei C, et al. (2014) Sphingomyelinase-Like Phosphodiesterase 3b Expression Levels Determine Podocyte Injury Phenotypes in Glomerular Disease. Journal of the American Society of Nephrology: ASN. 2013111213.
17. Spence MW, Byers D, Palmer F, Cook HW (1989) A new Zn2+-stimulated sphingomyelinase in fetal bovine serum. Journal of Biological Chemistry 264: 5358–5363.
18. Kolter T, Sandhoff K (2005) Principles of lysosomal membrane digestion: stimulation of sphingolipid degradation by sphingolipid activator proteins and anionic lysosomal lipids. Annu Rev Cell Dev Biol 21: 81–103.
19. Genin S, Denny TP (2012) Pathogenomics of the Ralstonia solanacearum species complex. Annual review of phytopathology 50: 67–89.
20. Poueymiro M, Genin S (2009) Secreted proteins from *Ralstonia solanacearum*: a hundred tricks to kill a plant. Current opinion in microbiology 12: 44–52.
21. Salanoubat M, Genin S, Artiguenave F, Gouzy J, Mangenot S, et al. (2002) Genome sequence of the plant pathogen Ralstonia solanacearum. Nature 415: 497–502.
22. Aldon D, Brito B, Boucher C, Genin S (2000) A bacterial sensor of plant cell contact controls the transcriptional induction of Ralstonia solanacearum pathogenicity genes. The EMBO journal 19: 2304–2314.
23. Peeters N, Guidot A, Vailleau F, Valls M (2013) Ralstonia solanacearum, a widespread bacterial plant pathogen in the post-genomic era. Molecular plant pathology 14: 651–662.
24. Huseby M, Shi K, Brown CK, Digre J, Mengistu F, et al. (2007) Structure and biological activities of beta toxin from Staphylococcus aureus. Journal of bacteriology 189: 8719–8726.
25. Titball RW, Leslie DL, Harvey S, Kelly D (1991) Hemolytic and sphingomyelinase activities of Clostridium perfringens alpha-toxin are dependent on a domain homologous to that of an enzyme from the human arachidonic acid pathway. Infection and immunity 59: 1872–1874.
26. Oda M, Hashimoto M, Takahashi M, Ohmae Y, Seike S, et al. (2012) Role of sphingomyelinase in infectious diseases caused by Bacillus cereus. PloS one 7: e38054.
27. Jenkins RW, Canals D, Idkowiak-Baldys J, Simbari F, Roddy P, et al. (2010) Regulated Secretion of Acid Sphingomyelinase IMPLICATIONS FOR SELECTIVITY OF CERAMIDE FORMATION. Journal of Biological Chemistry 285: 35706–35718.
28. Ferlinz K, Hurwitz R, Moczall H, Lansmann S, Schuchman EH, et al. (1997) Functional Characterization of the N-glycosylation Sites of Human Acid Sphingomyelinase by Site-Directed Mutagenesis. European Journal of Biochemistry 243: 511–517.
29. Bartelsen O, Lansmann S, Nettersheim M, Lemm T, Ferlinz K, et al. (1998) Expression of recombinant human acid sphingomyelinase in insect Sf 21 cells: purification, processing and enzymatic characterization. Journal of biotechnology 63: 29–40.
30. Altschul SF, Madden TL, Schäffer AA, Zhang J, Zhang Z, et al. (1997) Gapped BLAST and PSI-BLAST: a new generation of protein database search programs. Nucleic acids research 25: 3389–3402.
31. Ago H, Oda M, Takahashi M, Tsuge H, Ochi S, et al. (2006) Structural basis of the sphingomyelin phosphodiesterase activity in neutral sphingomyelinase from Bacillus cereus. Journal of Biological Chemistry 281: 16157–16167.
32. Openshaw AE, Race PR, Monzó HJ, Vázquez-Boland J-A, Banfield MJ (2005) Crystal structure of SmcL, a bacterial neutral sphingomyelinase C from Listeria. Journal of Biological Chemistry 280: 35011–35017.
33. Lansmann S, Schuette CG, Bartelsen O, Hoernschemeyer J, Linke T, et al. (2003) Human acid sphingomyelinase. European Journal of Biochemistry 270: 1076–1088.
34. Simonaro CM, Desnick RJ, McGovern MM, Wasserstein MP, Schuchman EH (2002) The demographics and distribution of type B Niemann-Pick disease: novel mutations lead to new genotype/phenotype correlations. The American Journal of Human Genetics 71: 1413–1419.
35. Pittis M, Ricci V, Guerci V, Marcais C, Ciana G, et al. (2004) Acid sphingomyelinase: Identification of nine novel mutations among Italian Niemann Pick type B patients and characterization of in vivo functional in-frame start codon. Human mutation 24: 186–187.
36. Pavlů-Pereira H, Asfaw B, Poupčtová H, Ledvinova J, Sikora J, et al. (2005) Acid sphingomyelinase deficiency. Phenotype variability with prevalence of intermediate phenotype in a series of twenty-five Czech and Slovak patients. A multi-approach study. Journal of inherited metabolic disease 28: 203–227.
37. Desnick JP, Kim J, He X, Wasserstein MP, Simonaro CM, et al. (2010) Identification and characterization of eight novel SMPD1 mutations causing types A and B Niemann-Pick disease. Molecular Medicine 16: 316.
38. Pavlů H, Elleder M (1997) Two novel mutations in patients with atypical phenotypes of acid sphingomyelinase deficiency. Journal of inherited metabolic disease 20: 615–616.
39. Rodríguez-Pascau L, Gort L, Schuchman EH, Vilageliu L, Grinberg D, et al. (2009) Identification and characterization of SMPD1 mutations causing Niemann-Pick types A and B in Spanish patients. Human mutation 30: 1117–1122.
40. Takahashi T, Akiyama K, Tomihara M, Tokudome T, Nishinomiya F, et al. (1997) Heterogeneity of liver disorder in type B Niemann-Pick disease. Human pathology 28: 385–388.
41. Sikora J, Pavlu-Pereira H, Elleder M, Roelofs H, Wevers R (2003) Seven Novel Acid Sphingomyelinase Gene Mutations in Niemann-Pick Type A and B Patients. Annals of human genetics 67: 63–70.
42. Ricci V, Stroppiano M, Corsolini F, Di Rocco M, Parenti G, et al. (2004) Screening of 25 Italian patients with Niemann-Pick a reveals fourteen new mutations, one common and thirteen private, in SMPD1. Human mutation 24: 105–105.

43. Dardis A, Zampieri S, Filocamo M, Burlina A, Bembi B, et al. (2005) Functional in vitro characterization of 14 SMPD1 mutations identified in Italian patients affected by Niemann Pick Type B disease. Human mutation 26: 164–164.

44. Atilla-Gokcumen GE, Muro E, Relat-Goberna J, Sasse S, Bedigian A, et al. (2014) Dividing Cells Regulate Their Lipid Composition and Localization. Cell.

45. Fornoni A, Sageshima J, Wei C, Merscher-Gomez S, Aguillon-Prada R, et al. (2011) Rituximab targets podocytes in recurrent focal segmental glomerulosclerosis. Science Translational Medicine 3: 85ra46–85ra46.

46. Noto PB, Bukhtiyarov Y, Shi M, McKeever BM, McGeehan GM, et al. (2012) Regulation of Sphingomyelin Phosphodiesterase Acid-Like 3A Gene (SMPDL3A) by Liver X Receptors. Molecular pharmacology 82: 719–727.

47. Sabourdy F, Selves J, Astudillo L, Laurent C, Brousset P, et al. (2011) Is active acid sphingomyelinase required for the antiproliferative response to rituximab? Blood 117: 3695–3696.

48. Kato K, Nishimasu H, Okudaira S, Mihara E, Ishitani R, et al. (2012) Crystal structure of Enpp1, an extracellular glycoprotein involved in bone mineralization and insulin signaling. Proceedings of the National Academy of Sciences 109: 16876–16881.

49. Nishimasu H, Okudaira S, Hama K, Mihara E, Dohmae N, et al. (2011) Crystal structure of autotaxin and insight into GPCR activation by lipid mediators. Nature structural & molecular biology 18: 205–212.

50. Hausmann J, Kamtekar S, Christodoulou E, Day JE, Wu T, et al. (2011) Structural basis of substrate discrimination and integrin binding by autotaxin. Nature structural & molecular biology 18: 198–204.

51. Stefan C, Jansen S, Bollen M (2005) NPP-type ectophosphodiesterases: unity in diversity. Trends in biochemical sciences 30: 542–550.

52. Schissel SL, Jiang X-c, Tweedie-Hardman J, Jeong T-s, Camejo EH, et al. (1998) Secretory Sphingomyelinase, a Product of the Acid Sphingomyelinase Gene, Can Hydrolyze Atherogenic Lipoproteins at Neutral pH IMPLICATIONS FOR ATHEROSCLEROTIC LESION DEVELOPMENT. Journal of Biological Chemistry 273: 2738–2746.

53. Choi J, Tanaka K, Cao Y, Qi Y, Qiu J, et al. (2014) Identification of a Plant Receptor for Extracellular ATP. Science 343: 290–294.

54. Song CJ, Steinebrunner I, Wang X, Stout SC, Roux SJ (2006) Extracellular ATP induces the accumulation of superoxide via NADPH oxidases in Arabidopsis. Plant physiology 140: 1222–1232.

## 3

Flagella-Mediated Adhesion and Extracellular DNA Release Contribute to Biofilm Formation and Stress Tolerance of *Campylobacter jejuni*

Sarah L. Svensson, Mark Pryjma, Erin C. Gaynor*

Department of Microbiology and Immunology, University of British Columbia, Vancouver, British Columbia, Canada

Abstract

Campylobacter jejuni is a leading cause of foodbourne gastroenteritis, despite fragile behaviour under standard laboratory conditions. In the environment, *C. jejuni* may survive within biofilms, which can impart resident bacteria with enhanced stress tolerance compared to their planktonic counterparts. While *C. jejuni* forms biofilms *in vitro* and in the wild, it had not been confirmed that this lifestyle confers stress tolerance. Moreover, little is understood about molecular mechanisms of biofilm formation in this pathogen. We previously found that a Δ*cprS* mutant, which carries a deletion in the sensor kinase of the CprRS two-component system, forms enhanced biofilms. Biofilms were also enhanced by the bile salt deoxycholate and contained extracellular DNA. Through more in-depth analysis of Δ*cprS* and WT under conditions that promote or inhibit biofilms, we sought to further define this lifestyle for *C. jejuni*. Epistasis experiments with Δ*cprS* and flagellar mutations (Δ*flhA*, Δ*pflA*) suggested that initiation is mediated by flagellum-mediated adherence, a process which was kinetically enhanced by motility. Lysis was also observed, especially under biofilm-enhancing conditions. Microscopy suggested adherence was followed by release of eDNA, which was required for biofilm maturation. Importantly, inhibiting biofilm formation by removal of eDNA with DNase decreased stress tolerance. This work suggests the biofilm lifestyle provides *C. jejuni* with resilience that has not been apparent from observation of planktonic bacteria during routine laboratory culture, and provides a framework for subsequent molecular studies of *C. jejuni* biofilms.

Editor: Roy M. Roopll, East Carolina University School of Medicine, United States of America

Funding: S. L. S. is supported by a Senior Graduate Traineeship from the Michael Smith Foundation for Health Research. E. C. G. is supported by a Canada Research Chair Award. This work was funded by a Burroughs Wellcome Fund Career Development Award, Canadian Institutes of Health Research MOP-68981, and Natural Sciences and Engineering Research Council of Canada grant F11-05378 to E. C. G. The funders had no role in study design, data collection and analysis, decision to publish, or preparation of the manuscript.

Competing Interests: The authors have declared that no competing interests exist.

* Email: egaynor@mail.ubc.ca

Introduction

Campylobacter jejuni is a prevalent food- and waterborne gastrointestinal pathogen. Infection commonly presents as an acute gastroenteritis, marked by fever, stomach cramps, and diarrhea. Although illness is usually self-limiting, the high incidence of infection, together with the significant subset of cases that go on to manifest as serious autoimmune sequelae (such as Guillain-Barré syndrome), contributes to the significant burden of *C. jejuni* infection. Preventative strategies that limit *C. jejuni* exposure and infection may thus greatly reduce its impact.

Mechanisms by which *C. jejuni* causes disease are relatively enigmatic, suggesting it may use unique strategies compared to more extensively characterized enteric pathogens. In fact, many factors identified as critical to pathogenesis include those related to survival of stress and basic biology of the organism, including the stringent response, motility, and surface carbohydrates [1,2]. In the gastrointestinal tract of commensal or susceptible animals, *C. jejuni* tolerates insults such as bile salts. Moreover, *C. jejuni* is zoonotic, with sporadic cases associated with consumption of undercooked poultry and outbreaks arising from contaminated water, and thus survives transmission environments characterized by a range of nutrient availabilities, temperatures, oxygen tensions, and osmolarities.

Understanding how *C. jejuni* survives in such environments may help direct strategies to limit its impact. However, inspection of the genome of numerous strains suggests lacks many classical stress tolerance factors, including the RpoS stationary phase sigma factor [3]. *C. jejuni* is also relatively fragile and fastidious in the laboratory, with specific atmospheric and nutrient requirements for growth, bringing into question how it can adapt to environments both inside and outside the host. *C. jejuni* may serve as a model for understanding how pathogens with limited regulatory repertoires adapt to pathogenesis-related environments.

The paradox of *C. jejuni*'s success may be explained by a tendency to persist in distinct lifestyles in natural environments. Phenotypes displayed during logarithmic growth in rich broth may not be representative of phenotypes expressed in nature. In fact, most bacteria exist naturally in sessile biofilms: adhered communities of microorganisms encased in a polymeric extracellular matrix. Formation of a biofilm proceeds in a set of distinct steps (adherence, microcolony formation, matrix release, dispersal) that have been proposed to represent 'microbial development' [4]. The mechanisms and factors that underlie each step appear distinct for each bacterial species. For example, attachment may be mediated by flagella, pili, carbohydrates, or protein adhesins, and the biofilm matrix can be a unique mixture of hydrated extracellular polymeric substances, such as carbohydrates, proteins, and lipids [5]. The significant contribution of extracellular DNA (eDNA) to biofilm structure and function, including structural integrity, recombination, and antibiotic resistance, has been also recently become apparent [6,7]. Autolysis can underlie either biofilm formation or dispersal, and can release eDNA [6,8,9].

Biofilms have been proposed to contribute to survival of *C. jejuni* in the food chain, from farm to fork [10]. However, the contribution of biofilms to *C. jejuni* resilience is unclear. Observation of *C. jejuni* in the wild, such as within biofilms in aquatic environments suggests that biofilm residents display several phenotypic differences from their planktonic counterparts, including enhanced stress tolerance. There is some evidence that *C. jejuni* cells residing within biofilms in aquatic environments survive better than their planktonic counterparts [11–15]. Strains residing within chicken house drinking water biofilms have also been found to colonize broiler flocks [16], and *C. jejuni* forms biofilms *in vitro* under conditions that may mimic environments encountered during pathogenesis. Biofilm formation is affected by nutrients, temperature, oxygen tension, and osmolarity [17,18] and notably, enhanced in the presence of the bile salt sodium deoxycholate (DOC) [19]. Although biofilm-enhanced mutants have been identified, including $\Delta ppk1$, $\Delta ppk2$, $\Delta spoT$, $\Delta peb4$, $\Delta kpsM$, and $\Delta waaF$ [20–24], demonstration of increased tolerance by such strains is hampered by the pleiotropic effects of such mutations. Molecular factors that mediate *C. jejuni* biofilm formation are also poorly understood. Flagella are required [17,25,26] and may mediate adhesion - both bacteria-bacteria and biofilm-host cell [27–29]. Whether motility or the flagellar structure itself is important, and at which stage each is required, is unclear. Moreover, the biofilm matrix of such a "sugary" bug remains surprisingly ill-defined: while carbohydrate changes correlate with biofilm formation [22,23], a carbohydrate component has not been definitively identified. Instead, extracellular DNA has been observed in *C. jejuni* biofilms [19,30].

We previously identified a two-component regulatory system, CprRS (*Campylobacter* planktonic growth regulation), which may control phenomena central to biofilm formation [19]. A $\Delta cprS$ sensor kinase mutant forms markedly enhanced biofilms compared to the parental strain, but has no obvious differences in carbohydrate production. Here, we extend characterization of this strain to provide insight into mechanisms of *C. jejuni* biofilm formation through exploration of the temporal development of *C. jejuni* biofilms using confocal microscopy. We show that flagella are required for initial attachment of biofilms, and that eDNA is dispensable for this step. We have also find evidence of a lytic process that correlates with biofilm maturation and releases eDNA. Finally, we provide evidence that inhibition of biofilm formation affects the fitness of *C. jejuni*, and that biofilm formation may contribute to long-term survival of *C. jejuni* populations by contributing to genetic diversity.

Materials and Methods

Bacterial strains and routine culture conditions

C. jejuni strain 81–176, a highly invasive isolate from a raw milk outbreak [31], was used as the WT parental strain. All other strains are listed in **Table 1**. Targeted deletion mutants, such as $\Delta cprS$, $\Delta flhA$, and $\Delta flgR$, have been described previously [19,32]. The $\Delta pflA$ mutant was isolated from a transposon mutant screen using the Mariner system developed for *C. jejuni* [33]. The double $\Delta cprS$ $\Delta flhA$ mutant was constructed by naturally transforming genomic DNA from $\Delta flhA$ into $\Delta cprS$ and selecting for Kan^R Cm^R colonies. *C. jejuni* strains were routinely cultured in Müller-Hinton (MH) broth (Oxoid, Hampshire, England) or on MH agar (1.7%) plates at 37°C under microaerobic conditions (6% O_2, 12% CO_2) in a Sanyo tri-gas incubator (plates and standing liquid cultures/biofilms) or generated using the Oxoid CampyGen system (shaking broth cultures). All media were supplemented with 10 μg mL^{-1} vancomycin and 5 μg mL^{-1} trimethoprim (Sigma, Oakville, ON). Where appropriate, antibiotics kanamycin (Kan), chloramphenicol (Cm), and streptomycin (Str) were added to a final concentration of 40 μg mL^{-1}, 15 μg mL^{-1}, and 100 μg mL^{-1}, respectively.

Crystal violet biofilm assay

Biofilm formation was assessed as described previously [19,20,22,23,34,35]. Values shown are the mean A_{570} +/- standard error of three biofilm cultures for each strain/condition. To determine statistical significance, an unpaired *t*-test was performed with significance set at $p<0.05$. Where indicated, DNase I (Invitrogen) was added to a final concentration of 90 U mL^{-1}, and sodium deoxycholate (DOC, Sigma) was included at 0.05%. In some experiments, 50 μL of WT culture supernatant, isolated as described below, was added to each mL of fresh MH broth. Genomic DNA was isolated from WT *C. jejuni* grown for 24 h on MH agar using the Qiagen genomic tip 100/G kit and was added, where indicated, at a concentration of 500 ng μL^{-1}.

Standing culture growth

Standard biofilm cultures of each strain (WT, $\Delta cprS$, $\Delta flhA$, and $\Delta cprS$ $\Delta flhA$) were inoculated to an OD_{600} of 0.002 in either MH broth only, MH/DOC (0.05%), MH/DNase (90 U mL^{-1}), or MH/DOC/DNase. Following 2 days of growth under microaerobic conditions, tubes were either stained with crystal violet to assess biofilm formation, or vortexed for one min., followed by measuring OD_{600}.

Detection of bacterial cell lysis

Lysis was assessed by SDS-PAGE and Western blot analysis of culture supernatants. Following growth in shaking broth culture (10 mL) for 24 h, a 1 mL sample of culture was harvested for analysis of total cellular protein expression. Cells from the rest of the culture were removed by centrifugation at 10,000×g for 5 min. and discarded. Any cells remaining in this clarified supernatant were removed by filtration through a 0.22 μM filter. Supernatants were then concentrated approximately 10-fold from 2.5 mL to 250 μL using 3 kDa cutoff Amicon Ultra centrifugal filter units (Millipore, Billerica, MA) by centrifugation for 60 min. at 4,000 x g. Samples were then analyzed by SDS-PAGE/Western blotting with an anti-CosR antibody (a gift from Dr. Stu Thompson).

Quantification of eDNA

The amount of DNA present in culture supernatants was measured by QPCR. Supernatants, prepared as above, were used as templates for qPCR using primers *cprR*-QPCRFWD/REV (5'-GACCTTTCTTTGCCAGGGCTTGAT and 5'-GGTAGG-TAATCATCTGCTCCAAGCTC, respectively). QPCR was performed in triplicate on equal volumes (0.5 μL) of supernatant as template using IQ SYBR Green Supermix and the MyIQ Real-time PCR Detection System (Biorad, Mississauga, ON) according to the manufacturer's specifications.

Table 1. Strains used in this study.

Strain	Genotype	Relevant characteristics	Source
C. jejuni strains			
81–176	Wild type (WT)		Korlath *et al.* 1985
DRH461	Δ*astA*	Str[R]	
Δ*cprS*	Δ*cprS::aph3*	biofilm-enhanced, Kan[R]	Svensson *et al.* 2009
Δ*flhA*	*flhA::cat-rpsL*	aflagellate, non-motile, Cm[R]	Hendrixson and DiRita 2003
Δ*cprS* Δ*flhA*	Δ*cprS::aph3 flhA::cat-rpsL*	aflagellate, non-motile, Kan[R] Cm[R]	this study
Δ*pflA*	*pflA::solo*	paralyzed flagella, non-motile, Kan[R]	laboratory collection
Δ*flgR*	*flgR::kan-rpsL*	aflagellate, non-motile, Kan[R]	Hendrixson and DiRita 2003
Plasmids			
P$_{atpF}$-*gfp* pRY112	GFP under control of the *atpF'* promoter in pRY112	Cm[R]	Apel *et al.* 2012

Confocal microscopy of biofilms

For confocal microscopy, a plasmid encoding green fluorescent protein (GFP) expressed from the *atpF'* promoter [36] was introduced into strains by natural transformation. Biofilm cultures were set up as in above, except a glass coverslip was included standing upright in each tube. MH broth was supplemented with Cm for plasmid selection. Following 12, 24, or 36 h, culture medium was removed and biofilms were fixed by the addition of 4% paraformaldehyde in PBS, pH 7.4 for 15 min. Fixing solution was removed and replaced with PBS, and coverslips were stored at 4°C. Samples were mounted using Prolong Gold Antifade with DAPI (4',6-diamino-2-phenylindol; Invitrogen), and imaging was performed with an Olympus Fluoview FV1000 laser scanning confocal microscope with FV10-ASW 2.0 Viewer software to adjust images.

Measurement of recombination

Transfer of resistance markers between strains was measured in mixed-strain shaking broth cultures. Each strain was marked (on the chromosome) with a different antibiotic resistance marker and recombination was determined by measuring appearance of doubly resistant recombinant clones. Briefly, WT (marked with Str[R]) was grown in mixed culture (1:1) with either an isogenic WT strain (marked with Kan[R]; insertion into an rRNA spacer via pRRK) or the Δ*cprS* hyper-biofilm mutant (marked with Kan[R]; allelic replacement of the *cprS* locus). Cultures were grown in either MH alone or MH/DOC (0.05%). Cells were removed A) immediately following inoculation and B) following 8 h growth and plated on MH agar with the appropriate antibiotics (Kan, Cm, and/or Str) for determination of colony-forming units (CFUs).

Results

Strains grown under conditions that promote biofilms show enhanced and accelerated appearance of eDNA

We previously noted extracellular DNA (eDNA) in *C. jejuni* biofilms, and thatthe amount of eDNA appeared to be qualitatively increased in strains forming enhanced biofilms, such as Δ*cprS* and WT in MH/deoxycholate (DOC; a bile salt) [19]. Furthermore, Production of a specific surface polysaccharide does not appear to correlate with *C. jejuni* biofilm formation, unlike in other bacteria. We thus hypothesized that eDNA could instead be a marker for *C. jejuni* biofilm formation. To this end, the temporal relationship between biofilm formation and appearance of eDNA was examined by confocal microscopy observation of biofilms in a time course experiment (**Fig. 1**). GFP-expressing bacteria were inoculated into MH broth standing cultures (WT and Δ*cprS*). WT grown in MH/DOC was also included for comparison. At different time points post-inoculation (12, 24, 36 h), biofilms were fixed, stained with DAPI, and samples were observed by confocal microscopy.

For WT under routine laboratory conditions (top panels, 'WT; MH alone'), green bacteria first adhered to the coverslip in small microcolonies (12 h). This was followed, at 24 h, by the appearance of blue DNA. DNA, likely extracellular due to its mucoid appearance, was more prevalent in regions closer to the interior of the biofilm. As time progressed to 36 h, the amount of eDNA and the apparent thickness of the biofilm increased further. This suggested that DNA not only was a quantitative marker for biofilm formation, but was also a temporal marker as it was not present in appreciable amounts upon biofilm initiation.

We next compared kinetics for a strain and condition previously shown to enhance biofilm formation (middle and bottom panels, 'Δ*cprS*' and 'WT in MH/DOC'). A similar temporal relationship between initiation and DNA appearance was seen for both biofilm-enhanced cultures. However, the process appeared both accelerated and markedly enhanced. For example, at 12 h, we generally observed that Δ*cprS* microcolonies were markedly larger than those observed for WT. The heterogeneous nature of the biofilms formed on coverslips precluded quantification of thickness for comparison. However, Δ*cprS* and WT in MH/DOC generally appeared thicker, compared to WT in MH alone, at this time point. Furthermore, foci of blue DNA were already visible at 12 h in some regions of Δ*cprS* biofilms. By 36 h, even larger strands of eDNA were observed. WT in MH/DOC appeared more accelerated and enhanced than Δ*cprS*. For example, at 12 h, had significant amounts of blue eDNA present. For both Δ*cprS* and WT in MH/DOC at 36 h, less overall coverage of both bacteria and DNA on the coverslip likely reflects some sloughing off of very thick biofilms. Together, observations of both WT and enhanced *C. jejuni* biofilms suggest that the appearance of eDNA coincides with maturation and may be a temporal, and potentially quantitative, marker for biofilm formation in *C. jejuni*.

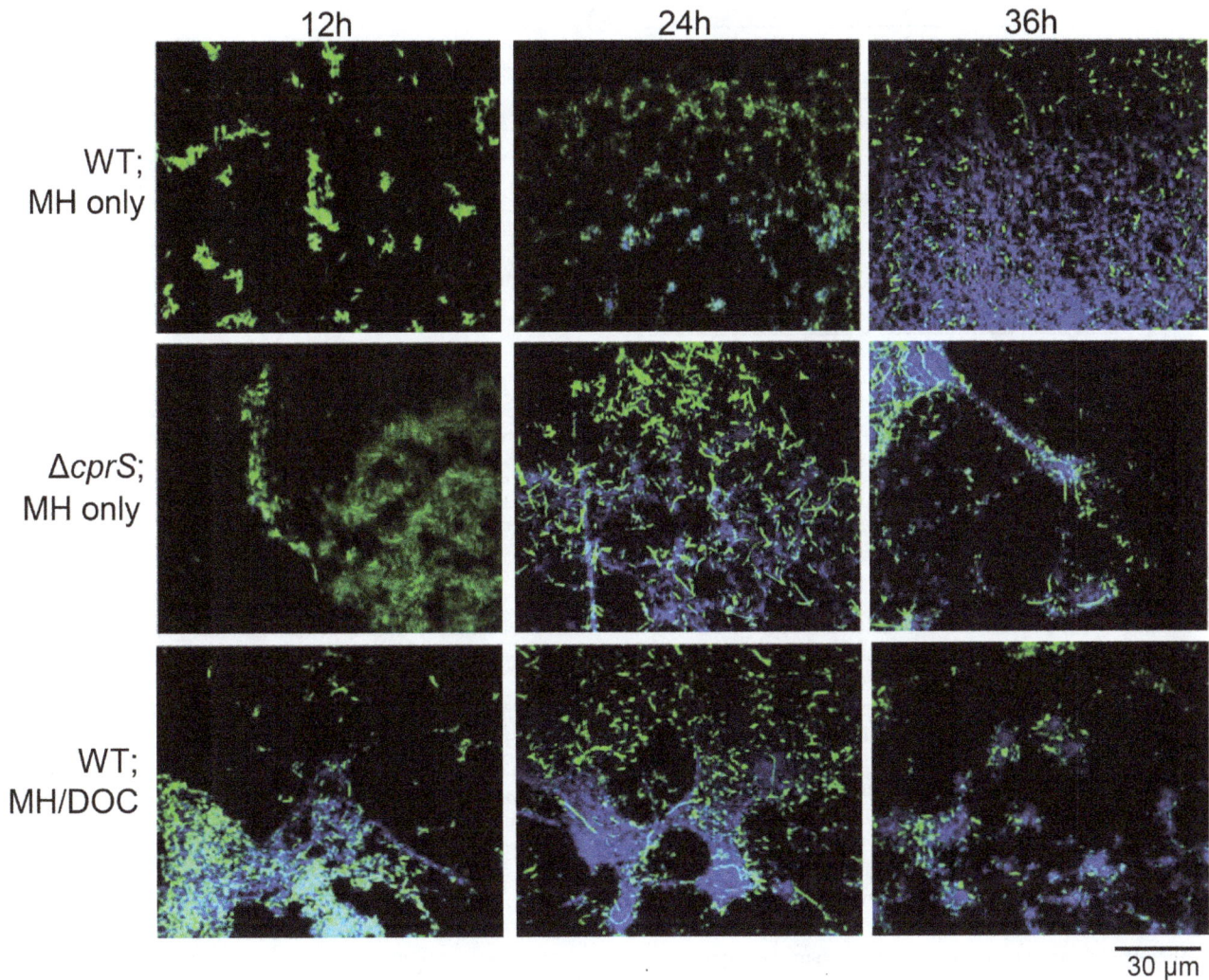

Figure 1. DNA appears in WT biofilms following attachment and is more pronounced under conditions that promote biofilm formation. Biofilms of WT or ΔcprS were grown on glass coverslips in MH broth alone or MH/DOC (0.05%). At indicated times post-inoculation, coverslips were fixed, stained with DAPI, and visualized by confocal microscopy. Green: GFP-expressing bacteria; Blue: DAPI-stained DNA.

Conditions that increase extracellular DNA also correlate with bacterial lysis

To confirm microscopy observations that suggested more eDNA was present under biofilm-enhanced conditions, the relative DNA concentration of culture supernatants was measured (**Fig. 2A**). Strains were grown overnight, and supernatants were harvested and subjected to qPCR with primers specific for the chromosomal *cprR* gene. For supernatants of biofilm-enhanced bacteria (ΔcprS and WT in MH/DOC), an approximately 3-4-fold increase in DNA was measured compared to WT in MH only (**Fig. 2A**; left three bars). Supernatants harvested from ΔcprS contained 3.2-fold more DNA than WT (p<0.0001). Similarly, when WT was grown in MH/DOC, bacteria released 3.8-fold higher levels of DNA than in MH alone (p = 0.0015). These values were consistent with microscopy observations (**Fig. 1**). We also did not observe an effect of flagellar mutation on the amount of DNA in the media (see below). This suggests that eDNA could be a quantitative marker for *C. jejuni* biofilm formation.

Microscopy suggested the mechanism underlying appearance of eDNA may be lytic, as can be the case in other bacteria [9,37]. *C. jejuni* often displays a 2-log reduction in colony-forming units

following log phase. This is enhanced in the ΔcprS mutant and was previously hypothesized to occur as a result of lysis [19]. The ΔcprS mutant also displays increased protein species in the media compared to WT during routine culture [19]. We hypothesized that lysis may underlie release of eDNA. Like ΔcprS, we observed increased protein in the media when WT was grown in the presence of sub-MIC levels of DOC (data not shown). We thus used Western blotting to detect the presence the cytosolic regulatory protein CosR [24] in culture supernatants (**Fig. 2B**). We did not observe differences in CosR expression in total cell protein samples. Supernatants from both ΔcprS and WT in MH/DOC contained significant levels of CosR compared to WT in MH only, which had undetectable levels. Detection of cytosolic proteins in supernatants strongly suggested lysis in ΔcprS, and WT in MH/DOC. However, the flagellar export apparatus has been reported to act as a type III secretion system-like machine in *C. jejuni*, secreting non-flagellar proteins into the medium and host cells. [38,39]. Furthermore, DOC stimulates secretion of some flagellar-secreted proteins [40], and expression of FlaA is upregulated in ΔcprS [19]. Thus, we wanted to rule out flagella-dependent secretion of CosR under biofilm and DNA release-

A

B

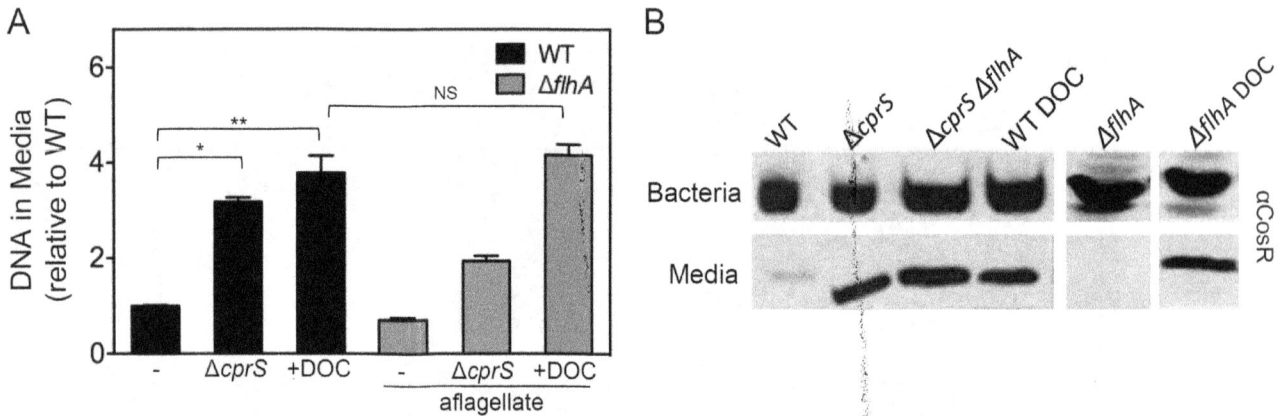

Figure 2. Increased extracellular DNA and lysis occur in biofilm-enhanced cultures. A) Cell-free supernatants contain more DNA under biofilm-enhancing conditions. Supernatants were isolated from cultures (at a similar OD_{600}) of WT, $\Delta cprS$, $\Delta flhA$, or $\Delta cprS \, \Delta flhA$ grown in either MH alone or MH/DOC. Equal volumes were used as templates for qPCR. DNA amounts are normalized to WT in MH alone. Error bars represent the mean of three separate cultures. *p<0.0001; **p=0.0015; NS p=0.42, unpaired t-test. **B)** Lysis, independent of flagella, occurs under conditions that promote biofilms. Cell-free supernatants were isolated as in above. Both total cellular protein ('Bacteria') and supernatants ('Media') were analyzed by Western blotting with an antibody specific for the cytoplasmic response regulator CosR.

promoting conditions. As such, WT and $\Delta cprS$ strains harboring a deletion of the flagellar export apparatus gene $flhA$ [39] were also included in Western blot analyses (**Fig. 2B**; lanes denoted $\Delta flhA$). Importantly, we observed CosR in supernatants when $\Delta flhA$ was introduced into either $\Delta cprS$ or WT in DOC. Together, these observations suggested that a lytic mechanism, occurring independently of flagella-mediated export, underlies the appearance of both eDNA (**Fig. 2A**, right three bars) and protein (**Fig. 2B**) in supernatants of biofilm-enhanced *C. jejuni*.

Addition of exogenous DNA enhances biofilms; removal of eDNA inhibits biofilm formation

Although DNA release correlated with biofilm phenotype, it was still unclear whether eDNA contributed mechanistically to biofilm formation. Pre-formed *C. jejuni* biofilms can be disrupted with DNase I, suggesting that like for other bacteria, eDNA plays a functional role in these structures [19]. To determine if eDNA affects the *C. jejuni* biofilm formation process, we used a crystal violet assay standard to our laboratory [19,20,22,23,34,35] to test the effect of adding exogenous *C. jejuni* genomic DNA to standing cultures (**Fig. 3A**). The effect of adding cell-free culture supernatants, which contain eDNA, was also examined. We consistently observed that culture supernatants modestly enhanced biofilm formation (MH/sup), although this difference did not reach statistical significance (p = 0.08). The spent media (1/20 volume 10-fold concentrated supernatants was added) also appeared to partially inhibit growth. We thus determined the effect of added DNA alone. Purified *C. jejuni* genomic DNA was added to WT biofilm cultures (MH/gDNA), and these cultures showed a significant increase in biofilm formation compared to those grown in broth alone (p = 0.003). We also performed the complementary experiment and asked whether endogenous DNA was required for biofilm formation (**Fig. 3B**). Biofilm cultures of WT, $\Delta cprS$, and WT in MH/DOC were grown in the presence or absence of DNase I. Biofilm formation by WT in DNase (MH/DNase) was reduced compared to that of WT in MH alone (p = 0.0013). Furthermore, when DNase was included in biofilm-enhanced cultures ($\Delta cprS$; WT in MH/DOC), they formed significantly less biofilm than their counterparts grown without DNase (p = 0.0017, p = 0.0025, respectively). Collectively, these experiments suggested

that eDNA release during biofilm formation was not simply a consequence of, but required for, biofilm formation.

DNA is required for maturation of the *C. jejuni* biofilm

The relatively low sensitivity of the crystal violet assay prevented us from determining if DNase completely inhibited biofilm formation, or if it arrested it at a very early stage of development. To determine at which stage biofilms were arrested by DNase, and to confirm that eDNA was in fact being degraded by the addition of DNase, we used confocal microscopy to observe biofilm formation in the presence and absence of the enzyme (**Fig. 4**). For WT (left panels, 12 h, 24 h, and 36 h), DAPI-stained DNA surrounding cells observed in MH alone was not observed when DNase was included, suggesting that the enzyme was sufficiently active. However, a few cells that were not expressing green GFP were blue, as the DAPI was presumably able to enter the cells and stain chromosomal DNA, but DNase was too large to enter. Unlike what was suggested by the low-resolution crystal violet assay above (**Fig. 3B**), confocal microscopy showed that DNase did not completely eliminate biofilm formation. Closer observation suggested that DNase arrested WT biofilms following adherence. Cultures with DNase included were still adhered to the coverslip (bottom three panels), but they remained in a monolayer and did not progress to more elaborate structures observed in MH only in the same time period (top three panels). Thus, the crystal violet assay, used extensively to measure biofilm formation in multiple bacterial species, may not be sensitive enough to detect the initial adherence step in *C. jejuni*.

Data in Fig. 3 showed that, by the crystal violet assay, biofilm-enhanced cultures ($\Delta cprS$, WT in MH/DOC; right panels) displayed a similar inhibition of biofilm formation upon inclusion of DNase as WT under routine conditions. By confocal microscopy, however, addition of DNase not only inhibited biofilm formation in $\Delta cprS$ and WT (MH/DOC), but very few adherent bacteria were observed. Based on this observation, together with poor growth of $\Delta cprS \, \Delta flhA$ during the above experiment (**Fig. 2A**), we hypothesized the biofilm may provide *C. jejuni* with fitness that is especially required by $\Delta cprS$ and WT in MH/DOC. This was addressed in subsequent experiments (below). Nonetheless, our observations are consistent with removal

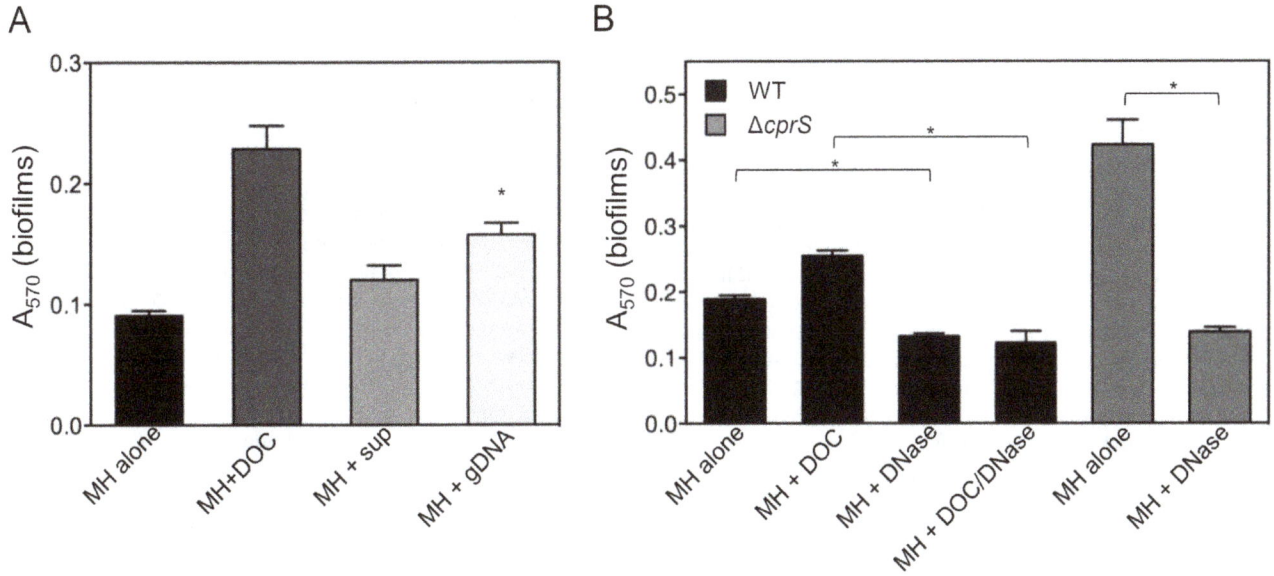

Figure 3. Cell-free supernatants and exogenous DNA promote biofilms and DNA is necessary for biofilm formation. A) Exogenous DNA enhances biofilms. Culture supernatants, concentrated for >3 kDa size components, or gDNA isolated from WT *C. jejuni* grown for 24 h on MH plates (500 ng) were included in fresh MH broth. Tubes were then inoculated with WT, and following 2 days growth, biofilms were quantified with crystal violet. *p = 0.08; **p = 0.003 (vs. MH alone). **B)** Biofilm formation is inhibited by DNase I. Biofilms (WT/black bars or ΔcprS/grey bars) were grown in either MH alone, MH/DOC, (0.05%) MH/DNase (90 U mL⁻¹), or MH/DOC/DNase, followed by CV staining after 2 days growth. Error bars represent the mean of three biological replicates. *p<0.005 vs. counterpart without DNase.

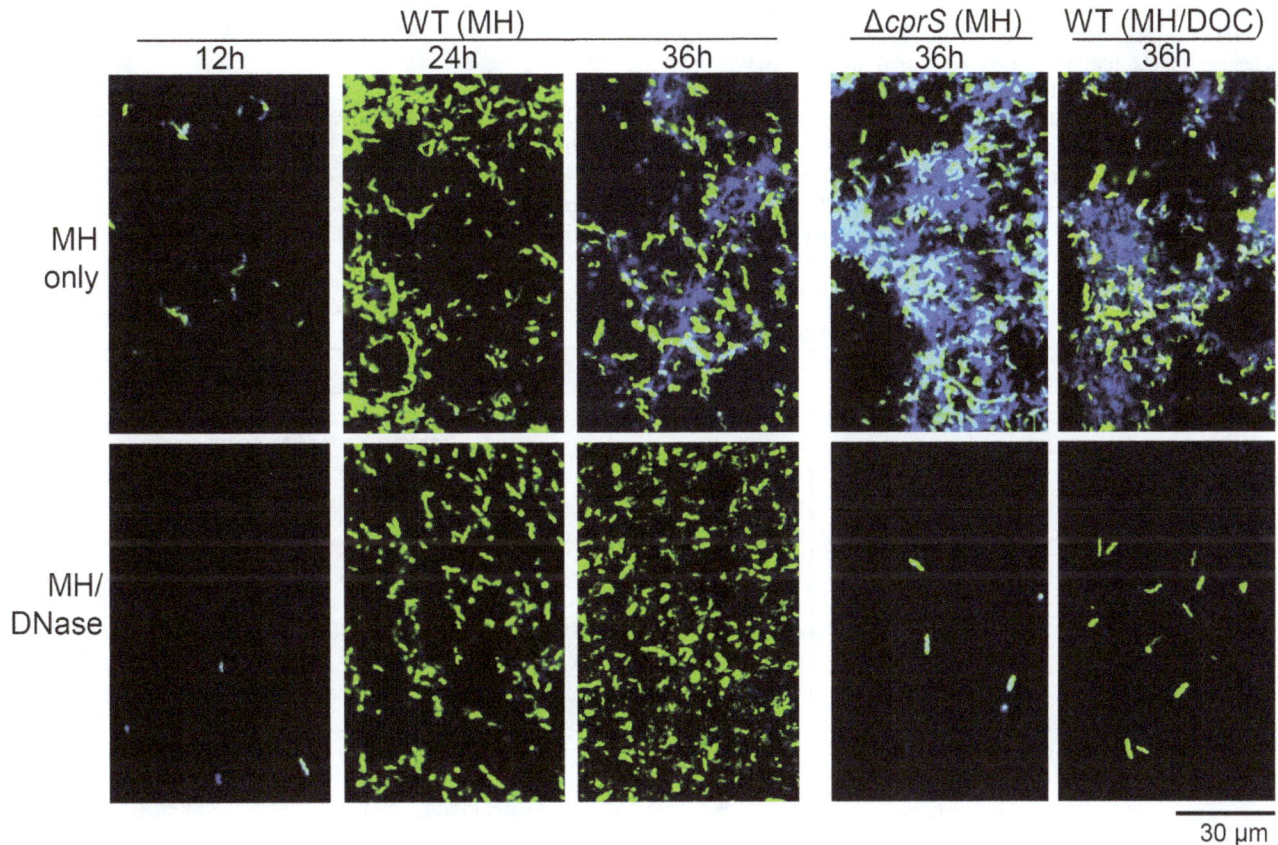

Figure 4. DNase arrests biofilms following adherence. Biofilms of WT, ΔcprS, or WT in MH/DOC (0.05%) were grown on coverslips in the presence (top panels) or absence (bottom panels) of DNase (90 U mL⁻¹). After the indicated times, biofilms were fixed, stained with DAPI, and visualized by confocal microscopy. Green: GFP-expressing bacteria; Blue: DAPI-stained DNA.

of DNA arresting biofilm formation following attachment, and having little effect on adherence of WT.

Flagella are necessary for biofilm formation by both WT and biofilm-enhanced *C. jejuni*

As we observed release of DNA at later time points, following adherence, we next sought to identify a factor required for the initial attachment step. Mutations that cause an aflagellate phenotype have consistently been reported to cause defective biofilm formation in *C. jejuni* [17,25,26]. As mutation of flagella did not affect lysis or release of eDNA (**Fig. 2**), we hypothesized that flagella may be required in a step prior to the stage where eDNA is relevant – specifically, adherence. To test this, biofilm formation by flagellar mutants was assessed (**Fig. 5A**). Deletion of *flhA* caused severely defective biofilm formation. When introduced into a Δ*cprS* background, the Δ*flhA* mutation also resulted in defective biofilm formation, suggesting flagella are epistatic to Δ*cprS*. Thus, we confirmed the biofilm-defective phenotype upon loss of flagella, and also showed that flagella were absolutely required for biofilm formation, even under conditions that can enhance biofilms.

The flagellar filament is required for attachment; motility aids kinetics of biofilm formation

In addition to motility, in the absence of structures such as pili, *C. jejuni* flagella also appear to mediate adhesion [28]. We therefore sought to determine if motility or the flagellar structure was required. As a Δ*flhA* mutant is aflagellate [26] a *pflA* mutant, that expresses paralyzed flagella [41], was included in our analyses. In MH only (**Fig. 5B**, left), Δ*flhA* and Δ*pflA* were both markedly defective for biofilm formation compared to WT and Δ*cprS*. However, in MH/DOC (**Fig. 5B**, right), these strains displayed distinct behaviour. While the Δ*flhA* mutant remained biofilm-defective in MH/DOC, Δ*pflA* was not as defective, displaying a significant 3-fold increase in biofilm formation in MH/DOC compared to MH only (p<0.0001). The "+DOC" observations suggested that loss of motility could be partially rescued in

conditions that promote biofilm formation in *C. jejuni*, but only in the presence of the flagellin adhesin.

Microscopy was used to determine at which stages flagella and motility might contribute to biofilm formation. An aflagellate Δ*flgR* mutant (Kan[R]), that is also biofilm defective (data not shown), was used in place of Δ*flhA* to allow introduction of GFP on a Cm[R] plasmid for microscopy. In MH only, the aflagellate Δ*flgR* mutant adhered poorly to coverslips compared to WT, with very few green bacteria observed attached to the slide (**Fig. 6**, top left and top middle panels, respectively). In contrast, more adherent Δ*pflA* bacteria were seen than for Δ*flgR* (top right panel). However, in contrast to WT, Δ*pflA* still appeared defective and/or delayed for both adherence and biofilm formation. Fewer adhered bacteria and little DNA were observed compared to WT at the same time point. Although very few bacteria were observed for Δ*flgR*, DNA was still observed attached to Δ*flgR*-incubated slides, confirming the above observations (**Fig. 2A**) that DNA release was not abolished by loss of flagella. Consistent with crystal violet results, inclusion of DOC in MH broth appeared to allow the flagellate, but non-motile Δ*pflA* mutant (bottom right panel) to form better biofilms, although still not to the levels of WT. Significantly more DNA were observed surrounding Δ*pflA* in MH/DOC compared to MH only. Thus, the defect observed in non-motile bacteria can be partially rescued by stimulating biofilm formation with conditions that enhance lysis and eDNA release, such in MH/DOC. Thus, this suggests that *C. jejuni* absolutely requires the flagellar structure to initiate biofilm formation, presumably to mediate adherence, and that eDNA is not sufficient to mediate adherence. While motility is dispensable under certain conditions, we conclude that it aids the kinetics of biofilm formation.

Biofilms contribute to fitness under adverse conditions *in vitro*

In experiments described above (**Fig. 4**), WT/DOC or Δ*cprS* biofilm cultures incubated with DNase displayed very few bacteria adhering to the coverslip. Closer inspection of biofilm cultures

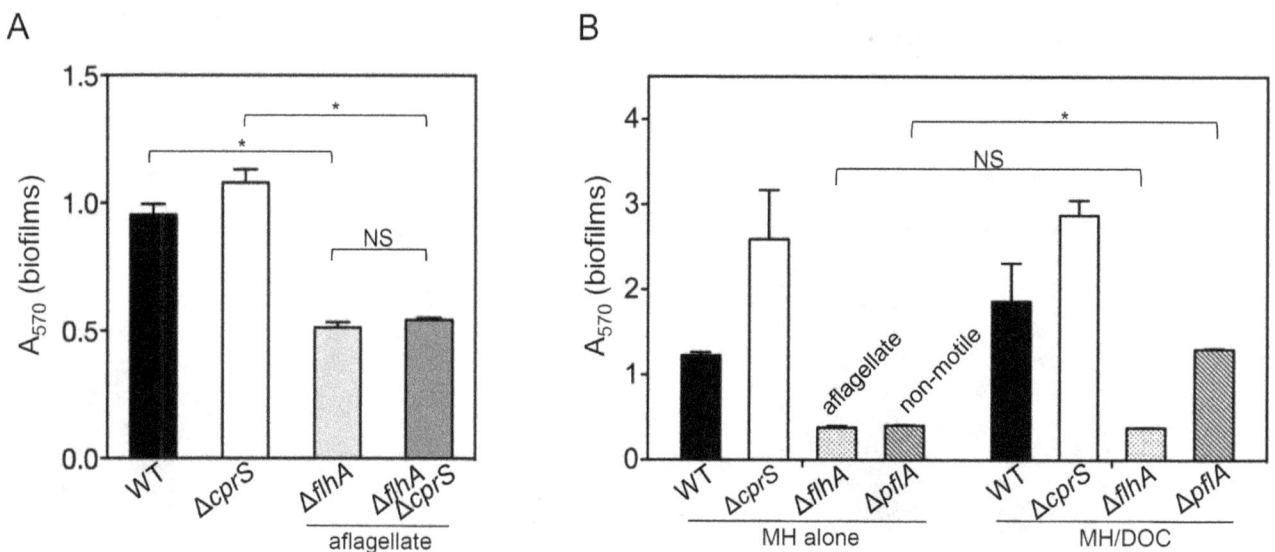

Figure 5. The flagellum, but not motility, is absolutely required for biofilm formation. A) Aflagellate mutants are defective for biofilm formation in WT and Δ*cprS* backgrounds. *p<0.0001; NS p>0.1 **B)** Only non-flagellate bacteria remain completely defective in biofilm-promoting media. *p<0.0001; NS, p = 1. Indicated strains were grown in static culture for 2 days in either MH broth alone or MH/DOC, followed by staining and quantification with crystal violet.

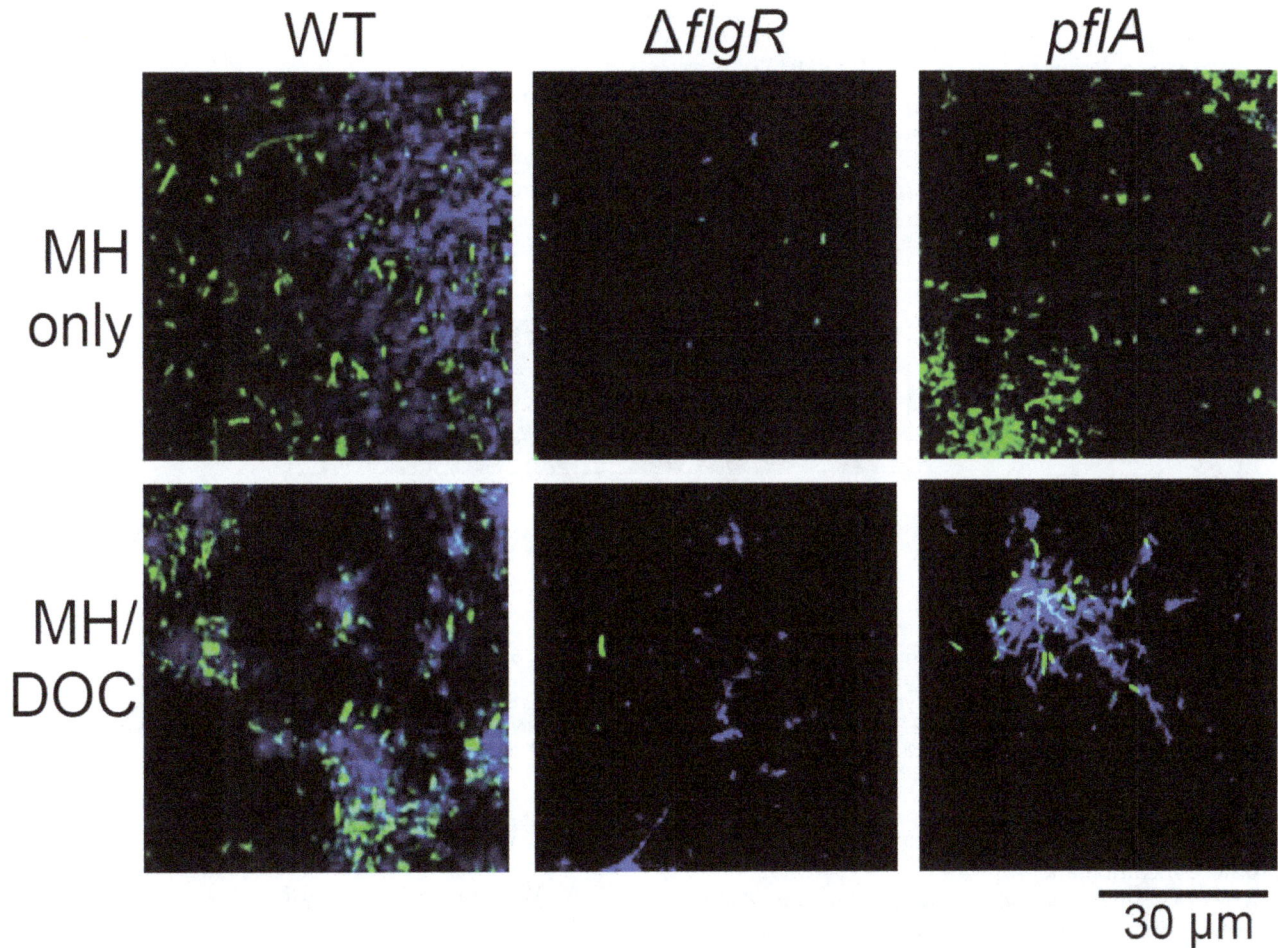

Figure 6. Aflagellate bacteria are defective for adherence; kinetics of biofilm formation is delayed in bacteria expressing paralyzed flagella. Biofilms of WT, ΔflgR (aflagellate), and pflA (paralyzed flagella) were grown on coverslips for 36 h, fixed, stained with DAPI, and visualized by confocal microscopy. Green: GFP-expressing bacteria; Blue: DAPI-stained DNA.

suggested that while each strain was able to grow in the sub-MIC levels of DOC, there appeared to be a decrease in total biomass produced by cultures that were both biofilm-inhibited and experiencing 'stress,' compared to those that were only biofilm-inhibited (for example, WT in DNase vs. WT in both DOC and DNase). There have also been reports in the literature of *C. jejuni* flagellar mutant strains (such as Δ*rpoN* and Δ*fliA*) that are likely biofilm-impaired that display growth and stress tolerance defects in standing culture [42,43]. Together, this suggested to us that *C. jejuni* requires biofilm formation to tolerate adverse conditions *in vitro*.

To test this, we determined if strains showed decreased fitness in the presence of a pathogenesis-related condition that normally did not markedly affect growth (sub-MIC DOC) if they were not able to form a biofilm, as measured above (**Figs. 3–6**). As a measure of fitness, we used the total biomass that was reached for each strain/condition during standing culture, similar conditions to our biofilm assay.

We first sought to determine the requirement of flagella-mediated adhesion for adaptation to DOC. However, aflagellate strains reached much lower total biomasses than flagellate strains, which made comparisons difficult (data not shown). We thus focused on the requirement for biofilm maturation (ie, the effect of DNase, which arrests biofilm formation following adherence). We

noted no difference in growth (OD_{600} of resuspended cultures) for WT in MH or MH/DNase (**Fig. 7**, black bars), suggesting that DNase did not appreciably affect fitness of WT under routine conditions. In contrast, when DOC was included along with DNase, we saw a significant decrease in final biomass ($p < 0.0001$) reached by WT. Sub-MIC levels of DOC increased total resuspended culture density, together with reduced density of flagellar mutants (see above) suggested that biofilm formation allows *C. jejuni* to reach higher bacterial loads.

We also included Δ*cprS*, which like WT in DOC is presumably experiencing stress due to absence of CprRS signaling [19], as a comparison (**Fig. 7**, grey bars). Like WT, Δ*cprS* reached a higher resuspended culture density in the presence of DOC, compared to MH alone. However, in contrast to WT, Δ*cprS* was significantly affected by inclusion of DNase in the culture media, even in the absence of DOC ($p = 0.0018$). As Δ*cprS* shows numerous *in vitro* stress-related phenotypes, including reduced tolerance of osmotic and oxidative stress [19], this suggests that its enhanced biofilm phenotype may be a compensatory stress response. Together, our *in vitro* observations suggest that *C. jejuni* requires biofilm formation for fitness in the face of challenging conditions.

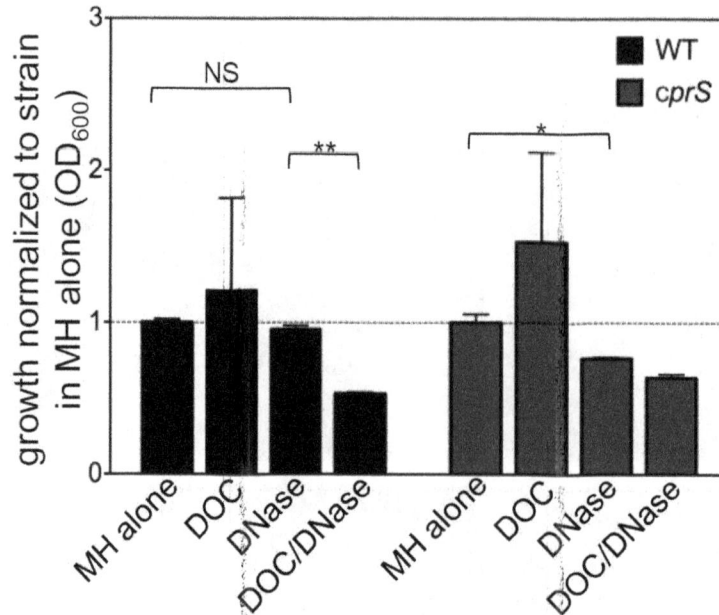

Figure 7. Biofilm formation confers stress tolerance *in vitro*. Standing cultures of the indicated strains (black bars, WT background; grey bars, Δ*cprS* background) were grown in MH broth with the indicated additions (labels below). Biofilm formation was impaired by addition of DNase (90 U mL^{-1}). Sub-MIC levels of DOC were included where indicated. Total OD$_{600}$ of three independent cultures, following 2 days growth and resuspension by vortexing was measured. Cultures were normalized to the strain background (WT or Δ*cprS*) in MH alone. Error bars represent the mean of three biological replicates. NS: not significant **p<0.0001 *p=0.0018.

Conditions that increase DNA release and biofilms also promote recombination

As eDNA was increased under conditions that promoted biofilm formation (**Fig. 1C**), we asked whether increased extracellular DNA could also increase horizontal gene transfer. Genetic exchange was measured under two conditions that promote biofilm formation and eDNA release: mutation of *cprS*, and growth in MH/DOC. Strains marked with antibiotic resistance (Δ*cprS*, Kan^R, WT, Str^R or Kan^R) on the chromosome were grown in mixed culture. When WT (Str^R) was grown with WT (Kan^R), the appearance of doubly resistant colonies, not present upon inoculation of the cultures, was observed (**Fig. 8**). When the same mixed cultures were grown in MH/DOC, appearance of more of these doubly-resistant clones was observed compared to cultures in MH alone (p = 0.09). Moreover, when WT (Str^R) was co-cultured with Δ*cprS* (Kan^R), we also recovered significantly more (p = 0.02) colonies on plates containing both Kan and Str compared to those from cultures of the two WT strains.

Discussion

Previous work had not identified dedicated virulence factors or specific stress response proteins that sufficiently explain why *C. jejuni* is such a successful zoonotic pathogen, surviving and thriving in numerous environments during transmission and pathogenesis. In this work, analysis of strains enhanced for biofilm formation (Δ*cprS*; WT in MH/DOC) identified stages and molecular factors involved in *C. jejuni* biofilm formation, a phenomenon that may explain the resilience of *C. jejuni* outside of the laboratory. Two specific phenomena that appear to be related to *C. jejuni* biofilm formation, at least *in vitro*, were determined: flagella and eDNA. Flagella appear to be necessary for initiation of biofilm formation on a surface by mediating adhesion. Furthermore, motility provided by flagella also aided kinetics of biofilm

formation. We also observed a lytic phenomenon that correlates with biofilm formation and appears to be responsible for release of eDNA. We also found that eDNA was then required for maturation from microcolonies into a three-dimensional biofilm. Finally, we observed that inhibition of biofilm formation lead to reduced fitness in the presence of DOC, a pathogenesis-related stress that also appears to trigger *C. jejuni* biofilm formation.

The process of biofilm formation in *C. jejuni*, like other bacteria, appears to proceed in discrete steps, starting with adhesion. We propose that flagella are required for adhesion, as aflagellate mutants were not observed to adhere to coverslips, even under conditions that normally enhance biofilm formation of WT (such as with DOC) (**Fig. 6**). This is consistent with two previous studies have noted that bacteria adhere to *ex vivo* tissue samples by flagella in microcolony-like structures [27,29]. Analysis of biofilm formation on abiotic surfaces also found microcolonies formed on glass coverslips with flagella forming bridges between organisms [30]. Moreover, autoagglutination, which is thought to be dependent on flagella and biofilm formation, also seem to be correlated in *C. jejuni* [18,44].

A central role for flagella in biofilm formation is also supported by previously reported expression data. Motility peaks during late log phase [45], and Class II and III flagellar genes exhibit sustained or increasing expression through stationary phase, suggesting components of the flagellum may be necessary for this transition. Biofilm cells often exhibit characteristics of stationary phase cells and share similar expression profiles [46], and *C. jejuni* biofilm cells also display higher expression of flagellar genes compared to stationary phase cells grown planktonically [26]. Finally, proteomic and microarray expression analysis of the Δ*cprS* hyper-biofilm mutant [[19]; S.L. Svensson and E.C. Gaynor, in preparation] also suggest expression of flagellar genes is increased in this strain.

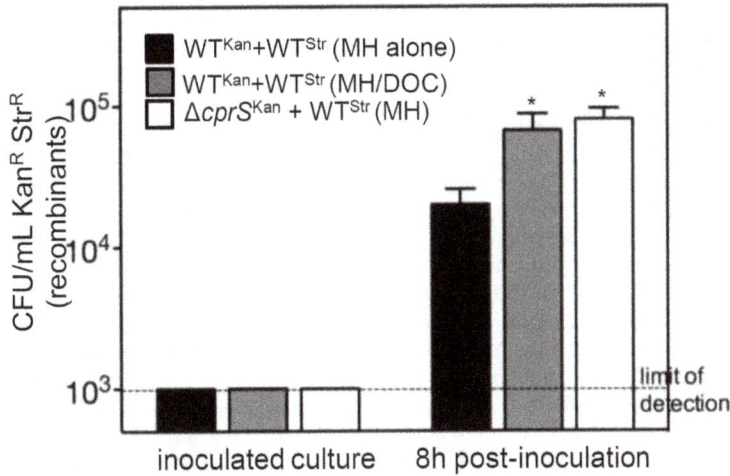

Figure 8. Conditions that promote DNA release and biofilms also increase genetic exchange and UV tolerance. Genetic exchange. WT bacteria, marked with Str^R, were grown in mixed culture (1:1) with either an isogenic WT strain marked with Kan^R or the $\Delta cprS$ mutant marked with Kan^R. Cultures were grown in either MH broth alone or MH/DOC. Cells were removed at indicated time points and CFUs were determined on the appropriate antibiotics. Error bars represent the mean of three biological replicates. *p<0.1 vs. WT+WT (MH).

It was initially unclear whether motility or the flagellum itself was required for adhesion. Further mutant analyses using a paralyzed flagellum mutant suggested that while motility might aid the kinetics of biofilm formation, it was not absolutely required. In contrast, the flagellum structure itself was. Our observations are consistent with behaviour of other mutants with a variety flagellar morphologies and motilities [17,26]. Biofilm formation is consistently severely defective in aflagellate mutants (such as $\Delta flhA$), but delayed in strains such as $\Delta flaA$, $\Delta flaB$, $\Delta fliA$, and $\Delta flaC$ [26]. These strains express either normal or morphologically aberrant flagella and have reduced (~20% of WT) or absent motility [26,47]. Interestingly, a $\Delta flaG$ mutant, which expresses long flagella but retains full motility, is completely defective for biofilm formation, even upon extended incubation [26]. This suggests that motility alone may be insufficient for biofilm formation, and that aspects of the flagellar structure itself are critical for biofilm formation.

We observed release of eDNA following adherence and found that it is required for further maturation of the biofilm. DNase did not appear to affect the initial adherence step. Consistent with this, *C. jejuni* has been proposed to use both flagellum-dependent and -independent mechanisms of biofilm formation [18]. In other bacteria, adherence is often followed by biogenesis or release of polymeric matrix components that encase the mature biofilm. Surface carbohydrates are common components of biofilm matrices, and the *C. jejuni* surface is highly glycosylated. It is therefore puzzling that a specific carbohydrate component of the *C. jejuni* matrix has yet to be identified. We previously noted that DNA surrounds *C. jejuni* biofilms, especially in $\Delta cprS$ and in WT bacteria under conditions favouring biofilm formation (MH/DOC), and treatment of pre-formed biofilms with DNase also disrupted them [19]. An extracellular material that binds Ruthenium Red [30], a dye that stains carbohydrate matrices, but also binds double-helical DNA, was previously observed [48]. The $\Delta cprS$ mutant carries no gross defects in surface polysaccharides [19]. We have now measured a 2-3-fold increase in eDNA for $\Delta cprS$ compared to WT after the same incubation period.

We have also shown that exogenous, purified *C. jejuni* gDNA enhances biofilms, and inclusion of DNase in standing cultures inhibits biofilm formation (**Fig. 3**). Thus, it appears that eDNA does in fact play an important role in *C. jejuni* biofilm formation, and does not simply correlate with the transition to a sessile lifestyle. Consistent with this, the presence and important role of eDNA in biofilms is now well-appreciated in many species. It is

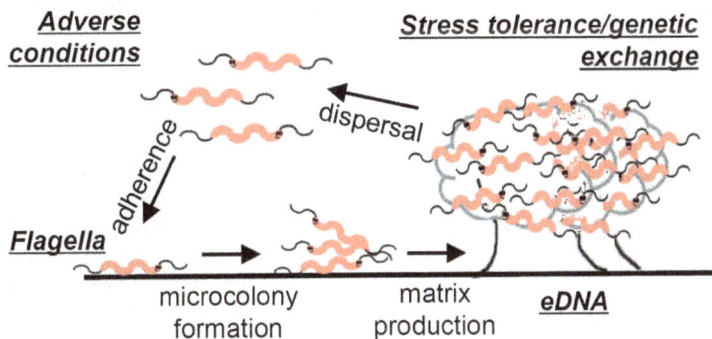

Figure 9. Model of *C. jejuni* biofilm formation. Evidence for the role of stress conditions, flagella and motility, eDNA release, and genetic exchange has been provided. Biofilm formation also appears to confer tolerance of specific stresses, such as those that may be encountered during pathogenesis.

interesting to note that deletion of *dps*, encoding a nucleoid-binding protein, reduces biofilm formation by 50% [49]. Unlike addition of *C. jejuni* gDNA, highly purified salmon DNA does not enhance biofilm formation (data not shown). It is possible that chromatin-like material, possibly containing proteins like Dps, may serve as an enucleating factor for biofilm maturation. We cannot rule out a potential contribution of other proteins released during lysis in *C. jejuni* biofilms. The enhanced biofilm phenotype of many loss-of-function mutants in surface carbohydrate loci of *C. jejuni* is intriguing and suggests that biofilm formation in this organism does not require a specific carbohydrate matrix component. Expression of a particular surface carbohydrate may instead be negatively correlated with biofilm formation, such as glycosylation of flagella or the major outer membrane protein [44,50], which would change surface hydrophobicity. Alternatively, DNA may fulfill the role played by exopolysaccharides in other bacteria, or a carbohydrate component, which may not be absolutely required under laboratory conditions, could be provided by a neighbouring organism in a multi-species biofilm in nature.

The source of the eDNA is unknown; however, we noted co-occurrence of increased eDNA with cytosolic proteins in culture supernatant (**Fig. 1**). An increase in many of protein species was previously noted in supernatants of Δ*cprS* [19]. The Δ*cprS* mutant displays a more marked loss of culturability following log phase compared to WT [19]. *C. jejuni* is thought to convert to a coccoid viable but non-culturable state; however, Δ*cprS* morphology is not consistent with an accelerated progression to this form [19]. Taken together, this implicates a lytic process. It is unknown whether the released DNA is chromosomal, consistent with lysis, or shows any enrichment for particular sequences. Furthermore, while DNA uptake appears to be mediated by a Type II secretion system, a putative DNA secretion apparatus has not been identified in *C. jejuni*. The pVIR plasmid carried by some strains, including the robust biofilm former 81–176, encodes a putative Type IV secretion system that could possibly mediate this, as in *Neisseria gonorrhoeae* [51]. However, mutation of *virB11*, encoding an essential component of this secretion system, does not affect biofilm formation in strain 81–176 (S.L. Svensson and E.C. Gaynor, unpublished observations). In the related gastric pathogen *H. pylori*, eDNA has also been identified as a component of the biofilm matrix [52]. DNA fingerprinting suggested a marked difference between eDNA and intracellular DNA, suggesting that a non-specific lytic mechanism does not release of DNA in this pathogen. However, DNase does not affect biofilm formation by *H. pylori*, and thus, it was concluded that the main function of eDNA in this bacterium was to contribute to genetic variation.

Our observations do not allow us to propose whether such a lytic mechanism is passive or autolytic. A connection between autolysis and biofilm formation exists in other bacteria. In *P. aeruginosa*, autolysis appears to contribute to dispersal of organisms from the biofilm, whereas in other bacteria such as *Enterococcus faecalis*, *Staphylococcus aureus*, and *N. meningitidis*, it appears to be involved in both eDNA release and biofilm development [8,9,53,54]. Lytic transglycosylases in *Salmonella* Typhimurium also link cell wall turnover to biofilm formation [55]. Unfortunately, we did not observe any accessory autolysins in the *C. jejuni* that may provide support for a lytic mechanism. Instead, 'housekeeping' peptidoglycan modification enzymes may be involved. Such enzymes are only now being identified and characterized in *C. jejuni* [34,35]. While a regulated autolysis program has not yet been described in *C. jejuni*, a decrease in CFUs (approximately 2 logs) is often observed in WT cultures after exponential phase of growth.

Biofilm formation by *C. jejuni* appears to be triggered under particular stress conditions. It was recently reported that aerobic conditions stimulate biofilm formation in *C. jejuni* [18], and bile upregulates the *flaA* flagellin promoter [56]. We previously reported that DOC, and other detergents, upregulate biofilm formation in *C. jejuni* [19]. Furthermore, there is a positive correlation between envelope perturbations, such as in Δ*kpsS*, Δ*waaF*, and Δ*spoT*, as well as WT grown in polymyxin B and ampicillin, and a tendency to form enhanced biofilms [[22–24] (S.L. Svensson and E.C. Gaynor, unpublished observations)]. A close relationship between envelope stress and biofilm formation exists in other pathogens. For example, it has been proposed that the Cpx-controlled envelope stress response of Gram-negative bacteria mediates biofilm formation [57]. Similar to *C. jejuni* in DOC, bile stimulates biofilm formation in *Vibrio cholerae* [58]. Interestingly, it has been shown that deletion of oxidative stress genes such as *ahpC* or catalase increases biofilm formation, where as overexpression of *ahpC* correlated with decreased biofilm formation [59]. Thus, biofilm formation may be a more general response to adverse conditions.

In support of observations that suggest biofilm formation is a stress response, we have also shown that inhibition of biofilm formation in *C. jejuni* increases the inhibitory effect of sub-MIC levels of DOC (**Fig. 7**). In general, we observed that bacteria that could not form a mature biofilm, either by genetic lesion of flagellar genes (data not shown) or enzymatic removal of eDNA, were less able to grow in standing culture with added DOC than those that could form a biofilm. Consistent with our observation, other work has shown that flagellar mutants (Δ*rpoN* and Δ*fliA*) exhibit growth differences and/or stress sensitivity in standing culture [38,43]. Cultures that could form biofilms also reached higher total biomass than those growing solely planktonically, even in MH broth alone, suggesting that biofilms could presumably increase the burden of this pathogen in the environment. The mechanism by which biofilms conferred *C. jejuni* with increased stress tolerance in this work is currently unknown. In general, the contribution of biofilms to stress tolerance in other bacteria is thought to be multi-factorial, and may include altered metabolism, induction of stress response genes, changes in the cell envelope, decreased penetration of O_2 or inhibitory compounds (such as DOC), or specific contributions of the properties of matrix components, such as eDNA. Nonetheless, it appears that the biofilm provides a niche well-suited to growth and/or survival of this pathogen, and conditions that promote biofilm formation may contribute to high bacterial loads in infection reservoirs. It also follows that antimicrobial agents may, to some extent, contribute to persistence of this pathogen by stimulating biofilm formation.

While our *in vitro* observations suggest that biofilm-residing *C. jejuni* are more stress tolerant, the role of biofilms *in vivo* has thus far been unclear. *C. jejuni* encounters numerous stresses in both commensal and susceptible hosts, and has been observed to form microcolonies on intestinal epithelial tissue *in vitro* [29]. Moreover, species of *Campylobacter* have been identified within biofilms in the upper gastrointestinal tract of patients with Barrett's esophagus [60], and *H. pylori* also forms biofilm-like structures in the gastric mucosa [61,62]. Indirect evidence suggests biofilms may partially protect otherwise sensitive mutants of *C. jejuni*. A Δ*spoT* stringent response mutant forms enhanced biofilms[22,63] and retains its capacity to colonize animal hosts, even though it displays specific *in vitro* stress-related defects [E. Gaynor, unpublished observations]. In addition, a Δ*ppk1* mutant, which also exhibits stress tolerance defects *in vitro*, displays a dose-dependent trend for both *in vitro* biofilm formation and chick colonization

[20]. Collectively, this suggests that biofilms do confer stress-sensitive mutants with *in vivo* resilience.

In addition to tolerance of acute instances of stress, our observations suggest that the mechanism of *C. jejuni* biofilm formation support its high genetic diversity, which could contribute to longer-term adaptation to varying environmental conditions. *C. jejuni* exhibits phase variation of genes relating to its cell surface – genes that are critical to its interaction with the host environment - and this has been shown to occur during colonization of chicks [64]. Exchange of genetic markers has also been observed in chicks [65]. eDNA released under biofilm-promoting conditions has the potential to serve as a substrate for horizontal gene transfer, and we observed an increased rate of marker exchange under biofilm-promoting conditions. However, the reason for this may be multi-factorial, and it remains to be demonstrated whether processes such is DNA uptake and recombination may also be upregulated during biofilm-enhancing conditions. Autolysis can in fact be a trigger for natural transformation in other bacteria [66]. Importantly, we observed increased recombination in conditions that may be encountered during both colonization of commensal hosts and pathogenic infection of humans (i.e., presence of DOC). This suggests that such a mechanism may occur *in vivo*.

In the absence of the large repertoire of survival factors expected for a zoonotic pathogen, global changes in physiology may underlie adaptation of *C. jejuni* to stressful environments. Phenotypes required for rapid growth are often expressed at the expense of stress tolerance [67]. Thus, some of the resilience of *C. jejuni* may not be observed planktonic broth culture, explaining the apparent fastidiousness of *C. jejuni* in the lab. In this work, we have extended understanding of the steps and molecular mechanisms of *C. jejuni* biofilm formation, a process that provides this pathogen with stress tolerance, providing a framework for future studies (**Fig. 9**). Further characterization of these mechanisms will contribute to our knowledge of how *C. jejuni* navigates environments encountered during pathogenesis.

Acknowledgments

The authors thank Dmitry Apel for the pRY112-*gfp* plasmid and Dr. Emilisa Frirdich and Dr. David Hendrixson for providing flagellar mutant strains.

Author Contributions

Conceived and designed the experiments: SLS ECG. Performed the experiments: SLS MP. Analyzed the data: SLS MP ECG. Wrote the paper: SLS ECG.

References

1. Gilbreath JJ, Cody WL, Merrell DS, Hendrixson DR (2011) Change is good: variations in common biological mechanisms in the epsilonproteobacterial genera *Campylobacter* and *Helicobacter*. Microbiol Mol Biol Rev 75: 84–132.
2. Szymanski CM, Gaynor EC (2012) How a sugary bug gets through the day: Recent developments in understanding fundamental processes impacting *Campylobacter jejuni* pathogenesis. Gut Microbes.
3. Parkhill J, Wren BW, Mungall K, Ketley JM, Churcher C, et al. (2000) The genome sequence of the food-borne pathogen Campylobacter jejuni reveals hypervariable sequences. Nature 403: 665–668.
4. O'Toole G, Kaplan HB, Kolter R (2000) Biofilm formation as microbial development. Annu Rev Microbiol 54: 49–79.
5. Flemming HC, Wingender J (2010) The biofilm matrix. Nat Rev Microbiol 8: 623–633.
6. Montanaro L, Poggi A, Visai L, Ravaioli S, Campoccia D, et al. (2011) Extracellular DNA in biofilms. Int J Artif Organs 34: 824–831.
7. Mulcahy H, Charron-Mazenod L, Lewenza S (2008) Extracellular DNA chelates cations and induces antibiotic resistance in *Pseudomonas aeruginosa* biofilms. PLoS Pathog 4: e1000213.
8. Ma L, Conover M, Lu H, Parsek MR, Bayles K, et al. (2009) Assembly and development of the *Pseudomonas aeruginosa* biofilm matrix. PLoS Pathog 5: e1000354.
9. Thomas VC, Thurlow LR, Boyle D, Hancock LE (2008) Regulation of autolysis-dependent extracellular DNA release by *Enterococcus faecalis* extracellular proteases influences biofilm development. J Bacteriol 190: 5690–5698.
10. Nguyen D, Joshi-Datar A, Lepine F, Bauerle E, Olakanmi O, et al. (2011) Active starvation responses mediate antibiotic tolerance in biofilms and nutrient-limited bacteria. Science 334: 982–986.
11. Sanders SQ, Boothe DH, Frank JF, Arnold JW (2007) Culture and detection of *Campylobacter jejuni* within mixed microbial populations of biofilms on stainless steel. J Food Prot 70: 1379–1385.
12. Trachoo N, Frank JF, Stern NJ (2002) Survival of *Campylobacter jejuni* in biofilms isolated from chicken houses. J Food Prot 65: 1110–1116.
13. Trachoo N, Frank JF (2002) Effectiveness of chemical sanitizers against *Campylobacter jejuni*-containing biofilms. J Food Prot 65: 1117–1121.
14. Rollins DM, Colwell RR (1986) Viable but nonculturable stage of *Campylobacter jejuni* and its role in survival in the natural aquatic environment. Appl Environ Microbiol 52: 531–538.
15. Buswell CM, Herlihy YM, Lawrence LM, McGuiggan JT, Marsh PD, et al. (1998) Extended survival and persistence of *Campylobacter* spp. in water and aquatic biofilms and their detection by immunofluorescent-antibody and -rRNA staining. Appl Environ Microbiol 64: 733–741.
16. Zimmer M, Barnhart H, Idris U, Lee MD (2003) Detection of *Campylobacter jejuni* strains in the water lines of a commercial broiler operation and their relationship to the strains that colonized the chickens. Avian Dis 47: 101–107.
17. Reeser RJ, Medler RT, Billington SJ, Jost BH, Joens LA (2007) Characterization of *Campylobacter jejuni* biofilms under defined growth conditions. Appl Environ Microbiol 73: 1908–1913.
18. Reuter M, Mallett A, Pearson BM, van Vliet AH (2010) Biofilm formation by Campylobacter jejuni is increased under aerobic conditions. Appl Environ Microbiol 76: 2122–2128.
19. Svensson SL, Davis LM, MacKichan JK, Allan BJ, Pajaniappan M, et al. (2009) The CprS sensor kinase of the zoonotic pathogen *Campylobacter jejuni* influences biofilm formation and is required for optimal chick colonization. Mol Microbiol 71: 253–272.
20. Candon HL, Allan BJ, Fraley CD, Gaynor EC (2007) Polyphosphate kinase 1 is a pathogenesis determinant in *Campylobacter jejuni*. J Bacteriol 189: 8099–8108.
21. Gangaiah D, Liu Z, Arcos J, Kassem, II, Sanad Y, et al. (2010) Polyphosphate kinase 2: a novel determinant of stress responses and pathogenesis in *Campylobacter jejuni*. PLoS One 5: e12142.
22. McLennan MK, Ringoir DD, Frirdich E, Svensson SL, Wells DH, et al. (2008) *Campylobacter jejuni* biofilms up-regulated in the absence of the stringent response utilize a calcofluor white-reactive polysaccharide. J Bacteriol 190: 1097–1107.
23. Naito M, Frirdich E, Fields JA, Pryjma M, Li J, et al. (2010) Effects of sequential *Campylobacter jejuni* 81–176 lipooligosaccharide core truncations on biofilm formation, stress survival, and pathogenesis. J Bacteriol 192: 2182–2192.
24. Rathbun KM, Thompson SA (2009) Mutation of PEB4 alters the outer membrane protein profile of *Campylobacter jejuni*. FEMS Microbiol Lett 300: 188–194.
25. Joshua GW, Guthrie-Irons C, Karlyshev AV, Wren BW (2006) Biofilm formation in *Campylobacter jejuni*. Microbiology 152: 387–396.
26. Kalmokoff M, Lanthier P, Tremblay TL, Foss M, Lau PC, et al. (2006) Proteomic analysis of *Campylobacter jejuni* 11168 biofilms reveals a role for the motility complex in biofilm formation. J Bacteriol 188: 4312–4320.
27. Grant AJ, Woodward J, Maskell DJ (2006) Development of an *ex vivo* organ culture model using human gastro-intestinal tissue and *Campylobacter jejuni*. FEMS Microbiol Lett 263: 240–243.
28. Guerry P (2007) *Campylobacter* flagella: not just for motility. Trends Microbiol 15: 456–461.
29. Haddock G, Mullin M, MacCallum A, Sherry A, Tetley L, et al. (2010) *Campylobacter jejuni* 81–176 forms distinct microcolonies on *in vitro*-infected human small intestinal tissue prior to biofilm formation. Microbiology 156: 3079–3084.
30. Moe KK, Mimura J, Ohnishi T, Wake T, Yamazaki W, et al. (2010) The mode of biofilm formation on smooth surfaces by *Campylobacter jejuni*. J Vet Med Sci 72: 411–416.
31. Korlath JA, Osterholm MT, Judy LA, Forfang JC, Robinson RA (1985) A point-source outbreak of campylobacteriosis associated with consumption of raw milk. J Infect Dis 152: 592–596.
32. Hendrixson DR, DiRita VJ (2003) Transcription of σ^{54}-dependent but not σ^{28}-dependent flagellar genes in *Campylobacter jejuni* is associated with formation of the flagellar secretory apparatus. Mol Microbiol 50: 687–702.
33. Hendrixson DR, Akerley BJ, DiRita VJ (2001) Transposon mutagenesis of *Campylobacter jejuni* identifies a bipartite energy taxis system required for motility. Mol Microbiol 40: 214–224.
34. Frirdich E, Biboy J, Adams C, Lee J, Ellermeier J, et al. (2012) Peptidoglycan-Modifying Enzyme Pgp1 Is Required for Helical Cell Shape and Pathogenicity Traits in *Campylobacter jejuni*. PLoS Pathog 8: e1002602.
35. Frirdich E, Vermeulen J, Biboy J, Soares F, Taveirne ME, et al. (2014) Peptidoglycan LD-Carboxypeptidase Pgp2 Influences Campylobacter jejuni

Helical Cell Shape and Pathogenic Properties, and Provides the Substrate for the DL-Carboxypeptidase Pgp1. J Biol Chem.

36. Apel D, Ellermeier J, Pryjma M, Dirita VJ, Gaynor EC (2012) Characterization of *Campylobacter jejuni* RacRS reveals a role in the heat shock response, motility, and maintenance of cell length population homogeneity. J Bacteriol.

37. Fujita Y, Yamaguchi K, Kamegaya T, Sato H, Semura K, et al. (2005) A novel mechanism of autolysis in *Helicobacter pylori*: possible involvement of peptidergic substances. Helicobacter 10: 567–576.

38. Barrero-Tobon AM, Hendrixson DR (2012) Identification and analysis of flagellar coexpressed determinants (Feds) of *Campylobacter jejuni* involved in colonization. Mol Microbiol.

39. Konkel ME, Klena JD, Rivera-Amill V, Monteville MR, Biswas D, et al. (2004) Secretion of virulence proteins from *Campylobacter jejuni* is dependent on a functional flagellar export apparatus. J Bacteriol 186: 3296–3303.

40. Rivera-Amill V, Kim BJ, Seshu J, Konkel ME (2001) Secretion of the virulence-associated *Campylobacter* invasion antigens from *Campylobacter jejuni* requires a stimulatory signal. J Infect Dis 183: 1607–1616.

41. Yao R, Burr DH, Doig P, Trust TJ, Niu H, et al. (1994) Isolation of motile and non-motile insertional mutants of *Campylobacter jejuni*: the role of motility in adherence and invasion of eukaryotic cells. Mol Microbiol 14: 883–893.

42. Barrero-Tobon AM, Hendrixson DR (2012) Identification and analysis of flagellar coexpressed determinants (Feds) of *Campylobacter jejuni* involved in colonization. Mol Microbiol 84: 352–369.

43. Hwang S, Jeon B, Yun J, Ryu S (2011) Roles of RpoN in the resistance of *Campylobacter jejuni* under various stress conditions. BMC Microbiol 11: 207.

44. Howard SL, Jagannathan A, Soo EC, Hui JP, Aubry AJ, et al. (2009) *Campylobacter jejuni* glycosylation island important in cell charge, legionaminic acid biosynthesis, and colonization of chickens. Infect Immun 77: 2544–2556.

45. Wright JA, Grant AJ, Hurd D, Harrison M, Guccione EJ, et al. (2009) Metabolite and transcriptome analysis of *Campylobacter jejuni in vitro* growth reveals a stationary-phase physiological switch. Microbiology 155: 80–94.

46. Beloin C, Valle J, Latour-Lambert P, Faure P, Kzreminski M, et al. (2004) Global impact of mature biofilm lifestyle on *Escherichia coli* K-12 gene expression. Mol Microbiol 51: 659–674.

47. Golden NJ, Acheson DW (2002) Identification of motility and autoagglutination *Campylobacter jejuni* mutants by random transposon mutagenesis. Infect Immun 70: 1761–1771.

48. Karpel RL, Shirley MS, Holt SR (1981) Interaction of the ruthenium red cation with nucleic acid double helices. Biophys Chem 13: 151–165.

49. Theoret JR, Cooper KK, Zekarias B, Roland KL, Law BF, et al. (2012) The Campylobacter jejuni Dps homologue is important for in vitro biofilm formation and cecal colonization of poultry, and may serve as a protective antigen for vaccination. Clin Vaccine Immunol.

50. Mahdavi J, Pirinccioglu N, Oldfield NJ, Carlsohn E, Stoof J, et al. (2014) A novel O-linked glycan modulates Campylobacter jejuni major outer membrane protein-mediated adhesion to human histo-blood group antigens and chicken colonization. Open Biol 4: 130202.

51. Hamilton HL, Dominguez NM, Schwartz KJ, Hackett KT, Dillard JP (2005) *Neisseria gonorrhoeae* secretes chromosomal DNA via a novel type IV secretion system. Mol Microbiol 55: 1704–1721.

52. Grande R, Di Giulio M, Bessa LJ, Di Campli E, Baffoni M, et al. (2011) Extracellular DNA in *Helicobacter pylori* biofilm: a backstairs rumour. J Appl Microbiol 110: 490–498.

53. Fournier B, Hooper DC (2000) A new two-component regulatory system involved in adhesion, autolysis, and extracellular proteolytic activity of *Staphylococcus aureus*. J Bacteriol 182: 3955–3964.

54. Lappann M, Claus H, van Alen T, Harmsen M, Elias J, et al. (2010) A dual role of extracellular DNA during biofilm formation of *Neisseria meningitidis*. Mol Microbiol 75: 1355–1371.

55. Monteiro C, Fang X, Ahmad I, Gomelsky M, Romling U (2011) Regulation of biofilm components in *Salmonella enterica* serovar Typhimurium by lytic transglycosylases involved in cell wall turnover. J Bacteriol 193: 6443–6451.

56. Allen KJ, Griffiths MW (2001) Effect of environmental and chemotactic stimuli on the activity of the Campylobacter jejuni flaA sigma(28) promoter. FEMS Microbiol Lett 205: 43–48.

57. Dorel C, Lejeune P, Rodrigue A (2006) The Cpx system of *Escherichia coli*, a strategic signaling pathway for confronting adverse conditions and for settling biofilm communities? Res Microbiol 157: 306–314.

58. Hung DT, Zhu J, Sturtevant D, Mekalanos JJ (2006) Bile acids stimulate biofilm formation in *Vibrio cholerae*. Mol Microbiol 59: 193–201.

59. Oh E, Jeon B (2014) Role of Alkyl Hydroperoxide Reductase (AhpC) in the Biofilm Formation of Campylobacter jejuni. PLoS One 9: e87312.

60. Macfarlane S, Furrie E, Macfarlane GT, Dillon JF (2007) Microbial colonization of the upper gastrointestinal tract in patients with Barrett's esophagus. Clin Infect Dis 45: 29–38.

61. Carron MA, Tran VR, Sugawa C, Coticchia JM (2006) Identification of *Helicobacter pylori* Biofilms in Human Gastric Mucosa. J Gastrointest Surg 10: 712–717.

62. Coticchia JM, Sugawa C, Tran VR, Gurrola J, Kowalski E, et al. (2006) Presence and density of *Helicobacter pylori* biofilms in human gastric mucosa in patients with peptic ulcer disease. J Gastrointest Surg 10: 883–889.

63. Gaynor EC, Wells DH, MacKichan JK, Falkow S (2005) The *Campylobacter jejuni* stringent response controls specific stress survival and virulence-associated phenotypes. Mol Microbiol 56: 8–27.

64. Bayliss CD, Bidmos FA, Anjum A, Manchev VT, Richards RL, et al. (2012) Phase variable genes of Campylobacter jejuni exhibit high mutation rates and specific mutational patterns but mutability is not the major determinant of population structure during host colonization. Nucleic Acids Res.

65. de Boer P, Wagenaar JA, Achterberg RP, van Putten JP, Schouls LM, et al. (2002) Generation of *Campylobacter jejuni* genetic diversity *in vivo*. Mol Microbiol 44: 351–359.

66. Lewis K (2000) Programmed death in bacteria. Microbiol Mol Biol Rev 64: 503–514.

67. Ferenci T, Spira B (2007) Variation in stress responses within a bacterial species and the indirect costs of stress resistance. Ann N Y Acad Sci 1113: 105–113.

## 4

Francisella novicida Pathogenicity Island Encoded Proteins Were Secreted during Infection of Macrophage-Like Cells

Rebekah F. Hare[1], Karsten Hueffer[2]*

1 Department of Biology and Wildlife, Institute of Arctic Biology, University of Alaska Fairbanks, Fairbanks, Alaska, United States of America, 2 Department of Veterinary Medicine, University of Alaska Fairbanks, Fairbanks, Alaska, United States of America

Abstract

Intracellular pathogens and other organisms have evolved mechanisms to exploit host cells for their life cycles. Virulence genes of some intracellular bacteria responsible for these mechanisms are located in pathogenicity islands, such as secretion systems that secrete effector proteins. The *Francisella* pathogenicity island is required for phagosomal escape, intracellular replication, evasion of host immune responses, virulence, and encodes a type 6 secretion system. We hypothesize that some *Francisella novicida* pathogenicity island proteins are secreted during infection of host cells. To test this hypothesis, expression plasmids for all *Francisella novicida* FPI-encoded proteins with C-terminal and N-terminal epitope FLAG tags were developed. These plasmids expressed their respective epitope FLAG-tagged proteins at their predicted molecular weights. J774 murine macrophage-like cells were infected with *Francisella novicida* containing these plasmids. The FPI proteins expressed from these plasmids successfully restored the intramacrophage growth phenotype in mutants of the respective genes that were deficient for intramacrophage growth. Using these expression plasmids, the localization of the *Francisella* pathogenicity island proteins were examined via immuno-fluorescence microscopy within infected macrophage-like cells. Several *Francisella* pathogenicity island encoded proteins (IglABCDEFGHIJ, PdpACE, DotU and VgrG) were detected extracellularly and they were co-localized with the bacteria, while PdpBD and Anmk were not detected and thus remained inside bacteria. Proteins that were co-localized with bacteria had different patterns of localization. The localization of IglC was dependent on the type 6 secretion system. This suggests that some *Francisella* pathogenicity island proteins were secreted while others remain within the bacterium during infection of host cells as structural components of the secretion system and were necessary for secretion.

Editor: Yousef Abu Kwaik, University of Louisville, United States of America

Funding: Research reported in this publication was supported by the National Institute of General Medical Sciences of the National Institutes of Health under Award Number P20GM103395. The content is solely the responsibility of the authors and does not necessarily represent the official views of the National Institutes of Health. The funders had no role in study design, data collection and analysis, decision to publish, or preparation of the manuscript.

Competing Interests: The authors have declared that no competing interests exist.

* Email: khueffer@alaska.edu

Introduction

Pathogenicity islands exist in many pathogenic bacteria, are acquired via horizontal gene transfer, and encode genes that facilitate interactions with host cells [1]. Secretion systems in bacteria involve the transport or translocation of effector molecules from the interior of a bacterial cell through its membranes to the exterior. Protein secretion is an important mechanism for bacteria to adapt and survive in their environment, including within an infected host [2]. Effector proteins are enzymes or toxins that facilitate infection and are secreted by these secretion systems [3].

Francisella tularensis is an intracellular pathogen that possesses the *Francisella* pathogenicity island (FPI) [4]. The FPI is found in all *Francisella* species and strains, and is duplicated in all human-virulent biovars of *F. tularensis*. *F. novicida* and *F. philomiragia* harbor only one copy of the FPI, which makes these species attractive for creating isogenic FPI gene deletion mutants [4,5]. The molecular mechanisms contributing to the intracellular survival of *Francisella* are poorly understood, and FPI mutagenesis approaches are useful in identifying genes required for intracellular replication and virulence [4,6,7,8,9,10,11].

The FPI contains genes with homology to genes encoding type 6 secretion systems (T6SS) in other bacteria [12,13,14,15]. Bioinformatics, genetics, biochemical, and cell biology approaches provide evidence the FPI encodes a functional secretion system [12,13]. Homologues of *iglAB*, *pdpB*, *dotU*, and *vgrG* are found in most T6SS identified to date; therefore, some suspect the secretion system of the FPI is a T6SS, although this is debatable [15,16]. DotU and PdpB are inner membrane components that are homologous with the T6SS proteins DotU and IcmF, respectively [15]. IglA and IglB are IcmF-associated homologous proteins seen in *Rhizobium leguminosarum*, *Salmonella enterica*, and *Vibrio cholerae* [4,12,13,15,16,17,18]. The solubility properties of IglABC suggest these proteins could be part of the needle spanning through the bacterial membranes, and the protein-protein interactions of IglAB also suggests the auxiliary roles within *F. novicida* as described in other species [13]. Mutations in IglA and IglB result in bacteria that are unable to escape the phagosome and unable to replicate intracellularly [4,6,12,19,20]. In some species, these homologues are responsible for secretion of proteins, including Hcp and VgrG [16,18,21,22,23]. Recent studies suggest the T6SSs constitute important *Francisella* virulence, intracellular growth, or survival factors; however, only basic aspects of this system have been characterized [13,24,25].

Although the ability of *F. tularensis* to replicate within macrophages is multifactorial, our working hypothesis is that *F. tularensis* secretes FPI-encoded proteins that facilitate the organism's ability to escape the vacuole, enter the cytoplasm to replicate intracellularly, and down regulate the host immune cytokine response. If this is correct, then FPI-encoded proteins should be secreted during infection within host macrophages. Currently available genetic tools for studying the FPI-encoded proteins consist of green fluorescent protein (GFP) tags [8] and more recently reporter fusion tag systems [11]. Secretion of FPI-encoded proteins have previously been examined in the *Francisella* live vaccine strain (LVS) with a fusion β-lactamase, however, this system is not applicable to wild type *F. novicida* and was assessed in a β-lactamase gene mutant because *F. novicida* possesses native β-lactamase genes that exhibit the same activity toward the TEM substrate and interfere with the assay [11]. In the current study, FPI-encoded proteins were expressed as fusion proteins with the small triple FLAG tag and tracked within infected macrophage-like cells. The localization of IglC in a T6SS mutant was also assessed.

Materials and Methods

Bacterial and Cell Cultures

Bacterial strains and cell lines used in this study are listed in Table 1. *Escherichia coli* D10 (Invitrogen) was grown aerobically at 37°C on Luria-Bertani (LB) media, containing 50 µg/ml of ampicillin (LBA) when appropriate for selection and maintenance. *F. novicida* U112 (ATTC 15482) was cultured aerobically at 37°C on tryptic soy agar (TSA) or in tryptic soy broth (TSB) supplemented with 0.1% cysteine. When selecting for or maintaining transformants, *Francisella* was cultured on TSA containing 15 µg/ml of kanamycin. J774-1A murine macrophage-like cells were obtained from the American Type Culture Collection (ATCC, TIB 67, BALB/C macrophage). J774 cells were grown in flasks in Dulbecco's Modified Eagle Medium (DMEM) (GIBCO Invitrogen Grand Island, NY USA) supplemented with 10% newborn calf serum (NCS) and maintained at 37°C in a humidified 6.5% CO_2 incubator. The mosquito hemocyte like cells Sua-1B were grown in Schneider's Insect Medium (Sigma Aldrich St. Louis, MO USA) supplemented with 20% fetal bovine serum at 28°C and flasks were capped tightly [26].

DNA Manipulations

Restriction enzyme digests, sub-cloning, cloning, and DNA electrophoresis for *E. coli* was performed using standard cloning techniques [27] and the (Invitrogen Carlsbad, CA. USA) E-Gel Clonewell 0.8% SYBR Safe gel system. By cutting the *Francisella* expression plasmid groE-GFP- pFNLTP6 [8] with BamHI, the GFP insert was removed, leaving the groE promoter and the multiple cloning site (MCS) in place. Within the MCS a fragment containing *sopB* and a triple FLAG tag from the plasmid pSB2598 was inserted [28]. Next through quick-change mutagenesis, the second NcoI site in the plasmid's kanamycin resistance gene was removed without changing the coding sequence. By modifying this second NcoI site, most of the FPI genes have been inserted into plasmids using the restriction enzymes EcoRI and NcoI within the MCS; this allows for easy primer design and cloning of C-terminus triple FLAG plasmids. Primers used to construct the *Francisella* expression plasmids with the epitope tag on the C-terminus of FPI are listed in Table 2. After cloning all C-terminal tagged FPI genes, pKH8, IglA-FLAG, has been modified becoming the backbone for N-terminus tagged plasmids. These modifications involved removing *iglA*, leaving the triple FLAG tag down stream of the groE promoter and the Shine-Delgarno sequences of *iglA* yet upstream of the MCS. Primers were designed for N-terminus triple FLAG-tagged FPI genes to be inserted with the restriction enzymes Xma1 and Xho1 (Fig. S1). Primers used to construct the *Francisella* expression plasmids with the epitope tag on the N-terminus of FPI are listed in Table 3. PCR for cloning was done using Phusion High-Fidelity PCR (Finnzymes). Restriction enzyme digest was performed as described in New England Biolabs Catalog and Technical Reference (Ipswich, MA. USA). PCR products and restriction enzyme digest products were purified via Wizard SV Gel and PCR clean up System (Promega Madison, WI, USA). Ligations using T4 DNA ligase (Fisher Scientific) were done at 16°C for 14–16 h. Plasmids were recovered from *E. coli* through PureYield Plasmid Miniprep System (Promega Madison, WI, USA) for screening and PureYield Plasmid Midiprep System (Promega Madison, WI, USA) to collect a stock for transformations. Plasmids generated in this study are listed in Table 4. Plasmids were initially screened using restriction enzymes that were used for cloning and when applicable another restriction enzyme that would cut within the specific FPI gene. Plasmids were also screened for correct gene and triple FLAG sequence using standard Taq PCR. After plasmids and gene inserts were ligated, they were transformed into *E. coli* D10 using electroporation. Plasmids collected from *E. coli* were sequenced to confirm that the FPI gene sequence was not altered before being transformed into *F. novicida*.

Transforming *Francisella*

These newly constructed *Francisella* expression plasmids containing the FPI encoded ORFs were chemically transformed into *F. novicida* strain U112. A sub-culture of bacteria was grown aerobically at 37°C, shaking at 200 rpm until mid log phase or an $OD_{600 nm}$ of 0.3–0.5. Cells were pelleted at $5,000 \times G$ for 5 min at room temperature. Cells were suspended in *Francisella* transformation buffer or transformation medium [29] and then 400 µl of cell suspension was mixed with DNA and incubated aerobically at 37°C with shaking at 90 rpm for 1 h. 56 µl of 10% glucose and one ml of TSB was then added per transformation and incubated overnight aerobically at 37°C with shaking at 150 rpm. Cells were plated in 100 µl aliquots on freshly prepared TSA containing 15 µg/ml of kanamycin. Colonies were picked, isolated, and then screened by PCR, restriction enzyme digest, and Western blotting for confirmation of successful transformation.

SDS-PAGE and Western Blotting

SDS-PAGE was performed using standard techniques [27]. Proteins were transferred to Immobilon-P membrane (Millipore Billerica, MA USA), and then blocked in 5% non-fat dry milk (NFDM) in Tris-Buffered Saline and Tween 20 (Fisher BioReagents Fair Lawn, NJ US) solution (TBST) containing 1 mM Tris, 15 mM NaCl, 2 mM KCl, and 0.1% Tween 20 for 1 h. To detect FLAG-tagged proteins, the blots were incubated with (1/5000) monoclonal M2 anti-FLAG antibodies (Sigma Aldrich St. Louis, MO USA) in 5% NFDM in TBST. For the detection of IglA, IglC, PdpA, and PdpC, polyclonal rabbit anti IglA, IglC,

Table 1. List of Strains and Plasmids.

Strain	Description	Reference
J774-1A	Murine Macrophages cell lines	ATTC
Sua-1B	Mosquito hemocytes cell lines	[26]
E. coli DH5α	Sub cloning competent cells	Invitrogen
F. novicida U112	Francisella novicida prototype strain	ATTC
F. novicida U112R 2008	U112 Δ restriction genes	[40]
Jlo	U112 with deletion of gene FTN1758	[36]
ΔpdpA	U112 ΔpdpA	[32]
ΔpdpB	U112 ΔpdpB	[13]
ΔiglE	U112 ΔiglE	[13]
ΔvgrG	U112 ΔvgrG	[13]
ΔiglF	U112 ΔiglF	[13]
ΔiglG	U112 ΔiglF	[13]
ΔiglH	U112 ΔiglH	[13]
ΔdotU	U112 ΔdotU	[13]
ΔiglI	U112 ΔiglI	[13]
ΔiglJ	U112 ΔiglJ	[13]
ΔpdpC	U112 ΔpdpC	[13]
ΔpdpE	U112 ΔpdpE	[13]
ΔiglD	U112 ΔiglD	[13]
ΔiglC	U112 ΔiglC	[12]
ΔiglB	U112 ΔiglB	[13]
ΔiglA	U112 ΔiglA	[13]
ΔpdpD	U112 ΔpdpD	[36]

PdpA, and PdpC antibodies were used 1/2000. To detect bound antibodies, blots were incubated with Peroxidase-Goat Anti-Mouse or Peroxidase-Goat Anti-Rabbit secondary antibodies (1/5000) (Zymed Laboratories Invitrogen Immundetection San Francisco, CA USA) in 5% NFDM in TBST. To visualize protein bands, blots were incubated with SuperSignal West Pico Chemiluminescent Substrate (Thermo Scientific Rockford, IL USA) prior to exposing and developing film.

Macrophage Growth Assay and Analysis

J774A.1 mouse macrophage like cells (ATCC TIB-67) were seeded in 24-well cell culture plates at 1.4×10^5 cells/well for 24 h in complete Dulbecco's Modified Eagle Medium (cDMEM) containing 10% newborn calf serum (NCS). In 4 independent experiments, cells were infected in triplicate with F. novicida strains at a multiplicity of infection (MOI) of 1:50 (bacterium-to-macrophage). To help promote bacterial uptake after bacteria have been added, the 24-well dishes containing infected macrophages were centrifuged at $600 \times G$ for 10 min. Infected monolayers were incubated for 2 h in DMEM to allow for phagocytosis to occur, washed five times in Hank's Phosphate Buffered Saline (HPBS) (GIBCO Invitrogen Grand Island, NY USA). At this time, the infection is at 0 h, and infected macrophages were then either lysed at this time or incubated at 37°C in 5% CO_2 for 24 and 48 h. To determine bacterial replication, infected macrophages were lysed in 0.1% dexoycholate in HPBS at 0, 24, and 48 h post infection. The lysates were serially diluted in HPBS and plated on TSA and incubated at 37°C for 24 or 48 h. The colony forming units (cfu) were enumerated, and used to plot growth curves. To perform statistical analysis, the fold replication at 48 h was first determined (cfu 48 h/cfu 0 h), and then the log of the 48 h fold replication was used in a two-way ANOVA with XLSTAT to compare the means of each group in Tukey multiple comparisons ($\alpha = 0.05$).

Immuno-fluorescence Microscopy and Analysis

J774 murine macrophage-like cells or Sua-1B mosquito hemocyte-like cells were grown on coverslips and infected with F. novicida strains as indicated in each figure (Table 1) [13]. Cells were infected for 30 min at an MOI of 50:1 (bacteria per macrophage), washed with phosphate buffered saline (PBS), and incubated until the desired time point in DMEM containing 10% NCS. Cells were then fixed in 4% paraformaldehyde for 15 min at room temperature and rinsed three times with PBS. FLAG-tagged proteins and Francisella were detected with anti-FLAG M2 monoclonal antibodies and rabbit anti-Francisella novicida serum, respectively. The antibodies were diluted (1/500) in PBS containing 0.5% BSA and 0.1% saponin to permeate host cell membranes, while leaving the bacterial cell membranes intact [30]. Primary antibodies were detected with goat anti-mouse and goat anti-rabbit serum conjugated to Alexa Fluor 488 and 594, respectively (Invitrogen MOLECULAR PROBES Eugene, OR US). DNA was detected with DAPI (Invitrogen MOLECULAR PROBES Eugene, OR US). Coverslips were mounted using Prolong Gold Antifade reagent (Invitrogen MOLECULAR PROBES Eugene, OR US) and examined using an Olympus TE81 inverted fluorescent microscope with spinning disc confocal capabilities.

Table 2. C-term Primers Used in this Study.

C-terminal FLAG *Francisella* expression plasmid primers
pdpA_C_terminalFLAG_F_Nde1: ggcagCATATGctaattaagtagacaatgatagc
pdpA_C_terminalFLAG_B_Nco1: ggcagCCATGGgatttccttttgatttatat
pdpB_C_terminalFLAG_F_KpnI1: agGGTACCcaaaaggaaattaaaagtatg
pdpB_C_terminalFLAG_B_Nco1: agCCATGGttgtacattaacttctccttg
iglE_C_terminal_F_EcoR1: aggaGAATCCggcaaaacaaggagaagttaatg
iglE_C_terminal_B_Nco1: gacgCCATGGcatcttttctatgctgctatc
vgrG_C_terminal_F_EcoR1: gagaGAATTCgattaagggataттcttatg
vgrG_C_terminal_B_Nco1: agagCCATGGctccaaccattgttgctgcggaacc
iglF_C_terminal_F_Nde1: gcagCATATGcaatggttggataataatatg
iglF_C_terminal_B_Nco1: agcaCCATGGcattttccataagcttcttgcttgc
iglG_C_terminal_F_EcoR1: agagGAATTCgaagcttattggaaaatttaaatg
iglG_C_terminal_B_Nco1: agagCCATGGcagatgtttttacatttatttg
iglH_C_terminal_F_EcoR1: agagGAATTCcttagaaggtcattatcatg
iglH_C_terminal_B_Nco1: agagCCATGGctatagagttatttaaaacaatc
dotU_C_terminal_F_EcoR1: aggaGAATTCgctatataaaggatattagaaatg
dotU_C_terminal_B_Nco1: aggaCCATGGccagcttaataaaattag
iglI_C_terminal_F_EcoR1: cgagGAATTCgggtaagaggagatttatatg
iglI_C_terminal_B_Nco1: gaagCCATGGctatgtcaaaaagatcttc
iglJ_C_terminal_F_EcoR1: caagGAATTCcaaatgagatagatg
iglJ_C_terminal_B_Nco1: agcgCCATGGctaaattaaaataacc
pdpC_C_terminal_F_EcoR1: ggcgaGAATTCgataaattaaggaagtacatatg
pdpC_C_terminal_B_Nco1: ggcagCCATGGtgacgatatttttttaaaaaagtc
pdpE_C_terminal_F_EcoR1: gaagGAATTCcttaaggatgcaaaaatatg
pdpE_C_terminal_B_Nco1: ggccCCATGGctattatagtaattttctttttc
iglD_C_terminal_F_EcoR1: ggcagGAATTCaagatcggagttgattctaatg
iglD_C_terminal_B_Nco1: ggcgaCCATGGagaaaaggctataaagaaatc
iglC_C_terminal_F_EcoR1: gaGAATTCaaaggagaatgattatgagtgag
iglC_C_terminal_B_Nco1: gaCCATGGGtgcagctgcaatatatcc
iglB_C_terminal_F_Nde1: gagaCATATGgtagagaggattтttgttatg
iglB_C_terminal_B_Nco1: agagCCATGGcgttattatttgtacc
iglA_C_terminal_F_EcoR1: agagGAATTCgtaaaaaaaggacaataagatg
iglA_C_terminal_B_Nco1: ggaaCCATGGccttatcatctacttg
pdpD_C_terminal_F_EcoR1: ggcgaGAATTCgtaagagtagtaagtatggatcaag
pdpD_C_terminal_B_Nco1: ggcgaCCATGGgaacccagatcattggtctatac
anmK_C_terminal_F: agagGAATTCgaatataaaatattgtgtaggaatcatg
anmK_C_terminal_B_Nco1: aggaCCATGGCaaagaaatttatttggacc

The description contains the FPI ORF nomenclature, which terminal the tag was fused to, direction of the primer, restriction enzyme used in cloning the respective ORF, the sequence of the primer in 5'-3' direction, underlining represents in frame codon.

Table 3. N-term Primers Used in this Study.

N-terminal FLAG *Francisella* expression plasmid primers
pdpA_N_terminal_F_Xma1: acggCCCGGGgaatagcagtaaaagatataac
pdpA_N_terminal_B_Xho1: acggCTCGAGttaatttccttttgatttatatc
pdpB_N_terminal_F_Xma1: acggCCCGGGgaaattttattaaaaatcatc
pdpB_N_terminal_B_Xho1: acggCTCGAGttattgtacattaacttctccttg
iglE_N_terminal_F_Xma1: acggCCCGGGgatacaataaattattgaaaaatc
iglE_N_terminal_B_Xho1: acggCTCGAGttaatcttttctatgctgc
vgrG_N_terminal_F_Xma1: acggCCCGGGgatcaaaagcagaccatatttc
vgrG_N_terminal_B_Not1: caggGCGGCCGCttatccaaccattgttgctgcgg
iglF_N_terminal_F_Xma1: acggCCCGGGgaaataatgatattgataaatgg
iglF_N_terminal_B_Xho1: acggCTCGAGttaaattttccaataagcttcttgc
iglG_N_terminal_F_Xma1: acggCCCGGGgattaaatattataaatgactcc
iglG_N_terminal_B_Xho1: acggCTCGAGctaagatgtttttacatttatttgtcc
iglH_N_terminal_F_Xma1: acggCCCGGGatgaaaaaagaaaagatttaag
iglH_N_terminal_B_Xho1: acggCTCGAGttatatagagttatttaaaacaatc
dotU_N_terminal_F_Xma1: acggCCCGGGgaaaagactttaaagagatag
dotU_N_terminal_B_Xho1: acggCTCGAGttaccagcttaataaaattagtaagc
iglI_N_terminal_F_Xma1: acggCCCGGGgaagtcagataatatctacac
iglI_N_terminal_B_Xho1: acggCTCGAGttatatgtcaaaaagatcttc
iglJ_N_terminal_F_Xma1: acggCCCGGGgaaagactattttgaagatcttttg
iglJ_N_terminal_B_Xho1: acggCTCGAGtcataaattaaaataacctagatatatc
pdpC_N_terminal_F_Xma1: acggCCCGGGgaaacgacaaatatgaactaaatatc
pdpC_N_terminal_B_Xho1: acggCTCGAGctatgacgatatttttttaaaaaag
pdpE_N_terminal_F_Xma1: acggCCCGGGgaagtaaaaaagtatttcaattattattaatatttg
pdpE_N_terminal_B_Xho1: acggCTCGAGttatattatagtaattttctttttc
iglD_N_terminal_F_Xma1: acggCCCGGGgatttctagaaaggatttattg
iglD_N_terminal_B_Xho1: acggCTCGAGttaagaaaaggctataaagaaatc
iglC_N_terminal_F_Xma1: acggCCCGGGgaattatgagtgagatgataacaag
iglC_N_terminal_B_Xho1: acggCTCGAGctatgcagctgcaatatatcc
iglB_N_terminal_F_Xma1: acggCCCGGGgaacaataaataaattaag
iglB_N_terminal_B_Xho1: acggCTCGAGttagttattatttgtaccg
iglA_N_terminal_F_Xma1: acggCCCGGGcaaaaaataaaatcccaaattc
iglA_N_terminal_B_Not1: caggGCGGCCGCctacttatcatctacttgttgattac
pdpD_N_terminal_F_Xma1: acggCCCGGGatcaagatatcaacgatttattatatg
pdpD_N_terminal_B_Xho1: acggCTCGAGttaaacccagatcattggtctatac
anmk_N_terminal_F_Xma1: acggCCCGGGgatctggaacatcactagatgg
anmk_N_terminal_B_Xho1: acggCTCGAGttaaaagaaatttatttggacc

The description contains the FPI ORF nomenclature, which terminal the tag was fused to, direction of the primer, restriction enzyme used in cloning the respective ORF, the sequence of the primer in 5'-3' direction, underlining represents in frame codon.

Images containing a total of 10–60 infected cells and 100–600 bacteria for each of the 3 independent experiments were collected as Z-stacks and a projection image was generated using the Intelligent Imaging SlideBook software package. Exposure time and settings were constant for all slides in each experiment. Using SlideBook software, masks were generated for infected macrophage-like cells, bacteria, and FLAG-tagged protein signals. The percentage of bacterial masks that overlapped with FLAG masks was used to determine the percentage of bacteria associated or co-localized with FLAG-tagged protein. However, this did not account for FLAG-tagged protein that dispersed away from bacteria, therefore the percentage of infected macrophage masks containing FLAG-tagged masks were also determined for every FPI protein with each the C-terminal and N-terminal FLAG-tag. Three independent experiments were performed. The data were analyzed in XLSTAT with an ANOVA paired with a left sided Dennett's test comparing each FLAG-tagged proteins' mean to the mean of bacteria not containing a FLAG expressing plasmid for either the percentage of bacteria co-localized with FLAG signal or the percentages of infected macrophage-like cells containing FLAG signal. Significant differences were determined with an $\alpha = 0.05$. We also tested the data for correlations between the same proteins with different tags

Table 4. Plasmids used in this study.

Plasmid	Description	Reference
pFNLTP6-gro-gfp	groE-gfp; Km^r Ap^r	[8]
pKH1	Km^r Ap^r	This study
pSB2598	sopB-FLAG	[28]
pKH2	Km^r Ap^r	This study
pKH3	gro-sopB-FLAG; Km^r Ap^r	This study
pKH22	gro-pdpA-FLAG; Km^r Ap^r	This study
pKH24	gro-pdpB-FLAG; Km^r Ap^r	[13]
pKH9	gro-iglE-FLAG; Km^r Ap^r	[13]
pKH10	gro-vgrG-FLAG; Km^r Ap^r	[13]
pKH26	gro-iglF-FLAG; Km^r Ap^r	[13]
pKH11	gro-iglG-FLAG; Km^r Ap^r	[13]
pKH12	gro-iglH-FLAG; Km^r Ap^r	[13]
pKH13	gro-dotU-FLAG; Km^r Ap^r	[13]
pKH14	gro-iglI-FLAG; Km^r Ap^r	[13]
pKH15	gro-iglJ-FLAG; Km^r Ap^r	[13]
pKH5	gro-pdpC-FLAG; Km^r Ap^r	This study
pKH16	gro-pdpE-FLAG; Km^r Ap^r	This study
pKH6	gro-iglD-FLAG; Km^r Ap^r	This study
pKH4	gro-iglC-FLAG; Km^r Ap^r	[13]
pKH18	gro-iglB-FLAG; Km^r Ap^r	This study
pKH8	gro-iglA-FLAG; Km^r Ap^r	This study
pKH7	gro-pdpD-FLAG; Km^r Ap^r	This study
pKH25	gro-anmK-FLAG; Km^r Ap^r	This study
pKH40	gro-FLAG-pdpA; Km^r Ap^r	This study
pKH50	gro-FLAG-pdpB; Km^r Ap^r	This study
pKH34	gro-FLAG-iglE; Km^r Ap^r	This study
pKH35	gro-FLAG-vgrG; Km^r Ap^r	This study
pKH44	gro-FLAG-iglF; Km^r Ap^r	This study
pKH36	gro-FLAG-iglG; Km^r Ap^r	This study
pKH45	gro-FLAG-iglH; Km^r Ap^r	This study
pKH37	gro-FLAG-dotU; Km^r Ap^r	This study
pKH39	gro-FLAG-iglI; Km^r Ap^r	This study
pKH47	gro-FLAG-iglJ; Km^r Ap^r	This study
pKH41	gro-FLAG-pdpC; Km^r Ap^r	This study
pKH38	gro-LAG-pdpE; Km^r Ap^r	This study
pKH48	gro-FLAG-iglD; Km^r Ap^r	This study
pKH46	gro-FLAG-iglC; Km^r Ap^r	This study
pKH43	gro-FLAG-iglB; Km^r Ap^r	This study
pKH27	gro-FLAG-iglA; Km^r Ap^r	This study
pKH42	gro-FLAG-pdpD; Km^r Ap^r	This study
pKH49	gro-FLAG-anmK; Km^r Ap^r	This study

This table lists all the plasmids used in designing the *Francisella* expression plasmids and all of the *Francisella* expression plasmids that were generated in this study.

using the Spearman correlation test in XLSTAT. Significant differences were determined with an $\alpha = 0.05$.

Results

FLAG-tagged FPI Protein Expression in *F. novicida*

Expression of C-terminal and N-terminal epitope tagged FPI proteins from *F. novicida* U112 was confirmed by Western

blotting (Fig. 1). Western blotting showed the C-terminal and N-terminal tags do not interrupt FPI protein expression at their expected sizes (Fig. 1 and Table S1), with the exception of FLAG-PdpE. In addition to the expected sizes, lower intensity bands of different sizes were detected for some proteins (Fig. 1). IglG-FLAG expression was lower than the other FPI proteins and was not visible here (Fig. 1C); expression of IglG-FLAG was confirmed with longer exposure times causing over exposure with the other proteins (data not shown). Antibodies towards IglA, IglC, PdpA, and PdpC detected proteins of the same size as western blots detecting the FLAG-tag (Fig. S2).

Intramacrophage Growth Complementation

Since several FPI genes are needed for intracellular growth, the C-FLAG and N-FLAG-tagged proteins ability to complement respective knock out mutant strains were assessed (Table 1). As previously described, the FPI deletion mutants of *iglABCDEFHIJ, pdpAB, dotU*, and *vgrG* were defective for intramacrophage growth [8,12,30,32] (Fig. 2). Expression of C-terminal and N-terminal tagged FPI proteins, IglABCDEFHIJ, PdpAB, DotU, and VgrG, in FPI mutants increased growth rates, indicating that *Francisella* expression plasmids complemented their mutants. Genetic complementation of each deletion mutant with the C-FLAG and N-FLAG-tagged *Francisella* expression plasmids restored intramacrophage growth, and the growth of complemented mutants were significantly higher compared to their parental mutant (Fig. 2) (P<0.05). Expression of the tagged proteins did not always completely restore growth to that of the wild type; growth of 22 of the 26 complements were equivalent to that of the wild type (P>0.05). However, growth of the complements, FLAG-IglB, IglC-FLAG, IglJ-FLAG, and FLAG-IglJ, were significantly different compared to wild type and their respective mutant (P<0.05). Together, these data indicated that most of the plasmids expressed biologically functional proteins.

Unique Patterns of FPI-Encoded Proteins Co-localized with *F. novicida* within Infected Cells

FPI proteins with consistent FLAG detection had varying patterns of distribution of FLAG signal when compared to each other within infected murine macrophage-like cells (Fig. 3) and mosquito hemocyte-like cells (Fig. 4). IglA was localized with bacteria (Fig. 3 and Fig. 4). IglCE and VgrG were co-localized with bacteria and also extending beyond, completely surrounding bacteria (Fig. 3 and Fig. 4). IglD and PdpA were also localized with bacteria, and on occasion surrounding the bacteria (Fig. 3 and Fig. 4). PdpC was also detected both co-localized with bacteria and dispersing away from the bacteria (not shown). IglI was distinctly localized to the bacterium, it surrounded the bacterium uniformly (Fig. 3 and Fig. 4). The C-FLAG PdpE was studded around the bacterium, while the N-terminal tagged protein was not detected by immuno-fluorescent microscopy (Fig. S3).

C-terminal FLAG-tagged FPI Proteins Co-Localization with Bacteria During Cell Infection

The localization of FPI-encoded proteins was examined via immuno-fluorescent microscopy of infected murine macrophage-like J774 cells with bacteria expressing C-terminal fusion proteins. The percent of bacteria co-localized with FLAG signal within infected macrophages was determined for all 18 FPI-encoded proteins. At 30 min post-infection, bacteria expressing FLAG-tagged IglABDEH, PdpE, VgrG, and DotU were significantly more often co-localized with fluorescent signal compared to control bacteria not expressing epitope-tagged protein (p≤0.038) (Fig. 5A and Table S2). Bacteria expressing the remaining FLAG-

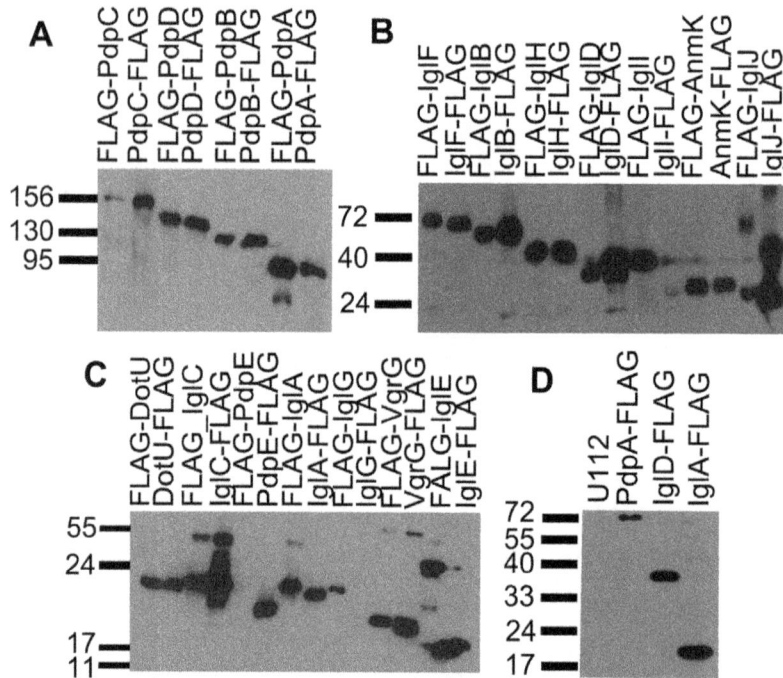

Figure 1. FPI FLAG-tagged Protein Expression in *F. novicida*. Whole cell lysates have been analyzed for production of N- terminal and C-terminal FLAG-tagged proteins by Western blotting. *F. novicida* U112 wild type and U112 expressing the respective FPI protein from the *Francisella* expression plasmid are labeled above each lane. C-terminally tagged proteins are referred to as protein-FLAG, and N-terminally tagged proteins are referred to as FLAG-protein. FPI encoded proteins have been grouped into similar predicted sizes (A) 156–95 kDa, (B) 67.6–30.9, (C) 24.6–14.5, and (D) a 10% gel for comparisons. FLAG-tagged proteins are detected with monoclonal mouse anti FLAG, goat anti mouse conjugated HRP, and chemiluminescent substrate.

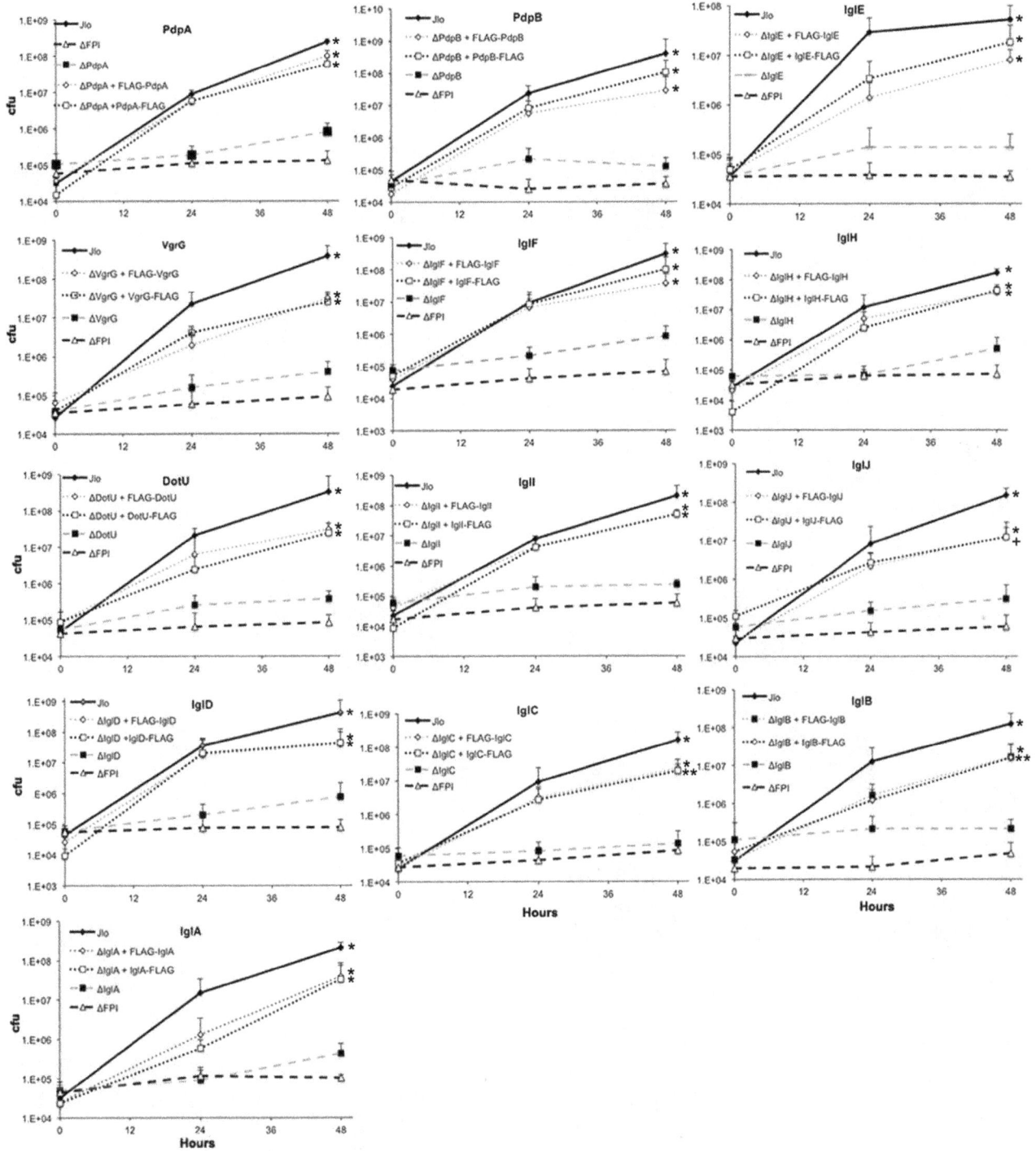

Figure 2. Intracellular Growth. J774 cells were infected with indicated strains and harvested at 0, 24 and 48 h post infection and the amount of intracellular bacteria at each time point was determined. The FPI mutant lacks all FPI genes and was defective for intracellular growth, as were each of the individual FPI gene mutants. All of the genetic complements express the cognate FPI gene as a C-terminal or an N-terminal triple FLAG-tagged version from a plasmid vector. Graphs are the average of four independent experiments and error bars show standard deviations, * indicate significance from mutants, ** partial complementation that was different form mutants and wild type, and + approaching significance at 0.054. P values generated by Tukey multiple comparisons via a two-way ANOVA.

tagged FPI-encoded proteins were not statistically different from the control bacteria at 30 min (p≥0.368) (Fig. 5A and Table S2). Bacteria expressing FLAG-tagged IglACEI, PdpA, VgrG, and DotU all had significantly more bacteria co-localized with

fluorescent signal compared to control bacteria not expressing epitope-tagged protein (p≤0.008) at 4 h post-infection (Fig. 5B and Table S2). Also at 4 h, bacteria expressing tagged IglH were approaching statistical significance when compared to the control

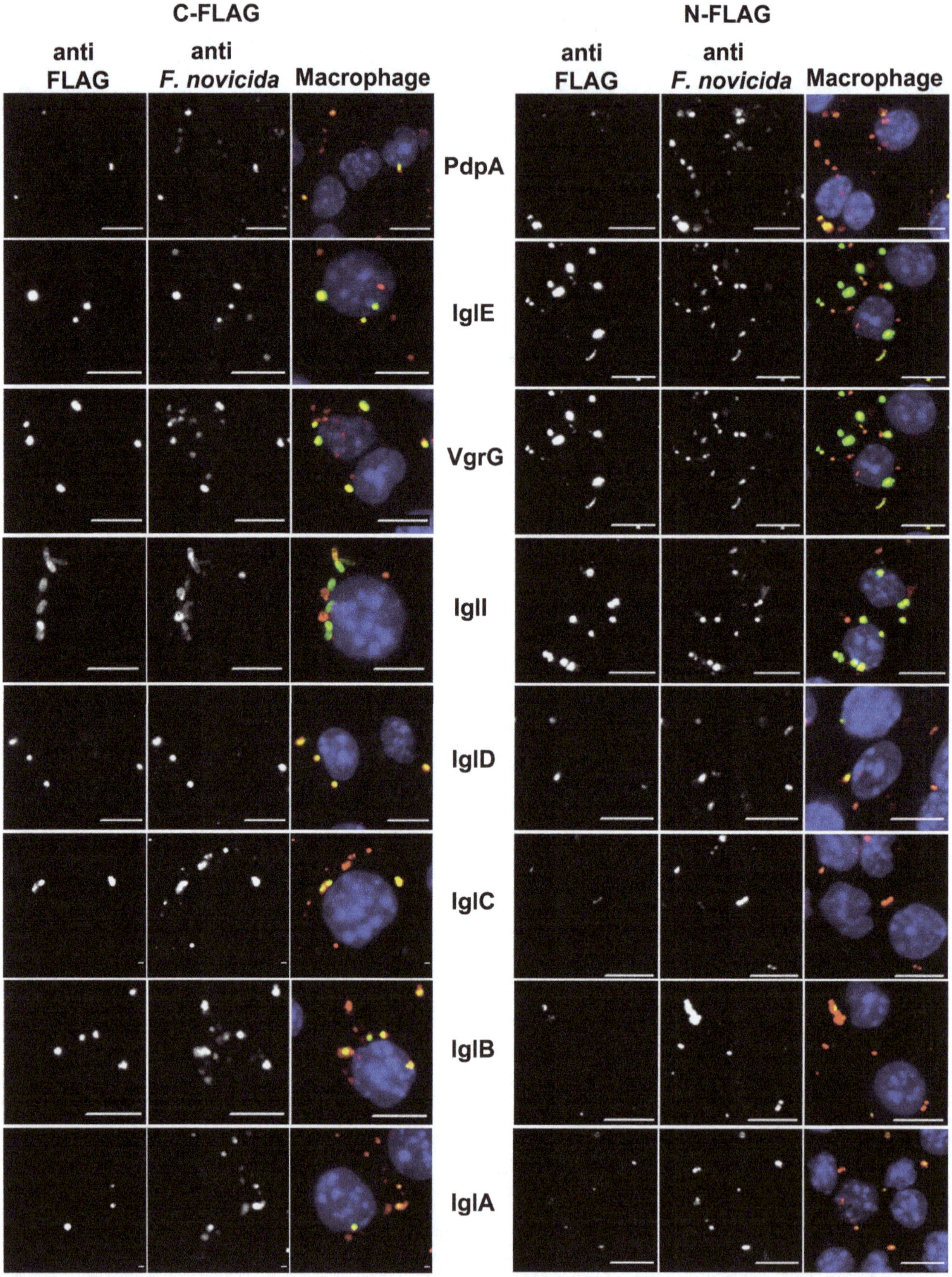
C-FLAG
anti FLAG
anti F. novicida
Macrophage
N-FLAG
anti FLAG
anti F. novicida
Macrophage
PdpA
IglE
VgrG
IglI
IglD
IglC
IglB
IglA

Figure 3. Localization of FPI Proteins within Infected Macrophages. J774 cells were infected with wild type bacteria without a plasmid and wild type containing the *Francisella* expression plasmids that express PdpA, IglABCDEDI and VgrG with either C-term FLAG tag (left) or N-term FLAG tag (right) at 4 h post-infection. Middle columns and the red in the merged right columns indicate bacteria. Left columns and the green in the merged right columns indicate FLAG-tagged proteins. Host cell nuclei are indicated by blue in merged right columns.

(p = 0.054) (Fig. 5B and Table S2). The bacteria expressing the rest of the FPI-tagged proteins were not different from control bacteria, when examining the bacteria co-localized with FLAG signal at 4 h (p≥0.814) (Fig. 5B and Table S2). 8 h into infection, bacteria expressing IglEHI and DotU had significantly more bacteria co-localized with fluorescent signal than the control bacteria (p≤0.029) (Fig. 5C and Table S2). Also at 8 h, bacteria expressing IglA had 55% of bacteria co-localized with FLAG signal, approaching statistical significance when compared to the control (p = 0.054) (Fig. 5C and Table S2). The other FLAG-tagged FPI-encoded proteins were not different from the controls (p≥0.274) (Fig. 5C and Table S2). More bacteria were co-localized with fluorescent signal from tagged IglE, DotU, PdpE, and IglA (p≤0.012) with bacteria expressing those tagged proteins compared to control bacteria at 12 h, while bacteria expressing

the other FPI-tagged proteins were not different than the control bacteria (p≥0.129) (Fig. 5D and Table S2).

C-term FLAG-tagged FPI Protein Localization within Infected Cells

The localization of C-term FLAG-tagged FPI proteins were alternatively assessed by calculating the percent of infected cells containing fluorescent signal to account for proteins that were secreted into the infected cells but did not co-localize with the bacteria expressing the epitope tagged proteins. Within 30 min of infection, the cells infected with bacteria expressing tagged IglABCDEHI, PdpACDE, VgrG, DotU, and Amnk all had significantly more infected cells containing fluorescent signal when compared to the control cells infected with bacteria not expressing FLAG-tagged proteins (p≤0.036) (Fig. 6A and Table S2). Also at 30 min into infection, the cells infected with bacteria expressing tagged PdpB and IglF were not different from cells infected with control bacteria (p≥0.232) (Fig. 6A and Table S2). At 4 h post infection, cells infected with bacteria expressing tagged IglEHI, VgrG, and DotU, all had significantly more infected cells containing fluorescent signal (p≤0.029) than the cells infected with bacteria not expressing FLAG-tagged proteins. However, the cells infected with tagged IglBCDFGJ, PdpABCE, and Anmk were not significantly different from cells infected with bacteria not expressing FLAG-tagged proteins (Fig. 6B and Table S2). When comparing the amount of infected cells containing FLAG signal at 8 h after infection, the infected cells that expressed IglABC-DEFGI, PdpACE VgrG, and DotU had significantly more infected cells containing FLAG signal (p≤0.009). While the cells infected with bacteria expressing tagged IglG, PdpBD and Anmk were not different from the control cells containing bacteria not expressing FLAG-tagged proteins (p≥0.142) (Fig. 6C and Table S2). More infected cells contained FLAG signal at 12 h within cells infected with bacteria expressing tagged PdpACE, IglABCDEJ, VgrG, DotU, and Anmk (p≤0.042). Additionally at 12 h, cells infected with bacteria expressing FLAG-tagged PdpBD and IglFGH were not statistically different (p≥0.103) when compared to cells infected with bacteria not expressing FLAG-tagged proteins (Fig. 6D and Table S2).

Bacteria Co-Localization with N-terminal FLAG-tagged FPI Proteins During Cell Infection

The localization of FPI-encoded proteins was also examined via immuno-fluorescent microscopy of infected murine macrophage-like cells with bacteria expressing N-terminal fusion proteins in order to assess effects of tags on the termini of FPI encoded proteins. The percent of bacteria co-localized with FLAG signal within infected macrophages was determined for all 18 FPI proteins with a N-terminus FLAG tag. At 30 min post-infection, bacteria expressing FLAG-tagged IglEI, VgrG, DotU, and PdpC had significantly more bacteria co-localized with FLAG signal compared to bacteria not expressing epitope-tagged protein (p≤ 0.047 (Fig. 7A and Table S2). Also at 30 min, bacteria expressing tagged IglD were approaching statistical significance with the amount of bacteria co-localized with FLAG-tagged protein when compared to the control bacteria (p = 0.072) (Fig. 7A and Table S2). The bacteria expressing the other tagged proteins at 30 min were not statistically different from the control bacteria not

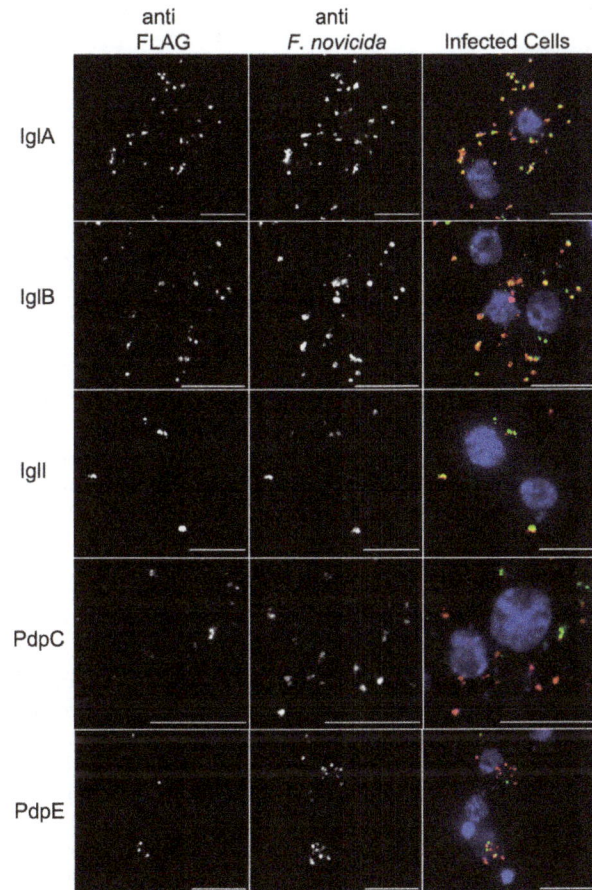

Figure 4. Localization of FPI Proteins During Infection of Sua-1B Cells. J774 cells infected with wild type containing the *Francisella* expression plasmids that express IglA, IglC, IglI, and PdpE for 4 h (A). Sua-1B cells infected with wild type containing the *Francisella* expression plasmids that express IglA, IglC, IglI, and PdpE for 4 h (B). Middle columns and red in the right column indicate bacteria. C-terminally FLAG-tagged proteins are shown in the left columns and are also green in the right column. Host cell nuclei are displayed as blue in the merged, right columns.

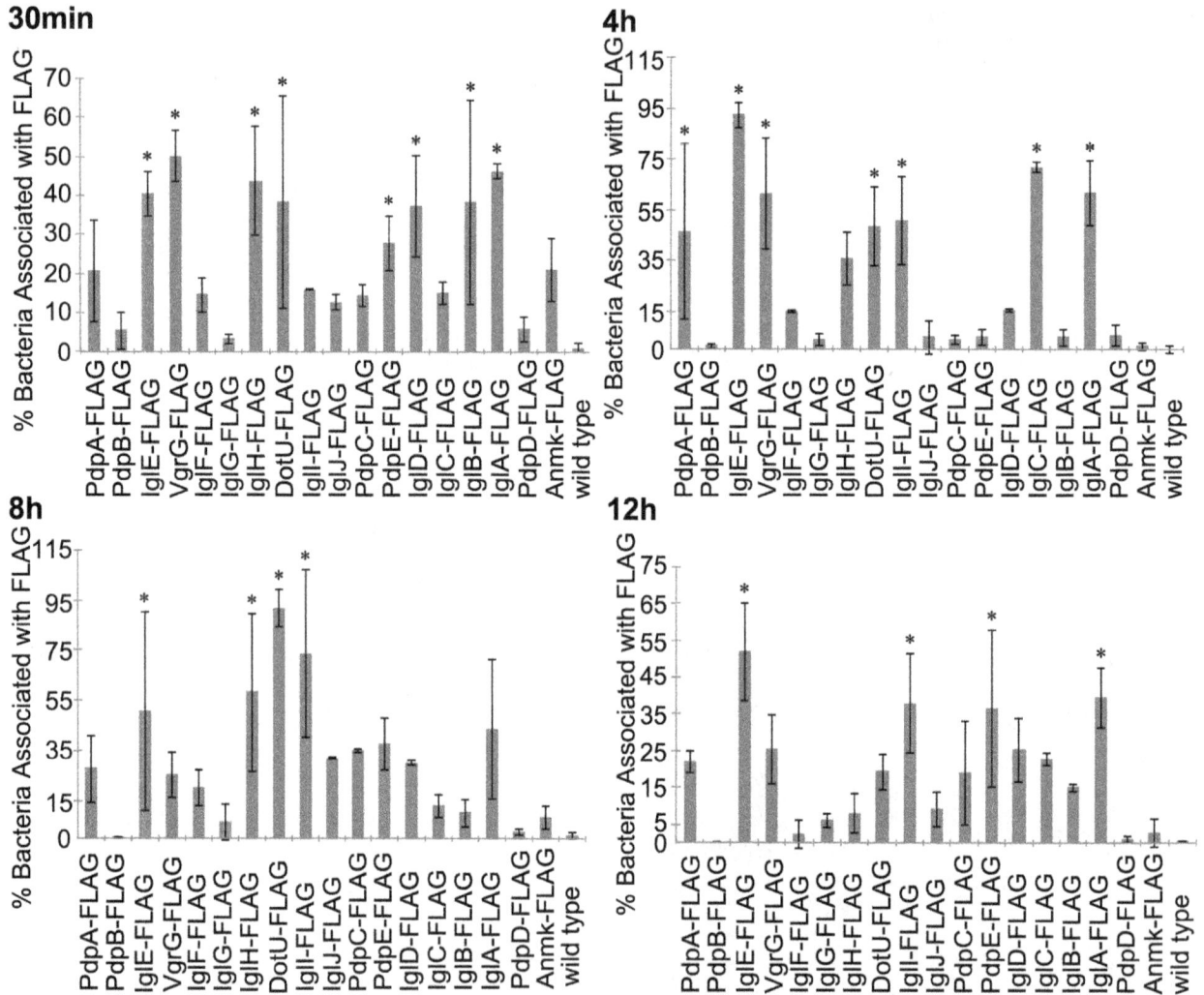

Figure 5. Bacteria Associated with FPI C-tagged Proteins. The percent of bacteria associated with C-term FLAG-tagged FPI proteins within J774 cells at various time-points during infection were determined. This graph represents the mean of three independent experiments. Error bars indicate the standard deviations. Asterisks indicate significance (p≤0.05) and plus signs indicate approaching significance (p = 0.051–0.099) when compared to the no plasmid wild type control.

expressing tagged protein (p≥0.166) (Fig. 7A and Table S2). Bacteria expressing tagged IglID and PdpC had significantly more bacteria co-localized with FLAG signal compared to cells infected with control bacteria not expressing epitope-tagged protein (p≤ 0.021) at 4 h post-infection (Fig. 7B and Table S2). Additionally at 4 h into infection, bacteria expressing the rest of the FPI-tagged proteins were not different from control bacteria when examining the bacteria co-localized with FLAG signal (p≥0.347) (Fig. 7B and Table S2). Bacteria expressing tagged IglEI and DotU had significantly more bacteria co-localized with FLAG signal than cells infected with the control bacteria at 8 h post infection (p≤ 0.037) (Fig. 7C and Table S2). Bacteria expressing the other FPI proteins were not different from the control (p≥0.161) (Fig. 7C). More bacteria were co-localized with FLAG signal from tagged VgrG, DotU, IglACEI, and PdpC (p≤0.032) when compared to control bacteria at 12 h, while bacteria expressing the other FPI-tagged proteins were not different than the control bacteria (p≥ 0.115) (Fig. 7D and Table S2).

N-term FLAG-tagged FPI Protein Localization within Infected Cells

Similar to the analysis of C-terminally tagged proteins, the localization of FPI proteins with N-term FLAG tags were also assessed by calculating the percent of infected cells containing FLAG signal. At 30 min into the infection, cells infected with bacteria expressing tagged PdpAC, IglAEI, DotU, and PdpC all had significantly more infected cells containing FLAG signal when compared to control cells infected with bacteria not expressing FLAG-tagged proteins (p≤0.008) (Fig. 8A and Table S2). Also at 30 min, cells infected with bacteria expressing tagged IglBCDFGHJ PdpB, VgrG, and PdpDE, and Amnk were not significant (p≥0.135) (Fig. 8A and Table S2). Cells infected with bacteria that expressed tagged IglCDE and PdpC all had significantly more infected cells containing FLAG signal (p≤ 0.045) than cells infected with bacteria not expressing FLAG-tagged proteins at 4 h post-infection (Fig. 8B and Table S2). Additionally at 4 h post infection, the cells infected with bacteria expressing tagged PdpABD, IglABIFGHJ, VgrG, DotU, and Amnk were not significantly different from the cells infected with

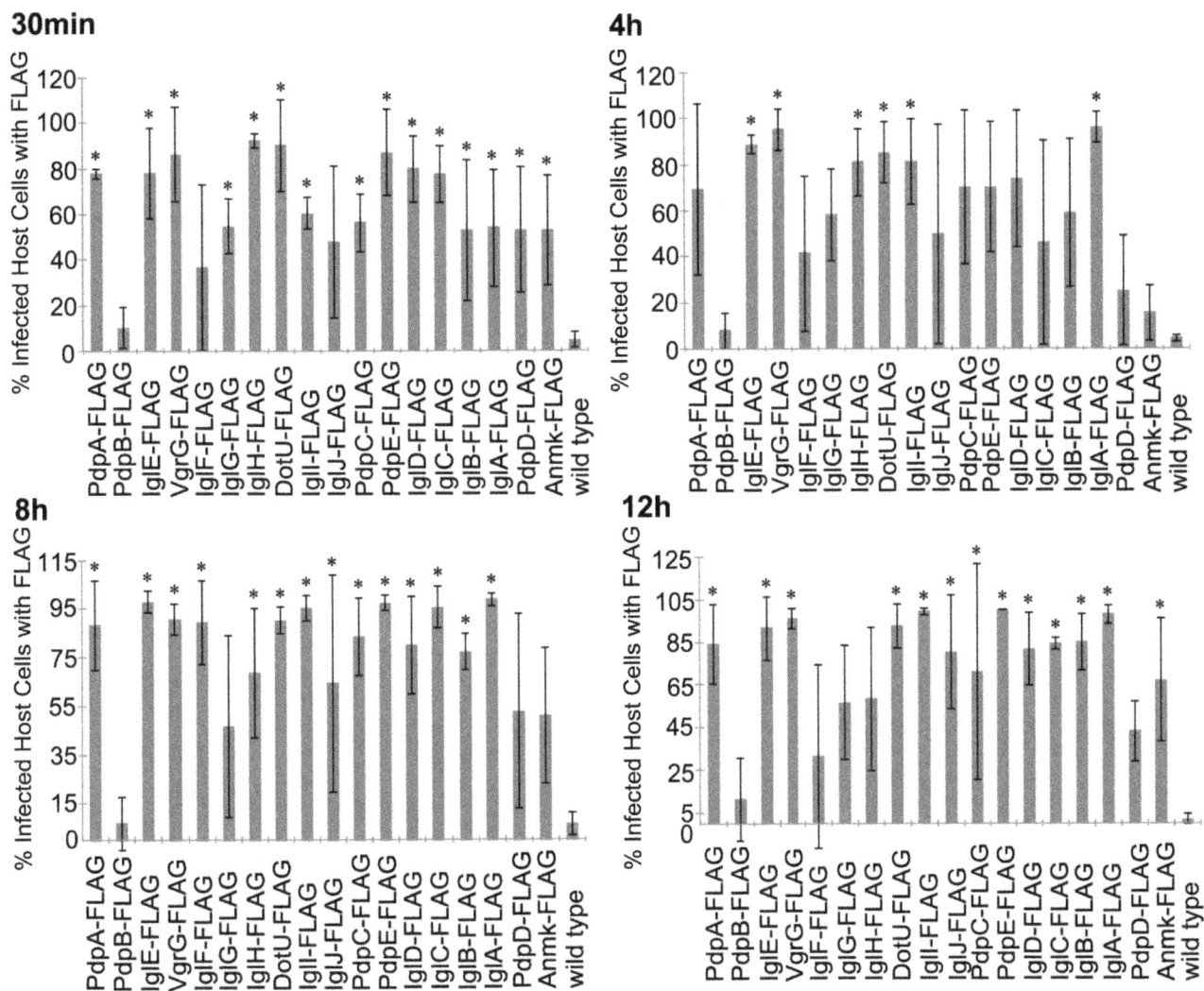

Figure 6. Infected Macrophages with FPI C-tagged Proteins. The percent of infected macrophages containing C-term FLAG-tagged FPI proteins within J774 cells at various time-points during infection were determined. This graph represents the mean of three independent experiments. Error bars indicate the standard deviations. Asterisks indicate significance (p≤0.05) and plus signs indicate approaching significance (p = 0.051–0.099) when compared to the no plasmid wild type control.

bacteria not expressing FLAG-tagged proteins (p≥0.173) (Fig. 8B and Table S2). When comparing the amount of infected cells containing FLAG signal at 8 h after infection, the cells infected with bacteria that expressed tagged FPI proteins were not different (p≥0.132) than infected cells containing bacteria not expressing FLAG-tagged proteins (Fig. 8C and Table S2). None of the infected cells were significant for containing FLAG signal in cells infected (p≥0.105) when compared to cells infected with bacteria not expressing FLAG-tagged proteins at 12 h into the infection (Fig. 8D and Table S2).

IglC Localization was Dependent on PdpB

Less FLAG was detected when IglC was expressed by the ΔpdpB mutant when compared to expression of tagged IglC by wild type bacteria (Fig. 9). Over 20% of wild type bacteria were associated with IglC, while less than 7% of ΔpdpB bacteria were associated with IglC (Fig. 9B). The differences between IglC association with wild type or ΔpdpB mutant bacteria were statistically significant. IglC was examined by Western blot to

determine if the mutant expressed the tagged proteins. IglC-FLAG was expressed at similar levels in the wild type and ΔpdpB mutant backgrounds (Fig. 9C). Therefore, extracellular co-localization of IglC with bacteria was dependent on the T6SS.

Discussion

Genes within the FPI are required for a T6SS, intracellular growth, and virulence [13,24,25,32,33]. Many of the FPI-encoded proteins are part of a T6SS, therefore we hypothesized that some of the FPI encoded proteins would be directed for secretion by that secretion system, such as effector proteins or chaperones [11,33]. A recent study that utilized a fusion Temoniera (TEM) β-lactamase reporter in LVS identified IglCEFIJ, PdpAE, and VgrG as secreted proteins, and determined that secretion is dependent on the core components of the T6SS: IglCG, DotU, and VgrG [11]. Also within that same study, only IglCE and PdpAE are secreted from *F. novicida*, suggesting differences in the secretion of FPI proteins among subspecies of *Francisella* [11]. In another

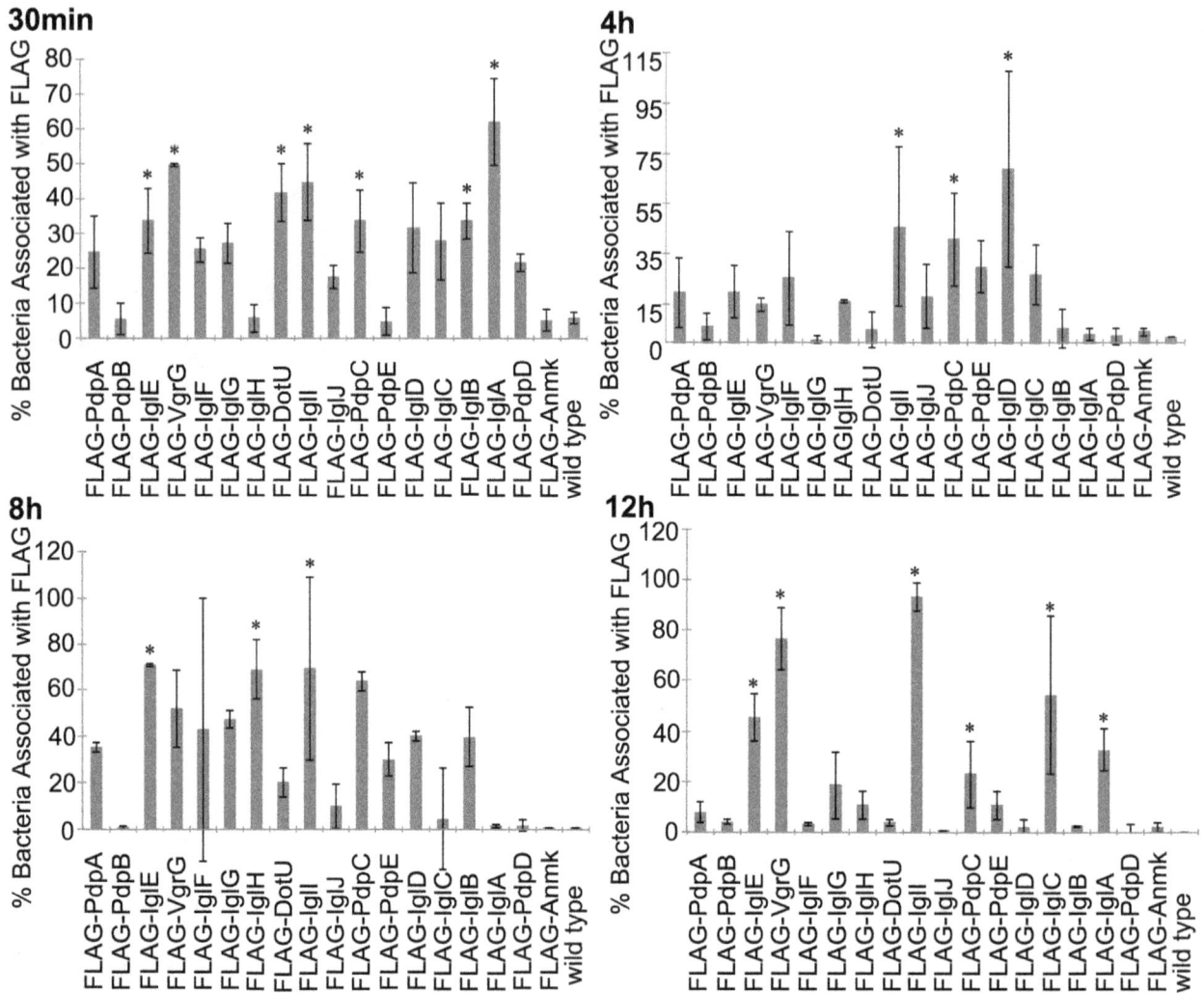

Figure 7. Bacteria Associated with FPI N-tagged Proteins. The percent of bacteria associated with N-term FLAG-tagged FPI proteins within J774 cells at various time-points during infection were determined. This graph represents the mean of three independent experiments. Error bars indicate the standard deviations. Asterisks indicate significance (p≤0.05) and plus signs indicate approaching significance (p = 0.051–0.099) when compared to the no plasmid wild type control.

study a CyaA reporter was used to show IglI and VgrG are secreted in both LVS and *F. novicida* [33].

Although these studies were the first to identify secreted FPI-encoded proteins, there were some limitations to the tools used in those experiments. The TEM β-lactamase reporters have adverse effects including low levels of protein expression; therefore results were only based on one time point at 18 h post-infection [11]. In addition to low expression, the TEM β-lactamase expresses functionless proteins that are unable to complement an intracellular growth phenotype [11]. The TEM β-lactamase assay was used in a *F. novicida* β-lactamase mutant because *F. novicida* possesses native β-lactamases that interfered with the fusion tag [11].

In the present study, we examined the localization of FPI proteins in *F. novicida* during infection of macrophage-like cells. We used *Francisella* expression plasmids that express all 18 of the FPI-encoded proteins from *F. novicida* with a C-terminal epitope FLAG tag as well as a N-terminal epitope FLAG tag. The *Francisella* expression plasmids were the first tools described for *Francisella novicida* that have epitope tags for both termini for the

entire set of FPI proteins. Some secreted proteins possess a secretion signal on either the N- or C-terminus [16]. Consequently, adding amino acids at either end may block the secretion signal, disrupting the protein's localization and function. It was unlikely that both the N- and C- termini of the proteins were required for localization. Therefore, the *Francisella* expression plasmids were developed to contain epitope FLAG tags at both the N- and C-termini. Moreover, the triple FLAG tag is short; it was less likely to alter protein folding and function. Triple repeating sequences increase affinity of FLAG monoclonal antibodies and reduce background in Western blotting, immuno-fluorescence microscopy, and many other commercially available biochemical products specific for localizing the FLAG tag sequence. However, there were also some disadvantages associated with the *Francisella* expression plasmids. Adding the triple FLAG tag does increase the molecular weight and could alter protein processing as evident by altered migration pattern in SDS gels (Fig. 1 and Fig. S2). These proteins encoded by the plasmids were constitutively expressed, which may lead to over expression and could also explain the apparent altered processing for some proteins. Additionally the

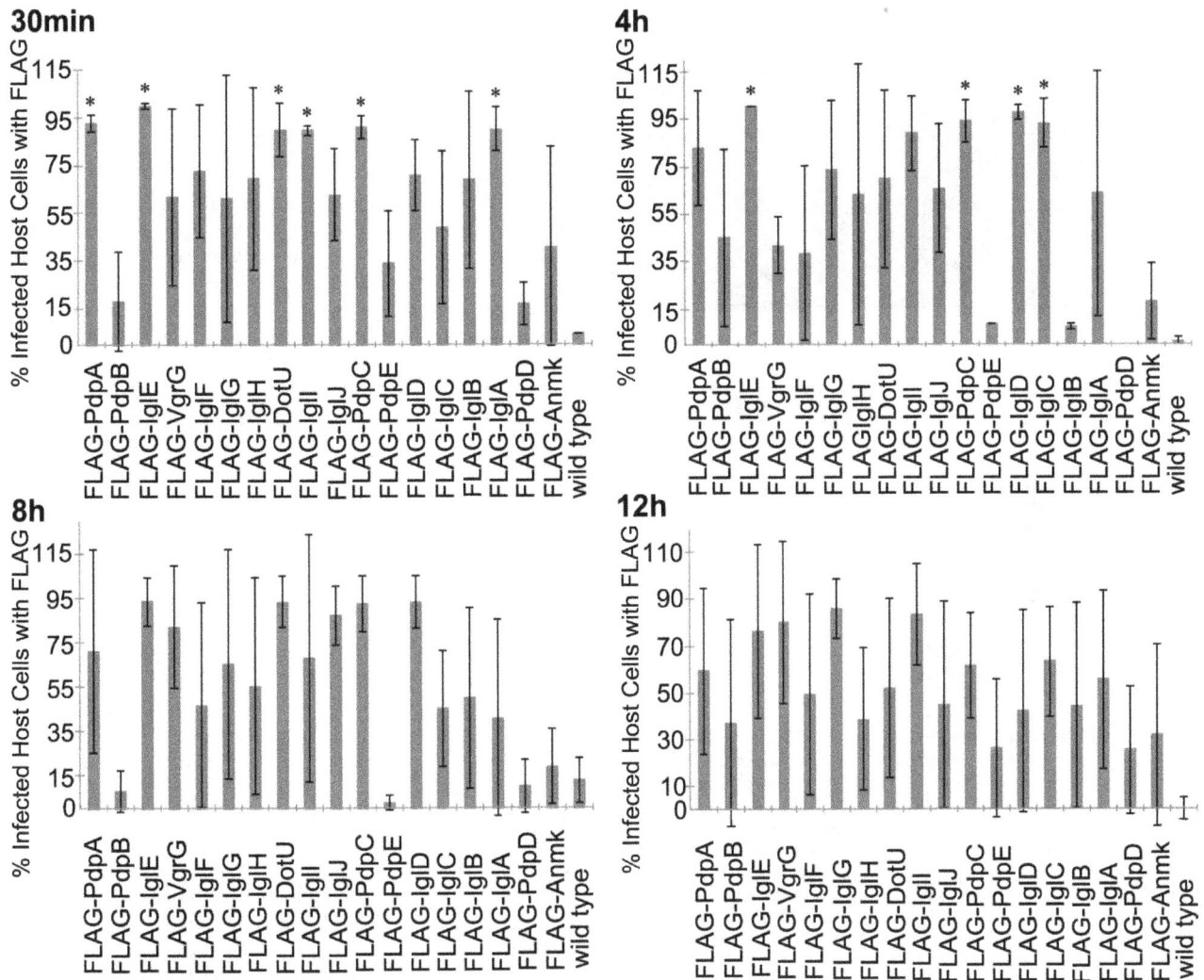

Figure 8. Infected Macrophages with FPI N-tagged Proteins. The percent of infected macrophages containing N-term FLAG-tagged FPI proteins within J774 cells at various time-points during infection were determined. This graph represents the mean of three independent experiments. Error bars indicate the standard deviations. Asterisks indicate significance ($p \leq 0.05$) and plus signs indicate approaching significance ($p = 0.051–0.099$) when compared to the no plasmid wild type control.

FPI blots were performed with similar amounts of bacteria and not adjusted for optimal visualization of individual proteins, which lead to over exposure of some proteins (Fig. 1). The lack of a wild type PdpC band visible with the PdpC antibody could be explained by low expression levels of native PdpC compared to the overexpression of PdpC-FLAG. Also it should be noted that different antibodies were used for detection of PdpC; the FLAG-tagged proteins contain three repeats of the FLAG sequences, which increases the avidity of the FLAG antibodies compared to the antibody raised against native PdpC (Fig. S2B).

To help discern the functions of FPI-encoded proteins, their localization within host cells were examined via microscopy. There are limitations associated with microscopy. Microscopy permeabilization techniques can result in false positives due to cell death or leakiness of membranes. In this study we used saponin to permeate host cell membranes, while leaving the bacterial cell membranes intact [30].

If the localization of a FPI protein was dependent on the expression of other proteins at a specific time, then we would have detected it because we examined four different time points during

infection. These time points were chosen according to *Francisella's* intracellular life cycle [34,35]. By 30 min post-infection, internalized bacteria escape the phagosome and enter the host cell's cytoplasm [20,31]. Bacteria replicate intracellularly by four and 8 h bacteria post-infection [35]. By 12h post-infection bacteria manipulate host cells by avoiding immune responses and initiating autophagy [37]. However, since the plasmids constitutively express proteins inferences cannot be made on the timeline of secretion under natural transcriptional control.

In this study the secretion and localization of FPI tagged proteins during an infection of a macrophage-like cell line was determined by assessing the amount of fluorescence per cell. First, we generated a complete set of plasmids that contain either N- or C-terminal tags. These plasmids were used to examine the entire FPI for potentially secreted proteins, and internal controls within the FPI were used for this study. IglE is described as an outer-membrane protein, PdpB is an inner-membrane protein, and several FPI proteins have previously been identified as secreted in *F. tularensis* [11,13,33]. As an initial screen of the entire FPI, we hope this study inspires future investigations that further charac-

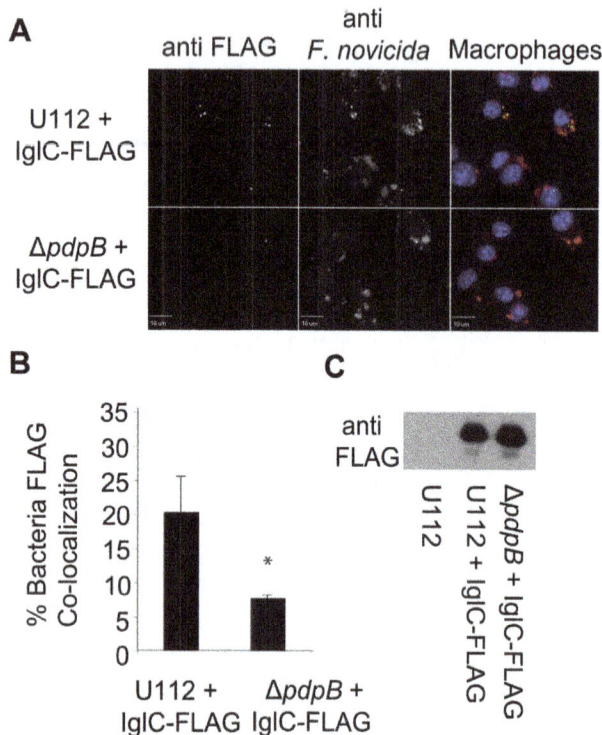

Figure 9. IglC Secretion was Dependent on T6SS. Microscopy of IglC-FLAG expressed in both wild type and Δ*pdpB* strains of U112 within J774 cells at 4 h post infection (A). Automated image analysis of bacterial cells associated with FLAG signal in wild type and Δ*pdpB* backgrounds shown in (B). Each bar represents the mean from three independent experiments of the percent of bacteria associated with FLAG signal, and error bars represent the standard deviation. An asterisk indicates the value was significantly different from that of the wild type in a Two-tailed t test (p≤0.05). Western blot shows FLAG expression of IglC-FLAG in wild type and Δ*pdpB* strains of U112 (C).

terize individual FPI proteins and their secretion pattern and localization within infected cells.

Using the *Francisella* expression plasmids in *F. novicida* IglABCDEFGHIJ, PdpACE, DotU, and VgrG were localized within macrophage-like cells on the outside of bacterial cell membranes. Proteins that were consistently detected in the cytoplasm of host cells were IglABCDEI and VgrG. There were high degrees of variation observed among time points, in the localization of proteins within infected host cells, among the N-terminal tagged proteins at, and between different tags of the same proteins. The detection of IglFGHJ, PdpACE, and DotU varied depending on time point. Since these proteins were constitutively expressed inferences based on time would not be valid. Although if some of the FPI proteins require the expression of other proteins for their delivery, secretion, or localization this may have been missed with only examining one time point. The variance in the cellular analyses can be explained by the fact that more area was assessed in the cell level analysis increasing total non-specific background fluorescence compared to bacterial association. Additionally, it was not possible to address how bacteria that were co-localized with FLAG were dispersed among infected cells. In addition to the increase in background signal in the analysis of host cells infected with FLAG, the variation among the N-tagged proteins can be explained by the fact that proteins are translated from N-terminus to C-terminus; therefore, the additional amino

acid sequences from the FLAG tag may have an affect on folding and protein stability, however these proteins were able to restore the growth phenotype (Fig. 2). It was also not surprising that N-terminally tagged PdpE was not expressed, as PdpE is predicted to have an export signal sequence on the N-terminus [11]. In addition IglE also may possess an N-terminal signal and is an outer-membrane protein [11].

Since the C-terminal and the N-terminal-flagged proteins were individually analyzed for each protein, the expression levels of the two variants were statistically compared for correlation (Fig. S4). The two tags were significantly correlated at half of the time points; overall there was always a positive correlation trend. Reasons that detection of N- and C-terminally tagged proteins were not always correlated could include effects from the tag such as alterations in expression and processing, post translation modifications, masking transportation of proteins, or stability. To avoid artifacts from tagging one terminus of the individual FPI proteins, this study used two tags and examined several time points. It should also be noted that for a given protein there was also a wide variation from one time point to another (Fig. 5–8). This is not surprising since the stabilities of these proteins are not known and previous studies of stability of pathogen proteins in host cells can be altered by tagging effects or degraded by host cell [37].

Some proteins, while expressed, were not detected in the cytoplasm of host cells. The lack of detection of fluorescence signal for PdpBD and Anmk possessing FLAG tags were similar to that of wild type cells in bacteria, and within infected host cells. This suggests PdpBD and Anmk were not secreted from bacteria, which is consistent with other studies [13,38]. These non-secreted proteins were expressed at similar levels compared to most of the secreted proteins (Fig. 1). Our inability to detect PdpB confirms the appropriateness of using saponin-permeabilized cells to detect FLAG epitopes outside of bacteria while leaving the bacterial cell wall intact [30]. Previous fractionation of *F. novicida* expressing FLAG-tagged PdpB show this PdpB localized to the inner-membrane [13]. The current model for the T6SS in *Francisella* suggests PdpB is a transmembrane anchor protein, which spans the inner-membrane with parts extending into the periplasmic space [13]. DotU is an inner-membrane component of the secretion system of *Francisella* and all T6SS's where it interacts with PdpB [13,38]. Solubility properties have identified DotU as predominantly membrane-associated, partially soluble, and localized to the inner-membrane and periplasmic space where it stabilizes the secretion system [13,39]. The localization of DotU has not been visualized before; it is interesting that in this study, microscopy detected DotU as extracellular. DotU could be temporarily exposed to the extracellular space of bacterial cells during the contraction of the tube of the secretion system as proteins were secreted. Also the extracellular localization of DotU could be from effects of the FLAG tag.

In the current model of the T6SS in *Francisella novicida* as described by de Bruin, the inner-tube of the T6SS is speculated to be a polymer of IglC, which lies within the IglA and IglB polymer that contracts and drives IglC through the host cell membrane [13]. This contraction of IglAB could temporarily expose components of the secretion system (IglABC, DotU, and potentially other proteins) to extracellular staining. Also, IglA-IglB polymers span both the inner- and outer-membrane of *Francisella*, and thus were exposed extracellularly but not necessarily secreted. VgrG and PdpE were located on the point of the secretion channel-forming tube and would therefore appear outside of bacteria, as shown in this study [13].

Several secreted proteins were identified in this study that have previously been identified as secreted from LVS, including PdpAE, IglCEFIJ, and VgrG [11,33,39]. This study also identified DotU, IglABDGH, and PdpC as being localized to the outside of bacteria within infected host cells. Fractionation of *F. novicida* show IglABCD localized in all fractions of the bacterial cell [13], which supports their detection, outside of bacteria within macrophages. IglH, and PdpC might be secreted proteins, as it is not clear whether these proteins were secreted, localized to the outer-membrane of *Francisella*, or temporarily localized to the outer-membrane as components of the secretion system during transport of other secreted proteins. An alternative explanation is their detection in this study is leakage due to over expression from the *Francisella* expression plasmids. Detection observed in microscopy is not likely from dying cells that leak tagged proteins; since we have shown these tagged, plasmid-expressed proteins restored the intracellular growth phenotype (Fig. 2). In addition, PdpB, a protein localized to the inner bacterial membrane was not detected in microscopy while it was expressed (Fig. 1) and restored growth (Fig. 2). If leakage were a systematic problem with this study, we would expect PdpB to be detected despite its localization to the inner-membrane. However, we cannot exclude this possibility for other proteins. In any type of secretion assay, leakage from dead or dying bacteria is always a possibility that has to be considered in data interpretation.

To further examine localization of IglC, the IglC-FLAG plasmid was transformed into a $\Delta pdpB$ strain to test if localization was dependent on the T6SS. PdpB is homologous to IcmF, which is an inner-membrane component of the T6SS in *V. cholera* and is required for the secretion of Hcp [14]. PdpB was not detected through microscopy because it is an inner-membrane protein of *Francisella* [13]. The co-localization of IglC-FLAG with bacterial cells was significantly lower in $\Delta pdpB$ bacteria compared to wild type cells (Fig. 9B). The expression of IglC was examined to determine if the mutant was expressing IglC-FLAG, and both wild type and $\Delta pdpB$ expressed IglC at similar levels (Fig. 9C).

The relevance of the findings may not be generally applicable to other, virulent subspecies of *Francisella* since there are previous data showing that the secretion patterns differ between *F. novicida* and LVS [11,33]. Although, due to the similarities of the secreted proteins between LVS and *F. novicida*, this study confirms *F. novicida* as a valuable model to study the molecular mechanism employed by *F. tularensis* during infection of host cells.

This study describes the development of genetic tools to assess and elucidate the function of the complete set of FPI encoded proteins. These genetic tools include plasmids that contain an entire set of both N- and C-terminus epitope triple FLAG-tagged FPI genes that express FPI proteins. Western blotting of bacterial lysates reveals expression of 35 full-length epitope tagged FPI proteins. The *Francisella* expression plasmid expresses full-length functional proteins that restore the intramacrophage growth phenotype in respective mutants. Therefore the *Francisella* expression plasmids were genetically viable tools that can be used to further understand the intracellular life cycle of *F. tularensis* and elucidate potential intervention strategies. Overall these plasmids will contribute to a better understanding of the molecular mechanisms involved in the intracellular life cycle of *F. tularensis*.

Supporting Information

Figure S1   *Francisella* Expression Plasmids. Representative diagram of the *Francisella* expression plasmids, pKH4 containing *iglC.* with a C-terminal FLAG tag and pKH46 containing *iglC* with a N-terminal FLAG tag, are shown as examples of all 36 plasmids. All of the *Francisella* expression plasmids contain a groE promoter, a multiple cloning site (MCS), triple FLAG epitope tag, antibiotic cassettes, and an origin of replication. The MCS shows restriction enzyme sites used for insertion of FPI genes. Each of the FPI genes was individually inserted where *iglC* is depicted in the diagram. Arrows represent the direction of transcription and size of gene products.

Figure S2   Native and FLAG-tagged FPI Proteins. Western blot of *F. novicida* U112 wild type and U112 expressing the respective C-terminally tagged FPI proteins from the *Francisella* expression plasmid are labeled above each lane. (A) 15% gel with IglAC, (B) 8% gel with PdpAC. IglAC and PdpAC proteins were detected with polyclonal rabbit anti IglAC and PdpAC. FLAG-tagged proteins are detected with monoclonal mouse anti FLAG, goat anti mouse conjugated HRP, and chemiluminescent substrate.

Figure S3   Three-dimensional Reconstruction of PdpE. Three-dimensional reconstructions were comprised from a series of images that were taken through the macrophage cell infected with wild type containing *Francisella* expression plasmids. Bacteria in red, FLAG-tagged protein in green, and host cell nuclei in blue.

Figure S4   N- and C-tag Correlation. The data for N- and C-tagged proteins in both bacterial and cellular analyses were plotted against each other for each protein, at each time point. The data were subjected to a Spearman correlation tests. A best-fit trend line was inserted along with the slope, R^2 values, and the P value. Asterisks indicate significance ($p \leq 0.05$) of the Spearman's test for correlation. Specific analysis and times points are indicated on graphs.

Table S1   Molecular weights of FPI proteins. The molecular weights of *F. novicida* FPI encoded proteins.

Table S2   Means of FLAG with bacteria and FLAG within cells and significance. Each of the FPI proteins were examined for their localization with bacteria or within infected host cells, which is indicated as Bacterial or Cellular in the analysis column. Within each analysis proteins were examined via the FLAG tag on the N-terminus and the C-terminus. Values indicate the mean percentage of bacterial-FLAG co-localization or the mean percentage of infected cells containing FLAG from 3 independent experiments. Significance was determined with a left sided Dunett's test, * p<0.05, **p<0.001, and ***p<0.0001.

Acknowledgments

Thank you Lisa McGilvary for providing excellent technical support. Thank you Thomas Zhart for pFNLTP-GFP-6. Thank you Fran Nano for supplying mutant strains and anti FPI protein antibodies.

Author Contributions

Conceived and designed the experiments: KH RFH. Performed the experiments: RFH. Analyzed the data: KH RFH. Contributed reagents/materials/analysis tools: KH RFH. Contributed to the writing of the manuscript: RFH KH.

References

1. Gal-Mor O, Finlay BB (2006) Pathogenicity islands: a molecular toolbox for bacterial virulence. Cell Microbiol 8(11): 1707–1719.
2. Galan JE, Wolf-Watz H (2006) Protein delivery into eukaryotic cells by type III secretion machines. Nature 444(7119): 567–573.
3. Mattoo S, Lee YM, Dixon JE (2007) Interactions of bacterial effector proteins with host proteins. Curr Opin Immunol 19(4): 392–401.
4. Nano FE, Zhang N, Cowley SC, Klose KE, Cheung KK, et al (2004) A *Francisella tularensis* pathogenicity island required for intramacrophage growth. J Bacteriol 186(19): 6430–6436.
5. Owen CR, Burker EO, Jellison WL, Lackman DB, Bell JF (1964) Comparative studies of *Francisella tularensis* and *Francisella novicida*. J Bacteriol 87: 676–83.
6. Gray CG, Cowley SC, Cheung KK, Nano FE (2002) The identification of five genetic loci of *Francisella novicida* associated with intracellular growth. FEMS Microbiol Lett 215(1): 53–56.
7. Tempel R, Lai XH, Crosa L, Kozlowicz B, Heffron F (2006) Attenuated *Francisella novicida* transposon mutants protect mice against wild type challenge. Infect Immun 74(9): 5095–5105.
8. Maier TM, Havig A, Casey M, Nano FE, Frank DW, et al (2004) Construction and characterization of a highly efficient *Francisella* shuttle plasmid. Appl Environ Microbiol 70(12): 7511–7519.
9. Nix EB, Cheung KK, Wang D, Zhang N, Burke RD, et al (2006) Virulence of *Francisella* spp. in chicken embryos. Infect Immun 74(8): 4809–4816.
10. Santic M, Molmeret M, Barker JR, Klose KE, Dekanic A, et al (2007) A *Francisella tularensis* pathogenicity island protein essential for bacterial proliferation within the host cell cytosol. Cell Microbiol 9(10): 2391–2403.
11. Bröms JE, Meyer L, Sun K, Lavander M, Sjöstedt A (2012) Unique substrates secreted by the type VI secretion system of *Francisella tularensis* during intramacrophage infection. PLOS ONE 7(11).
12. de Bruin OM, Ludu JS, Nano FE (2008) The *Francisella* pathogenicity island protein IglA localizes to the bacterial cytoplasm and is needed for intracellular growth. BMC Microbiol 7: 1.
13. de Bruin OM, Duplantis BN, Ludu JS, Hare RF, Nix EB, et al (2011) The biochemical properties of the *Francisella* pathogenicity island (FPI)-encoded proteins IglA, IglB, IglC, PdpB, and DotU suggest roles in type VI secretion. Microbiol 157: 3483–3491.
14. Barker JR, Chong A, Wehrly TD, Yu JJ, Rodriguez SA, et al (2009) The *Francisella tularensis* pathogenicity island encodes a secretion system that is required for phagosome escape and virulence. Mol Microbiol 74(6): 1459–1470.
15. Bingle LE, Bailey CM, Pallen MJ (2008) Type VI secretion: a beginner's guide. Curr Opin Microbiol 11(1): 3–8.
16. Filloux A, Hachani A, Bleves S (2008) The bacterial type VI secretion machine: yet another player for protein transport across membranes. Microbiol 154(2008): 1570–1583.
17. Bladergroen MR, Badelt K, Spaink HP (2003) Infection-blocking genes of a symbiotic *Rhizobium leguminosarum* strain that are involved in temperature-dependent protein secretion. Mol Plant Microbe Interact 16(1): 53–64.
18. Pukatzki S, Ma AT, Sturtevant D, Krastins B, Sarracinot D, et al (2006) Identification of a conserved bacterial protein secretion system in *Vibrio cholerae* using the *Dictyostelium* host model system. Proc Natl Acad Sci U S A 103(5): 1528–1533.
19. Lindgren H, Golovliov I, Baranov V, Ernst RK, Telepnev M, et al (2004) Factors affecting the escape of *Francisella tularensis* from the phagolysosome. J Med Microbiol 53(10): 953–958.
20. Santic M, Molmeret M, Klose KE, Jones S, Kwaik YA (2005) The *Francisella tularensis* pathogenicity island protein IglC and its regulator MglA are essential for modulating phagosome biogenesis and subsequent bacterial escape into the cytoplasm. Cell Microbiol 7(7): 969–979.

21. Mougous JD, Cuff ME, Raunser S, Shen A, Zhou M, et al. (2006) A virulence locus of *Pseudomonas aeruginosa* encodes a protein secretion apparatus. Science 312(5779): 1526–30.
22. Schell MA, Ulrich RL, Ribot WJ, Brueggemann EE, Hines HB, et al. (2007) Type VI secretion is a major virulence determinant in *Burkholderia mallei*. Mol Microbiol 64(6): 1466–1485.
23. Zheng J, Leung KY (2007) Dissection of a type VI secretion system in *Edwardsiella tarda*. Mol Microbiol 66(5): 1192–1206.
24. Lindgren M, Bröms J, Meyer L, Golovilov I, Sjöstedt A (2013) The *Francisella tularensis* LVS Delta-*pdpC* mutant exhibits a unique phenotype during intracellular infection. BMC Microbiol 13(20).
25. Lindgren M, Eneslatt K, Bröms J, Sjöstedt A (2013) Importance of PdpC, IglC, IglI, and IglG for modulation of a host cell death pathway induced by *Francisella tularensis*. Infect Immun 81(6): 2076–2084.
26. Read A, Sigrid J, Hueffer K, Gallagher L, Happ G (2008) *Francisella* genes required for replication in mosquito cells. J Med Entomol 45(6): 1108–1116.
27. Sambrook J, Fritsch EF, Maniatis T (1989) Molecular cloning: a laboratory manual, 2nd ed. Cold Spring Harbor Laboratory, Cold Spring Harbor, N.Y.
28. Patel JC, Hueffer K, Lam TT, Galan JE (2009) Diversification of a *Salmonella* virulence protein function by ubiquitin-dependent differential localization. Cell 137: 283–294.
29. Tyeryar FJ, Lawton WD (1970) Factors affecting transformation of *Pasteurella novicida*. J Bacteriol 104(3): 1312–1317.
30. Johnson MB, Criss AK (2013) Fluorescence microscopy methods for determining the viability of bacteria in association with mammalian cells. J Vis Exp 5(79).
31. Chong A, Wehrly TD, Nair V, Fischer ER, Barker JR, et al. (2008) The early phagosomal stage of *Francisella tularensis* determines optimal phagosomal escape and *Francisella* pathogenicity island protein expression. Infect Immun 76(12): 5488–5499.
32. Schmerk CL, Duplantis BN, Howard PL, Nano FE (2009) A *Francisella novicida pdpA* mutant exhibits limited intracellular replication and remains associated with the lysosomal marker LAMP-1. Microbiol 155(5): 1498–1504.
33. Bröms JE, Sjöstedt A, Lavander M (2010) The role of the *Francisella tularensis* pathogenicity island in type VI secretion, intracellular survival, and modulation of host cell signaling. Front Microbiol 1(136).
34. Clemens DL, Lee BY, Horwitz MA (2005) *Francisella tularensis* enters macrophages via a novel process involving pseudopod loops. Infect Immun 73(9): 5892–5902.
35. Santic M, Molmeret M, Klose KE, Kwaik YA (2006) *Francisella tularensis* travels a novel, twisted road within macrophages. Trends Microbiol 14(1): 37–44.
36. Checroun C, Wehrly TD, Fischer ER, Hayes SF, Celli J (2006) Autophagy-mediated reentry of *Francisella tularensis* into the endocytic compartment after cytoplasmic replication. Proc Natl Acad Sci U S A 103(39): 4578–4583.
37. Kubori T, Galan J (2003) Temporal regulation of *Salmonella* virulence effector function by proteasome-dependent protein degradation. Cell 115: 333–342.
38. Ludu JS, de Bruin OM, Duplantis BN, Schmerk CL, Chou AY, et al (2008) The *Francisella* pathogenicity island protein is PdpD is required for full virulence and associates with homologues of the type VI secretion system. J Bacteriol 190(13): 4584–4595.
39. Leiman PG, Basler M, Ramagopal UA, Bonanno JB, Sauder, et al (2009) Type VI secretion apparatus and phage tail-associated protein complexes share a common evolutionary origin. Proc Natl Acad Sci U S A 106(11): 4154–4159.
40. Robertson G, Child R, Ingle C, Celli J, Norgard M (2013) IglE is an outer membrane-associate lipoprotein essential for intracellular survival and murine virulence of Type A *Francisella tularensis*. Infect Immun 81(11): 4026–4040.

## 5

The Conserved *nhaAR* Operon Is Drastically Divergent between B2 and Non-B2 *Escherichia coli* and Is Involved in Extra-Intestinal Virulence

Mathilde Lescat[1,2]*, Florence Reibel[1], Coralie Pintard[1], Sara Dion[1,3], Jérémy Glodt[1,3], Cecile Gateau[1], Adrien Launay[1], Alice Ledda[1], Stephane Cruvellier[4], Jérôme Tourret[1,5], Olivier Tenaillon[1]

1 Institut National de la Santé et de la Recherche Médicale (INSERM), Unité Mixte de Recherche (UMR) 1137, Paris, France, 2 Laboratoire de Microbiologie, Hôpital Jean Verdier, Assistance Publique-Hôpitaux de Paris, Bondy, France et Université Paris Nord, Sorbonne Paris Cité, Paris, France, 3 UMR 1137, Université Paris Diderot, Sorbonne Paris Cité, Paris, France, 4 Laboratoire de Génomique Comparative, Centre national de la Recherche Scientifique (CNRS) UMR 8030, Institut de Génomique, Commissariat à l'énergie atomique et aux énergies alternatives (CEA), Genoscope, Evry, France, 5 Département d'Urologie, Néphrologie et Transplantation, Hôpital Pitié-Salpêtrière, Assistance Publique-Hôpitaux de Paris et Université Pierre et Marie Curie, Paris, France

Abstract

The *Escherichia coli* species is divided in phylogenetic groups that differ in their virulence and commensal distribution. Strains belonging to the B2 group are involved in extra-intestinal pathologies but also appear to be more prevalent as commensals among human occidental populations. To investigate the genetic specificities of B2 sub-group, we used 128 sequenced genomes and identified genes of the core genome that showed marked difference between B2 and non-B2 genomes. We focused on the gene and its surrounding region with the strongest divergence between B2 and non-B2, the antiporter gene *nhaA*. This gene is part of the *nhaAR* operon, which is in the core genome but flanked by mobile regions, and is involved in growth at high pH and high sodium concentrations. Consistently, we found that a panel of non-B2 strains grew faster than B2 at high pH and high sodium concentrations. However, we could not identify differences in expression of the *nhaAR* operon using fluorescence reporter plasmids. Furthermore, the operon deletion had no differential impact between B2 and non-B2 strains, and did not result in a fitness modification in a murine model of gut colonization. Nevertheless, sequence analysis and experiments in a murine model of septicemia revealed that recombination in *nhaA* among B2 strains was observed in strains with low virulence. Finally, *nhaA* and *nhaAR* operon deletions drastically decreased virulence in one B2 strain. This effect of *nhaAR* deletion appeared to be stronger than deletion of all pathogenicity islands. Thus, a population genetic approach allowed us to identify an operon in the core genome without strong effect in commensalism but with an important role in extra-intestinal virulence, a landmark of the B2 strains.

Editor: Christophe Beloin, Institut Pasteur, France

Funding: This work was supported by the European Research Council under the European Union's Seventh Framework Program (FP7/2007-2013)/ERC Grant 310944. The funder had no role in study design, data collection and analysis, decision to publish, or preparation of the manuscript.

Competing Interests: The authors have declared that no competing interests exist.

* Email: mathilde.lescat@inserm.fr

Background

Comparative genomics has unraveled the dynamics of microbial genome evolution [1]. The extent of lateral gene transfer has appeared to be one of the most striking characteristics of this dynamics. These transfers impact the most studied phenotypes of bacteria: antibiotic resistance and virulence. For example, horizontally acquired clusters of genes found in pathogenicity islands (PAI) have been shown to be involved in virulence [2]. Yet adaptation may also occur through mutations in genes present in the whole species, or core genes. This topic has been far less studied, despite the large potential for adaptation through mutations in core genes that experimental evolution has revealed [3]. The principal reason is that, because of the limited amount of recombination, most mutations are linked and therefore identifying the ones that are involved in adaptation is challenging. Nevertheless, if selective pressure is strong enough as in the case of antibiotic resistance or some cases of virulence, a few mutations in core genes have been found to be involved in adaptation [4,5]. In the present paper, we want to extend such an approach and try to identify some core genes that may contribute to the functional divergence between phylogroups in the *Escherichia coli* species.

E coli is a versatile bacterium, both retrieved in the environment and known as a widespread gut commensal of vertebrates, especially humans. *E. coli* is also a pathogen which is responsible for more than 1 million deaths a year due to intra and extra-intestinal diseases. In the wild, its population size has been estimated to more than 10^{20} bacteria [6]. The species has a clonal structure, and is subdivided in seven phylogenetic groups A, B1, B2, C, D, E and F [7]. These groups are not randomly distributed. Indeed, previous studies have shown a correlation between phylogeny and virulence in *E. coli*, with most extra-intestinal pathogenic *E. coli* (including urinary tract infection and meningitis associated strains) belonging to phylogenetic group B2 [8,9]. Moreover the prevalence of the different groups among commensal strains varies largely across host species and even across populations of a given host species. For instance, B2 strains not only are commonly isolated from extra-intestinal infections, but also appear to be efficient commensals frequently retrieved in wild animals and humans [10]. In humans, the prevalence of B2 commensals varies drastically according to populations, being low in tropical countries and high in developed countries [6]. It appears that the frequency of B2 carriage has increased over the last 30 years (e.g. from 20 to 40% in France) a worrying observation knowing their extra-intestinal pathogenic potential as well as their implication in colon cancer [11,12]. Unraveling the bases of this success in the commensal habitat is of medical relevance. Moreover, the inactivation of some of the virulence factors have been shown to reduce the ability to colonize the gut [13], comforting the idea that extra-intestinal virulence is a by product of commensalism [14].

Whether an *E. coli* strain behaves as a commensal or a pathogen is determined by an extremely complex balance between many factors: immune status of the host, production of virulence factors by the bacterium, portal of entry, inoculum, and the genetic background of the bacterium to cite some important ones. The latter appears to be essential for acquisition and expression of virulence factors [15]. Yet, the alleles involved in the specificities of the different group of strains remains largely unknown, especially in the primary habitat, the gut of vertebrates where *E. coli* is mainly a commensal strain.

To perform comparative genomics on a large scale, *E. coli* is an organism of choice with 128 complete genomes available. Based on that collection, we identified several candidate genes showing the highest divergence between B2 strains and the rest of the species. Our aim was not just to provide a list of genes but also to perform functional tests. Therefore, we focused our attention on the region centered on the gene *nhaA* as this was the candidate with the highest divergence opening the path to functional assays. *nhaA* is part of an operon coding for a sodium proton antiporter which is known to be responsible for pH and sodium homeostasis in *E. coli* [16] (Figure 1A). The aims of this study were (i) to identify markers of differentiation of the B2 phylogenetic group (ii) to perform population genetic analysis on the sequences of the candidate (iii) to identify a potential biological role for this marker *in vitro*, and (iv) to test its potential role *in vivo* in a mouse colonization assay and a mouse septicemia model.

Materials and Methods

Ethics statements

All *in vivo* experiments were realized in accordance with the ARRIVE guidelines. The murine septicemia was conducted following European and National regulations for housing and care of laboratory animals after pertinent review and approval by the Bioethics Committee at Santiago de Compostela University and by the French Veterinary Services (certificate number A 75-18-05). The murine gut colonization model was conducted after approval by the Debre-Bichat Ethics Committee for Animal Experimentation (Protocol Number 2012-17/722-0076) in accordance to the European Decret and French law on the protection of animals. All possible measures were taken to minimize animal suffering and to ensure animal welfare. When necessary, animals were sacrificed by lethal intra-peritoneal injection of phenobarbital after volatile anesthesia with sevoflurane.

Bacterial strains

All *E. coli* strains and plasmids are listed in Table S1. Strains have been chosen for their representativity of the phylogeny and their wide array of phenotypes as they were isolated from commensal, extra-intestinal and intra-intestinal pathogenic situations.

Inactivation of the *nhaAR* region and control experiment

Inactivation of *nhaAR* was performed using the modified method described by Datsenko *et al.* [17]. We first obtained a PCR product using the K-12, TA249 and IAI1 strains, with primers WanF_nonB2_nhaAR for all strains and WanR_K12-TA249_nhaAR for K-12 and TA249, and WanR_IAI1_nhaAR for IAI1. The same PCRs were done using CFT073, 536 and TA014 strains with primers WanF_B2_nhaAR and WanR_B2_nhaAR. We also performed the inactivation of *nhaA* and *nhaR* genes in 536 strain using the primers WanF_B2_nhaA with WanR_B2_nhaA and WanF_B2_nhaR with WanR_B2_nhaAR for *nhaA* and *nhaR* disruption, respectively. PCR products contained (i) the FLP recognition target FRT-flanked chloramphenicol resistance gene (*cat*) and (ii) the 50-bp sequences homologous to the 5′ and 3′ flanking regions of *nhaAR*, *nhaA* and *nhaR* for each corresponding strain. Inactivation of the *nhaAR* operon, *nhaA* and *nhaR* genes were confirmed by PCR using the following primers: verifWanF_nonB2_nhaAR and verifR_nonB2_nhaAR for non-B2 strains and the primers verifWanF_B2_nhaAR and verifR_B2_nhaAR for B2 strains, targeting sequences upstream and downstream from the *nhaAR* operon; and c1 and c2, targeting sequences within *cat* gene; All strains obtained and primers used are listed in Tables S1 and S2.

Complementation of the *nhaAR* region and control experiment

Complementation of the strain 536Δ*nhaAR* by the *nhaAR* region was performed using the GC Cloning & Amplification Kit (pSMART GC LK vector) (Lucigen, Middleton, WI). Briefly the *nhaAR* region including the promoter region were amplified from the 536 strain using the primers cp_536F and cp_536R targeting sequences 200 bp upstream *nhaA* start and 200 bp downstream *nhaR* stop, respectively. The *nhaA* gene was amplified using cp_536F and cp_536DnhaA_R targeting sequence 200 bp downstream *nhaA* stop. The fragments were then separately cloned into the blunt cloning site in the pSMART GC LK vector. The plasmids bearing the *nhaAR* region or *nhaA* gene were then electroporated in 536Δ*nhaAR* strain and or 536Δ*nhaA*. The complementation experiments were confirmed by PCR using the primers cp_536F and cp_536R. We also incorporated the pSMART GC LK vector in 536Δ*nhaAR* and 536Δ*nhaA* strains as controls. All strains obtained and primers used are listed in Tables S1 and S2.

Genomic environment analysis

The MicroScope platform [18] was used for comparative analysis of genetic sequences surrounding *nhaAR*. The MicroScope platform allows comparative analysis of available *E. coli* and closely related genomes, with visualization of *E. coli* genome annotations enhanced by a synchronized display of synteny groups in the other genomes chosen for comparison.

Reconstruction of the phylogenetic tree

The 121 *E. coli* genomes from the MicroScope website were included [23]. A maximum-likelihood phylogenetic tree was reconstructed with the PHYML software [19] using the concatenated multi locus sequence typing (MLST) sequences on the one hand, and *nhaA* sequences on the other hand. We used the MLST Pasteur scheme [7].

Sequence alignments and study of recombination

We compared 128 *E. coli*/*Escherichia* clade [20] *nhaAR* sequences of 2303 bp by sequence alignment using ClustalW software [21]. Observation of traces of recombination was

A

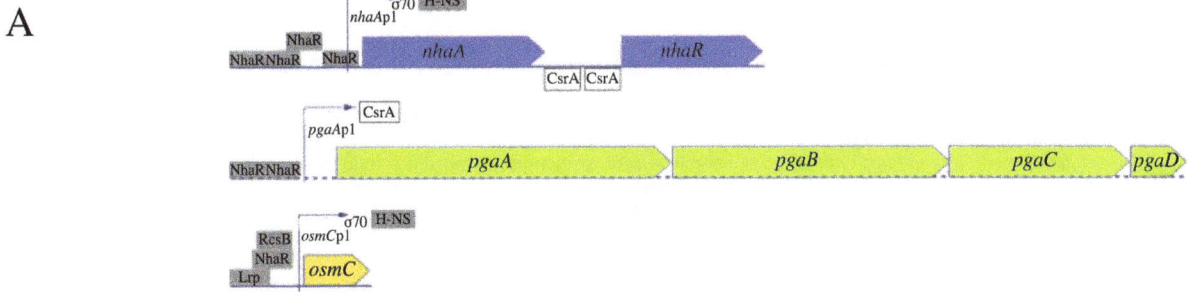

B

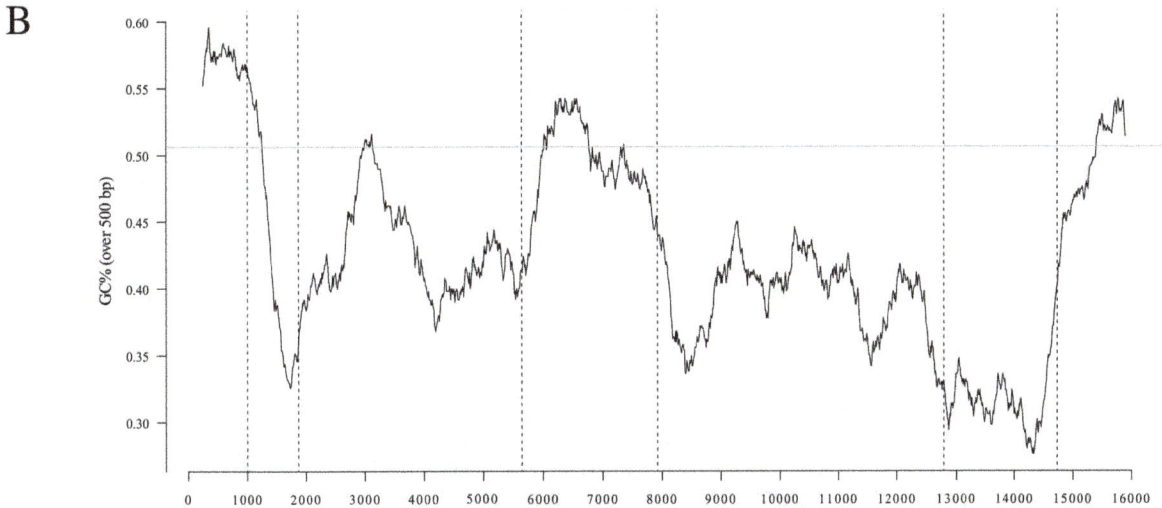

C

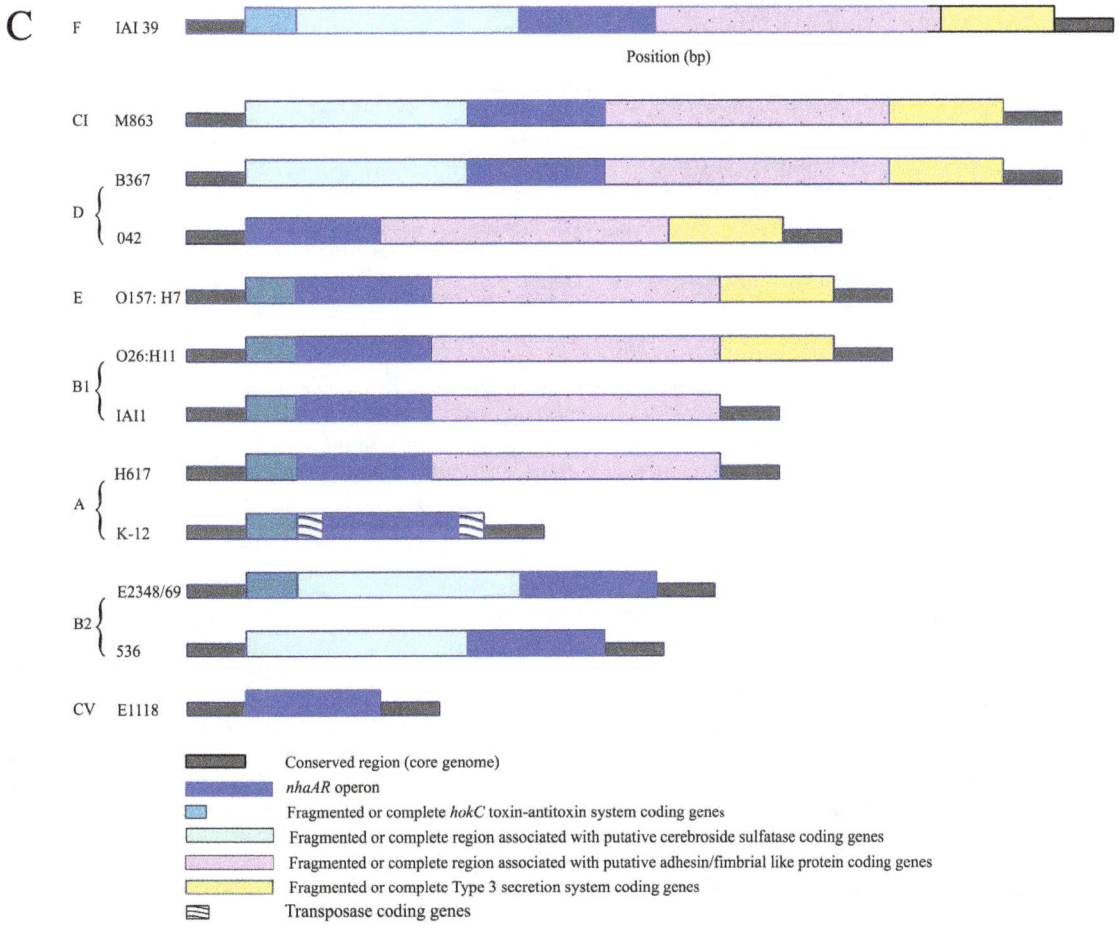

Figure 1. Genomic organization of *nhaAR* region. (A) Genomic representation of *nhaAR* and other operons under NhaR regulation in K-12 *E.coli* strain from http://www.ecocyc.org. All transcription or translation regulators are indicated. (B) GC percent along IAI39 *nhaAR* region (black line) and mean core genome GC percent (gray line) (C) Organization in various modules of *nhaAR* region. The modular organization of the region was defined using synteny breaks between 10 pathogenic and commensal *E. coli* and 2 clades strains from various phylogenetic groups including K-12 and H617 (two group A commensal strains), IAI1 and O26:H11 (two group B1 commensal strains), B367 (a group D commensal strain), 042 (an enteroaggregative group D, E2348/69 (an enteropathogenic group B2 strain), 536 (an extra-intestinal pathogen group B2 strain), O157:H7 Sakaï (an enterohemorrhagic group E strain). The two *Escherichia* clades (C) used were M863 (CI) and E1118 (CV). Five homologous modules have been defined, *nhaAR* operon being the third. Dark green: Fragmented or complete *hokC* toxin-antitoxin system coding genes; turquoise: Fragmented or conserved region associated with putative cerebroside sulfatase coding genes; blue: *nhaAR* region; pink: Fragmented or conserved region associated with putative adhesin/fimbrial like protein coding genes; yellow: Fragmented or conserved Type 3 secretion system coding genes. IS are also indicated.

performed on the 1167 bp of *nhaA* sequences by comparison between the sequences of B2 strains showing long branches in the phylogenetic tree sequences, other B2 strain consensus sequence, non-B2 strain consensus sequence and *Escherichia* clade sequences. Amino-acid sequences inferred from the nucleotide sequences of the *nhaAR* region were also analyzed. After the generation of the maximum likelihood tree (see above), amino-acid substitutions for each branch of the *nhaAR* tree were identified by comparison of consensus sequences between different branches using the BIOE-DIT software [22].

Analysis of genomic environment of *nhaAR*

The genomic environment was observed using the synteny breaks between two clades and 10 *E. coli*. Of these, 5 were pathogenic strains, including E2348/69 (an enteropathogenic group B2 strain), 536 (an extra-intestinal pathogen group B2 strain), 0157:H7 Sakaï (an entero-hemorrhagic group E strain), 042 (an entero-aggregative group D strain) and IAI39 (an extra-intestinal group F strain), whereas 5 were commensal strains including K-12 and H617 (two group A strains), IAI1 and O26:H11 (two group B1 strains) and B367 (a group D strain). The two clades were M863 (*Escherichia* clade I) and E1118 (*Escherichia* clade V). It appeared that this region had a composite structure, *i.e.* it is made up of five modules that are present or absent in the different strains. We then classified the strains according to the maximum number of regions present to retrace a parsimonious history of loss and gains of these modules.

Flow cytometry

The wild type and their Δ*nhaAR* isogenic mutants strains K-12, IAI1, TA249, CFT073, 536 and TA014 in which the plasmids from the Zaslaver collection have been introduced (Table S1) were compared using the wild type strains K-12, IAI1, TA249, CFT073, 536 and TA014 as controls. The Zaslaver collection is a bank of *E. coli* K-12 strains in which reporter plasmids bearing the Gfp protein under control of the promoter regions of each gene available was introduced [23] Plasmids were extracted using the plasmid min kit extraction kit (Sigma). The experiments were conducted as described elsewhere [24].

Growth curves

For the comparative growth assays, K-12, IAI1, TA249, CFT073, 536 and TA014 wild type strains and their mutants, K-12Δ*nhaAR*:Cm, IAI1Δ*nhaAR*:Cm, TA249Δ*nhaAR*:Cm, CFT 073Δ*nhaAR*:Cm, 536Δ*nhaAR*:Cm and TA014Δ*nhaAR*:Cm were grown at 37°C in 2 media: Luria Bertani (LB) and Davies minimum medium (DM) with glucose (NaH$_2$PO$_4$ 33.9 mmol/L, Na$_2$HPO$_4$ 31.1 mmol/L, (NH4)$_2$SO$_4$ 20 mmol/L, MgSO4 7 H2O 0.3 mmol/L, KCl 40,2 mmol/L, FeCl3 70 µmol/L, glucose 20 mmol/L), each media was adjusted at pH 7 and pH 8.5 with MOPS and TAPS, respectively (Sigma), DM was also adjusted to pH 8 and at NaCl: 170 mmol/L and 300 mmol/L (Table S3). LB is a complex medium, whereas DM is a minimal medium with

only one source of carbon. All the studied strains were grown overnight (O/N) in LB medium in deep-weel plates at 37°C with constant shaking at 280 rpm. O/N cultures were pre-diluted at 1/100 in saline buffer and strains were inoculated in four different wells each at 1/100 in a Costar 96 flat-bottomed well plate. Growth was recorded by an Infinite 200 Tecan, which measured the OD600 in each well every 5 minutes at 37°C, while shaking for 24 hours. Growth assays were repeated 3 times. The maximum growth rate (MGR in s^{-1}) was computed from growth curves obtained by Tecan. Briefly, OD600 were collected, log-transformed, and smoothed with a spline function. The MGR was defined as the maximum value of the derivative of the smoothed growth curve. The doubling times (DT) (in mn) have then been computed as followed. DT = Log2/(MGR*60). All DTs were compared by strain and by medium using the Welch test.

Murine Septicemia model

A mouse model of systemic infection [8] was used to assess the intrinsic virulence of strains SE15, H001, TA103 and TA435 which showed traces of recombination. To compare intrinsic virulence of B2 strains with a recombinant *nhaAR* operon and B2 strains without trace of recombination at the operon locus we used previous results of intrinsic virulence of strains CFT073, 536, F11, S88, APEC01, UTI89, LF82 and B2S [14,25]. In order to avoid a day-of-experiment bias, K-12 and 536 were included in all experiments as negative and positive controls of intrinsic virulence, respectively. To test the effect of the deletion of the *nhaAR* operon on intrinsic virulence of *E. coli* B2 strains in different genomic backgrounds, we tested CFT073, CFT073Δ*nhaAR*:Cm and CFT073Δ*nhaAR* strains, a mixture of equal quantities of CFT073 and CFT073Δ*nhaA*:Cm and a mixture of equal quantities of CFT073Δ*nhaAR* and CFT073Δ*nhaAR*:Cm to test for the cost of the antibiotic resistance. We also tested 536 and 536Δ*nhaAR* strains, a mixture of equal quantities of 536 and 536Δ*nhaAR*:Cm. To decipher which gene was responsible for virulence attenuation in the operon, we tested the deleted mutant strains 536Δ*nhaA* and 536Δ*nhaR*. Finally, we tested the complemented strains 536Δ*nhaAR* pGC*nhaAR*, 536Δ*nhaAR* pGC*nhaA*, 536Δ*nhaA* pGC*nhaAR* 536Δ*nhaA* pGC*nhaA*. The complemented strains 536Δ*nhaAR* pGC and 536Δ*nhaA* pGC in which the deleted mutant strains have been complemented with an empty vector were used as control of empty vector cost in the murine model of septicemia. The experiments were conducted as described elsewhere [8]. Briefly, the ability of bacterial strains to cause sepsis was determined using 5-wk old female OF1 mice (Charles River, L'Arbresle, France). 10 mice per strain or mixture of strains tested were used. A total of 200 µl of a suspension of 109 bacteria/ml in saline buffer was inoculated by subcutaneous injection in the neck, and mortality was recorded during the following 7 days. For competition assays, spleens were aseptically collected after death, homogenized in 1 ml of saline buffer, and plated in serial dilutions on LB agar with or without appropriate antibiotic. For assays where strains were tested alone, spleens were aseptically collected after death, homogenized in 1 ml of saline buffer, and plated in serial dilutions on LB agar with or without appropriate antibiotic, colony were verified by PCR using cp_536F and cp_536R primers (Table S2).

Mouse model of intestinal colonization

Intestinal colonization was assessed using a mouse model as described elsewhere [13]. Briefly, 6-wk old CD1 female mice (Charles River, L'Arbresle, France) treated with streptomycin were used. Five days before inoculation, was added to the sterile drinking water at a final concentration of 5 g/liter. Streptomycin was maintained throughout the whole experiment. Coliform-free mice were inoculated through oral gavage with 10^6 bacteria in 200 μl of saline buffer. Every day post-inoculation, dilutions of weighed fresh feces resuspended in 1 ml of saline buffer were plated on LB agar with or without appropriate antibiotic. We studied 536 wild type strain and isogenic mutants 536ΔnhaAR:Cm, and 536ΔnhaAR. For each strain two mice were used. We also performed competition assays using a mixture of equal quantities of 536 wild type and 536ΔnhaAR:Cm to test the effect of deletion of the region on the gut colonization ability and also a competition between 536ΔnhaAR:Cm and 536ΔnhaAR to test the cost of *cat* resistance gene. For each competition four mice were used once.

Statistical analysis

Population genetics analyses were performed using libsequence [26]. For phenotypic analysis, the values are given as medians (interquartile range) and, comparisons between strains were performed using either the Wilcoxon signed-rank test or the Kruskal-Wallis equality-of-populations rank test, unless specified otherwise. All statistics were computed using STATA (v10.0, College Station, TX, USA) or R (R Development Core Team, 2009, Vienna, Austria) and statistical significance was determined at a p-value of less than 0.05.

RNA isolation

Total RNA extraction was performed on 536, 536ΔnhaAR, 536ΔnhaR and 536ΔnhaA after O/N culture during 18 h at 37°C in LB medium. Each culture for each bacteria was repeated three time. Total RNA was extracted using the hot phenol method. Residual chromosomal DNA was removed by treating samples with a Ambion TURBO DNA-free Kit DNase-treated RNA samples were quantified using a NanoDrop 1000 spectrophotometer (Thermo Scientific).

Quantitative RT-PCR (qRT-PCR)

qRT-PCR experiments were performed using a KAPA SYBR One-Step qRT-PCR Kit (Kapa Biosystems) and a Lightcycler 480 (Roche) instrument with the program recommended by Kapa Biosystems. We applied the comparative CT quantification (ΔΔCt method) of qRT-PCR for comparing changes in gene expression of *nhaR* in the 536 deleted mutant strains. Relative quantification was performed using *16SrRNA* as endogenous control gene. Each experiment was performed in duplicate.

Results, Discussion and Conclusion

Results

Genomic analyses. To identify genetic markers in the core genome that would differentiate the B2 phylogenetic group from the other groups, we scanned all genes in the core genome of 128 *E. coli/Escherichia clade* genomes. For each gene, we computed the number of fixed mutations between B2 and non-B2 and compared it with a Fisher test to the pooled core genome number.

We studied the proportion of fixed sites compared to the total gene length or to the total number of polymorphism found in that gene. In both cases, the gene with the lowest p-value was *nhaA*, (p< 1e-62), the next gene being *ygbE* (p< 1e-33) a conserved gene of unknown function (Table S5). In this paper, we focused on the *nhaAR* operon, as *nhaA* is the first gene of the list, but also because it can be functionally assessed as it is a sodium proton antiporter involved in pH and sodium homeostasis [16].

We first compared the genomic environment of the *nhaAR* operon in 10 *E. coli* strains and two *Escherichia* clades (clade I and clade V) (Figure 1BC). We defined 5 homologous fragments composing this region excluding fragments of transposases. Apart from the fragment including exclusively *nhaAR* operon, none of the other fragments were found in all strains, yet they were all present in strain IAI39. The GC content in the region was on average 42.67%, and differed significantly from the average genome GC content of 50.63% (p<0.05) (Figure 1B). This suggests that the region might have been acquired through horizontal gene transfer. Nevertheless, *nhaAR* had a GC content compatible with the genomic one. The pattern of gain/loss of the fragments surrounding *nhaAR* appeared to be quite dynamic (Figure 2). All the fragments seemed to have been lost and or gained multiple times along the phylogeny. *nhaAR* operon has therefore been maintained in the core genome despite highly dynamic surrounding regions, an observation that suggests an important contribution of this operon to *E. coli* niche adaptation.

We reconstructed the phylogenetic tree of the *nhaA* gene from the 121 genomes of *E. coli* available in data banks (Table S4). Consistent with the screen used to identify *nhaA* region, we found that in the phylogenetic tree based on *nhaA* the branch leading to the B2 group of strains was much longer than what was found using the MLST genes of the Pasteur scheme [7] (Figure 3). There were 56 mutations that were fixed between the B2 and the non-B2 strains on a 1167 bp gene, or 4.8% of sites which contrast quite drastically with the whole genome average of 0.14%. This could be due to an accelerated evolution at this locus or to horizontal gene transfer or both. Yet, when we measured Ka/Ks (corresponding to the ratio of non synonymous mutation rate on the synonymous mutation rate) between B2 and non-B2 strains, a value between 0.01 and 0.02 was found. This means that synonymous mutations were in large excess compared to non-synonymous mutations. As we can exclude that selection of a succession of non-synonymous mutations was responsible for the long-branch, we favor horizontal gene transfer as the most likely explanation. The 5 to 10% divergence observed between B2 and non-B2 *nhaA* genes suggests that the transfer originated from a close species like an *Escherichia clade* and that this transfer may have been quite recent such that little recombination might have occurred subsequently between the B2 and the other strains. Accordingly, visual inspection of the *nhaA* sequences revealed that a few B2 strains (ED1a, SE15, E2348/69, H001, TA103, M605, and TA435) had three or more consecutive mutations that differed from the other B2, which can be considered as a trace of recombination. Similarly in the *nhaR* region, a long recombinant segment in strain M605 was responsible for most of the diversity within B2. When recombining strains were excluded from the analysis, *nhaR* appeared with an even stronger B2/non-B2 differentiation than *nhaA* with 13.0% of fixed differences compared to 5.8% for *nhaA*. Therefore the whole *nhaAR* operon and not just *nhaA* harbors a strong divergence between B2 and non-B2. Interestingly, while B2 strains are commonly isolated from extra-intestinal infections, none of the strains with sign of recombination have been isolated in extra-intestinal conditions. ED1a, SE15, H001, TA103, M605, TA103, TA435 were sampled

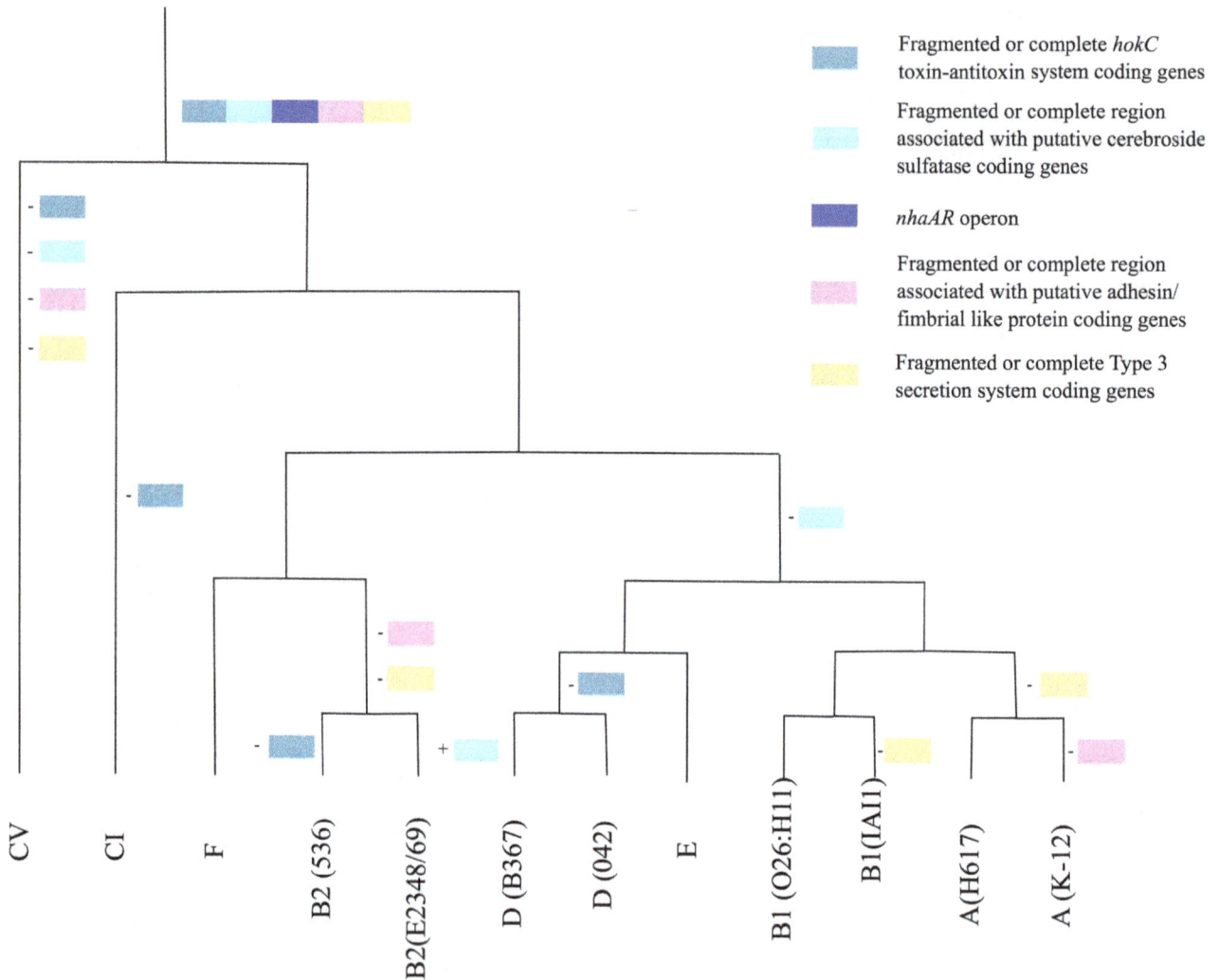

Figure 2. Multiple gains and losses of modules around *nhaAR* **operon.** A parsimonious scenario of gains and losses of the 5 modules defined in figure 1 is presented along the phylogenetic tree of the strains. − indicates losses and + acquisitions (colors of modules as in figure 1).

in commensal conditions [14,27,28]. E2348/69 is an entero-pathogenic strain [29]. Even more strikingly, some of these strains are quite atypical B2 strains in terms of extra-intestinal virulence as they are non-killer in a mouse model of septicemia. For instance, ED1a strain belongs to the B2 subgroup VIII, a specific commensal subgroup never retrieved in extra-intestinal virulence conditions, specific to the human digestive track [30], and avirulent.

We compared consensus sequences of the 1167 bp *nhaA* gene and the 172 bp promoter region between B2 and non-B2 strains. We identified 98 mutations. Of these, 3 were non-synonymous, 2 deletions and 3 indels, and none of them were in a position described as important for the protein. The promoter region was highly conserved among the B2 with a single polymorphism out of 172 bp among the 28 B2 strains (Watterson estimate per base: 0.0015), and much more diverse in the non-B2 (28 polymorphic sites among 90 strains; Watterson estimate per base: 0.0321). Five of the mutations that differentiated the B2 and non-B2 were found in the NhaR1 and NhaR4 binding sites of NhaR regulator (Figure 1A). These mutations suggest variable level of expression between B2 and non-B2 strains.

We also looked at the *nhaR* coding region (905 bp) and the inter-genic region between *nhaA* and *nhaR* (60 bp), which is involved in post-transcriptional regulation of *nhaR* by CsrA [31] (Figure 1A). Thirteen non-synonymous mutations were found between B2 and non-B2 strains. Moreover, the hairpin-loop binding site used by CsrA to modulate NhaR regulation harbored 3 mutations. This lead us to hypothesize a differential expression of the *nhaAR* operon with potential consequences on the genes regulated by NhaR, *i.e. nhaA, pgaA* and *osmC*.

Phenotypic results. To assess whether *nhaAR* region is implicated in virulence or commensalism, we tested different phenotypes linked with pH and osmolarity that could differentiate B2 and non-B2 strains. We first wanted to investigate the expression level of the operon in the two backgrounds. We used 3 strains of each group and introduced reporter plasmids bearing the Gfp protein under control of the promoter regions of *nhaA* and *osmC*. Both promoters are under NhaR control and can be used to monitor repression it imposes by flow cytometry. However, Gfp expression under *nhaA* promoter was too low in all tested conditions, and fluorescence controlled by *osmC* promoter did not show any B2/non-B2 difference across all the conditions tested (data not shown).

A

B

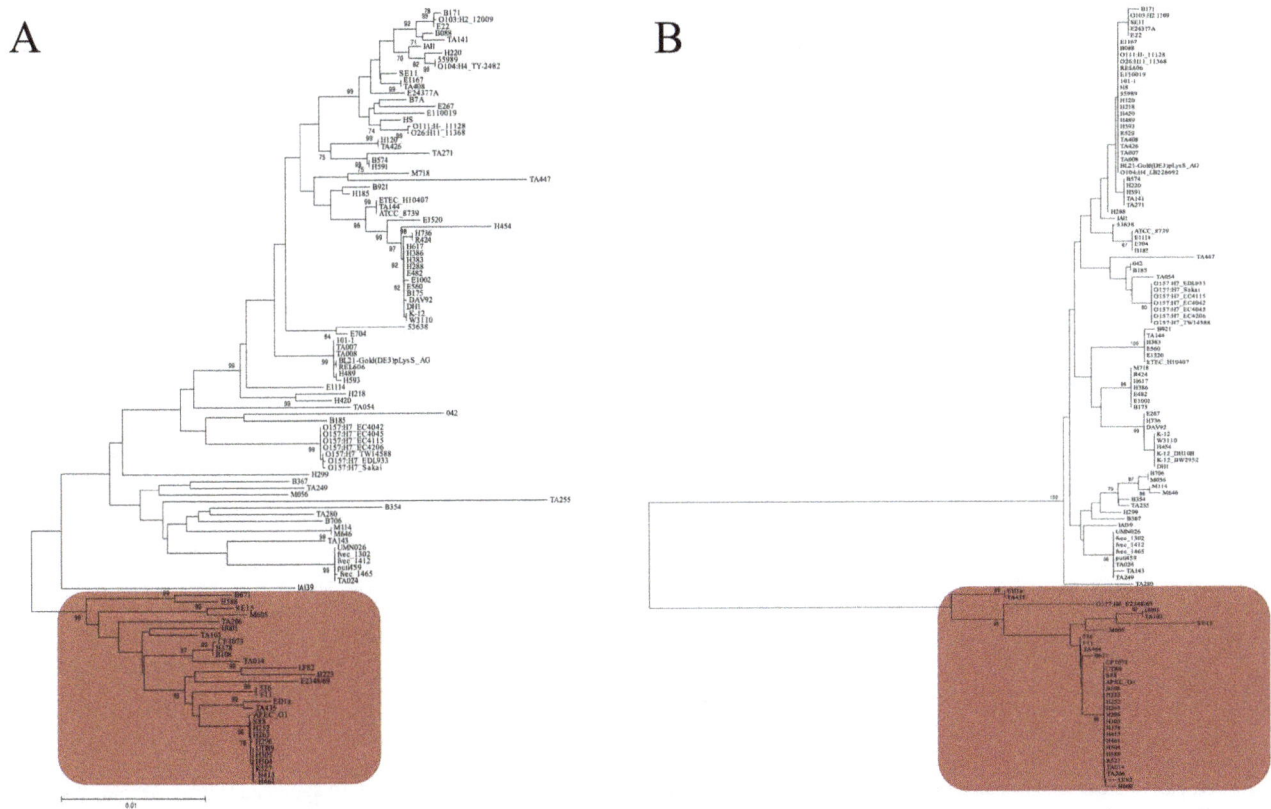

Figure 3. MLST and *nhaA* phylogenetic trees for 121 strains of *E. coli*. The trees were reconstructed from (A) multi-locus sequence typing of 8 partial housekeeping genes from the Pasteur scheme [7] representing the species phylogeny and (B) from the *nhaA* sequences using PHYML [19]. Bootstraps values are indicated. Strains studied and belonging to phylogenetic group B2 (red boxed) are indicated. Branches separating the B2 phylogenetic group strains from the other group strains are indicated in blue.

We then focused on growth curves in different media. Doubling times of 9 B2, 10 non-B2 and 3 B2 strains showing traces of recombination in *nhaAR* region were cultured in different media (LB and minimum growth medium with glucose as carbon source) at different pH (7, 8 and 8.5) and at different sodium concentrations (170 and 350 mmolL) (Figure 4). A statistical difference in growth between B2 and non-B2 strains was observed in LB at pH 8.5 and concentration of sodium of 350 mmol/L (p = 0.001). Interestingly, these conditions are the ones in which *nhaA* is induced [32–34]. The 4% difference observed in division time seems modest but corresponds to a drastic selective advantage: the ratio of non-B2 to B2 would double every 10 hours of competition in that media and result in a 150 fold increase of non-B2 over B2 in 3 days. However, this does not prove a direct contribution of *nhaAR* to this difference. We therefore looked for some direct implication of *nhaAR* operon by studying knock out mutants.

Effect on growth of the deletion of the operon *nhaAR* in 3 strains of B2 group (CFT073, 536 and TA014) and 3 strains from other groups (K-12, IAI1 and TA249) was studied in the same media. Most of mutants were not able to grow with minimum media at pH 8.5 as observed by others [33], we then used pH 8 to analyze growth in the minimum media. We analyzed several statistics of growth (MGR and maximal optical density) and compared the deletion mutants in absolute terms (MGR) or relative to their wild-type strain (change in MGR). The 3 B2 strains had a very comparable growth. In contrast, the non-B2 strains had very different growth characteristics, and one of the

three strains had a pattern similar to the B2. As a result, there was no significant differential effect of the *nhaAR* deletion between B2 and non-B2 on growth in the tested conditions.

Mouse models. Because some of the B2 strains with sign of recombination were known to be avirulent, we decided to study the intrinsic virulence of several B2 strains in the murine septicemia model. Among strains showing traces of recombination, we observed a significant decrease in lethality for ED1a [14], E2348/63 [14], SE15 (this work) but not for H001 (this work), TA435 (this work), and TA103 (this work) compared to other B2 strains responsible for extra intestinal infections. Indeed, we then compared the mean survival rate between strains with a recombinant *nhaAR* operon (ED1a, E2348/69, SE15, H001, TA103 and TA435) and strains with a non-recombinant *nhaAR operon* (CFT073, 536, F11, S88, APEC01, UTI89, LF82 and B2S) [14,25]. We found a significant decrease in the intrinsic virulence of recombinant strains (p<0.0001) (Figure 5).

We further tested the effect of the *nhaAR* deletion in this mice model, using 2 virulent B2 strains. 536Δ*nhaAR* showed a dramatically decreased lethality compared to wild-type 536 strain (p<0.001) (Figure 6). To confirm these results, we reproduced them in CFT073, another highly lethal B2 strain, and found similar results (data not shown). We also tested in this model the complemented strain 536Δ*nhaAR* pGC*nhaAR* and 536Δ*nhaAR* pGC. The comparison of 536Δ*nhaAR* pGC*nhaAR* with 536 strain (p = 0.48) and 536Δ*nhaAR* (p = 4.3e-06) proved that *nhaAR* operon was implicated in virulence and the comparison of 536Δ*nhaAR* pGC with 536Δ*nhaAR* (p<0.01) indicated a cost of

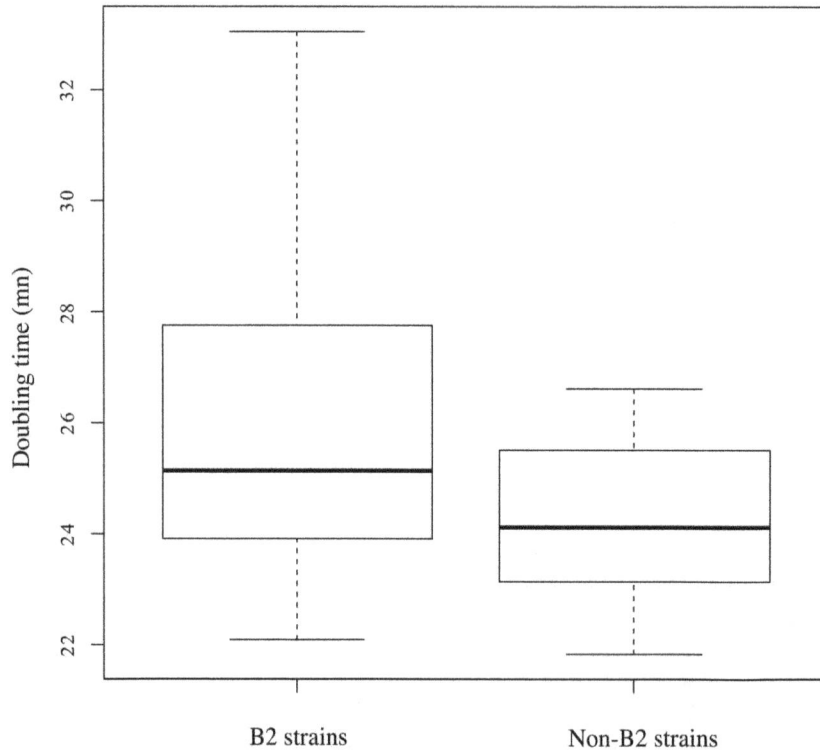

Figure 4. Non B2 grew faster than B2 in high pH high osmolarity. Boxplots of the doubling times (DT) in minutes of 12 B2 and 10 non-B2 representative strains of *E. coli* in LB, pH 8.5 with 350 mmol/L of sodium. We found a significant difference between B2 strains and non B2 strains using a Welch test (p = 0.001).

the empty plasmid in this model, reinforcing the implication of *nhaAR* in the extra-intestinal virulence (Figure 6B). We also performed competition assays between wild type CFT073 and 536 strains and their isogenic mutant counterparts, CFT073Δ*nha AR*:Cm and 536Δ*nhaAR*:Cm in the murine septicemia model. We used 5 mice in each group and the experiment was repeated one

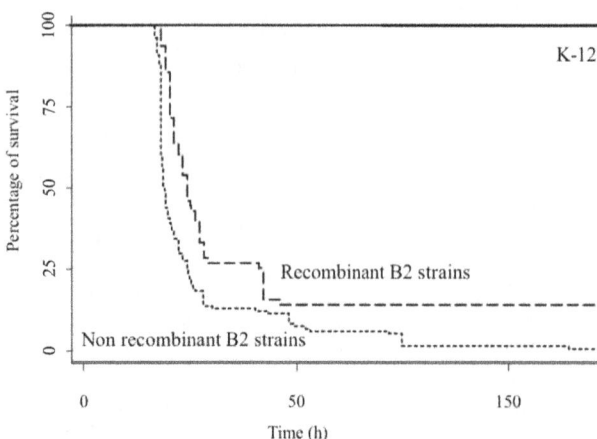

Figure 5. Recombinant *nhaA* B2 strains have a lower virulence. Lines represent the mean survival of OF1 mice after subcutaneous injection of 10^8 cells of the following strains: solid line: K-12 MG1655; dotted line: B2 strains lacking recombination in *nhaA* (*i.e* CFT073, 536, S88, RS218) and dashed line: strains showing evidence of recombination in *nhaA* region (*i.e* ED1a, E2348/65, SE15, B671, H001, TA103, TA435).

time (10 mice in each group, total). 536 and CFT073Δ*nhaAR* chloramphenicol-resistant cells were 5.33±1.14 and 3.06±1.4 orders of magnitude less numerous in the spleen than their wild type counterparts, respectively. In contrast, when Δ*nhaAR* and Δ*nhaAR*:Cm strains were injected together to the mice, no difference in spleen bacterial counts were noted, which is an indirect evidence for the absence of cost of the resistance marker *in vivo* (p = 0.51). To determine the implication of each gene of the operon *nhaAR*, we constructed and then tested the deleted mutant strains 536Δ*nhaA* and 536Δ*nhaR* in this model (Figure 6). We first determined using qRT-PCR of *nhaR* gene that in 536Δ*nhaR* strain, the gene *nhaR* was no longer expressed, and that, in 536Δ*nhaA*, *nhaR* was expressed (data not shown). In the mouse model 536Δ*nhaA* showed a significant attenuation of lethality compared to wild type 536 strains (p<0.001) whereas 536Δ*nhaR* did not show significant difference with 536 (Figure 6A). Complementation of deleted strains 536Δ*nhaA* and 536Δ*nhaAR* strains with pGC*nhaA* and pGC*nhaAR* allowed us to observe significant differentiations between 536Δ*nhaA* and 536Δ*nhaA* pGC*nhaAR* (p = 2.8E-5), 536Δ*nhaA* and 536Δ*nhaA* pGC*nhaA* in the 30 first hours (p<0.01), 536Δ*nhaAR* and 536Δ*nhaAR* pGC*nhaAR* as described above, 536Δ*nhaAR* and 536Δ*nhaAR* pGC*nhaA*. These observations lead us to conclude that complementation of *nhaA* deletion by achieved by either *nhaA* or *nhaAR* restored a high virulence (Figure 6B and C).

Hence, despite the fact that we could not find strong evidence of *nhaAR* phenotypic implication *in vitro*, it seems that the operon is critical in the mouse model of septicemia, and that the presence of the recombination in the operon is associated with a lower virulence. To go further in the *in vivo* characterization of *nhaAR* role, we finally tested strain 536 and 536Δ*nhaAR* in the murine

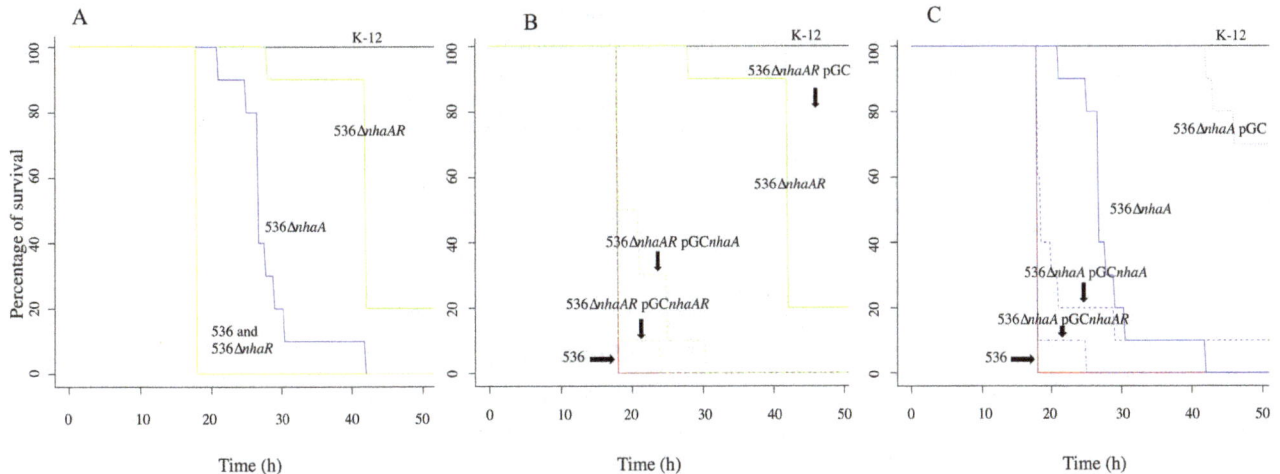

Figure 6. Impact of *nhaAR* operon on virulence. Lines represent the survival of OF1 mice after subcutaneous injection of 10^8 cells of the following strains. In (A), (B) and (C) black solid lines, K-12 MG1655 and red solid lines, strain 536. In (A) orange, blue and green solid lines, mice injected with mutants 536Δ*nhaR*, 536Δ*nhaA* and 536Δ*nhaAR*, respectively. In B (C) dashed-dotted lines, complemented mutants 536Δ*nhaA* pGCnhaAR (536Δ*nhaAR* pGCnhaAR), dashed lines, complemented mutants 536Δ*nhaA* pGCnhaA (536Δ*nhaAR* pGCnhaA), solid lines, 536Δ*nhaA* (536Δ*nhaAR*) and dotted lines, complemented mutants 536Δ*nhaApGC* (536Δ*nhaAR* pGC).

model of gut colonization. No effect of the mutation was found in the colonization model after 7 days, both when strains were added separately (mean p value during 7 days 0.2) or together in competition (mean p value during 7 days 0.63). Here again controls showed no cost of the chloramphenicol marker (mean p value during 7 days 0.74). The effect of *nhaAR* deletion seems to be restricted to virulence conditions.

Discussion

In modern biology, genomics is used to identify candidate genes associated with some phenotypes of interest. In microbiology, the genomic plasticity has lead scientists to use genomics-phenotype association mostly to focus on presence and absence of genes. In this paper, we tried to use the genomic approach to uncover some molecular determinants of B2 non-B2 differentiation using the sequence of the core genes. In that case, we lack a clear understanding of the phenotypes that may explain the difference of prevalence of these groups of strains as human commensals, and we thought that finding genes with marked difference between these groups might provide some hints. Using the fraction of fixed sites between the two groups, we identified *nhaA* as a clear outlier from the distribution. When we excluded some strains involved in recombination in the *nhaA* region, we found that this extreme pattern could be extended to the whole *nhaAR* operon.

How can such a pattern have emerged? When we looked at the *nhaAR* genomic environment we saw that the operon was flanked by highly volatile modules (Figure 1C). While the operon was conserved in all strains, flanking regions required multiple acquisition and loss to be compatible with the species phylogeny (Figure 2). Therefore, the large diversity between B2 and non-B2 could be due to the acquisition of an *nhaAR* operon by horizontal transfer. Several other observations support these hypotheses. The branch between B2 and non-B2 is not enriched in non-synonymous mutations as would be expected in the case of a strong selection. Moreover the diversity within B2 and non-B2 is not high (it is rather low indeed) which rejects a high local mutation rate.

Further investigation on the *nhaAR* sequences revealed some traces of recombination among the B2 strains (Figure 3).

Interestingly some of the strains involved in these recombinations were atypical B2 in term of virulence. None of them were isolated in extra-intestinal virulence conditions and the ones tested in a mouse model of septicemia were remarkable by their lack of virulence. We therefore decided to investigate functionally the role of *nhaAR* operon diversity.

nhaA is coding for a sodium proton antiporter which is know to be responsible for pH and sodium homeostasis in *E. coli* [16], particularly it allows growth of bacterial cells in high pH and high sodium concentrations [33]. NhaA protein is a membranous protein allowing the exchange of 2 protons against a sodium ion [34]. Padan *et al.* also described that *nhaA* mutants were not able to growth at high pH, underlining the importance of this protein in these conditions [35]. Transcription of *nhaA* is dependent of two regulators, Hns acting as a repressor and NhaR that activates the expression of *nhaA*, but also other genes, *i.e.*, *pgaA* and *osmC* [36,37]. *nhaR* is a central regulator of genes involved in stress responses (Figure 1A), *nhaA*, *pgaA* and *osmC* in high pH, high osmolarity, exposure to organic hydroperoxide or biofilm conditions [38–40]. *nhaR* is also regulated by the pleiotropic regulator CsrA in its upstream region. As high concentrations of Li + and Sodium ions [33] and high values of pH [32,34] promote the activity of NhaA, we studied the expression patterns of *nhaAR* and *osmC* at different pH and osmolarities, but failed to detect significant differences between B2 and non-B2 strains using reporter plasmids.

As changes of expression may be too small to be detected or may occur at a specific timing, we then focused on integrated phenotypes such as growth curves in different media (complex and minimum medium) at different pH (neutral and high pH) and concentration of sodium (170 and 350 mmoL/L). We found a significant differentiation between B2 and non-B2 strains at high pH and high concentration of sodium (Figure 4). These conditions are the ones in which NhaA is supposed to be expressed. However, when we looked for the specific implication of *nhaAR* region using wild type and deleted mutant strains of B2 and non-B2 group, we did not observed any significant differentiations between B2 and non-B2 strains.

Because our laboratory conditions may not be the ones in which a differentiation is strongly expressed, we tried some *in vivo* assays. As a significant number of recombinant strains for *nhaAR* region (ED1a and E2348/69) are known to be avirulent in a murine model of septicemia, we tested this model for some other recombinants identified (SE15, TA103 and TA435). We found that among the recombinant strains half of the strains tested showed decreased or absence of virulence in the murine model of septicemia which is significantly different from the other B2 strains (Figure 5) which are known to be virulent in this model [14,25]. This observation lead us to hypothesize that *nhaAR* could have implication in virulence or colonization process as it is now known that extra-intestinal and commensalism are linked [14]. We then tested two Δ*nhaAR* mutant strains belonging to the B2 phylogenetic group (CFT073 and 536) and observed an important decrease of the intrinsic virulence of the strains in this model completely complemented by a vector bearing *nhaAR* operon (Figure 6). Though, when we tested 536Δ*nhaAR* mutant in competition in a mouse gut colonization, we did not find any impact of the deletion. Hence, the effect of the mutation is only marked in the virulence model. Interestingly, the deletion of *nhaAR* operon seems to have a stronger impact on virulence than the deletion of the pathogenicity island (PAI) of strain 536 in isolations or in combination [13,41]. While the single PAI deletions had no effect with similar inoculum as the one used here, the mutant with all 7 PAI deleted killed 50% of mice in 28 hours compared to 42 hours in Δ*nhaAR* and 18 h in 536.

How could *nhaAR* contributes to virulence? NhaR is a central regulator of expression of the genes *nhaA*, *osmC* and the operon *pgaABCD* involved in stress responses such as high salinity, high pH or biofilm formation [38–40]. Implication of these genes in the virulence process is not clear, except for *pgaABCD* which has been proved to be implicated in urinary tract ascending infections [42]. However we were not able to prove specific *nhaR* implication in this model. But we clearly showed implication of *nhaA* gene in this attenuation of virulence using deleted and complemented strains with this gene. NhaA is known to be responsible for growth of bacterial cells in high pH and high sodium concentrations [33], yet such conditions are not the ones that seem to prevail during sepsis where low pH seem to be dominant [43]. Further investigation will therefore be needed to fully understand the contribution of *nhaAR* to virulence.

Conclusions

Through a bioinformatics approach we identified a candidate core gene involved in B2, non-B2 genetic differentiation. Many assays were performed to test some phenotypic expression of this diversity *in vitro* without a clear success. However, when we used *in vivo* experiments, though we only focused on the analysis of knock-outs, we found a strong and so far unnoticed implication of *nhaA* gene in virulence, despite a lack of effect in commensalism. This whole process illustrates that bioinformatics approaches may identify genes of interest whose effect is mostly if not only visible in complex *in vivo* environments.

Supporting Information

Table S1 Strains and plasmids used in the *in vitro* and *in vivo* assays in this study.

Table S2 List of primers used in this study.

Table S3 List of conditions used in the growth curves experiments.

Table S4 List of 128 genomes used in the study to identify markers of differentiation of the B2 phylogenetic group from other group.

Table S5 List of genes classified by the proportions of fixed differences between B2 and non-B2 between each gene of the core and the whole set of genes pooled together using libsequence [26].

Acknowledgments

We thank Olivier Clermont for technical assistance, Damien Roux for discussion and Erick Denamur for critical reading of the manuscript.

Author Contributions

Conceived and designed the experiments: ML OT. Performed the experiments: ML FR CP JG SD A. Launay A. Ledda SC CG JT. Analyzed the data: ML JT OT. Contributed reagents/materials/analysis tools: ML A. Ledda JT OT. Wrote the paper: ML JT OT.

References

1. Touchon M, Hoede C, Tenaillon O, Barbe V, Baeriswyl S, et al. (2009) Organised genome dynamics in the *Escherichia coli* species results in highly diverse adaptive paths. PLoS Genet 5: e1000344. doi:10.1371/journal.pgen.1000344

2. Hacker J, Kaper JB (2000) Pathogenicity islands and the evolution of microbes. Annu Rev Microbiol 54: 641–679. doi:10.1146/annurev.micro.54.1.641

3. Tenaillon O, Rodríguez-Verdugo A, Gaut RL, McDonald P, Bennett AF, et al. (2012) The molecular diversity of adaptive convergence. Science 335: 457–461. doi:10.1126/science.1212986

4. Hommais F, Gouriou S, Amorin C, Bui H, Rahimy MC, et al. (2003) The FimH A27V mutation is pathoadaptive for urovirulence in *Escherichia coli* B2 phylogenetic group isolates. Infect Immun 71: 3619–3622.

5. Sokurenko EV, Chesnokova V, Dykhuizen DE, Ofek I, Wu XR, et al. (1998) Pathogenic adaptation of *Escherichia coli* by natural variation of the FimH adhesin. Proc Natl Acad Sci USA 95: 8922–8926.

6. Tenaillon O, Skurnik D, Picard B, Denamur E (2010) The population genetics of commensal *Escherichia coli*. Nat Rev Microbiol 8: 207–217. doi:10.1038/nrmicro2298

7. Jaureguy F, Landraud L, Passet V, Diancourt L, Frapy E, et al. (2008) Phylogenetic and genomic diversity of human bacteremic *Escherichia coli* strains. BMC Genomics 9: 560. doi:10.1186/1471-2164-9-560

8. Picard B, Garcia JS, Gouriou S, Duriez P, Brahimi N, et al. (1999) The link between phylogeny and virulence in *Escherichia coli* extraintestinal infection. Infect Immun 67: 546–553.

9. Bingen E, Picard B, Brahimi N, Mathy S, Desjardins P, et al. (1998) Phylogenetic analysis of *Escherichia coli* strains causing neonatal meningitis suggests horizontal gene transfer from a predominant pool of highly virulent B2 group strains. J Infect Dis 177: 642–650.

10. Lescat M, Clermont O, Woerther PL, Glodt J, Dion S, et al. (2013) Commensal *Escherichia coli* strains in Guiana reveal a high genetic diversity with host-dependant population structure. Environ Microbiol Rep 5: 49–57. doi:10.1111/j.1758-2229.2012.00374.x

11. Nowrouzian FL, Oswald E (2012) *Escherichia coli* strains with the capacity for long-term persistence in the bowel microbiota carry the potentially genotoxic pks island. Microb Pathog 53: 180–182. doi:10.1016/j.micpath.2012.05.011

12. Secher T, Samba-Louaka A, Oswald E, Nougayrède J-P (2013) *Escherichia coli* Producing Colibactin Triggers Premature and Transmissible Senescence in Mammalian Cells. PLoS ONE 8: e77157. doi:10.1371/journal.pone.0077157

13. Diard M, Garry L, Selva M, Mosser T, Denamur E, et al. (2010) Pathogenicity-associated islands in extraintestinal pathogenic *Escherichia coli* are fitness elements involved in intestinal colonization. J Bacteriol 192: 4885–4893. doi:10.1128/JB.00804-10

14. Le Gall T, Clermont O, Gouriou S, Picard B, Nassif X, et al. (2007) Extraintestinal virulence is a coincidental by-product of commensalism in B2 phylogenetic group *Escherichia coli* strains. Mol Biol Evol 24: 2373–2384. doi:10.1093/molbev/msm172

15. Escobar-Páramo P, Clermont O, Blanc-Potard A-B, Bui H, Le Bouguénec C, et al. (2004) A specific genetic background is required for acquisition and

expression of virulence factors in *Escherichia coli*. Mol Biol Evol 21: 1085–1094. doi:10.1093/molbev/msh118

16. Padan E, Bibi E, Ito M, Krulwich TA (2005) Alkaline pH homeostasis in bacteria: new insights. Biochim Biophys Acta 1717: 67–88. doi:10.1016/j.bbamem.2005.09.010

17. Datsenko KA, Wanner BL (2000) One-step inactivation of chromosomal genes in *Escherichia coli* K-12 using PCR products. Proc Natl Acad Sci USA 97: 6640–6645. doi:10.1073/pnas.120163297

18. Vallenet D, Belda E, Calteau A, Cruveiller S, Engelen S, et al. (2013) MicroScope–an integrated microbial resource for the curation and comparative analysis of genomic and metabolic data. Nucleic Acids Res 41: D636–647. doi:10.1093/nar/gks1194

19. Guindon S, Lethiec F, Duroux P, Gascuel O (2005) PHYML Online–a web server for fast maximum likelihood-based phylogenetic inference. Nucleic Acids Res 33: W557–559. doi:10.1093/nar/gki352

20. Walk ST, Alm EW, Gordon DM, Ram JL, Toranzos GA, et al. (2009) Cryptic lineages of the genus *Escherichia*. Appl Environ Microbiol 75: 6534–6544. doi:10.1128/AEM.01262-09

21. Chenna R, Sugawara H, Koike T, Lopez R, Gibson TJ, et al. (2003) Multiple sequence alignment with the Clustal series of programs. Nucleic Acids Res 31: 3497–3500.

22. Bioedit website. Available: http://www.mbio.ncsu.edu/BioEdit/. Accessed 2012 October 10.

23. Zaslaver A, Bren A, Ronen M, Itzkovitz S, Kikoin I, et al. (2006) A comprehensive library of fluorescent transcriptional reporters for *Escherichia coli*. Nat Methods 3: 623–628. doi:10.1038/nmeth895

24. Bleibtreu A, Gros P-A, Laouénan C, Clermont O, Le Nagard H, et al. (2013) Fitness, Stress Resistance, and Extraintestinal Virulence in *Escherichia coli*. Infect Immun 81: 2733–2742. doi:10.1128/IAI.01329-12

25. Tourret J, Aloulou M, Garry L, Tenaillon O, Dion S, et al. (2011) The interaction between a non-pathogenic and a pathogenic strain synergistically enhances extra-intestinal virulence in *Escherichia coli*. Microbiology (Reading, Engl) 157: 774–785. doi:10.1099/mic.0.037416-0

26. Thornton K (2003) Libsequence: a C++ class library for evolutionary genetic analysis. Bioinformatics 19: 2325–2327.

27. Toh H, Oshima K, Toyoda A, Ogura Y, Ooka T, et al. (2010) Complete genome sequence of the wild-type commensal *Escherichia coli* strain SE15, belonging to phylogenetic group B2. J Bacteriol 192: 1165–1166. doi:10.1128/JB.01543-09

28. Broad institute website. Available: http://www.broadinstitute.org. Accessed 2011 June 15.

29. Reid SD, Herbelin CJ, Bumbaugh AC, Selander RK, Whittam TS (2000) Parallel evolution of virulence in pathogenic *Escherichia coli*. Nature 406: 64–67. doi:10.1038/35017546

30. Clermont O, Lescat M, O'Brien CL, Gordon DM, Tenaillon O, et al. (2008) Evidence for a human-specific *Escherichia coli* clone. Environ Microbiol 10: 1000–1006. doi:10.1111/j.1462-2920.2007.01520.x

31. Pannuri A, Yakhnin H, Vakulskas CA, Edwards AN, Babitzke P, et al. (2012) Translational repression of NhaR, a novel pathway for multi-tier regulation of biofilm circuitry by CsrA. J Bacteriol 194: 79–89. doi:10.1128/JB.06209-11

32. Maes M, Rimon A, Kozachkov-Magrisso L, Friedler A, Padan E (2012) Revealing the ligand binding site of NhaA Na+/H+ antiporter and its pH dependence. J Biol Chem 287: 38150–38157. doi:10.1074/jbc.M112.391128

33. Padan E, Maisler N, Taglicht D, Karpel R, Schuldiner S (1989) Deletion of ant in *Escherichia coli* reveals its function in adaptation to high salinity and an alternative Na+/H+ antiporter system(s). J Biol Chem 264: 20297–20302.

34. Taglicht D, Padan E, Schuldiner S (1991) Overproduction and purification of a functional Na+/H+ antiporter coded by nhaA (ant) from *Escherichia coli*. J Biol Chem 266: 11289–11294.

35. Padan E (2008) The enlightening encounter between structure and function in the NhaA Na+-H+ antiporter. Trends Biochem Sci 33: 435–443. doi:10.1016/j.tibs.2008.06.007

36. Sturny R, Cam K, Gutierrez C, Conter A (2003) NhaR and RcsB independently regulate the osmCp1 promoter of *Escherichia coli* at overlapping regulatory sites. J Bacteriol 185: 4298–4304.

37. Goller C, Wang X, Itoh Y, Romeo T (2006) The cation-responsive protein NhaR of *Escherichia coli* activates pgaABCD transcription, required for production of the biofilm adhesin poly-beta-1,6-N-acetyl-D-glucosamine. J Bacteriol 188: 8022–8032. doi:10.1128/JB.01106-06

38. Dover N, Higgins CF, Carmel O, Rimon A, Pinner E, et al. (1996) Na+-induced transcription of nhaA, which encodes an Na+/H+ antiporter in *Escherichia coli*, is positively regulated by nhaR and affected by hns. J Bacteriol 178: 6508–6517.

39. Wang X, Preston JF 3rd, Romeo T (2004) The pgaABCD locus of *Escherichia coli* promotes the synthesis of a polysaccharide adhesin required for biofilm formation. J Bacteriol 186: 2724–2734.

40. Lesniak J, Barton WA, Nikolov DB (2003) Structural and functional features of the *Escherichia coli* hydroperoxide resistance protein OsmC. Protein Sci 12: 2838–2843. doi:10.1110/ps.03375603

41. Tourret J, Diard M, Garry L, Matic I, Denamur E (2010) Effects of single and multiple pathogenicity island deletions on uropathogenic *Escherichia coli* strain 536 intrinsic extra-intestinal virulence. Int J Med Microbiol 300: 435–439. doi:10.1016/j.ijmm.2010.04.013

42. Subashchandrabose S, Smith SN, Spurbeck RR, Kole MM, Mobley HLT (2013) Genome-wide detection of fitness genes in uropathogenic *Escherichia coli* during systemic infection. PLoS Pathog 9: e1003788. doi:10.1371/journal.ppat.1003788

43. MacKenzie IM (2001) The haemodynamics of human septic shock. Anaesthesia 56: 130–144.

## 6

GABenchToB: A Genome Assembly Benchmark Tuned on Bacteria and Benchtop Sequencers

Sebastian Jünemann[1,2]*, Karola Prior[1], Andreas Albersmeier[3], Stefan Albaum[4], Jörn Kalinowski[3], Alexander Goesmann[5], Jens Stoye[2,6], Dag Harmsen[1]

1 Department for Periodontology, University of Münster, Münster, Germany, 2 Institute for Bioinformatics, Center for Biotechnology, Bielefeld University, Bielefeld, Germany, 3 Technology Platform Genomics, Center for Biotechnology, Bielefeld University, Bielefeld, Germany, 4 Bioinformatics Resource Facility, Center for Biotechnology, Bielefeld University, Bielefeld, Germany, 5 Bioinformatics and Systems Biology, Justus-Liebig-Univeristy Gießen, Gießen, Germany, 6 Genome Informatics Group, Faculty of Technology, Bielefeld University, Bielefeld, Germany

Abstract

De novo genome assembly is the process of reconstructing a complete genomic sequence from countless small sequencing reads. Due to the complexity of this task, numerous genome assemblers have been developed to cope with different requirements and the different kinds of data provided by sequencers within the fast evolving field of next-generation sequencing technologies. In particular, the recently introduced generation of benchtop sequencers, like Illumina's MiSeq and Ion Torrent's Personal Genome Machine (PGM), popularized the easy, fast, and cheap sequencing of bacterial organisms to a broad range of academic and clinical institutions. With a strong pragmatic focus, here, we give a novel insight into the line of assembly evaluation surveys as we benchmark popular *de novo* genome assemblers based on bacterial data generated by benchtop sequencers. Therefore, single-library assemblies were generated, assembled, and compared to each other by metrics describing assembly contiguity and accuracy, and also by practice-oriented criteria as for instance computing time. In addition, we extensively analyzed the effect of the depth of coverage on the genome assemblies within reasonable ranges and the k-mer optimization problem of de Bruijn Graph assemblers. Our results show that, although both MiSeq and PGM allow for good genome assemblies, they require different approaches. They not only pair with different assembler types, but also affect assemblies differently regarding the depth of coverage where oversampling can become problematic. Assemblies vary greatly with respect to contiguity and accuracy but also by the requirement on the computing power. Consequently, no assembler can be rated best for all preconditions. Instead, the given kind of data, the demands on assembly quality, and the available computing infrastructure determines which assembler suits best. The data sets, scripts and all additional information needed to replicate our results are freely available at ftp://ftp.cebitec.uni-bielefeld.de/pub/GABenchToB.

Editor: Christophe Antoniewski, CNRS UMR7622 & University Paris 6 Pierre-et-Marie-Curie, France

Funding: This work was supported in parts by grants of the German Federal Ministry of Education and Research (BMBF) in the framework of the FBI-Zoo project (FKZ 01KI1012B), of the Technology Platform Bioinformatics (TPB) project (FKZ 031A190) and by the European Commission's Seventh Framework Programme (EU PathoNGenTrace project agreement no. 278864). The funders had no role in study design, data collection and analysis, decision to publish, or preparation of the manuscript.

Competing Interests: The authors have declared that no competing interests exist.

* Email: jueneman@cebitec.uni-bielefeld.de

Introduction

With the introduction of massively parallel high-throughput next generation sequencing (NGS) platforms, fast and cost-effective whole genome shotgun sequencing of the full variety of organisms has been enabled. The rapid advancement in this field is best represented by the recently introduced small scaled benchtop sequencers (BS), e.g., the MiSeq by Illumina (San Diego, California) and the Ion Torrent PGM by Life Technologies (Carlsbad, California). Albeit providing a lower throughput than their conventional non-benchtop counterparts (e.g., Illumina's HiSeq and Ion Torrent's Proton), they still provide sufficient genomic coverage and sequencing accuracy to be efficiently used for sequencing bacterial genomes [1–2].

One crucial step in genome based analysis is the attempt to *de novo* assemble raw sequencing reads into a bacterial chromosome. *De novo* genome assembly is the process of reconstructing a whole genome sequence from short sequencing reads by finding common subsequences and assembling overlapping reads to longer continuous sequences, i.e. contigs, under the assumption that such reads originate from the same genomic location. If special pairs of reads with a known pairing distance are available, i.e. mate-pair (MP) or paired-end (PE) reads, this information can be used to arrange individual contigs in an ordered sequence consisting of contigs and gaps of known sizes (scaffolds). This is in particular useful to span gaps related to long repetitive elements which are hard to be resolved solely by overlapping reads of limited length. In general, most current assembly algorithms can be assigned to one of two classes based on their underlying data structure: de Bruijn graph (DBG) and overlap layout consensus (OLC) assemblers. Both approaches utilize a graph structure built upon the sequencing reads and algorithms for graph traversal in order to deduce overlapping sequences and to generate contigs. Very briefly, OLC assemblers build a graph by connecting nodes which represent the sequencing reads by edges representing the specific overlaps. For the DBG approach reads are initially partitioned into substrings of the reads of a fixed length (k-mers) and a graph is built by connecting nodes symbolizing sub-reads that share a specific prefix and suffix, respectively. See Compeau et al. 2011 [3], Li et al. 2012 [4] and Nagarajan and Pop 2013 [5] for detailed descriptions on the principles of DBG and OLC assemblers.

The rapid progress in the field of NGS as well as diverse sequencing procedures and protocols had also impacts on genome assembly algorithms and assembly software solutions. The number of sequence assemblers steadily increased in recent years with currently several dozens of assemblers available (22 genome assemblers compared in 2008 [6], additional 13 assemblers in 2010 [7], further ten assemblers in 2011 [8]). Although this variety becomes somewhat more limited as some assemblers require a particular kind or set of sequencing data as an input, researchers are still confronted with a wide range of assembler candidates. The decision which *de novo* assembler to use is conditional to several aspects, notably the specification of the applied sequencing platform and protocol (e.g., single-end versus PE reads), characteristics of the sequencing results (read lengths and error profiles) and, if available, characteristics of the sequenced genome (e.g., number of repetitive genomic elements or the genomic %GC-content).

Therefore, systematic evaluations of assemblers are necessary to provide the research community with scientifically sound decision-making support. In the past, several efforts have been made to assess genome assembler efficiency on different scales and for different application scenarios [8–11]. Most prominently, two comparative studies with recently introduced remakes contributed greatly to this field, i.e. Assemblathon and Genome Assembly Gold-Standard Evaluations (GAGE). For the first large scale assembler competition Assemblathon 1 [12] and its recent successor Assemblathon 2 [13] assemblies were performed and submitted by different institutions and thereafter evaluated. In the first study, 17 participants generated 41 assemblies based on simulated short Illumina HiSeq reads covering 16 different genome assemblers. In Assemblathon 2, 21 participants assembled 43 genomes of three vertebrate species sequenced on different instruments while using eight genome assemblers. In contrast, for the GAGE [14] and GAGE-B [15] competitions all assemblies were generated under equal conditions. Whereas GAGE compared eight different genome assemblers using multiple Illumina libraries of four different pro- and eukaryotic data sets, GAGE-B concentrated on single library assemblies of nine Illumina sequenced bacterial strains using also eight different assemblers.

Albeit these studies give a very comprehensive picture of the efficiency and applicability of state-of-the-art assembler algorithms, several questions remain unanswered. First and foremost, NGS is evolving fast and consumables, protocols, and technical specifications of BS differ in comparison to conventional NGS instruments. For instance, the MiSeq offers an improved read length of 2×300 base pairs (bp) compared to the maximum 2×150 bp of the HiSeq system. The Ion Torrent PGM, although comparable to Roche's 454 in terms of the error profile, produces

a different read length distribution (current maximum read length of about 400 bp for the PGM compared to the 1,000 bp at max. for the 454 GS−FLX+ system). Up to now, only the GAGE-B evaluation took into account MiSeq data sets while evaluations of assemblies originating from PGM instruments are missing. Therefore, methods need to be reconsidered for deciding how to assemble data originating from BS instruments. Secondly, bacterial genomes are underrepresented, particularly in the Assemblathon surveys. Even if bacterial genomes are lightweights regarding their assembly difficulty compared to e.g. vertebrates, they nevertheless have different requirements on sequencing procedures and the assembler algorithms. GAGE-B recently tried to close this gap, yet for the majority of the used data sets a high quality genome reference was not available. A third aspect of an assembler evaluation is more of a practice-oriented nature, e.g. the run time and memory usage of an assembly and the demands each assembler has on the compute infrastructure. An assembler recommendation would have little practical value if best performing assemblers cannot be operated due to impractical hardware requirements. However, this aspect was covered insufficiently in the past. Finally, a comparison study should be transparent in order to sustain reproducibility. This means that all steps beginning from the raw sequencing data to the final results should be sufficiently documented. Especially the Assemblathon competitions could not fully satisfy this requirement owed to the fact that the assemblies were performed at different institutions and documented at varying extent.

Here, we present a *de novo* assembler evaluation with a strong focus on practical aspects and which addresses the aforementioned unanswered questions. To this end, the main objectives of this Genome Assembly Benchmark Tuned On Bacteria and benchtop sequencers (GABenchToB) are:

- to use real (no synthetic) data originating from benchtop sequencers,
- to use single libraries only of bacterial genomes with an available high-quality reference,
- to consider open licensed as well as commercial assemblers,
- to select assemblers covering different assembly strategies,
- to not perform extensive assembly fine tuning but to rest on default parameters for the different sequencing platforms where possible and to use unprocessed raw reads,
- to include run time benchmark parameters, and
- to ensure for equal executing conditions by using dedicated computing hosts and to deposit all information necessary for reproduction at an open repository.

Results

Evaluation data sets

We have chosen three different bacterial strains for our assembly evaluation: *Escherichia coli O157:H7 Sakai* (American Tissue Culture Collection [ATCC] accession no. BAA-460), *Staphylococcus aureus COL* (Network on Antimicrobial Resistance in *Staphylococcus aureus* [NARSA] accession no. NRS100), and *Mycobacterium tuberculosis H37Rv* (ATCC accession no. 25618). For all three bacteria full reference sequences are available: NCBI Reference Sequence (RefSeq) accession no. NC_002695.1, NC_002128.1 and NC_002127.1 with modifications as described previously [2] were used as a reference for the *E. coli* genome and plasmids; NC_002951.2 and NC_006629.2 for the *S. aureus* genome and plasmid, and NC_000962.2 for the *M. tuberculosis* genome, respectively. The genomes cover a GC-content between 33 and 66% and have a genome size between 2.81 and 5.59 megabases (Mb). Samples of the three bacterial strains were sequenced on BS platforms Ion Torrent PGM (PGM) by Life Technologies and MiSeq (MIS) by Illumina. Both platforms are not only represented by all three organisms but also with different chemistries, i.e. 2×150 base pairs (bp) PE and 2×250 bp PE sequencing for the MiSeq as well as 200 bp and 400 bp sequencing for the PGM, resulting in a total of ten single library data sets (**Table 1**). Except for the four *E. coli* data sets, which were used and published previously [2], all other libraries were generated newly for the purpose of this study.

De novo assembler selection

To enable an assembler evaluation we aimed to select a set of *de novo* assemblers regarding the following criteria: assemblers which are (i) representing DBG or OLC approaches; (ii) free-to-use or open source as well as commercial products; (iii) unbiased in terms of the supported sequencing technology or the required sequencing library (processing single and paired-end reads, not relying on mate-pair libraries and no requirement on multiple libraries); (iv) up-to-date regarding to the date of their release or the latest software update; and (v) established and widely used either in recent *de novo* sequencing projects or in other assembler evaluations. Based on these criteria we selected nine representative assemblers: AbySS [16], Celera [17], CLC Assembly Cell (CLC bio A/S, Aarhus, Denmark) [18], GS De Novo Assembler (454 Life Sciences, Branford, CT) [19], MIRA [20], SeqMan Ngen (DNASTAR Inc., WI, USA) [21], SOAPdenovo2 [22], SPAdes [23], and Velvet [24] (**Table 2**). Some assemblers we did not include in our study, but which performed best in either one or several categories in previous evaluations [11–15] are: Allpaths-LG [25] because its algorithm requires input data consisting of at least two different libraries (one PE and one MP), ARACHNE [26] because it combines only long Sanger with MP reads, MaSuRCA [27] and SGA [28] because both can not handle single-end data, and Phusion [29] because it builds solely upon MP reads.

Assembly evaluation metrics

The evaluation of genome assemblers is a complex problem and single metric based evaluations using e.g., the N50 or NG50 value for comparisons examine only specific aspects of an assembly [9]. On the other hand, Haiminen et al. [9] criticize that providing tables full of different assembly metrics complicate the understanding and interpretation of an evaluation effort and limit their usability in practice. A single metric capturing the trade-off between contig contiguity and accuracy are feature response curves (FRC) as proposed by Narzisi and Mishra [11]. However, one limitation of FRC is their requirement of read based contig layouts, which are naturally not available for DBG assemblers. To overcome this limitation, Haiminen et al. [9] introduced FRCbam, which allows generating a read layout by aligning the sequencing reads to the assembled contigs by incorporating paired-end and mate-pair information. However, this is also a strong limitation for the evaluation of assemblies based on BS data, as no paired-end libraries are available (PGM) or are of limited use due to their small insert-size (MiSeq). Another single metric is the Log Average probability proposed by Ghodsi et al. [30] but it was designed to measure assemblies without knowledge of the true reference. Here, we assessed assembly accuracy using the recently published assembly evaluation software QUAST [31] following in essence the GAGE-B evaluation [15]. For that matter, assembly contiguity and accuracy are measured by two metrics introduced by QUAST and another two metrics are used to measure system workload: i.e. the NGA50 length, the number of mis-assemblies, the total run time (wall clock time), and the average system utilization (see the Methods section for a detailed definition of the metrics).

It is to be noted that we did not compare assemblies and platforms separately for contigs and scaffolds in order to preserve a focused evaluation. From a pragmatic point of view it is contradictory to use contigs for an evaluation even though scaffolds of the same assembly are available. Additionally, scaffolding requires PE or MP reads. As no PE-libraries have been available for the PGM platform scaffolding of PGM data was not possible but disabling scaffolding would unnecessarily disfavor the MiSeq data sets. Therefore, the term contig is used, with

Table 1. Overview of the data sets used in this study and their sequencing yield.

Platform	Library	Software version	Strain	Chip/Lane	Megabases	Coverage
PGM	200 bp[&] Ion Xpress Plus Fragment**	TSS v3.0[$]	*E. coli* (Sakai)	1×316 chip	733	133
PGM	400 bp Ion Xpress Plus Fragment**	TSS v3.4[$]	*E. coli* (Sakai)	1×318 chip	1,179	214
PGM	200 bp Ion Xpress Plus Fragment	TSS v3.0[$]	*S. aureus* (COL)	1×316 chip	555	197
PGM	400 bp Ion Xpress Plus Fragment	TSS v3.2[$]	*S. aureus* (COL)	1×318 chip	1,420	505
PGM	400 bp Ion Xpress Plus Fragment	TSS v3.6[$]	*M. tuberculosis* (H37)	1/3 ×318 chip	344	77
MiSeq	2×150 bp PE[+] Nextera*	MCS v1.2.3	*E. coli* (Sakai)	¼ multiplexed lane	565	102
MiSeq	2×250 bp PE Nextera*	MCS v2.0.5	*E. coli* (Sakai)	¼ multiplexed lane	776	141
MiSeq	2×150 bp PE Nextera	MCS v1.2.3	*S. aureus* (COL)	¼ multiplexed lane	445	158
MiSeq	2×250 bp PE Nextera	MCS v2.0.5	*S. aureus* (COL)	¼ multiplexed lane	509	181
MiSeq	2×250 bp PE Nextera	MCS v2.0.5	*M. tuberculosis* (H37)	15% multiplexed lane	340	77

[&]base pairs.
[+]Paired-end sequencing.
*Same raw data set as in Jünemann et al. [2].
**Same raw data set as in Jünemann et al. [2] but re-analyzed using a different sequencing software version. MiSeq, Illumina MiSeq; PGM, Ion Torrent Personal Genome Machine. MCS, MiSeq Control Software; TSS, PGM Torrent Suite Software.
[$]More stringent filter enabled.

Table 2. *De novo* assemblers used for comparison.

Assembler (acronym)	Software version	Type	Supports scaffolding	License	Supported operating systems
AbySS (ABYSS)	1.3.5	DBG$	yes	commercial*	Windows, Linux
Celera (CELERA)	7.0	OLC&	yes	open source	Windows, Linux
CLC Assembly Cell (CLC)	4.0.10	DBG	yes	commercial	Windows, Linux, Mac OS
GS De Novo Assembler (NEWBLER)	2.8	OLC	yes	commercial**	Linux
MIRA (MIRA)	3.9.9	OLC#	no	open source	Linux
SeqMan Ngen (SEQMAN)	11.0.0.172	OLC	no	commercial	Windows, Linux, Mac OS
SOAPdenovo2 (SOAP2)	2.04	DBG	yes	open source	Linux
SPAdes (SPADES)	2.5.0	DBG	yes	open source	Linux, Mac OS
Velvet (VELVET)	1.2.08	DBG	yes	open source	Linux, Mac OS

$de Bruijn Graph assembler,
&Overlap Layout Consensus assembler,
#MIRA is not a pure OLC assembler but uses also greedy assembler techniques.
*Free for non-commercial and academic applications.
**Freely available upon request.

Genome coverage

One major parameter that influences the result of a genome assembly is the amount of available data to cover the whole genome, i.e. the depth of coverage. Insufficient coverage may lead to an inferior assembly result either due to uncovered genomic regions or due to the impossibility to overrule randomly distributed sequencing errors. In contrast, each increase of the target coverage involves a higher load of the sequencing platform capacity, which comes at the price of an increased sequencing cost per library as well as of a growing computational effort due to the increased data volume. Also, non-randomly distributed sequencing errors (e.g., systematic errors related to homopolymer regions) can not be ruled out by redundant sequencing and accumulate with increasing coverage which can trick assemblers to handle them as true biological events [32]. Therefore, finding a balance between sequencing effort, cost, assembly turn-around time and assembly quality is crucial in advance to the sequencing itself. The observation that an ad infinitum increased depth of coverage does not necessarily improve assembly results was already reported by Lin et al. [10]. They found that for seven assemblers operated on six simulated data sets based on a eukaryotic chromosome the depth of coverage at which the N50 length plateau was reached never exceeded values of 50. Consequentially, an upper coverage threshold of 70x was used to compare the assembly results. Similar examinations in further studies yielded comparable results [9], [15], [33]. However, common to all these studies were shortcomings regarding the resolution, the range of assemblers, the amount of data at which this effect was studied, and most of all only Illumina reads were used. To address the question to what extent depth of coverage affects assembly results we have randomly sub-sampled all PGM and MiSeq data sets into a fix range of subsets. All subsets were subsequently assembled using three DBG (CLC, ABYSS, and SPADES) and three OLC assemblers (NEWBLER, CELRA, and MIRA). This reduced assembler set was chosen to cope with the high computational effort while maintaining meaningful results with regard to our assembler selection criteria.

Independent of the data set and the assembler, we found out that for each assembler and data set a window of reasonable coverage can be identified which is confined by a lower and upper bound, which mark the areas of inappropriate coverage (**Figure 1**). Given the *E. coli* Sakai data sets, too low coverages result in inferior NGA50 lengths progressively improving with increasing coverage, whereas above the upper bound a further increase in the depth of coverage does not necessarily effects superior assemblies. For the PGM platform oversampling, i.e. too high coverage, can even have a negative impact on the assembly especially if combined with OLC assemblers. Here, NGA50 lengths are constantly falling after reaching a specific maximum. This effect is apparent for both the PGM 200 bp and PGM 400 bp data, albeit the global coverage maxima are higher for the 400 bp (between 40- and 100-fold coverage) than for the 200 bp data sets (between 40- and 50-fold coverage). The MiSeq platform is not susceptible to oversampling. After reaching sufficient coverage, NGA50 lengths are mostly saturated and show only moderate improvement or degradation, marking the upper bound of appropriate coverage for the MiSeq data sets. These findings are in good accordance with the other data sets and with other metrics, as e.g. the number of mis-assemblies (local and non-local mis-assemblies) or the number of assembly errors (insertions, deletions and substitutions) and is further supported by another sub-sampling approach (see Supporting Information **Text S1** and **Figures S1–S5**). In order to achieve comparability between the data sets and the assemblers, we have selected two different cutoffs, reflecting the described upper coverage bounds, at which we sub-sample our data sets prior to any consecutive analysis. The 40-fold coverage limit was chosen for the PGM 200 bp data, whereas all MiSeq and the remaining PGM 400 bp data sets were sub-sampled at 75-fold depth of coverage. Even though these thresholds do not always represent individual global optima, they are not in favor or disfavor of any particular assembler or sequencing platform and in good accordance with previous findings [9], [15], [33].

K-mer parametrization of de Bruijn Graph assemblers

Generally stated, DBG assemblers break raw sequencing reads into a set of k-mers and construct a graph by connecting the suffix and prefix nodes of overlapping k-mers by edges. By that, the size of the DBG depends only on the length of the k-mer and not on the initial read length distribution. Albeit this has a beneficial effect on the memory footprint and graph traversing time, it comes at

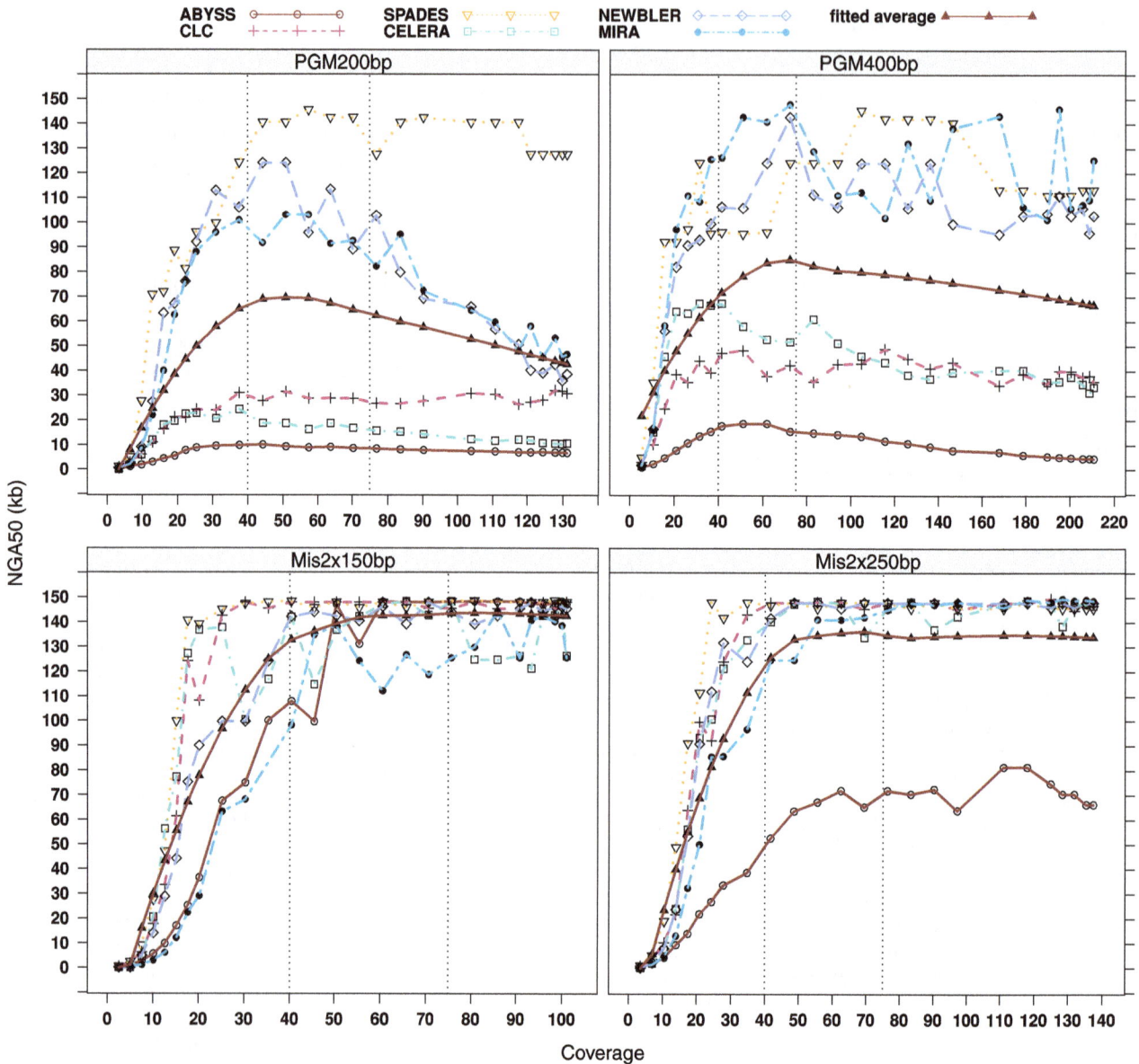

Figure 1. Effect of the depth of coverage on the assembly efficiency measured by NGA50 sizes based on randomly sub-sampled *E. coli* Sakai data sets. The coverage is referring to the average depth each genomic position is covered by the sequencing reads and not to the average depth of coverage the assemblies are actually reaching. The fitted average is, for each data set, the mean of all NGA50 lengths at each coverage fitted to a nonlinear local regression model. Sub-sampling was done in steps as a percentage of the original full sample size; hence, the x-axis ranges of the four sub-plots differ. The dotted vertical lines mark the finally used 40-fold (PGM 200 bp) and 75-fold coverage limits (PGM 400 bp, MiSeq 2×150 bp and MiSeq 2×250 bp).

the price for another parameter optimization step. Surprisingly, we are not aware of any study that tried to analyze this optimization problem extensively, nor were it a topic in previous assembly evaluation surveys, even though its existence is common knowledge in the assembly community. To meet this shortcoming, we have examined the effect of the k-mer parameter k on all data sets and all DBG assemblers for which a k needs to be specified (ABYSS, SOAP2 and VELVET) by iteratively running these assemblers at all supported values of k. Two main effects can be seen. First, the choice of an optimal k is depending on the underlying data set. Assemblages of different data sets using the same assembler provided different best performing values of k. Second, the optimal k also depends on the assembler and strongly

varies even though the same data set was assembled. An approximation of an optimal k is further complicated by the fact that already the results based on one assembler and one data set show more than one local optima along the full spectrum of k. Hence, best performing k-mer parameters for the consecutive assemblies were determined by optimizing over the whole possible k-mer spectrum. For PGM data sets a k-mer optimization is less reasonable. Even though comparable pattern as for the MiSeq data were observed, overall unsatisfactory results were obtained independent of k. For an in depth description on the K-mer parametrization of de Bruijn Graph assemblers see the Supporting Information (**Text S1** and **Figures S6** and **S7**).

Evaluation of *de novo* assemblies

Combining ten data sets and nine assemblers we have generated 90 *de novo* assemblies. For every single assembly, the assembly effort was measured using QUAST [31]. Results were transformed into a spread-sheet readable tabulator separated file format and stored in addition to the complete assembly output, log files, and run time benchmarking results. In **Figure 2** we present the results of each assembly using the following two key metrics: the NGA50 length and the number of mis-assemblies. As mentioned before, these figures show either scaffolds or contigs according to the data, i.e. scaffolds for all MiSeq assemblies except for MIRA and SEQMAN and contigs for all others including the PGM assemblies. Similarly, **Figure 3** shows the central aspect of the running time benchmark, i.e. the wall clock time and the average system utilization. Details about specific assembly parameters and the full execution pipeline are given in the Supporting Information (**Text S1**) or can be looked up at the following project site: ftp://ftp.cebitec.uni-bielefeld.de/pub/GABenchToB.

At this repository, also the 90 assembly results together with all computed assembly metrics can be accessed. In addition to this core set, each of the more than two thousand assemblies used for the coverage and k-mer parameter optimization are deposited here, with the limitation that not for all of those assembly and run time metrics are available.

Discussion

One of our main goals was to provide the research community with a practice-oriented *de novo* assembly evaluation of bacterial genomes sequenced by benchtop instruments. In this spirit, our study is intended to be more than a pure assembler comparison following the question which assembler performs best. Also, it is not a sequencing platform comparison trying to answer which platform allows for the best assemblies. Instead, given a diverse set of sequencing data, we identified those assemblers, which are best suited to handle individually different data sets and meet specific requirements such as the least amount of mis-assemblies or the fastest compute time.

Consequently, given the assembly results in **Figure 2** and **Figure 3**, none of the assemblers emerges as the overall winner. The individual assembler performance as given by the NGA50 length strongly depends on the nature of the data (**Figure 2A**). For MiSeq reads the assembler selection is less restricted than it is for PGM reads. Both, DBG and OLC assemblers are generally applicable on the MiSeq data sets. The *E. coli* MiSeq 2×150 bp and 2×250 bp assemblies, for example, yield NGA50 values of the same magnitude (about 155 kb) for both approaches. The highest MiSeq 2×150 bp *S. aureus* NGA50 lengths originate from DBG assemblers, whereas OLC assemblers performed better on the MiSeq 2×250 bp *S. aureus* and *M. tuberculosis* data. In contrast, assembled PGM data yielded better results more often when using OLC than DBG assemblers. Comparing the sum of all NGA50 lengths of the DBG assembled PGM 200 bp and 400 bp data sets with those using OLC, a more than 2-fold increase can be observed (719 megabases (Mb) compared to 1518 Mb). The only exception to this is the DBG SPADES assembler, which achieves NGA50 lengths on a par with the OLC assemblers. Omitting SPADES from this comparison, the proportion changes from a 2-fold increase to an over 6-fold increase (220 Mb compared to 1518 Mb), clearly showing the difficulties DBG assemblers have while dealing with PGM data. Considering the NGA50 length, we want to highlight the DBG assembler SPADES and the OLC assemblers NEWBLER and MIRA, as they repeatedly perform

above average in their category and are more robust with respect to the kind of sequencing data.

Laying focus on the number of mis-assemblies, a different picture arises (**Figure 2B**). The NGA50 is the NG50 length after contigs have been split at each observed mis-assembly position. Resting on this relation, the naïve assumption would be that lots of mis-assemblies are directly reflected by a low NGA50 and vice versa. However, here a high and low mis-assembly rates could be observed independent of the NGA50 length for both assembler approaches across all data sets. The highest rate of mis-assemblies originates from an assembly that can be, measured against the NGA50, declared as failed (*S. aureus* PGM200 bp SEQMAN with NGA50 of 2292 bp). In contrast, some assemblies with also a very low NGA50 show very few mis-assemblies (e.g., all PGM ABYSS assemblies). However, discussing rates of mis-assemblies for assemblies with extremely low NGA50 values has little meaning. Assemblies with a reasonable NGA50 length again show both, i.e. a high (e.g., the SOAP2 and VELVET MiSeq *E. coli* and *M. tuberculosis* assemblies) and low (e.g., the SPADES MiSeq *E. coli* assemblies) rate of mis-assemblies. Of those, three assemblers show the most consistent pattern of mis-assemblies. While SPADES produced the least mis-assemblies in total across all data sets, NEWBLER produced high rates of mis-assemblies only when dealing with MiSeq data, and MIRA had one of the most miss-assembled contigs across all platforms. For all other assemblers, the number of mis-assemblies depended on the processed data and varies from very high to very low. Moreover, the NGA50 metric shows no significant correlation with the mis-assembly rates (data not shown). This implies, that a low NGA50 value is not necessarily equivalent to many mis-assemblies. Instead, it simply indicates a comparably inferior assembly, either due to many mis-assemblies or low assembly contiguity. These findings are also consistent with other metrics, e.g. the amount of fully covered genes and the number of assembly errors (insertions, deletions, and substitutions; **Figure S8**).

Genomes of extreme GC-rich or GC-poor bacteria are known to be challenging for genome assemblers as amplification biases of GC-poor or GC-rich regions can result in uneven genome coverage. Here, a weak relationship between the genomic GC-content and assembly contiguity could be found. Assemblages of the GC-poor *S. aureus* genome (33% GC) tend to reach higher NGA50 lengths than the *E. coli* (51% GC) and the GC-rich *M. tuberculosis* (66% GC) genome, for which on average the lowest NGA50 lengths were achieved. Likewise, the *S. aureus* assemblies exhibited fewer mis-assemblies than those of the other two genomes. However, the relatively inferior *M. tuberculosis* assemblies cannot be entirely explained by increased mis-assemblies, as they are comparable to the ones based on the *E. coli* genome. Instead, this discrepancy might be explained by the finding that extreme GC-rich regions are especially difficult to amplify [34], possibly lowering the assembly completeness and by that the contiguity. Despite these findings, a general conclusion on the effect of the GC-content on genome assemblies cannot be drawn, as this would require a broader range of differing genomes to be analyzed. In addition, the differences shown here are not clear enough and for all genomes a successful assembly could be generated. This is also supported by a previous study, which reports that the degree of a GC bias, the factor most influencing the assembly contiguity, correlates neither with the mean GC content nor with the standard deviation of GC content of a genome [35]. In the same study it could also be shown that a sufficient depth of coverage can compensate for a GC bias, which may explain the comparatively low differences observed here with regard to the sub-sampling optimized approach used in our study.

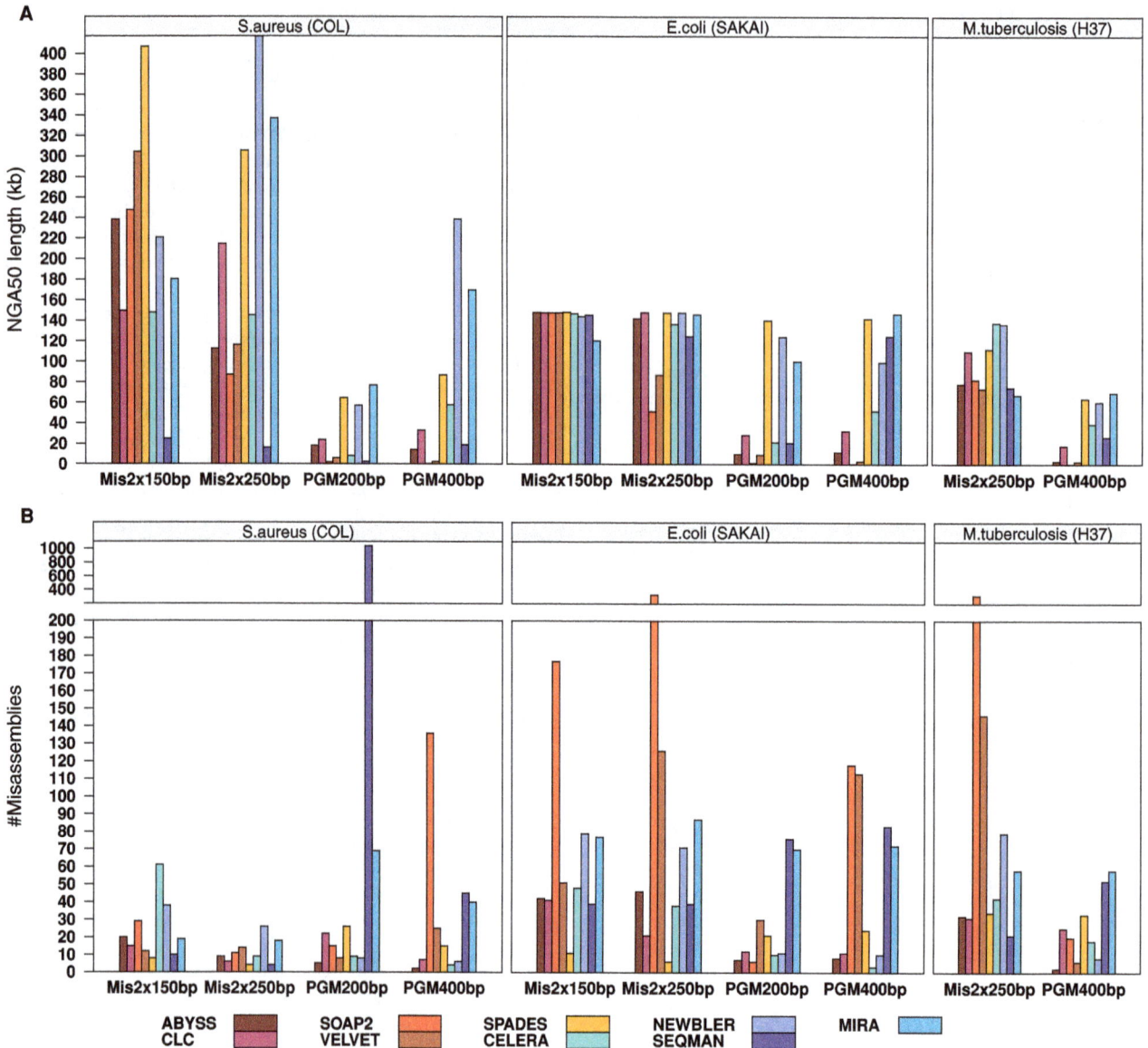

Figure 2. Comparison between the *de novo* **genome assemblies based on the NGA50 length and the number of mis-assemblies.** The NGA50 length (**A**, in kilobases) and the number of mis-assemblies (**B**, combining local and non-local mis-assemblies) on the y-axis are either contig or scaffold based, respectively. Scaffolds for MiSeq 2×150 bp and MiSeq 2×250 bp assemblies obtained by ABYSS, CELERA, CLC, NEWBLER, SOAP2, SPADES, and VELVET; contigs for MiSeq assemblies obtained by MIRA and SEQMAN as well as for all PGM assemblies. The second plot (**B**) is further divided into two plot rows where the upper row has an altered y-axis scale only showing high rates of mis-assemblies ranging from two hundred up to thousand.

One neglected aspect in preceding assembler evaluation studies has been the computational cost of an assembly. By measuring the wall clock time of all assembly processes we observed a great discrepancy between DBG and OLC assemblers (**Figure 3A**). With the exception of SPADES, all DBG assemblers finished within less than 20 minutes, the majority even within less than five minutes, whereas OLC assemblers took from eight minutes up to 20 hours. NEWBLER is to be highlighted positively by showing consistently the shortest run time of the OLC assemblers. CELERA and MIRA, on the contrary, repeatedly exceeded run times of three hours. However, for all DBG assemblers only the run time of a single assembly effort is shown, i.e. of the single assembly call resulting in the highest N50 length (**Table S1**). So the total wall clock time for an entire assembly project needs to be

adjusted with respect to the chosen k-mer optimization strategy and may markedly reduce the outstanding performance of the DBG assemblers. Exceptions to this are the assemblers CLC and SPADES as both do not rely on an external k-mer optimization step. Thus, their run times are equivalent to the overall needed run times. In this sense, CLC outperforms all other assemblers many times over. SPADES prolonged execution times, in contrast, can be explained by its operating procedures, as in the case of MiSeq PE data six and for PGM data five consecutive assemblies were performed internally.

A good indicator showing to which extent an assembler benefits from a parallelized computing environment is the average CPU utilization (**Figure 3B**). Surprisingly, only ABYSS is nearly full parallelized. All other assemblers have an average CPU utilization

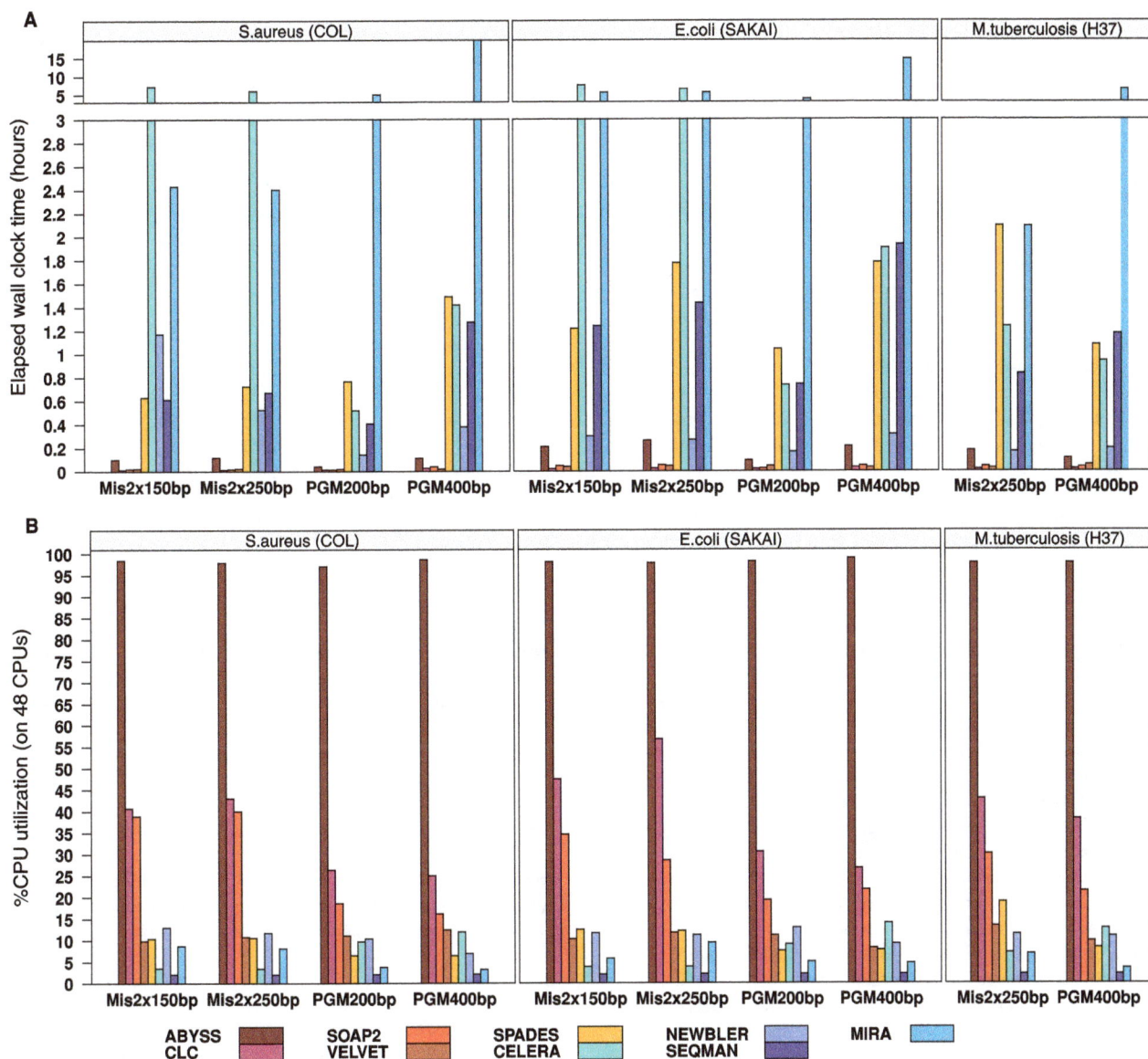

Figure 3. Computing time of *de novo* genome assemblies. Based on the elapsed wall clock time (**A**, in hours) and the total CPU utilization (**B**, in percent and relative to the 48 available CPU cores of the executing compute host). With regard to the CPU utilization, all assemblies have been instructed via proper parameterization to make maximal use of the 48 available CPU cores. The only exceptions to this were SEQMAN, which does not support parallelization, and CELERA, which due to configuration constraints has altering concurrency and multi-threading parameters for different internal processes. For DBG assemblers only run time and CPU utilization of the single assemblies with the best performing k-mer parameter are shown and not the summation of the full k-mer optimization procedure (for SPADES and CLC this is equivalent).

below 50%, which means that of the 48 available and requested CPUs, on average, only half of them were used during the full assembly procedure. Next to ABYSS, CLC and SOAP2 show the second and third best CPU utilizations, respectively. Also CLC and SOAP2 have a higher utilization if operating on MiSeq data than on PGM, showing that they are obviously optimized for Illumina data. All other assemblers have a total CPU utilization below 20%. This implies that the running time of those assemblers cannot be increased considerably by assigning more CPU cores. However, a low CPU utilization does not necessarily result in long running times, as demonstrated by VELVET, which always finished in less than four minutes but never exceeded a comparably low workload of 13.3%. One aspect negatively influencing the CPU utilization are input and output (I/O)

operations, causing processes to enter a waiting state. Here, most critical are I/O operations caused by swapping when system memory is insufficient. The memory usage of our assemblies was very different but system memory was always sufficient and no assembly process was waiting because of swapping (**Figure S9**). However, some assemblers (e.g., MIRA, SPADES and CELERA) utilize or demand considerably more memory than others, which should be considered before an assembly attempt in order to circumvent swapping. One possible solution, besides the extension of the system memory, is to make use of specific memory constraining parameters that memory intensive assemblers usually offer. With regard to the average system utilization, we want to clarify that the assemblers are different in terms of the implemented parallelization and the internal assembly pipeline.

Thus, a low CPU utilization does not mean that the entire assembly is inadequately parallelized, but that, for instance, parts of the assembly are constrained to system I/O (e.g., because of data pre- and post-processing), resulting in wait times of the depending processes. Therefore, this metric should not be considered as a fixed upper limit of the parallelization capabilities of each assembler. Instead, it reflects to what extent an assembler scales, in terms of run time, with the provided computing power, and helps to find a suitable combination between assemblers and the available hardware infrastructure.

One central aspect of this study was to compare the assemblies from a practical point of view. By that, we omitted additional data processing steps like, for instance, error correction methods. However, to respond to the question whether upstream error correction is reasonable and which assembler would profit from such methods we have exemplary used two different read correction methods on our data sets (BayesHammer from SPAdes [23] for MiSeq data and Coral version 1.4 [36] for PGM data) and compared the consecutive assemblies with those using uncorrected data (**Figure S10**). In essence, the observed differences are marginal. Even for those assemblers that do not include an error correcting pre-processing step (e.g., SOAP2, VELVET and NEWBLER) the beneficial effects are, albeit partially present, not altering their individual performance.

In this evaluation, we did not perform extensive parameter optimization, i.e. whenever possible we used default parameters. For mandatory or data specific parameters without default values we have chosen appropriate values (details given in **Text S1**). The only exception to this was the k-mer parameter optimization. We have shown that an optimal k-mer highly depends on both the assembler as well as the given data set. This implies that unless the assembler itself offers a suitable k-mer parameter estimation (as e.g., CLC) or default values (as e.g., SPADES), currently the best solution is to pursue a trial-and-error approach. Of course, this problem is not described for the first time and algorithms have been developed to predict optimal k-mer parameters *a priori*. KmerGenie, for instance, uses a heuristic to generate k-mer abundance histograms in order to estimate the best possible k-mer value [37]. However, for our data sets the optimal k-mer as predicted by KmerGenie did not matched the k-mer length at which the best assembly was achieved (data not shown). This may be caused by parameter estimations inferred only from the input data, which cannot take into account assembler peculiarities. Still, k-mer estimation may prove useful for other data sets and, especially due to quick heuristics, it is to be preferred over randomly chosen or alleged established k-mer lengths. Parameters proven successful in the past may not be adequate for new assembly problems. The downsides of the trial-and-error approach, in turn, are drastically increased running times countering the speed advantage of DBG assemblers. Besides the k-mer parameter, testing and comparing various parameter settings for each assembler is possible beyond the scope of every evaluation effort simply due to the high computational cost. Moreover, results of comparisons in which assemblies were highly optimized are hard to transfer to other application scenarios, weakening the overall conclusions of such a comparison. Therefore, good default parameters and standardized recipes are needed to support unbiased and useful comparisons. Every evaluation is affected by the applied evaluation procedure and the used metrics. Therefore, the results of this study cannot be interpreted as final principles to rule in or out individual assemblers, but to give a general advice which assemblers to shortlist in consideration of an upcoming bacterial genome assembly. Nevertheless, given the scope of our evaluation we want to highlight some promising combinations

between assemblers and bacterial BS data. The most obvious recommendation is to pair different BS platforms with specific assembler approaches, i.e. PGM data is better combined with OLC assemblers whereas MiSeq data shows stronger preferences towards DBG assemblers. The only exception here is the CELERA assembler, which consistently performs better on MiSeq than on PGM data. For MiSeq data, promising DBG assemblers to begin with are SPADES and CLC. Both assemblers offer good performing default k-mer parameters, are generally easy to execute and show one of the highest NGA50 and lowest mis-assembly rates among the DBG assemblers. In addition, SPADES also performed best on PGM data among DBG assemblers. The CLC assembler shows the least memory footprint and the notably fastest running time of all assemblers, which qualifies it for quick and reasonable draft assemblies. Given MiSeq data and a highly parallelized computing infrastructure, the ABYSS assembler might also be an option, as it generates good assembly results while offering the best CPU utilization. For PGM data, researchers should consider to sub-sample their data to a coverage between 40 and 80-fold prior to an assembly in order to prevent negative oversampling effects. The two OLC assemblers that are a good entry point for PGM data sets are MIRA and NEWBLER. MIRA is able to achieve very high NGA50 values particular at higher coverages but comes at the price of more mis-assemblies and a longer running time. NEWBLER, in contrast, convinces with the shortest running time of all OLC assemblers and a low rate of mis-assemblies. Finally, it is to be noted that there is still a great discrepancy between researchers who are developing or evaluating genome assemblers and researchers who simply want to use them. The former ones take great care to avoid exaggerated generalizations, like to conclude that a particular assembler provides the best assemblies. The latter ones, on the other hand, are confronted with concrete application scenarios and therefore require decision-making support without needing to perform extensive evaluations personally. Therefore, evaluations should consider realistic assembly scenarios and include assembly metrics that summarize several assembly features into easy to communicate metrics.

Methods

Library preparation and sequencing

Growing and DNA extraction of the entero-haemorrhagic *Escherichia coli* (EHEC) O157:H7 Sakai Japanese 1996 outbreak strain was done as previously described [38]. For *Staphylococcus aureus* COL, an early methicillin-resistant strain originally isolated in a hospital in Colindale (United Kingdom) [39], growing conditions were the same as for *E. coli Sakai*. For effective cell lysis of *S. aureus* an additional step to the before mentioned DNA isolation protocol was essential. It was performed with Lysostaphin (Sigma-Aldrich, Taufkirchen, Germany, final concentration100 µg/ml), for 30 min at 37°C. Isolated DNA from *Mycobacterium tuberculosis H37Rv* was kindly provided by the group of Stefan Niemann (Forschungszentrum Borstel, Borstel, Germany). Growing and isolation of *M. tuberculosis* high molecular genomic DNA was performed as described earlier [40].

As described previously, sequencing of the *S. aureus* and *M. tuberculosis* strains was carried out for both the MiSeq 2×150 bp and 2×250 bp sequencing runs, respectively [2]. A minor modification was applied to the *M. tuberculosis* library, which was pooled after gel-extraction with other lane samples according to their molarity such that a 15 percent ratio was reached. Sequencing on the Ion Torrent PGM was also done as described before for the *E. coli* 200 bp and 400 bp sequencing libraries [2]. The two PGM *E. coli* 200 bp and 400 bp data sets were re-analyzed using the

Torrent Sequencing Software (TSS) v3.0 and v3.4, respectively. PGM sequencing of *S. aureus* COL was performed in the same manner as it was described for *E. coli* Sakai with one minor modification, i.e. for the 400 bp sequencing run the TSS version was v3.2. Sequencing of *M. tuberculosis* was performed with the Ion PGM Template OT2 Kit (Life Technologies, Darmstadt, Germany) and the Ion PGM Sequencing 400 kit (Life Technologies) according to the manufacturer's instructions. Library preparation and quality controls were performed just as described for the *E. coli* Sakai 400 bp library. The TSS version was v3.6 for *M. tuberculosis* sequencing. Independent of the applied TSS version, the software parameters and quality filter criteria remained the same, i.e. the more stringent filter was enabled and the base recalibration was disabled.

De novo genome assemblies

According to the findings on the genome coverage analysis, all data sets were sub-sampled prior to the *de novo* genome assembly. For the PGM 200 bp data a 40-fold coverage threshold was chosen, whereas PGM 400 bp and all MiSeq data were sub-sampled to 75-fold coverage. For each MiSeq data set insert size distribution was determined by mapping the PE reads to their corresponding reference using the *aln* and *sampe* module of BWA v0.5.10 [41]. Thereafter, the low and high boundaries of the insert sizes, as well as mean and standard deviation were calculated using the same method as implemented in BWA (**Table S2**). MiSeq FASTQ files were additionally edited to cope with special file format prerequisites of individual assemblers having different standards for the pairing information of PE reads.

All assemblies were independently computed on ten identical compute hosts, i.e. A+ Server 2024G-TRF (Super Micro Computer Inc., San Jones, CA, USA) equipped with four AMD Opteron 6168 processors, each with 1.90 GHz and 12 cores and in total 128 GB system memory. These machines were chosen as they provide reasonable computing power and a high degree of parallelization for asset costs below 10,000 Euros. Custom shell scripts were used to submit assembly bulk jobs to a Sun Grid Engine instance and the scheduler was utilized to consecutively distribute individual jobs among the ten compute hosts. Custom wrapper scripts for each assembler were used to prepare and initiate the actual assembly. To prevent network file system influences on the benchmarking results all necessary input data was copied to local storage prior to the assemblies. To guarantee that per compute host only one individual assembly job was executed at a time, each job was configured in such a way that it consumed the entire system resources regardless of the actual workload.

For recording run time statistics, each plain assembly process was wrapped once again using the '/usr/bin/time -v ' system call. Additionally, at an interval of five seconds a 'ps -o "%cpu = %mem = cputime = etime = nlwp = size = vsize = sz = rss = lstart = psr = comm =' system call was invoked to monitor detailed processor and memory utilization of all assembly threads while the programs were executed. Details about used assembler versions are given in **Table 2**. After the assemblies were finished, QUAST v2.1 [31] was executed on the resulting contig or scaffold FASTA files with the minimum contig length filter '–min-contig' set to 200 and the parameters '–gage –genes –gene-finding ' enabled. QUAST utilizes MUMmer [42] to align contigs to a given reference and infer various assembly metrics. Regarding the measurement of the assembly accuracy, our evaluation effort follows in essence the GAGE-B evaluation [15]. In the following, the metrics used for our main comparison are explained briefly:

- NGA50 describes the length of the last contig that is taken from all assembled contigs sorted in descending order by contig length in such a way that the summarized length of this and all previously selected contigs have at least 50% of the size of the corresponding genome. Contigs were split at all mis-assembly positions prior to this calculation.

- The number of mis-assemblies, i.e. the combined number of relocations, translocations and inversions independent of the affected genomic area.

- Wall clock time refers to the time a plain assembly process takes in total from the very beginning to the very end. This measurement does not include any pre- or post-processing procedure and is not corrected for the system load, degree of parallelization or any software dependent idle time.

- The average system utilization U describes the degree of parallelization of an assembly process in percent and is defined as follows:

$$U = \left(\frac{t_u + t_{sys}}{t_{wc} \times n_{cpu}} \right) \times 100$$

Where t_u refers to the accumulated user time for which all CPUs were busy executing the assembly, t_{sys} to the amount of time the CPUs spent on system calls on behalf of the assembly process, t_{wc} to the wall clock time of the entire assembly and n_{cpu} to the number of available CPUs.

An average system load of 100% corresponds to 100% work load of all 48 available CPU cores for the entire execution, illustrating maximum possible parallelization of all parts of an assembly. In contrast, a system load of 2.08% means that for the entire execution only one of the 48 available cores was occupied, which is synonymous to no parallelization. Finally, assembly and run time metrics were collected and processed using custom Perl and shell scripts. Plots were generated using the statistical software suite R v 2.9.10 [43] and the lattice R-package [44]. For an in-detail description of the complete bioinformatic analysis it is referred to the Supporting Information (**Text S1**).

Supporting Information

Figure S1 Effect of the depth of coverage on NGA50 lengths using random sub-sampling. Shown are in rows the results of randomly sub-sampled *S. aureus*, *E. coli*, and *M. tuberculosis* data sets, respectively. The coverage is referring to the average depth each genomic position is covered by the sequencing reads and not on the average depth of coverage the assemblies are actually reaching. The dotted vertical lines mark the finally used 40-fold (PGM 200 bp) and 75-fold coverage limits (PGM 400 bp, MiSeq 2×150 bp and MiSeq 2×250 bp).

Figure S2 Effect of the depth of coverage on mis-assemblies using random sub-sampling. Shown are in rows the results of randomly sub-sampled *S. aureus*, *E. coli*, and *M. tuberculosis* data sets, respectively. Mis-assembly combines local and non-local mis-assemblies. The coverage is referring to the average depth each genomic position is covered by the sequencing reads and not on the average depth of coverage the assemblies are actually reaching. The dotted vertical lines mark the finally used 40-fold (PGM 200 bp) and 75-fold coverage limits (PGM 400 bp, MiSeq 2×150 bp and MiSeq 2×250 bp).

Figure S3 Effect of the depth of coverage on assembly errors using random sub-sampling. Shown are in rows the results of randomly sub-sampled *S. aureus*, *E. coli*, and *M. tuberculosis* data sets, respectively. Assembly error is summarizing substitutions, insertions, and deletions errors. The coverage is referring to the average depth each genomic position is covered by the sequencing reads and not on the average depth of coverage the assemblies are actually reaching. The dotted vertical lines mark the finally used 40-fold (PGM 200 bp) and 75-fold coverage limits (PGM 400 bp, MiSeq 2×150 bp and MiSeq 2×250 bp).

Figure S4 Effect of the depth of coverage on NGA50 lengths using progressive sub-sampling. Shown are in rows the results of progressively sub-sampled *E. coli* and *S. aureus* data sets, respectively. The coverage is referring to the average depth each genomic position is covered by the sequencing reads and not on the average depth of coverage the assemblies are actually reaching. The dotted vertical lines mark the finally used 40-fold (PGM 200 bp) and 75-fold coverage limits (PGM 400 bp, MiSeq 2×150 bp and MiSeq 2×250 bp).

Figure S5 Effect of the depth of coverage on mis-assemblies using progressive sub-sampling. Shown are in rows the results of progressively sub-sampled *E. coli* and *S. aureus* data sets, respectively. Mis-assembly combines local and non-local mis-assemblies. The coverage is referring to the average depth each genomic position is covered by the sequencing reads and not on the average depth of coverage the assemblies are actually reaching. The dotted vertical lines mark the finally used 40-fold (PGM 200 bp) and 75-fold coverage limits (PGM 400 bp, MiSeq 2×150 bp and MiSeq 2×250 bp).

Figure S6 Effect of the k-mer size parameter on the NGA50 length. Shown are values for the three de Bruijn Graph assembler ABYSS (**A**, **D**), SOAP2 (**B**, **E**), and VELVET (**C**, **F**). On the left side (**A**, **B**, **C**) using MiSeq 2×150 bp (dotted lines) and MiSeq 2×250 bp (solid lines); on the right side (**D**, **E**, **F**) using PGM 200 bp (dotted lines) and PGM 400 bp (solid lines) data sets of the *E. coli* (red), *M. tuberculosis* (blue), and *S. aureus* (green) genomes, respectively. For each line, the highest reached NGA50 length is indicated by a vertical arrow and the corresponding x- and y-values are given in the upper left legend.

Figure S7 Effect of the k-mer size parameter on the NGA50 length of the SPADES assembler. The assemblies were generated in two ways: using an increasing set of k-mer parameters where for each assembly process the NGA50 length of the last and final k-mer cycle is drawn (**A**); using the default set of k-mer parameters where the NGA50 length of all intermediate and the final k-mer cycle is drawn (**B**). MiSeq 2×150 bp (dark-red), MiSeq 2×250 bp (red), PGM 200 bp (green), and PGM 400 bp (dark-green) data sets of the *E. coli* (solid lines), *M. tuberculosis* (dot-dashed lines), and *S. aureus* (dashed lines) genomes are used, respectively. For each line, the highest reached NGA50 length is indicated by a vertical arrow and the corresponding x- and y-values are given in the upper left legend.

Figure S8 Gene coverage and assembly error rates of *de novo* genome assemblies. Based on the percentage of full covered genes (**A**) and the number of assembly errors (**B**, combining substitutions, insertions, and deletions). Full covered genes are completely covered positions in the reference genome where a gene annotation was provided (based on all chromosomal and plasmid genes). The numbers of assembly errors are either contig or scaffold based, respectively. Scaffolds for MiSeq 2×150 bp and MiSeq 2×250 bp assemblies obtained by ABYSS, CELERA, CLC, NEWBLER, SOAP2, SPADES, and VELVET; contigs for MiSeq assemblies obtained by MIRA and SEQMAN as well as for all PGM assemblies.

Figure S9 Memory usage of *de novo* genome assemblies. Shown is the maximum non-swapped physical memory, i.e. the peak resident size that an assembly process has used over the entire time. For assemblies running several processes or threads in parallel this value is calculated from the maximum summation of all concurrent processes at a specific time point. For the DBG assemblers ABYSS, SOAP2, and VELVET only the peak resident size of the best resulting k-mer parameter are shown and not the summation of all assemblies using different k-mer parameters.

Figure S10 Effect of upstream error correction on *de novo* genome assemblies. Compared are NGA50 lengths (in kilobase pairs) of assemblies without an upstream read based error correction (left side) with those based on error corrected reads (using BayesHammer on MiSeq data and Coral on PGM data; right side). The NGA50 length is either contig or scaffold based, respectively. Scaffolds for MiSeq 2×150 bp and MiSeq 2×250 bp assemblies obtained by ABYSS, CELERA, CLC, NEWBLER, SOAP2, SPADES, and VELVET; contigs for MiSeq assemblies obtained by MIRA and all PGM assemblies.

Table S1 Determined optimal k-mer sizes for DBG assemblies with mandatory k-mer parameterization.

Table S2 Calculated insert-sizes of MiSeq sequencing libraries.

Text S1 Supporting Information to GABenchToB: A Genome Assembly Benchmark Tuned on Bacteria and Benchtop Sequencers.

Acknowledgments

The authors gratefully thank the system administrators of the Bioinformatics Resource Facility at the CeBiTec for general technical support and in particular for providing the dedicated computing infrastructure. Furthermore, the authors thank DNASTAR Inc. (Madison, Wisconsin, USA), CLC bio A/S (Katrinebjerg, Aarhus, Denmark) and 454 Life Sciences (Branford, CT, USA) for providing extended trail or free licenses of their assembly software products.

Author Contributions

Conceived and designed the experiments: SJ DH. Performed the experiments: SJ KP AA. Analyzed the data: SJ SA AG JS. Contributed reagents/materials/analysis tools: SJ JK DH. Wrote the paper: SJ KP DH.

References

1. Loman NJ, Misra RV, Dallman TJ, Constantinidou C, Gharbia SE, et al. (2012) Performance comparison of benchtop high-throughput sequencing platforms. Nat Biotechnol 30: 434–439.
2. Jünemann S, Sedlazeck FJ, Prior K, Albersmeier A, John U, et al. (2013) Updating benchtop sequencing performance comparison. Nat Biotechnol 31: 294–296.
3. Compeau PE, Pevzner PA, Tesler G (2011) How to apply de Bruijn graphs to genome assembly. Nature biotechnology 29: 987–991.
4. Li Z, Chen Y, Mu D, Yuan J, Shi Y, et al. (2012) Comparison of the two major classes of assembly algorithms: overlap-layout-consensus and de-bruijn-graph. Brief Funct Genomics 11: 25–37.
5. Nagarajan N, Pop M (2013) Sequence assembly demystified. Nat Rev Genet 14: 157–167.
6. Scheibye-Alsing K, Hoffmann S, Frankel A, Jensen P, Stadler PF, et al. (2009) Sequence assembly Computational biology and chemistry 33: 121–136.
7. Paszkiewicz K, Studholme DJ (2010) De novo assembly of short sequence reads. Brief Bioinform 11: 457–472.
8. Zhang W, Chen J, Yang Y, Tang Y, Shang J, et al. (2011) A practical comparison of de novo genome assembly software tools for next-generation sequencing technologies. PLoS One 6: e17915.
9. Haiminen N, Kuhn DN, Parida L, Rigoutsos I (2011) Evaluation of methods for de novo genome assembly from high-throughput sequencing reveals dependencies that affect the quality of the results. PLoS One 6: e24182.
10. Lin Y, Li J, Shen H, Zhang L, Papasian CJ, et al. (2011) Comparative studies of de novo assembly tools for next-generation sequencing technologies. Bioinformatics 27: 2031–2037.
11. Narzisi G, Mishra B (2011) Comparing de novo genome assembly: the long and short of it. PLoS One 6: e19175.
12. Earl D, Bradnam K, St John J, Darling A, Lin D, et al. (2011) Assemblathon 1: a competitive assessment of de novo short read assembly methods. Genome Res 21: 2224–2241.
13. Bradnam KR, Fass JN, Alexandrov A, Baranay P, Bechner M, et al. (2013) Assemblathon 2: evaluating de novo methods of genome assembly in three vertebrate species. Gigascience 2: 10.
14. Salzberg SL, Phillippy AM, Zimin A, Puiu D, Magoc T, et al. (2012) GAGE: A critical evaluation of genome assemblies and assembly algorithms. Genome Res 22: 557–567.
15. Magoc T, Pabinger S, Canzar S, Liu X, Su Q, et al. (2013) GAGE-B: an evaluation of genome assemblers for bacterial organisms. Bioinformatics 29: 1718–1725.
16. Simpson JT, Wong K, Jackman SD, Schein JE, Jones SJM, et al. (2009) ABySS: a parallel assembler for short read sequence data. Genome Res 19: 1117–1123.
17. Miller JR, Delcher AL, Koren S, Venter E, Walenz BP, et al. (2008) Aggressive assembly of pyrosequencing reads with mates. Bioinformatics 24: 2818–2824.
18. CLC bio (2014) CLC Assembly Cell. Available: http://clcbio.com. Accessed 23 April 2014.
19. Life Sciences (2014) GS De Novo Assembler. Available: http://www.454.com/products/analysis-software/. Accessed 2014 Apr 23.
20. Chevreux B, Pfisterer T, Drescher B, Driesel AJ, Müller WEG, et al. (2004) Using the miraEST assembler for reliable and automated mRNA transcript assembly and SNP detection in sequenced ESTs. Genome Res 14: 1147–1159.
21. DNASTAR (2014) SeqMan Ngen. Available: http://www.dnastar.com/t-nextgen-seqman-ngen.aspx. Accessed 2014 Apr 23.
22. Luo R, Liu B, Xie Y, Li Z, Huang W, et al. (2012) SOAPdenovo2: an empirically improved memory-efficient short-read de novo assembler. Gigascience 1: 18.
23. Bankevich A, Nurk S, Antipov D, Gurevich AA, Dvorkin M, et al. (2012) SPAdes: a new genome assembly algorithm and its applications to single-cell sequencing. J Comput Biol 19: 455–477.
24. Zerbino DR, Birney E (2008) Velvet: algorithms for de novo short read assembly using de Bruijn graphs. Genome Res 18: 821–829.
25. Gnerre S, Maccallum I, Przybylski D, Ribeiro FJ, Burton JN, et al. (2011) High-quality draft assemblies of mammalian genomes from massively parallel sequence data. Proc Natl Acad Sci U S A 108: 1513–1518.
26. Batzoglou S, Jaffe DB, Stanley K, Butler J, Gnerre S, et al. (2002) ARACHNE: a whole-genome shotgun assembler. Genome Res 12: 177–189.
27. Zimin AV, Marçais G, Puiu D, Roberts M, Salzberg SL, et al. (2013) The MaSuRCA genome assembler. Bioinformatics 29: 2669–2677.
28. Simpson JT, Durbin R (2012) Efficient de novo assembly of large genomes using compressed data structures. Genome Res 22: 549–556.
29. Mullikin JC, Ning Z (2003) The phusion assembler. Genome Res 13: 81–90.
30. Ghodsi M, Hill CM, Astrovskaya I, Lin H, Sommer DD, et al. (2013) De novo likelihood-based measures for comparing genome assemblies. BMC Res Notes 6: 334.
31. Gurevich A, Saveliev V, Vyahhi N, Tesler G (2013) QUAST: quality assessment tool for genome assemblies. Bioinformatics 29: 1072–1075.
32. Hubisz MJ, Lin MF, Kellis M, Siepel A (2011) Error and error mitigation in low-coverage genome assemblies. PLoS One 6: e17034.
33. Illumina (2009) De novo assembly using Illumina reads. Available: http://www.illumina.com/Documents/products/technotes/technote_denovo_assembly_ecoli.pdf. Accessed 2014 Apr 23.
34. Arezi B, Xing W, Sorge JA, Hogrefe HH (2003) Amplification efficiency of thermostable DNA polymerases. Anal Biochem 321: 226–235.
35. Chen Y-C, Liu T, Yu C-H, Chiang T-Y, Hwang C-C (2013) Effects of GC bias in next-generation-sequencing data on de novo genome assembly. PLoS One 8: e62856.
36. Salmela L, Schröder J (2011) Correcting errors in short reads by multiple alignments. Bioinformatics 27: 1455–1461
37. Chikhi R, Medvedev P (2014) Informed and automated k-mer size selection for genome assembly. Bioinformatics 30: 31–37.
38. Mellmann A, Harmsen D, Cummings CA, Zentz EB, Leopold SR, et al. (2011) Prospective genomic characterization of the German enterohemorrhagic Escherichia coli O104:H4 outbreak by rapid next generation sequencing technology. PLoS One 6: e22751.
39. Jevons PM (1961) "Celbenin" - resistant Staphylococci. Br Med J 1: 124–125.
40. van Embden JD, Cave MD, Crawford JT, Dale JW, Eisenach KD, et al. (1993) Strain identification of Mycobacterium tuberculosis by DNA fingerprinting: recommendations for a standardized methodology. J Clin Microbiol 31: 406–409.
41. Li H, Durbin R (2010) Fast and accurate long-read alignment with Burrows-Wheeler transform. Bioinformatics 26: 589–595.
42. Kurtz S, Phillippy A, Delcher AL, Smoot M, Shumway M, et al. (2004) Versatile and open software for comparing large genomes. Genome Biol 5: R12.
43. Team RDC (2011) R: a language and environment for statistical computing. Available: http://www.R-project.org. Accessed 23 April 2014.
44. Sarkar D (2008) Lattice: multivariate data visualization with R. New York: Springer.

## 7

Production of Outer Membrane Vesicles by the Plague Pathogen *Yersinia pestis*

Justin L. Eddy[9], Lindsay M. Gielda[9], Adam J. Caulfield, Stephanie M. Rangel, Wyndham W. Lathem*

Department of Microbiology-Immunology, Northwestern University Feinberg School of Medicine, Chicago, Illinois, United States of America

Abstract

Many Gram-negative bacteria produce outer membrane vesicles (OMVs) during cell growth and division, and some bacterial pathogens deliver virulence factors to the host via the release of OMVs during infection. Here we show that *Yersinia pestis*, the causative agent of the disease plague, produces and releases native OMVs under physiological conditions. These OMVs, approximately 100 nm in diameter, contain multiple virulence-associated outer membrane proteins including the adhesin Ail, the F1 outer fimbrial antigen, and the protease Pla. We found that OMVs released by *Y. pestis* contain catalytically active Pla that is competent for plasminogen activation and α2-antiplasmin degradation. The abundance of OMV-associated proteins released by *Y. pestis* is significantly elevated at 37°C compared to 26°C and is increased in response to membrane stress and mutations in RseA, Hfq, and the major Braun lipoprotein (Lpp). In addition, we show that *Y. pestis* OMVs are able to bind to components of the extracellular matrix such as fibronectin and laminin. These data suggest that *Y. pestis* may produce OMVs during mammalian infection and we propose that dispersal of Pla via OMV release may influence the outcome of infection through interactions with Pla substrates such as plasminogen and Fas ligand.

Editor: Lisa A. Morici, Tulane University School of Medicine, United States of America

Funding: National Institutes of Health P30 CA060553 to RHLCCC; National Institutes of Health R01 AI093727 to WWL. The funders had no role in study design, data collection and analysis, decision to publish, or preparation of the manuscript.

Competing Interests: The authors have declared that no competing interests exist.

* Email: lathem@northwestern.edu

[9] These authors contributed equally to this work.

Introduction

Outer membrane vesicles (OMVs) are closed spherical portions of the bacterial outer membrane that contain phospholipids, outer membrane proteins, lipopolysaccharide (LPS), and periplasmic contents [1]. Produced by many Gram-negative bacteria such as *Escherichia coli*, *Pseudomonas aeruginosa*, and *Helicobacter pylori* [2–4], OMVs are formed when small portions of the outer membrane pinch off from the cell and are released as self-contained spherical structures that range from 20–250 nm in size [5]. While the biogenesis of OMVs is poorly understood, it is thought that expansion of the outer leaflet of the membrane relative to the inner leaflet induces membrane curvature that forces the outer membrane to bud away from the cell [5,6]. OMV production can be detected in bacterial communities growing under a variety of conditions, including planktonic cultures as well as in surface-attached biofilm communities [7,8].

OMVs are produced by both pathogenic and non-pathogenic bacteria [9–11]. OMVs released by pathogens can contain multiple components that interact with the host, including LPS, virulence factors, and other antigens. Pathogen-derived OMVs may contribute to virulence by modulating the innate immune response, delivering toxins to cells, dispersing antigens and virulence factors away from the bacterium, trafficking signaling molecules between bacteria, and more. Microscopic examination of tissues has detected the presence of OMVs near host cells or within host tissues, suggesting an interaction between OMVs and the host during infection [12–14]. Further, OMVs have been found to deliver active toxins to host cells, including the enterotoxigenic *E. coli* heat-labile enterotoxin (LT), the enterohemorrhagic *E.coli* pore-forming cytotoxin ClyA, and the *H. pylori* VacA protein [3,11,15]. Environmental stresses contribute to the production of OMVs [16], suggesting that, as bacteria encounter stressors such as those found within the infected host, the production of OMVs may not only manipulate interactions with the host but also aid in the survival of the bacterium.

The Gram-negative bacterium *Yersinia pestis*, a pathogen of both insects and mammals, can be transmitted to humans via the bite of hematophagous insects (typically fleas) or through the inhalation of respiratory droplets or aerosols containing the bacteria, and can cause bubonic, pneumonic, or septicemic plague [17]. Temperature is a major regulator of gene expression in *Y. pestis*, controlling both transcriptional and post-transcriptional responses [18,19]. At lower temperatures (<25°C), *Y. pestis* produces factors that maximize survival and colonization in the flea, such as biofilms [20], while at higher temperatures (>30°C), the bacterium expresses genes required for mammalian infection, including the adhesin Ail, the F1 fimbrial antigen (Caf1), the outer membrane protease Pla, and the Yop-Ysc type III secretion system (T3SS) [21–24]. Thus, *Y. pestis* possesses a variety of virulence factors, including a number of outer membrane-associated factors, which are necessary for interacting with its hosts to ultimately cause disease.

Among these, the Pla protease is necessary for the progression of both bubonic and pneumonic plague, but is dispensable during septicemic plague [21,25,26]. Pla is known to cleave a number of mammalian host proteins, including the zymogen plasminogen (plg), the plasmin inhibitor α2-antiplasmin, and the recently identified substrate Fas ligand (FasL), a major inducer of host cell death via apoptosis [27–31]. In addition, Pla has also been shown *in vitro* to act as an adhesin to extracellular matrices by binding laminin as well as promoting the bacterial invasion of HeLa cells [24,32,33].

As Pla is an insoluble outer membrane protein dependent on rough LPS for its protease activity, it is not thought to be secreted by *Y. pestis* [34–36]. However, we have detected active Pla in cell-free culture supernatants, suggesting that this cell-free form of Pla could be contained on OMVs. Here we investigate the ability of *Y. pestis* to produce native OMVs, characterize the presence and activities of various virulence factors carried on released OMVs, and propose a role for these OMVs during mammalian infection.

Results

Outer membrane protein activity in cell-free culture supernatants

Our laboratory has detected the activity of the outer membrane protein Pla in cell-free culture supernatants during the exponential growth phase of *Y. pestis*. To explore this further, 0.2 μm-filtered, cell-free culture supernatants from either wild-type *Y. pestis* or an isogenic mutant of *Y. pestis* lacking Pla (*Y. pestis* Δ*pla*) were grown in the rich media brain-heart infusion (BHI) at 37°C and tested for the ability to convert plg to the active plasmin form, an activity dependent on Pla. We found that filtered culture supernatants from wild-type but not *Y. pestis* Δ*pla* contained measurable levels of Pla activity (**Fig. 1**). This activity was lost when these 0.2 μm-filtered culture supernatants were further passed through a filter with a 100 kDa cutoff (**Fig. 1**). As the molecular weight of Pla is 37 kDa and Pla is not predicted to form multimers, these data suggest that the form of Pla found in cell-free culture supernatants may be contained on bacterial superstructures greater than 100 kDa [37]. While this could represent cellular lysis, the observation of OMV formation by other Gram-negative bacteria prompted the consideration of OMV production by *Y. pestis*.

Y. pestis produces OMVs

OMV-like structures have been previously observed on the surface of *Y. pestis* bacteria [38]. To examine whether *Y. pestis* produces OMVs under laboratory conditions, bacteria were cultured in BHI at 37°C and at various times during growth, aliquots of bacteria were removed, fixed, and examined via both scanning and transmission electron microscopy. Micrographs revealed round, vesicle-like structures attached to or affiliated with the surface of *Y. pestis* bacilli (**Fig. 2A & 2B**). While these structures could be artifacts of the fixation procedure, they are similar to those observed on the surfaces of other bacterial species, suggesting the formation of OMVs [39].

To determine whether these structures are truly natural products of *Y. pestis* and share characteristics with OMVs produced by other bacteria, we purified potential native OMVs released by *Y. pestis* using standard, established vesicle isolation techniques that do not require sonication, shearing, or chemical treatments to induce vesicle production [5,40]. Briefly, 0.2 μm-filtered, mid-log growth-phase culture supernatants were concentrated and ultracentrifuged to isolate outer membranes. To purify vesicles from cellular debris, the isolated material was subjected to Optiprep-based gradient ultracentrifugation, resulting in a sepa-

ration of contaminating cellular proteins and the OMVs based on lipid content into multiple independent fractions [3]. We analyzed the fractions by transmission electron microscopy and found characteristic OMV-like spherical structures (**Fig. 2C**), similar to vesicles isolated from other Gram-negative bacteria [1,8,41]. Together, these SEM and TEM images indicate that *Y. pestis* releases material under standard laboratory conditions that is consistent with that of bacterial OMVs. We determined the average size of these isolated OMVs to be 93.07+/−11.75 nm in diameter (**Fig. 2D**).

Characterization of proteins associated with *Y. pestis* OMVs

OMVs are known to carry a wide array of proteins associated with the outer membrane, periplasm, and cytoplasm of Gram-negative bacteria. Therefore, we examined if *Y. pestis* OMVs are enriched for protein subsets compared to whole bacteria and if specific outer membrane virulence factors are present on these OMVs. To minimize contamination from cellular lysis, OMVs were isolated from mid-log phase cultures without the use of sonication or chemical treatment in order to purify naturally occurring OMVs. First, we analyzed by reducing SDS-PAGE fractions 2–7 of the Optiprep gradient used to purify OMVs. We found that fractions from the Optiprep gradient contained proteins that were either enriched or reduced in abundance compared to *Y. pestis* whole cell lysates (WCL) (**Fig. 3A**). In *Y. pestis*, a number of proteins contained within or associated with the outer membrane are virulence determinants, and many of these are produced at 37°C, including Ail, Pla, and Caf1 [42]. Therefore, to determine if OMVs produced by *Y. pestis* at 37°C contain these specific outer membrane-associated proteins, immunoblot analyses were performed with antibodies to Ail, Caf1, and Pla. To determine the enrichment of these proteins compared to the cytoplasmic fraction, we also examined OMVs for the presence of Hfq, a cytoplasmic protein that serves as a chaperone for small RNAs, and RpoA, the alpha subunit of RNA polymerase that is also found in the cytoplasm. We consistently found that the outer membrane-associated proteins Ail and Caf1 were present in gradient fractions 4–6, and that these same fractions contained minimal RpoA and Hfq (**Fig. 3B**, left panels). On the other hand, we were unable to detect Pla in the individual pure OMV fractions; therefore, to increase protein abundance we combined and concentrated fractions 4–6 and repeated the same immunoblot analysis. Using this approach we could detect the presence of Pla on the OMVs (**Fig. 3B**, right panels). We also isolated OMVs from the Δ*pla* strain of *Y. pestis*, as Pla is known to cleave *Y. pestis* proteins and thus could alter the composition of the OMVs themselves [37,43]. OMVs from *Y. pestis* Δ*pla* contain the outer membrane proteins Ail and Caf1 and lack Pla as well as the cytoplasmic proteins RpoA and Hfq (**Fig. 3B**, left panels). This indicates that the loss of Pla does not impact the presence or absence of these other *Y. pestis* proteins contained on OMVs. To demonstrate the presence of Caf1 on the surface of OMVs, OMVs were immuno-labeled with antibodies to Caf1 using 6 nm-sized gold beads. Transmission electron micrographs of OMVs labeled with anti-Caf1 antibody demonstrates that Caf1 protein is indeed present on the surface of isolated OMVs (**Fig. 3C**).

Proteomic analysis of *Y. pestis* OMVs

In order to more thoroughly analyze the proteins associated with *Y. pestis* OMVs, we purified OMVs from bacteria cultured at 37°C in biological triplicate and analyzed the protein content by mass spectrometry. A total of 270 unique proteins present in at least 2 of the 3 replicates were identified and the subcellular

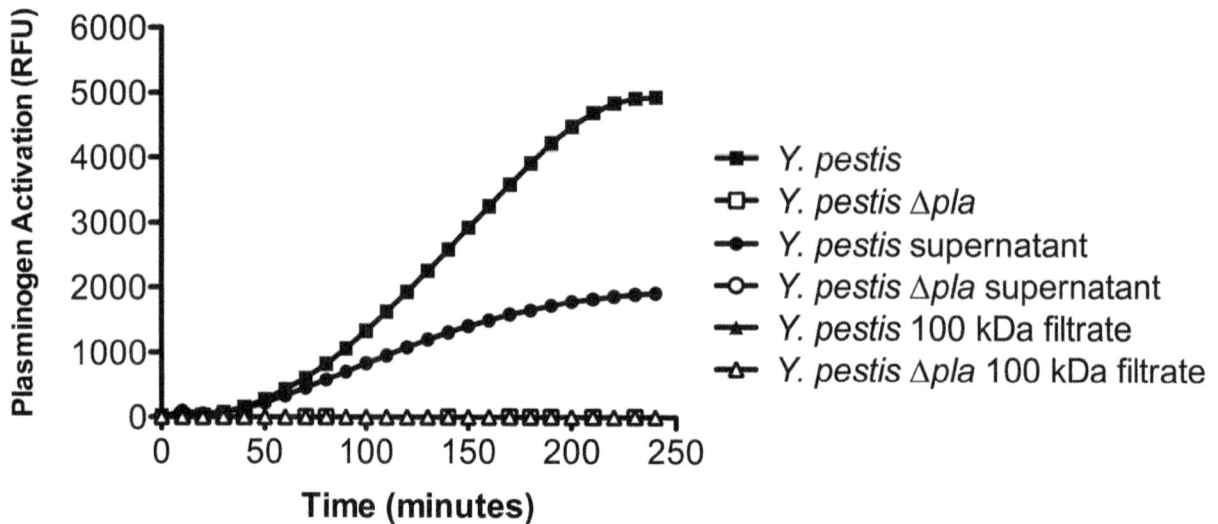

Figure 1. Y. pestis culture supernatants contain active Pla. (A) Plg-activating ability of whole bacteria, 0.2 μm-filtered culture supernatants, or the filtrate of 100 kDa-passed culture supernatants, from wild-type or Δpla Y. pestis, respectively. Materials were incubated with human glu-plg and a fluorescent substrate of plasmin for 3 h at 37°C. One experiment representative of 3 independent biological replicates is shown.

localization of each protein was predicted using the PSORTb algorithm (**Table S1**). This analysis indicated that of the 270 proteins identified, 15 (6%) are derived from the outer membrane (including Ail and Caf1), 68 (25%) are found in the periplasm, 5 (2%) are from the inner membrane, and 160 (58%) are cytoplasmic (**Fig. 3D**). Of note, we failed to detect Pla peptides by mass spectrometry, even though we are able to observe the presence of Pla and its activity by immunoblot and other assays (see below). In total, these results confirm that native OMVs produced by Y. pestis contain and display a significant number of proteins, including multiple virulence factors.

Increased production of OMVs in response to temperature and stress

Temperature is a major regulator of gene expression in Y. pestis, and OMV production by other bacteria has been observed at both low and higher temperatures [7,8,44]. With this in mind, we examined whether changes in temperature affect OMV production by Y. pestis. OMVs were isolated from Y. pestis cultured at either 26°C or 37°C and total protein content associated with the purified OMV fractions was measured. We found a significantly greater quantity of OMV-associated protein released into the culture media at 37°C compared to 26°C, suggesting that OMV production is more abundant at elevated temperatures (**Fig. 4A**). In addition, activation of bacterial stress response pathways has been shown to increase the formation of OMVs [40,45–48]. We first investigated the impact of cold shock, a well-established inducer of stress in Gram-negative bacteria [49], on OMV production by incubating cultures of Y. pestis grown at 37°C on ice for one hour. Quantification of OMV-associated proteins isolated from these cold-shocked bacteria demonstrated a significant increase in the release of OMVs compared to bacteria maintained at 37°C (**Fig. 4B**).

To test whether the loss of factors that respond to membrane stress contributes to or alters OMV production by Y. pestis, we employed deletions in the genes encoding RseA, Hfq, and the major Braun lipoprotein Lpp. The anti-sigma factor protein RseA is a negative regulator of SigmaE; deletion of rseA results in elevated activity of SigmaE, a regulator of the outer membrane

stress response [50–52]. Hfq is a chaperone for small, non-coding regulatory RNAs (sRNAs), and recent studies have shown that Hfq is necessary for resistance to multiple stressors by Yersinia species [53–55]. Lpp links the outer membrane to peptidoglycan, and the deletion of lpp disrupts membrane stability, contributing to increases in OMV formation in several bacteria [5,56,57]. Isogenic deletions of rseA, hfq, and lpp in Y. pestis results in 3.0, 2.4, and 3.6-fold increases in OMV-associated proteins present in the culture media, respectively, when compared to OMV production by wild-type bacteria at 37°C (**Fig. 4B**). In total, these results provide evidence that the production and release of OMVs by Y. pestis likely increases when undergoing both temperature and cell envelope stress in a manner similar to other Gram-negative bacteria.

Y. pestis OMVs contain active Pla

Proteins contained within bacterial OMVs often retain biological activity [5,58–60], therefore we hypothesized that OMV-bound Pla may remain catalytically active and able to cleave its substrates, such as plg. To test this, we isolated OMVs from wild-type and Δpla Y. pestis and then performed a plg-activation assay with these vesicles. Wild-type OMVs containing Pla activated plg in a dose-dependent manner, while OMVs from Y. pestis Δpla were unable to activate plg (**Fig. 5A**). We also examined the ability of purified OMVs to degrade α2-antiplasmin, another established substrate of Pla [28]. Incubation of OMVs with purified α2-antiplasmin resulted in a Pla-dependent loss of detectable α2-antiplasmin over time as determined by immunoblot analysis (**Fig. 5B**). Thus, these results demonstrate that OMV-bound Pla retains the ability to cleave known substrates of the protease.

Y. pestis OMVs adhere to the extracellular matrix

Both Ail and Pla facilitate binding of Y. pestis to components of the extracellular matrix (ECM) [61]. Since we have shown that Y. pestis OMVs contain these adhesins, we tested whether wild-type or Δpla OMVs are also able to bind to ECM components. We incubated Y. pestis-derived OMVs in 96-well plates coated with Matrigel (a 3-dimensional ECM composed of laminin, collagen

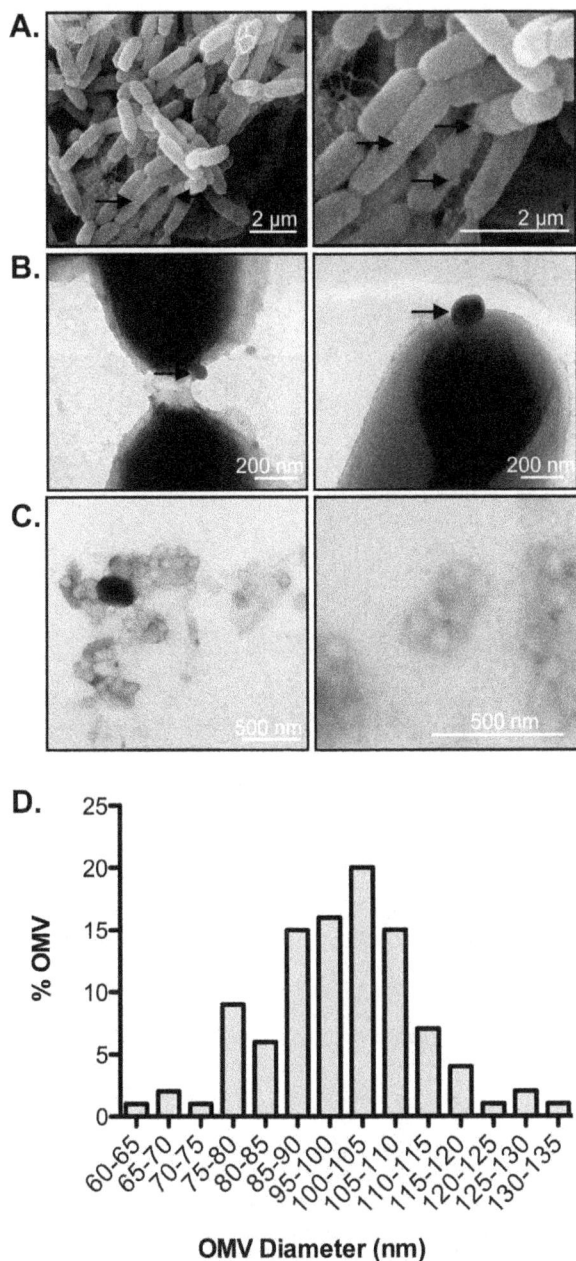

Figure 2. Y. pestis produces membrane blebs consistent with OMVs. Y. pestis bacteria cultured for 6 h at 37°C were fixed and imaged via SEM. (A and B) Images reveal round membrane protrusions on the bacterial surface (arrows) that are consistent with OMVs. (C) TEM of OMVs purified from Y. pestis supernatants. Bar represents size in nanometers. (D) Size distribution of OMVs purified from Y. pestis. One hundred OMVs were measured and diameters are shown as a percent of the total. The average OMV diameter is 93.07+/−11.75 nm. Bars represent size in nanometers as indicated.

type IV, heparan sulfate proteoglycan, and entactin) or individual components of the ECM such as fibronectin, laminin, and collagen, and assessed binding by ELISA using fluorescently labeled anti-Caf1 antibodies. We found that OMVs derived from wild-type Y. pestis were able to interact with Matrigel, fibronectin, and laminin to a significantly greater degree than to bovine serum albumin (BSA)-coated wells, demonstrating that Y. pestis OMVs

retain the ability to bind to ECM components (**Fig. 6**). Furthermore, the presence of Pla on these OMVs significantly contributes to the binding of OMVs to Matrigel and laminin, suggesting that in this context, Pla may also serve as an adhesin for OMVs to these components of the ECM (**Fig. 6**). Taken together, our data demonstrate that Pla retains both attributed biological functions (i.e adhesive and protease activities) when contained on OMVs.

Discussion

A growing body of evidence suggests that OMVs play critical roles in the physiology and life cycle of many bacteria, including the killing of competing species, transferring genetic material to other bacteria, delivering toxins and virulence factors to host cells, and modulating the immune response of the host [3,58–60]. While a large aspect of OMV research is aimed at understanding the host recognition of antigenic OMV-bound factors, particularly for vaccine development, the native activity of proteins on OMVs may also play distinct roles in pathogenesis and the modulation of host defense during bacterial infection. Indeed, this has been observed for enterotoxigenic E. coli via the OMV-mediated delivery of LT to host cells and for the OMV-mediated induction of IL-8 by H. pylori and P. aeruginosa [3,4,62].

Here we show that Y. pestis, the causative agent of plague, releases OMVs under physiological conditions. As expected based on their derivation, these OMVs carry multiple constituents of the outer membrane, although it is not yet known whether Y. pestis actively sorts specific proteins and/or modulates their abundance into OMVs during biogenesis. Our data suggest that a limited number of outer membrane proteins are associated with native OMVs produced by Y. pestis in vitro and that the F1 capsular antigen Caf1 represents the major constituent of the OMVs, as illustrated by immunoblot and a high MASCOT score determined by mass spectrometry. We hypothesize that the abundance of Caf1 might exclude other outer membrane proteins from associating with native OMVs. This is supported by our data indicating that there is less Ail per μg of protein in OMVs compared to the equivalent amount of protein derived from a Y. pestis whole cell lysate. This observation, coupled with the fact that Pla autoprocesses itself (potentially limiting the number of peptide fragments for detection), could explain why Pla protein levels were below the detectable limit in our mass spectrometry analysis, even though OMV-associated Pla activity can be measured in a variety of assays. Thus we cannot rule out that there may be additional proteins associated with OMVs that were not detected by mass spectrometry but could be identified through alternative protein isolation techniques.

Our analysis of the protein profile associated with native OMVs is consistent with a number of studies that find an abundance of both outer membrane and periplasmic proteins and an exclusion of inner membrane proteins [58,60,63–65]. Studies characterizing the outer membrane proteome of Y. pestis grown in vitro have shown that between 50 and 70 proteins are associated with the outer membrane and these are altered in a temperature shift between 26°C and 37°C [42,66]. Additionally these studies identified 31 outer membrane proteins associated with the outer membrane fraction isolated from Y. pestis and we identified 15 outer membrane proteins by mass spectrometry that are associated with OMVs when bacteria are grown at 37°C, representing approximately 20% of the total outer membrane proteome and 33% of the OM proteins previously identified associated with whole bacterial membranes. We found a number of cytoplasmic proteins associated with Y. pestis OMVs, including Elongation

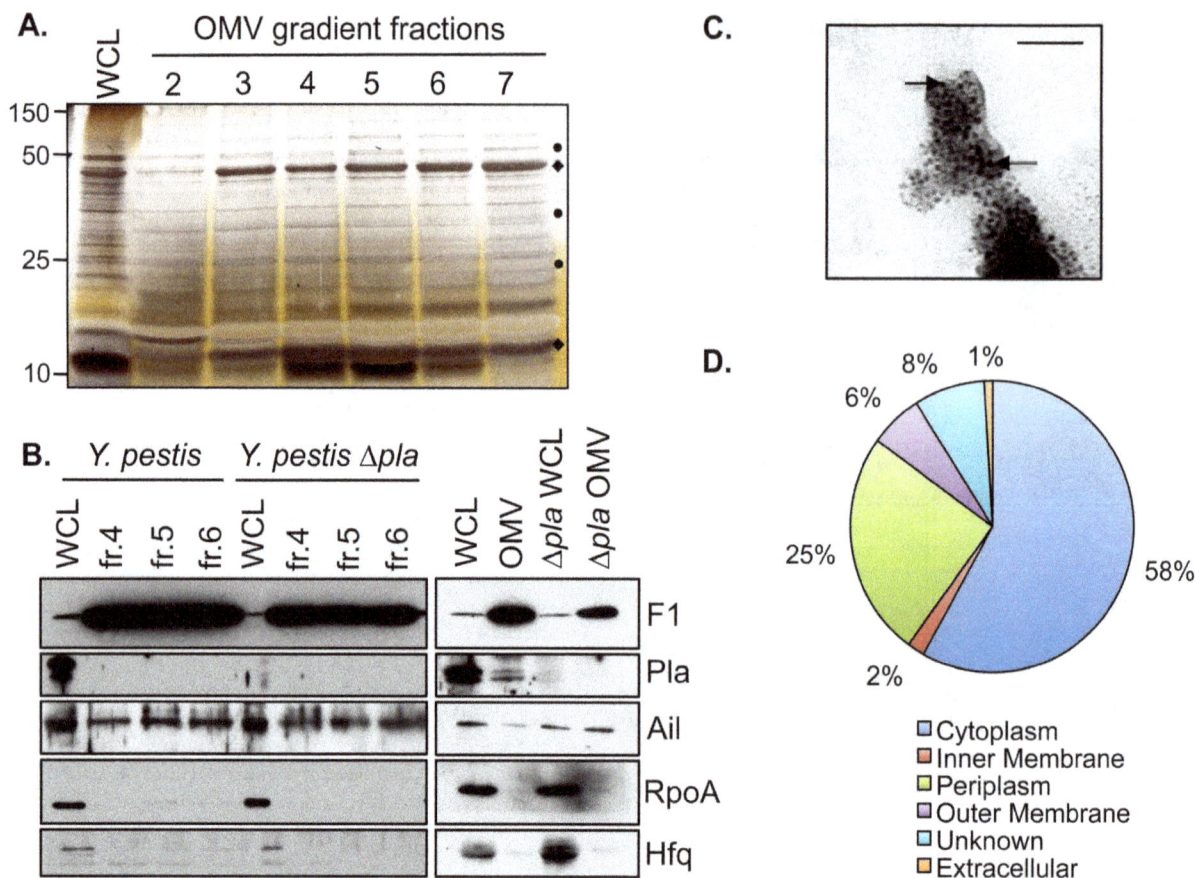

Figure 3. OMVs contain outer membrane-associated virulence factors. (A) Whole cell lysates (WCL) or density centrifugation gradient fractions from OMVs isolated from *Y. pestis* were separated by SDS-PAGE and gels were silver stained. (♦) denotes an enriched band and (●) denotes reduced bands. (B) WCL or gradient fractions (4–6) from OMVs isolated from *Y. pestis* and *Y. pestis* Δ*pla* were examined for the presence of the virulence factors Pla, Ail, and Caf1 by immunoblot. Immunoblots for RpoA and Hfq, two cytoplasmic proteins, are shown to demonstrate the absence of contaminating proteins from the OMV preparation. (C) OMVs were immuno-gold labeled with an anti-Caf1 antibody conjugated to gold beads and examined by TEM. Black arrows indicate representative gold particles. Bar represents 50 nm. (D) Subcellular distribution of proteins present in *Y. pestis* OMVs as a percentage of total proteins identified by mass spectrometry listed in (Table S3).

factor Tu, GroEL, RpsA, RplL, and DnaK, which is consistent with findings from the studies of multiple Gram-negative OMVs, including *Neisseria meningitidis* [67], *E. coli* [65], *Brucella melitensis* [60], and *Edwardsiella tarda* [63]. Some of the cytoplasmic contaminants we observed are also known virulence determinants or immune stimulants such as GroEL, Ymt, and a tellurium resistance protein. These proteins were also identified as associated with *Y. pestis* OMV-like outer membrane blebs [38]. It is unknown whether these common cytoplasmic proteins associated with Gram-negative OMVs are contaminants or represent cytoplasmic proteins either non-specifically or specifically targeted into OMVs during their biogenesis.

Based on studies of OMV production by other bacterial pathogens during infection, we hypothesize that OMV release by *Y. pestis* in the mammal could have multiple physiological consequences, such as influencing the immune response to the infection, altering host cell function, and aiding bacterial spread through the dysregulation of the host hemostatic and innate immune systems. For instance, OMV interaction with the ECM could facilitate disruption of the epithelial layer via Pla or other factors, thereby permitting development of the characteristic edema and fluid accumulation observed during pneumonic plague. Indeed, OMVs from a variety of pathogens have been

detected in the fluids of infected hosts, demonstrating their ability to spread from the site(s) of infection [68–70]. While it is not yet known if *Y. pestis* produces OMVs during infection, dispersal of OMVs may prove beneficial for the plague bacillus by delivering antigens and virulence factors, such as Pla or Ail, to sites distal to the bacterium. For instance, OMV-mediated dispersal of active Pla could expand the range of fibrin degradation near the bacteria, allowing for further bacterial spread in the tissue.

In addition, while the T3SS dampens immune cell activation around the bacteria themselves through the direct injection of T3SS Yop effectors into recruited hematopoietic cells, dispersal of OMVs could redirect the focus of polymorphonuclear cells away from the bacteria, prolonging bacterial survival. Furthermore, if OMVs interact with host cells that are not otherwise targeted by the T3SS, those host cells could themselves become activated in a manner that results in inflammation and further immune cell recruitment. We speculate that OMVs may allow for catalytically active Pla to act upon targets in the lungs during pneumonic plague, such as the newly discovered target Fas ligand, resulting in altered host cell apoptosis and innate immunity [29]. Thus, the production of OMVs by *Y. pestis* may provide an explanation for how a pathogen with a significant array of anti-inflammatory

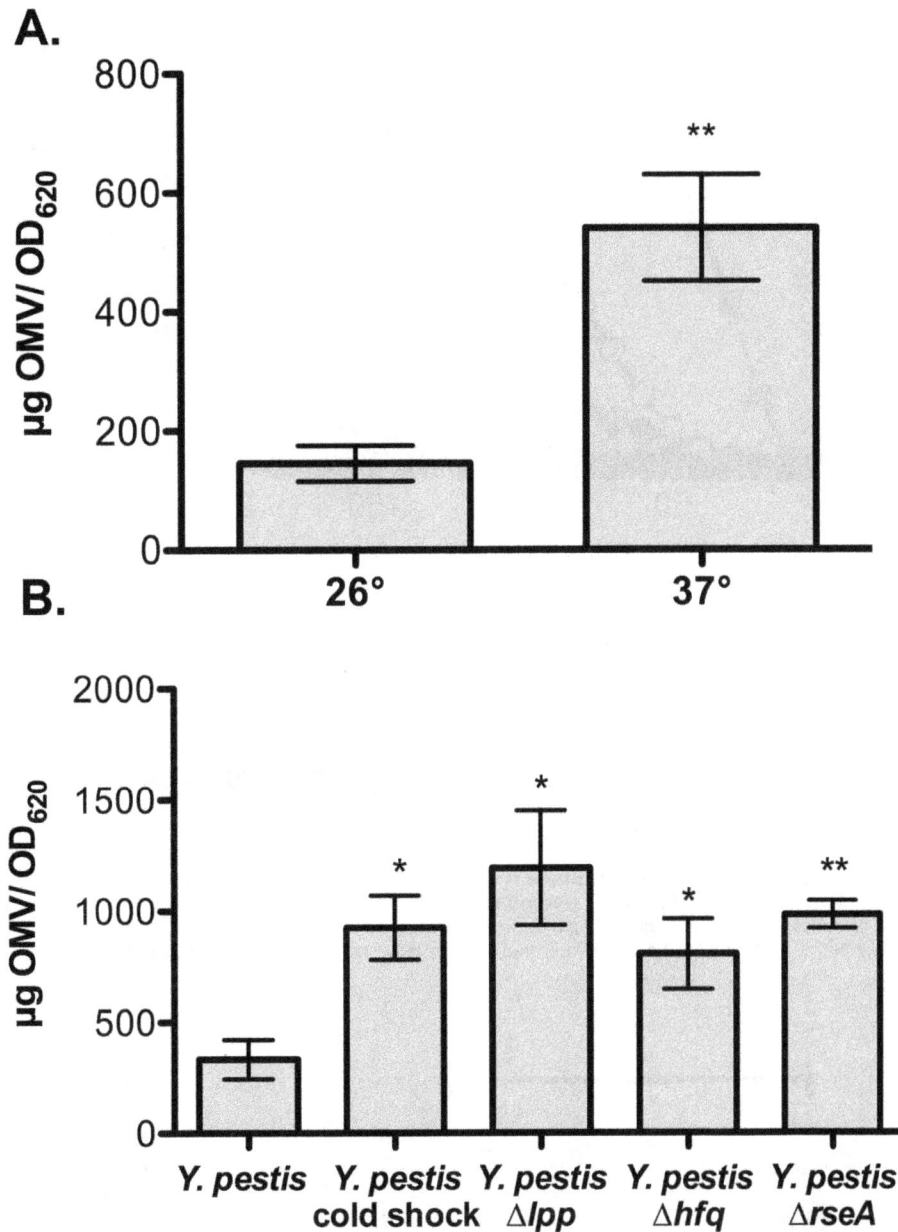

Figure 4. Effects of temperature and stress response factors on OMV production. (A) OMVs were isolated from *Y. pestis* cultured at either 26°C or 37°C, and total protein abundance associated with the OMVs was measured. Protein concentrations were normalized to the OD_{620} of the bacterial cultures. (B) Wild-type *Y. pestis* or strains lacking the genes for RseA, Hfq, or Lpp were cultured at 37°C as above and OMVs were isolated and total associated protein was measured and normalized to the OD_{620}. For cold shock experiments, bacteria were placed in an ice water bath for one h before proceeding. One experiment representative of two biological replicates is shown. The mean and SE are shown. *$p<0.05$, **$p<0.005$ (student's t–test).

virulence factors is able to induce a highly pro-inflammatory state during disease.

Y. pestis infection of mammals is generally extracellular in nature, and the only bacterial products thought to be delivered to the host cell cytoplasm are those injected by the T3SS. It has been repeatedly demonstrated, however, that OMVs released by other bacterial pathogens are capable of fusing with or are internalized by host cells [4,14]. Thus, OMV production by *Y. pestis* could potentially result in the delivery of otherwise extracellular or cell surface-associated bacterial factors directly to the eukaryotic cell cytosol. If *Y. pestis*-produced OMVs are capable of fusing with the

host cell during infection, this raises the possibility that extracellular virulence factors of the plague bacillus may also have intracellular activities. For instance, if Pla is internalized via OMV fusion or endocytosis, the Pla protease could alter host cell function by cleaving or degrading intracellular proteins. If these targeted proteins contribute to pathogen sensing, signaling, or basic biological processes, this could explain the diverse roles of Pla during pneumonic plague beyond its effects on fibrinolysis and apoptosis.

While the release of OMVs by *Y. pestis* may be playing a natural role during host infections, it is also tempting to speculate on the

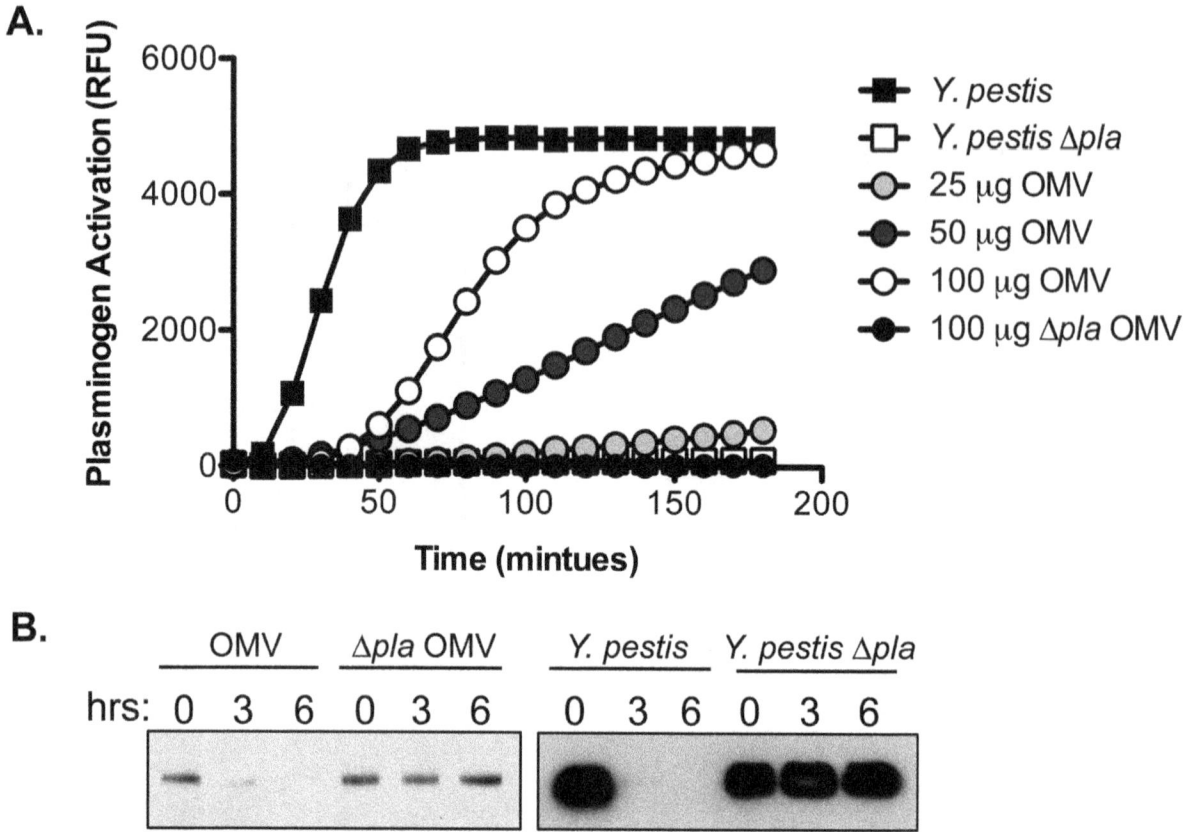

Figure 5. OMV-bound Pla is catalytically active and interacts with components of the ECM. (A) Plg-activating ability of wild-type or Δpla Y. pestis bacteria or OMVs. Whole bacteria or purified OMVs were incubated with human glu-plg and a fluorescent substrate of plasmin for 3 hours at 37°C. (B) Degradation of α2-antiplasmin by wild-type or Δpla Y. pestis bacteria or OMVs. Whole bacteria or purified OMVs were incubated with purified human α2-antiplasmin at 37°C and at the times indicated, the presence of uncleaved α2-antiplasmin was determined by immunoblot analysis.

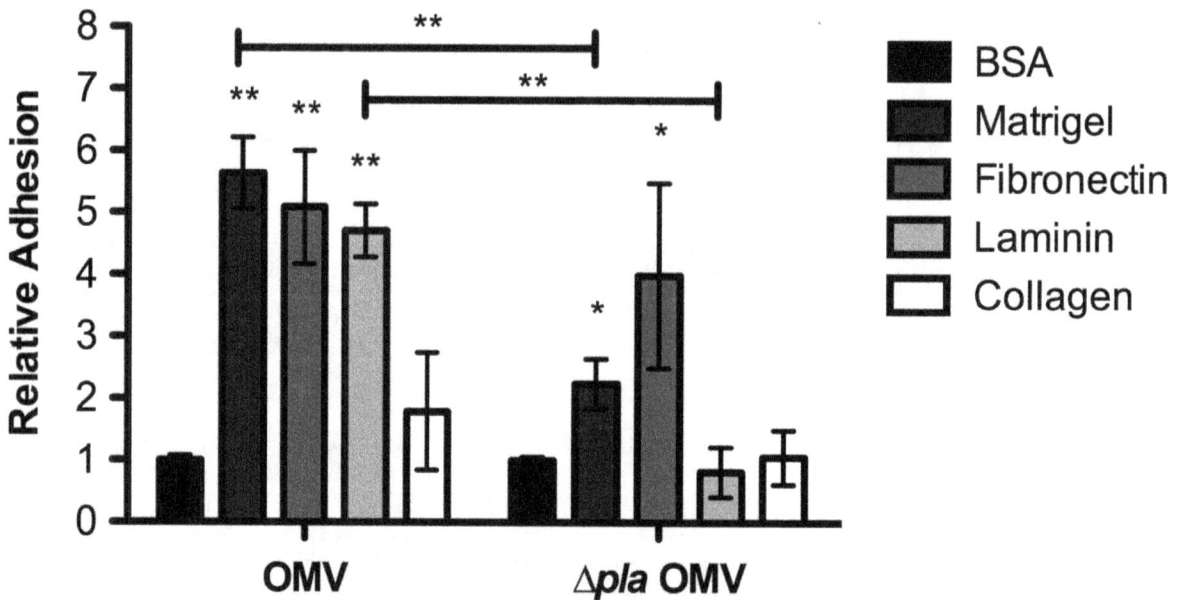

Figure 6. Binding of Y. pestis OMVs to components of the extracellular matrix. Wild-type or Δpla OMVs were examined by ELISA for the ability to bind the ECM components Matrigel, fibronectin, laminin, and collagen. BSA was used as a negative control for binding and differences in fluorescence are presented as fold change compared to BSA (set at 1). The combined mean and SE of 3 independent experiments are shown. *p< 0.05, **p<0.005 (two-way ANOVA).

use of purified OMVs as a tool to determine the specific roles of outer membrane virulence factors in the host independent of replicating, metabolically active, or secretion-competent bacteria. This could be particularly useful for the study of proteins that are otherwise intransigent to purification due to their structure or requirement for bacterial co-factors for full activity, such as Pla. Experiments using OMVs as a virulence factor "delivery system" have been performed with a variety of Gram-negative organisms including uropathogenic *E. coli* [59], *E. tarda* [63], *B. melitensis* [60], and *N. meningitidis* [71], and similar experiments with *Y. pestis* OMVs are likely to elaborate our understanding of the overall virulence strategy of this high-risk pathogen.

Materials and Methods

Reagents, bacterial strains and growth conditions

All reagents were obtained from Sigma-Aldrich or VWR unless otherwise indicated. All *Y. pestis* strains described in this study lack the pCD1 virulence plasmid, and bacterial strains used in this study are listed in **Table S2**. *Y. pestis* strains were routinely cultured on brain heart infusion (BHI) (Difco) agar or in liquid BHI broth at 26°C unless otherwise indicated. Media were supplemented with ampicillin (100 μg/ml) or kanamycin (50 μg/ml) as appropriate.

Deletion of *rseA* and *lpp*

The coding sequences for *rseA* and *lpp* were deleted from *Y. pestis* by lambda red recombination following procedures previously described [25,72]. Regions of homology upstream and downstream of the genes were amplified by PCR using the primer sets listed in **Table S3**. The kanamycin resistance cassette used for the selection of recombinants was excised using an FRT-based system as described [72].

OMV isolation

OMVs were isolated from *Y. pestis* strains based on previously published protocols [5,40]. Briefly, *Y. pestis* was cultured in Erlenmeyer flasks with shaking at 250 RPM for 10 h at 37°C or 26°C. For cold shock experiments, bacteria were cultured as above at 37°C and then placed in an ice slurry water bath for 1 h before proceeding with OMV isolation. Bacteria were centrifuged at 5,000 x g for 20 min and the supernatant, containing the released OMVs, was removed and filter sterilized through a 0.22 μm PES membrane (Millipore), and subsequently checked for sterility by plating on BHI agar. The sterilized supernatant was then concentrated using Centricon Plus-70 100 kDa centrifugation filters (Millipore) according to the manufacturer's recommendations. The concentrated filtrate was subjected to ultracentrifugation at 180,000 x g for 2 h at 4°C. The pelleted fraction containing OMVs was resuspended in 45% Optiprep solution, 10 mM HEPES, 0.85% NaCl, pH 7.4, and OMVs were subjected to density gradient centrifugation (40%, 35%, 30%, 25%, 20% Optiprep/Tris solutions) for 16 h at 100,000 x g at 4°C. Fractions were dialyzed against 50 mM Tris-HCl, pH 6.9 and analyzed for OMV recovery.

OMV protein quantification

To determine the total protein abundance associated with OMVs, the Bradford Assay (Bio-Rad) was performed according to the manufacturer's commendations as described previously for quantifying OMV abundance [9,63]. For those experiments in which OMV preparations from different conditions and/or strains were compared, OMV protein abundance was normalized to the optical density (OD_{620}) of the bacterial culture at the time of harvest.

Immunoblot analyses

The presence of Caf1, Ail, Pla, Hfq, and RpoA in OMV preparations were determined by immunoblot. Bacterial whole cell lysates were prepared by sonication as previously described [73]. OMVs or lysates (20 μg each) were mixed with reducing sample buffer (10% glycerol, 100 mM Tris-HCl, pH 6.8, 2% sodium dodecyl sulfate (SDS), 0.02% bromophenol blue, 5% β-mercaptoethanol) and separated by SDS-PAGE. Proteins were transferred to nitrocellulose membranes for immunblot analyses with antibodies against Pla [74], Ail (Eric Krukonis, University of Detroit Mercy School of Dentistry), Hfq [53], Caf1 (Abcam), and RpoA (Melanie Marketon, Indiana University).

Electron microscopy

For scanning and transmission electron microscopy of *Y. pestis* or purified OMVs, bacteria were cultured for 6 h at 37°C or OMVs were isolated as above. For transmission electron microscopy, 10 μl of each preparation were spotted on nickel grids and incubated at room temperature for 30 min. The grids were then dried, and a solution of 2% formaldehyde/0.5% glutaraldehyde was applied for 15 min. Grids were then rinsed with PBS and negatively stained using 1% uranyl acetate for one min. For immune-gold labeling of OMVs, prior to fixation the Caf1 antibody was incubated with OMVs for 30 min and then washed 3 x for 10 min each with PBS. Secondary antibody conjugated to 6 mM gold beads (Invitrogen) was incubated with OMVs for 30 min and then washed 3 x with PBS followed by negative staining as described above. Images were obtained using the FEI Tecnai Spirit G2 microscopy. For scanning electron microscopy, samples were prepared as described, fixed with 4% paraformaldehyde/1% glutaraldehyde for 30 min followed by sequential dehydration with 20%, 40%, 60%, 80%, 95%, and 3×100% ethanol for 10 min each. Dehydrated samples were sputter-coated using the Baltec coating system and imaged on the JEOL Neo Scope Benchtop SEM.

Plasminogen activation assay

Assessment of plg activation by *Y. pestis* bacteria, culture supernatants, or OMVs was performed as previously described [25]. Briefly, bacteria (8×10^6 CFU, cultured in BHI at 37°C for 6 h), 0.22 μm-filtered culture supernatant, 100 kDa-filtered culture supernatant (filtrate), or increasing concentrations of OMVs were incubated with purified human glu-plg (Hematologic Technologies) (4 μg) and the chromogenic substrate D-AFK-ANSNH-iC$_4$H$_9$-2HBr (SN5; Haematologic) (50 μM) in a total volume of 200 μl PBS. Reaction mixtures were incubated in triplicate for 3 h at 37°C, and the absorbance at 460 nm was measured every 10–11 min in a Molecular Devices SpectraMax M5 fluorescence microplate reader.

α2-antiplasmin degradation assay

Purified OMVs (100 μg) or *Y. pestis* bacteria (1×10^8 CFU, cultured in BHI broth for 6 hours at 37°C) were incubated with active α2-antiplasmin (1 μg, Abcam) at 37°C. At various times, bacteria were removed by centrifugation and proteins contained within the supernatant or the OMV-containing samples were precipitated with 10% trichloroacetic acid and resuspended in an excess of sample buffer. Samples were then separated by SDS-PAGE and transferred to nitrocellulose membranes for analysis with an antibody to α2-antiplasmin (Abcam).

ECM binding assay

To test OMV binding to various ECM components, purified BSA, Matrigel, laminin, collagen, or fibronectin (50 µg each) were added in triplicate to the wells of a 96-well plate overnight at 4°C. Unbound ECM components were removed and the wells washed 3 x with PBS. OMVs (50 µg) were then added to the wells for 18 h at 4°C. The wells were subsequently washed 3 x with PBS and then incubated with an anti-Caf1 antibody for 4 h (1:2,000 dilution). Wells were washed 3 x with PBS, incubated for one h with a FITC-conjugated secondary antibody, washed 3 x with PBS, and then 100 µl of PBS was added. Fluorescence was measured on a Tecan Safire² spectrophotometer with an excitation wavelength of 488 nm and an emission wavelength of 519 nm. Results are presented as fold change compared to the BSA wells.

LC-MS/MS analysis

OMVs were isolated as described and proteins were denatured at 50°C with 8 M urea for 60 min. After denaturation, proteins were reduced by adding DTT to a final concentration of 1 mM and incubating at 50°C for 15 min, and subsequently alkylated by adding iodoacetamide to a final concentration of 10 mM and incubating in the dark at room temperature for 15 min. The protein sample was then diluted by the addition of ammonium bicarbonate (100 mM) to a final concentration of 1 M urea and digested with trypsin at 37°C overnight. Samples were desalted using reverse phase C18 spin columns (Thermo Fisher Scientific), and the peptides were concentrated to dryness *in vaccuo*. After drying, the peptides were suspended in 5% acetonitrile and 0.1% formic acid, loaded directly onto a 15 cm-long, 75 µM reversed-phase capillary column (ProteoPep II C18, 300 Å, 5 µm size, New Objective), and separated with a 200 min gradient from 5% acetonitrile to 100% acetonitrile on a Proxeon Easy n-LC II (Thermo Scientific). The peptides were eluted into an LTQ Orbitrap Velos mass spectrometer (Thermo Scientific) with electrospray ionization at 350 nl/minute flow rate. The mass spectrometer was operated in data-dependent mode, and for each MS1 precursor ion scan the 10 most intense ions were selected from fragmentation by collision-induced dissociation. The other parameters for mass spectrometry analysis included: resolution of MS1 set at 60,000; normalized collision energy 35%; activation time 10 ms; isolation width 1.5; and +4 and higher charge states were rejected.

The data were processed using Proteome Discoverer (version 1.4, Thermo Scientific) and searched using embedded sequest HT search engine. The data were searched against the reference

proteome of *Y. pestis* downloaded from Uniprot.org. The other parameters were as follows: (i) enzyme specificity: trypsin; (ii) fixed modification: cysteine carbamidomethylation; (iv) variable modification: methionine oxidation and N-terminal acetylation; (v) precursor mass tolerance: ± 10 ppm; and (vi) fragment ion mass tolerance: ± 0.8 Da. All the spectra were searched against target/decoy databases and targeted false discovery rate of 1% was set to achieve high confidence assignment of peptides. Protein grouping was enabled in Proteome discoverer and proteins were grouped to satisfy the rule of parsimony. Further, in the final protein list, protein identification was considered only valid if supported by minimum of two peptides of which at least one has to be unique. The subcellular localization of identified proteins was predicted using PSORTb version 3.0 (http://www.psort.org).

Statistical analysis

Statistical analysis were performed using GraphPad Prism 5.0. For comparison between two groups a two-tailed student's t-test was performed. For comparison of multiple groups a two-way ANOVA was performed with a Bonferroni post-test. In all cases, significance was set to a p value of <0.05.

Supporting Information

Table S1 Proteins associated with *Yersinia pestis* OMVs identified by LC-MS/MS.

Table S2 Bacterial strains used in this study.

Table S3 Oligonucleotides used in this study.

Acknowledgments

We thank Drs. Eric Krukonis and Melanie Marketon for the kind gifts of the Ail and RpoA antibodies, respectively. We also wish to thank Drs. Meta Kuehn and Jason Huntley for helpful discussions, and Lauren Bellows, Jay Schroeder, and Dr. Dhaval Nanavati for technical assistance with this project. Imaging work was performed at the Northwestern University Cell Imaging Facility.

Author Contributions

Conceived and designed the experiments: JLE LMG WWL. Performed the experiments: JLE LMG AJC SMR. Analyzed the data: JLE LMG WWL. Contributed reagents/materials/analysis tools: JLE LMG AJC. Contributed to the writing of the manuscript: JLE LMG WWL.

References

1. Kesty NC, Kuehn MJ (2004) Incorporation of heterologous outer membrane and periplasmic proteins into *Escherichia coli* outer membrane vesicles. J Biol Chem 279: 2069–2076.
2. Parker H, Keenan JI (2012) Composition and function of *Helicobacter pylori* outer membrane vesicles. Microbes Infect 14: 9–16.
3. Horstman AL, Kuehn MJ (2000) Enterotoxigenic *Escherichia coli* secretes active heat-labile enterotoxin via outer membrane vesicles. J Biol Chem 275: 12489–12496.
4. Bauman SJ, Kuehn MJ (2006) Purification of outer membrane vesicles from *Pseudomonas aeruginosa* and their activation of an IL-8 response. Microbes Infect 8: 2400–2408.
5. Kulp A, Kuehn MJ (2010) Biological functions and biogenesis of secreted bacterial outer membrane vesicles. Annu Rev Microbiol 64: 163–184.
6. Schertzer JW, Whiteley M (2012) A bilayer-couple model of bacterial outer membrane vesicle biogenesis. mBio 3: e00297–00211.
7. Schooling SR, Beveridge TJ (2006) Membrane vesicles: an overlooked component of the matrices of biofilms. J Bacteriol 188: 5945–5957.
8. Kuehn MJ, Kesty NC (2005) Bacterial outer membrane vesicles and the host-pathogen interaction. Genes Dev 19: 2645–2655.
9. Wai SN, Takade A, Amako K (1995) The release of outer membrane vesicles from the strains of enterotoxigenic *Escherichia coli*. Microbiol Immunol 39: 451–456.
10. Lai CH, Listgarten MA, Hammond BF (1981) Comparative ultrastructure of leukotoxic and non-leukotoxic strains of *Actinobacillus actinomycetemcomitans*. J Periodontal Res 16: 379–389.
11. Horstman AL, Kuehn MJ (2002) Bacterial surface association of heat-labile enterotoxin through lipopolysaccharide after secretion via the general secretory pathway. J Biol Chem 277: 32538–32545.
12. Tan TT, Morgelin M, Forsgren A, Riesbeck K (2007) *Haemophilus influenzae* survival during complement-mediated attacks is promoted by *Moraxella catarrhalis* outer membrane vesicles. J Infect Dis 195: 1661–1670.
13. Hellman J, Warren HS (2001) Outer membrane protein A (OmpA), peptidoglycan-associated lipoprotein (PAL), and murein lipoprotein (MLP) are released in experimental Gram-negative sepsis. J Endotoxin Res 7: 69–72.
14. Fiocca R, Necchi V, Sommi P, Ricci V, Telford J, et al. (1999) Release of *Helicobacter pylori* vacuolating cytotoxin by both a specific secretion pathway and budding of outer membrane vesicles. Uptake of released toxin and vesicles by gastric epithelium. J Pathol 188: 220–226.

15. Gankema H, Wensink J, Guinee PA, Jansen WH, Witholt B (1980) Some characteristics of the outer membrane material released by growing enterotoxigenic *Escherichia coli*. Infect Immun 29: 704–713.

16. Manning AJ, Kuehn MJ (2011) Contribution of bacterial outer membrane vesicles to innate bacterial defense. BMC Microbiol 11: 258.

17. Butler T (2013) Plague gives surprises in the first decade of the 21st century in the United States and worldwide. Am J Trop Med Hyg 89: 788–793.

18. Schiano CA, Lathem WW (2012) Post-transcriptional regulation of gene expression in *Yersinia* species. Front Cell Infect Microbiol 2: 129.

19. Marceau M (2005) Transcriptional regulation in *Yersinia*: an update. Curr Issues Mol Biol 7: 151–177.

20. Hinnebusch BJ, Fischer ER, Schwan TG (1998) Evaluation of the role of the *Yersinia pestis* plasminogen activator and other plasmid-encoded factors in temperature-dependent blockage of the flea. J Infect Dis 178: 1406–1415.

21. Sodeinde OA, Subrahmanyam YV, Stark K, Quan T, Bao Y, et al. (1992) A surface protease and the invasive character of plague. Science 258: 1004–1007.

22. Burrows TW (1956) An antigen determining virulence in *Pasteurella pestis*. Nature 177: 426–427.

23. Cornelis GR (2002) The *Yersinia* Ysc-Yop virulence apparatus. Int J Med Microbiol 291: 455–462.

24. Cowan C, Jones HA, Kaya YH, Perry RD, Straley SC (2000) Invasion of epithelial cells by *Yersinia pestis*: evidence for a *Y. pestis*-specific invasin. Infect Immun 68: 4523–4530.

25. Lathem WW, Price PA, Miller VL, Goldman WE (2007) A plasminogen-activating protease specifically controls the development of primary pneumonic plague. Science 315: 509–513.

26. Sebbane F, Jarrett CO, Gardner D, Long D, Hinnebusch BJ (2006) Role of the *Yersinia pestis* plasminogen activator in the incidence of distinct septicemic and bubonic forms of flea-borne plague. Proc Natl Acad Sci U S A 103: 5526–5530.

27. Sodeinde OA, Goguen JD (1989) Nucleotide sequence of the plasminogen activator gene of *Yersinia pestis*: relationship to *ompT* of *Escherichia coli* and gene E of *Salmonella typhimurium*. Infect Immun 57: 1517–1523.

28. Kukkonen M, Lahteenmaki K, Suomalainen M, Kalkkinen N, Emody L, et al. (2001) Protein regions important for plasminogen activation and inactivation of alpha2-antiplasmin in the surface protease Pla of *Yersinia pestis*. Mol Microbiol 40: 1097–1111.

29. Caulfield AJ, Walker ME, Gielda LM, Lathem WW (2014) The Pla protease of *Yersinia pestis* degrades Fas ligand to manipulate host cell death and inflammation. Cell Host Microbe 15: 424–434.

30. Caulfield AJ, Lathem WW (2012) Substrates of the plasminogen activator protease of *Yersinia pestis*. Adv Exp Med Biol 954: 253–260.

31. Beesley ED, Brubaker RR, Janssen WA, Surgalla MJ (1967) Pesticins. 3. Expression of coagulase and mechanism of fibrinolysis. J Bacteriol 94: 19–26.

32. Lahteenmaki K, Virkola R, Saren A, Emody L, Korhonen TK (1998) Expression of plasminogen activator Pla of *Yersinia pestis* enhances bacterial attachment to the mammalian extracellular matrix. Infect Immun 66: 5755–5762.

33. Kienle Z, Emody L, Svanborg C, O'Toole PW (1992) Adhesive properties conferred by the plasminogen activator of *Yersinia pestis*. J Gen Microbiol 138 Pt 8: 1679–1687.

34. Eren E, van den Berg B (2012) Structural basis for activation of an integral membrane protease by lipopolysaccharide. J Biol Chem 287: 23971–23976.

35. Eren E, Murphy M, Goguen J, van den Berg B (2010) An active site water network in the plasminogen activator pla from *Yersinia pestis*. Structure 18: 809–818.

36. Pouillot F, Derbise A, Kukkonen M, Foulon J, Korhonen TK, et al. (2005) Evaluation of O-antigen inactivation on Pla activity and virulence of *Yersinia pseudotuberculosis* harbouring the pPla plasmid. Microbiology 151: 3759–3768.

37. Sodeinde OA, Sample AK, Brubaker RR, Goguen JD (1988) Plasminogen activator/coagulase gene of *Yersinia pestis* is responsible for degradation of plasmid-encoded outer membrane proteins. Infect Immun 56: 2749–2752.

38. Kolodziejek AM, Caplan AB, Bohach GA, Paszczynski AJ, Minnich SA, et al. (2013) Physiological levels of glucose induce membrane vesicle secretion and affect the lipid and protein composition of *Yersinia pestis* cell surfaces. Appl Environ Microbiol 79: 4509–4514.

39. Ellis TN, Leiman SA, Kuehn MJ (2010) Naturally produced outer membrane vesicles from *Pseudomonas aeruginosa* elicit a potent innate immune response via combined sensing of both lipopolysaccharide and protein components. Infect Immun 78: 3822–3831.

40. McBroom AJ, Kuehn MJ (2007) Release of outer membrane vesicles by Gram-negative bacteria is a novel envelope stress response. Mol Microbiol 63: 545–558.

41. Chutkan H, Macdonald I, Manning A, Kuehn MJ (2013) Quantitative and qualitative preparations of bacterial outer membrane vesicles. Methods Mol Biol 966: 259–272.

42. Pieper R, Huang ST, Robinson JM, Clark DJ, Alami H, et al. (2009) Temperature and growth phase influence the outer-membrane proteome and the expression of a type VI secretion system in *Yersinia pestis*. Microbiology 155: 498–512.

43. Lane MC, Lenz JD, Miller VL (2013) Proteolytic processing of the *Yersinia pestis* YapG autotransporter by the omptin protease Pla and the contribution of YapG to murine plague pathogenesis. J Med Microbiol 62: 1124–1134.

44. Yonezawa H, Osaki T, Woo T, Kurata S, Zaman C, et al. (2011) Analysis of outer membrane vesicle protein involved in biofilm formation of *Helicobacter pylori*. Anaerobe 17: 388–390.

45. Schwechheimer C, Sullivan CJ, Kuehn MJ (2013) Envelope control of outer membrane vesicle production in Gram-negative bacteria. Biochemistry 52: 3031–3040.

46. Schwechheimer C, Kuehn MJ (2013) Synthetic effect between envelope stress and lack of outer membrane vesicle production in *Escherichia coli*. J Bacteriol 195: 4161–4173.

47. Macdonald IA, Kuehn MJ (2013) Stress-induced outer membrane vesicle production by *Pseudomonas aeruginosa*. J Bacteriol 195: 2971–2981.

48. Galindo CL, Sha J, Moen ST, Agar SL, Kirtley ML, et al. (2010) Comparative global gene expression profiles of wild-type *Yersinia pestis* CO92 and its braun lipoprotein mutant at flea and human body temperatures. Comp Funct Genomics: 342168.

49. Phadtare S (2004) Recent developments in bacterial cold-shock response. Curr Issues Mol Biol 6: 125–136.

50. Rowley G, Spector M, Kormanec J, Roberts M (2006) Pushing the envelope: extracytoplasmic stress responses in bacterial pathogens. Nat Rev Microbiol 4: 383–394.

51. Missiakas D, Mayer MP, Lemaire M, Georgopoulos C, Raina S (1997) Modulation of the *Escherichia coli* sigmaE (RpoE) heat-shock transcription-factor activity by the RseA, RseB and RseC proteins. Mol Microbiol 24: 355–371.

52. Alba BM, Gross CA (2004) Regulation of the *Escherichia coli* sigma-dependent envelope stress response. Mol Microbiol 52: 613–619.

53. Schiano CA, Bellows LE, Lathem WW (2010) The small RNA chaperone Hfq is required for the virulence of *Yersinia pseudotuberculosis*. Infect Immun 78: 2034–2044.

54. Lathem WW, Schroeder JA, Bellows LE, Ritzert JT, Koo JT, et al. (2014) Posttranscriptional regulation of the *Yersinia pestis* cyclic AMP receptor protein Crp and impact on virulence. mBio 5: e01038–01013.

55. Geng J, Song Y, Yang L, Feng Y, Qiu Y, et al. (2009) Involvement of the post-transcriptional regulator Hfq in *Yersinia pestis* virulence. PLoS One 4: e6213.

56. Nikaido H, Bavoil P, Hirota Y (1977) Outer membranes of gram-negative bacteria. XV. Transmembrane diffusion rates in lipoprotein-deficient mutants of *Escherichia coli*. J Bacteriol 132: 1045–1047.

57. Cascales E, Bernadac A, Gavioli M, Lazzaroni JC, Lloubes R (2002) Pal lipoprotein of *Escherichia coli* plays a major role in outer membrane integrity. J Bacteriol 184: 754–759.

58. Kesty NC, Mason KM, Reedy M, Miller SE, Kuehn MJ (2004) Enterotoxigenic *Escherichia coli* vesicles target toxin delivery into mammalian cells. EMBO J 23: 4538–4549.

59. Davis JM, Carvalho HM, Rasmussen SB, O'Brien AD (2006) Cytotoxic necrotizing factor type 1 delivered by outer membrane vesicles of uropathogenic *Escherichia coli* attenuates polymorphonuclear leukocyte antimicrobial activity and chemotaxis. Infect Immun 74: 4401–4408.

60. Avila-Calderon ED, Lopez-Merino A, Jain N, Peralta H, Lopez-Villegas EO, et al. (2012) Characterization of outer membrane vesicles from *Brucella melitensis* and protection induced in mice. Clin Dev Immunol 2012: 352493.

61. Tsang TM, Felek S, Krukonis ES (2010) Ail binding to fibronectin facilitates *Yersinia pestis* binding to host cells and Yop delivery. Infect Immun 78: 3358–3368.

62. Ismail S, Hampton MB, Keenan JI (2003) *Helicobacter pylori* outer membrane vesicles modulate proliferation and interleukin-8 production by gastric epithelial cells. Infect Immun 71: 5670–5675.

63. Park SB, Jang HB, Nho SW, Cha IS, Hikima J, et al. (2011) Outer membrane vesicles as a candidate vaccine against edwardsiellosis. PLoS One 6: e17629.

64. Lee EY, Choi DS, Kim KP, Gho YS (2008) Proteomics in gram-negative bacterial outer membrane vesicles. Mass Spectrom Rev 27: 535–555.

65. Lee EY, Bang JY, Park GW, Choi DS, Kang JS, et al. (2007) Global proteomic profiling of native outer membrane vesicles derived from *Escherichia coli*. Proteomics 7: 3143–3153.

66. Pieper R, Huang ST, Clark DJ, Robinson JM, Alami H, et al. (2009) Integral and peripheral association of proteins and protein complexes with *Yersinia pestis* inner and outer membranes. Proteome Sci 7: 5.

67. Post DM, Zhang D, Eastvold JS, Teghanemt A, Gibson BW, et al. (2005) Biochemical and functional characterization of membrane blebs purified from *Neisseria meningitidis* serogroup B. J Biol Chem 280: 38383–38394.

68. Shah S, Miller A, Mastellone A, Kim K, Colaninno P, et al. (1998) Rapid diagnosis of tuberculosis in various biopsy and body fluid specimens by the AMPLICOR *Mycobacterium tuberculosis* polymerase chain reaction test. Chest 113: 1190–1194.

69. Dorward DW, Schwan TG, Garon CF (1991) Immune capture and detection of *Borrelia burgdorferi* antigens in urine, blood, or tissues from infected ticks, mice, dogs, and humans. J Clin Microbiol 29: 1162–1170.

70. Bjerre A, Brusletto B, Rosenqvist E, Namork E, Kierulf P, et al. (2000) Cellular activating properties and morphology of membrane-bound and purified meningococcal lipopolysaccharide. J Endotoxin Res 6: 437–445.

71. Saunders NB, Brandt BL, Warren RL, Hansen BD, Zollinger WD (1998) Immunological and molecular characterization of three variant subtype P1.14 strains of *Neisseria meningitidis*. Infect Immun 66: 3218–3222.

72. Koo JT, Alleyne TM, Schiano CA, Jafari N, Lathem WW (2011) Global discovery of small RNAs in *Yersinia pseudotuberculosis* identifies *Yersinia*-

specific small, noncoding RNAs required for virulence. Proc Natl Acad Sci U S A 108: E709–717.

73. Bellows LE, Koestler BJ, Karaba SM, Waters CM, Lathem WW (2012) Hfq-dependent, co-ordinate control of cyclic diguanylate synthesis and catabolism in the plague pathogen *Yersinia pestis*. Mol Microbiol 86: 661–674.

74. Houppert AS, Bohman L, Merritt PM, Cole CB, Caulfield AJ, et al. (2013) RfaL is required for *Yersinia pestis* type III secretion and virulence. Infect Immun 81: 1186–1197.

## 8

Fabrication of SWCNT-Ag Nanoparticle Hybrid Included Self-Assemblies for Antibacterial Applications

Sayanti Brahmachari, Subhra Kanti Mandal, Prasanta Kumar Das*

Department of Biological Chemistry, Indian Association for the Cultivation of Science, Kolkata, India

Abstract

The present article reports the development of soft nanohybrids comprising of single walled carbon nanotube (SWCNT) included silver nanoparticles (AgNPs) having superior antibacterial property. In this regard aqueous dispersing agent of carbon nanotube (CNT) containing a silver ion reducing unit was synthesised by the inclusion of tryptophan and tyrosine within the backbone of the amphiphile. The dispersions were characterized spectroscopically and microscopically using TEM, AFM and Raman spectroscopy. The nanotube-nanoparticle conjugates were prepared by the *in situ* photoreduction of AgNO$_3$. The phenolate residue and the indole moieties of tyrosine and tryptophan, respectively reduces the sliver ion as well as acts as stabilizing agents for the synthesized AgNPs. The nanohybrids were characterized using TEM and AFM. The antibacterial activity of the nanohybrids was studied against Gram-positive (*Bacillus subtilis* and *Micrococcus luteus*) and Gram-negative bacteria (*Escherichia coli* and *Klebsiella aerogenes*). The SWCNT dispersions showed moderate killing ability (40–60%) against Gram-positive bacteria however no antibacterial activity was observed against the Gram negative ones. Interestingly, the developed SWCNT-amphiphile-AgNP nanohybrids exhibited significant killing ability (~90%) against all bacteria. Importantly, the cell viability of these newly developed self-assemblies was checked towards chinese hamster ovarian cells and high cell viability was observed after 24 h of incubation. This specific killing of bacterial cells may have been achieved due to the presence of higher –SH containing proteins in the cell walls of the bacteria. The developed nanohybrids were subsequently infused into tissue engineering scaffold agar-gelatin films and the films similarly showed bactericidal activity towards both kinds of bacterial strains while allowing normal growth of eukaryotic cells on the surface of the films.

Editor: Vipul Bansal, RMIT University, Australia

Funding: P.K.D. is thankful to Department of Science and Technology, India (SR/S1/OC-25/2011) for financial assistance. S.B. and S.K.M. acknowledge the Council of Scientific and Industrial Research, India for Research fellowships. The funders had no role in study design, data collection and analysis, decision to publish, or preparation of the manuscript.

Competing Interests: The authors have declared that no competing interests exist.

* Email: bcpkd@iacs.res.in

Introduction

Over the decades, the development of novel antimicrobial agents has undergone a continuous process of evolution and still remains an important domain of research [1–13]. The growing resistance of microbes against the conventional antibiotics necessitated the restructuring of the antibiotic design and newer formulations have emerged with time. This drug resistance mostly arises as a natural process of adaptation and random selection through mutation. To this end, in addition to the conventionally known antibiotics, nanomaterials like silver nanoparticles (AgNPs) have emerged as a class of alternative antibiotics possessing a different mechanism of bacteria killing [14–19]. The mechanism of antibacterial activity of AgNPs is still not well understood, however there are theories like (i) membrane damage by free radicals, (ii) membrane structure degradation by "pits" in cell walls, and (iii) penetration of cell walls and dephosphorylation of key peptides in cellular signalling cycles [20–22]. To date there are few reports of bacterial resistance towards these nanoparticles and this antibacterial activity has enhanced the broad spectrum applications of these nanomaterials [23–24]. In fact recently gold

and silver nanoparticles were shown to have specific antibacterial and anticancer activities and AgNP is being used to incorporate antimicrobial activity in paints and biomedical implants [20,25–27].

Single walled carbon nanotube (SWCNTs) – the one dimensional allotrope of carbon, also belongs to the important class of nanomaterials because of its extraordinary optical, electronic, mechanical properties and high aspect ratio [28–31]. It is finding applications in almost all branches of sciences from energy research to biotechnology due to its unique intrinsic features. Amongst others, the huge surface area of carbon nanotubes (CNTs) makes it suitable to be utilized as cellular transporters [28–31]. However, studies on its interaction with the prokaryotic cells have received comparatively little attention. Few recent reports investigated the antibacterial activity of CNTs and its modified forms [32–41]. Size dependent antibacterial activity of CNTs was first reported by Kang et al [33]. Later on the findings were further supported by Liu et al where the membrane damage of prokaryotic cells resulting from direct contact with pristine SWCNTs was investigated [36]. However, the inherent insolubility of these nanostructures greatly bars its applications and the key towards exploiting this nanomaterial in the biomedicinal arena lies in designing judicious CNT dispersing agents [42–51]. To this end, we recently reported amino acid based biocompatible SWCNT dispersions for the delivery of biomolecules and drugs into eukaryotic cells [52–55]. Additionally we have also utilized AgNPs for the development of composite hydrogel matrices having antibacterial activity [56–58]. However it would be intriguing if the complementary properties of these two nanomaterials could be simultaneously exploited to develop superior antimicrobial agents. In particular, fabrication of AgNPs decorated CNT dissolution in aqueous medium by the assistance of amphiphilic dispersing agent could be used for developing antibacterial scaffolds. Such soft-nanocomposites would find wide range application in the biomedicinal arena including tissue engineering, drug delivery and so forth.

Herein, the present work reports the design and development of aqueous SWCNT dispersion by L-tyrosine and L-tryptophan based neutral amphiphiles (Figure 1) comprising of polyethylene glycol (PEG) unit. The AgNP was synthesized within these dispersions by *in situ* photo-reduction under sunlight [59]. The SWCNT dispersion and nanoconjugates were characterized using transmission elector microscopy (TEM), atomic force microscopy (AFM) and Raman spectroscopy. Encouragingly, more than 90% killing of both Gram-positive (*Bacillus subtilis* and *Micrococcus luteus*) and Gram-negative bacteria (*Escherichia coli* and *Klebsiella aerogenes*) was achieved using SWCNT-amphiphile-AgNP hybrid (6–10 μg/mL AgNP). Moreover, substantial cell viability of the nanohybrids was observed against Chinese Hamster Ovarian cells (CHO cells). Interestingly, normal growth of eukaryotic cells were noted on the surface of these nanocomposites infused agar-gelatin film (tissue engineering scaffold) while it was lethal toward bacteria.

Results and Discussion

Synthesis of dispersing agent

Development of SWCNT based antimicrobial agents is still at its infancy while its potential demands wider exploitation particularly in the backdrop of antibiotics resistance microbes. The first step towards developing biocompatible antibacterial dispersion of CNTs is to design dispersing agents that would facilitate the exfoliation of CNTs in water. Any non-covalent SWCNT dispersing agent typically contains a hydrophobic and a hydrophilic end. Generally, the hydrophobic unit binds to the surface of the nanotubes while the hydrophilic end assists its solubilization in water through the formation of supramolecular aggregates. To this end recently we reported the formation of electrostatically bound composite material of SWCNT and gold nanoparticles where the nanotubes were dispersed in water using cationic amphiphile and the nanoparticles were capped with anionic surfactants [60]. However in several previous instances the

cationic dispersing agents was found to be cytotoxic towards mammalian cells and hence those SWCNT dispersions were not suitable for developing biocompatible scaffolds [54,56]. On the other hand, nanotubes dispersing with anionic surfactants may result in overall repulsion with the negatively charged membrane of prokaryotic and eukaryotic cells. Hence, instead of cationic or anionic hydrophilic head groups, we designed amino acid based neutral dispersing agent comprising of polyethylene glycol (PEG) as the hydrophilic unit. The cetyl (C-16) chain generally acts as a good surface anchoring unit for the CNTs and it was taken as the hydrophobic unit. Additionally, the presence of a nanoparticle capping residue was mandatory to ensure the binding of AgNP on the nanotube surface. Therefore, tryptophan and tyrosine amino acids were integrated into the backbone of dispersing agent because of the well known capping and nanoparticle stabilizing ability of the indole residue of tryptophan and phenolate residue of tyrosine [56–58]. Thus, amphiphilic dispersing gents **1** and **2** were synthesized by coupling of the C-16 chain to the acid terminal of the amino acid and PEG to the amine terminal via a succinic acid linker (Figure 1).

Quantification and characterization of SWCNT-dispersions

Aqueous suspensions of SWCNT using **1** and **2** were prepared following the previously reported protocol [52]. Briefly, SWCNT (1 mg) was taken in 4 mL of amphiphile solution (2.5 mg/mL) and tip sonicated followed by bath sonication. The suspension was then centrifuged at 2500 g for 90 min. The supernatant was collected and the amount of dispersed SWCNT was calculated using the previously reported calibration plot prepared using commercially available surfactant sodium dodecyl benzene sulfonate (SDBS) [53]. Both neutral amphiphiles dispersed SWCNTs with an efficient of 75% and 78%, respectively for **1** and **2**, which indicates most of the nanotubes remain in the suspension with respect to its initial weight.

These aqueous SWCNT suspensions (SWCNT-**1** and SWCNT-**2**) were stable for months and showed no sign of aggregation/ precipitation. In case of positively/negatively charged dispersing agents, the electrostatic repulsion between the nanotubes hindered their re-aggregation while in case of dispersion with neutral PEG based amphiphiles, the steric bulk of the PEG facilitated the dispersion of the nanotubes. Also the hydrogen-bonded water molecules with glycol chain further aided dissolution of SWCNT possibly by preventing the coagulation of nanotubes [61]. These SWCNT dispersions were characterized by microscopic and spectroscopic analysis. In accordance to the atomic force microscopy (AFM) and transmission electron microscopy (TEM) images the diameter of debundled SWCNT-**1** and SWCNT-**2** was found to be ~ 5 nm (Figure 2). Statistical AFM analysis was performed in order to compare the dimension of individualized

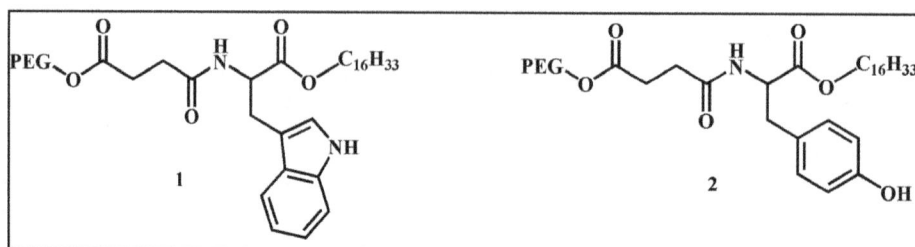

Figure 1. Structure of the dispersing agents. PEG = -(CH$_2$CH$_2$O)$_{12}$CH$_3$.

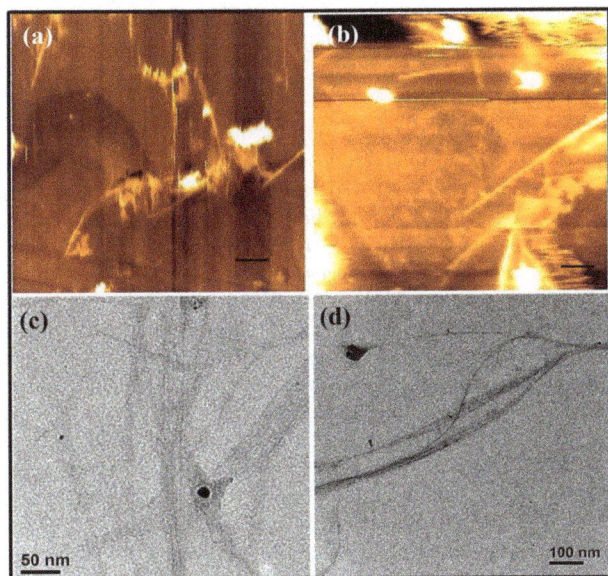

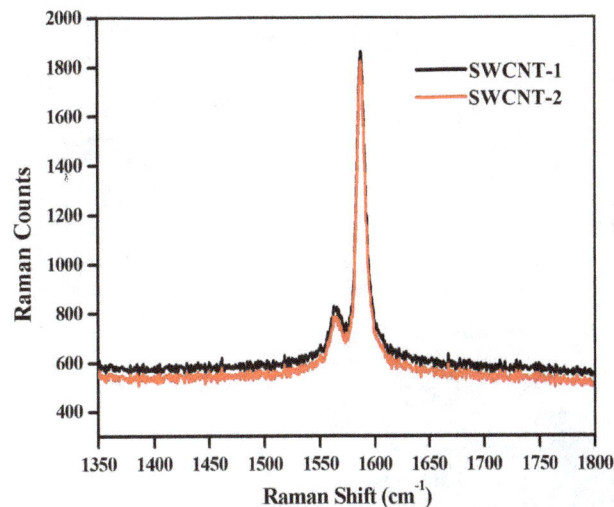

Figure 3. Raman Spectra of dispersed SWCNT-1 and SWCNT-2 using 514.5 nm excitation.

Figure 2. AFM images of dispersed (a) SWCNT-1 and (b) SWCNT-2 (the scale bars in the AFM images indicate 200 nm) and TEM images of dispersed (c) SWCNT-1 (d) SWCNT-2.

nanotubes in an aqueous suspension of **1** and **2**. From several AFM images the average bundle diameter and length of the exfoliated nanotubes were found to between 4–6 nm and 400–500 nm, respectively (Figure S1, S2).

Zeta (ζ)-potential is another well established parameter of assessing the colloidal stability of any dispersion [52,61]. It gives an idea about the interplay of the different forces operating to inhibit the aggregation of the nanomaterials. Conventionally, ζ-potential values higher and lower than $+/-15$ mV indicate greater stability of a colloidal dispersion. In the present study, the ζ-potential values were found to be -23 mV and -30 mV, respectively for the aqueous suspension of SWCNT-**1** and SWCNT-**2** [61]. Considering the neutral nature of dispersing agents, the high ζ-potential value clearly indicates substantial stability of these SWCNT dispersions. The quality of nanotube dispersions was also studied using Raman spectroscopy. Primarily, the G-band of SWCNT originates from several tangential C–C stretching transitions of the SWCNT carbon atoms whereas the D-band is generally associated with defects in the SWCNT structure [62–63]. The Raman spectra of SWCNT-**1** and SWCNT-**2** was acquired by excitation of 514.5 nm laser and a sharp peak corresponding to the G-band at 1590 cm^{-1} was observed (Figure 3). In a control experiment the Raman spectrum of solid SWCNT was recorded (Figure S3). Importantly no change in the spectral nature was observed when the nanotubes were dispersed using amphiphiles **1** and **2**. Also the area under the G-band directly corresponds to the amount of dispersed SWCNT. The comparable area under the G-band in case of both SWCNT-**1** and SWCNT-**2** indicated similar quantity of nanotube dispersion, which was in concurrence with above mentioned quantification.

In situ synthesis of AgNPs and characterization of nanohybrids

These SWCNT-amphiphile suspensions were subsequently used for the synthesis of AgNPs [58]. A green technique was adopted where AgNPs were synthesized in situ under sunlight from AgNO$_3$ at physiological pH. The ratio of AgNO$_3$: capping agent

(dispersing agents **1** and **2**) was maintained at 1:10 and the suspensions were exposed to sunlight for 15 min. This photo reduction method does not involve the use of any harmful chemicals [16–17]. The formation of AgNPs was monitored by UV-vis spectroscopy. In case of SWCNT-**1**, a surface plasmon resonance (SPR) peak generation was observed after 15 min of sunlight exposure at 425 nm, which gets intensified after 30 min (Figure S4). Similar peak formation due to the synthesis of AgNP was observed at 410 nm in case of SWCNT-**2**. The developed nanoconjugates (SWCNT-**1**-AgNP and SWCNT-**2**-AgNP) were then subjected to microscopic investigations. In the AFM images, AgNPs of 8–10 nm diameters were found to be decorated on the surface of SWCNT-**1** and SWCNT-**2** (Figure 4a,b). Importantly very little amount of unbound nanoparticle was observed. Similarly nanoparticles having a diameter of about 8 to 10 nm were observed on the walls of the nanotubes in the TEM images of SWCNT-**1**-AgNP and SWCNT-**2**-AgNP (Figure 4c,d). Thus the microscopic studies clearly delineated the formation of the nanotube-nanoparticle conjugates. The tryptophan and tyrosine amino acids within the amphiphilic dispersing agents act as capping and stabilizing agent for the synthesized AgNPs that facilitated the binding of the nanoparticles to the nanotube surface [56–60]. Interestingly, no loss in colloidal stability of the conjugate was observed when it was kept in dark for a week.

Quantification of the SWCNT-AgNP nanohybrids

In order to quantify the amount of amphiphile, SWCNT and AgNP present in the nanohybrids, the samples were subjected to thermo gravimetric analysis (TGA). Firstly, SWCNT (30 mg) was sonicated and dispersed using **1** and **2**, respectively, followed by centrifuged at 2500 g to remove heavier bundles. The supernatant was collected and half of the obtained supernatant was taken for in situ synthesis of AgNP. The AgNO$_3$ solution was added to this supernatant and AgNP was synthesized in a similar way as described above. Next, all suspensions were centrifuged twice at 45000 rpm to remove excess amphiphiles and unconverted AgNO$_3$. The obtained pellets were lyophilized for 6 h to remove trace amount of water. Finally, each conjugate (SWCNT-**1**, SWCNT-**2**, SWCNT-**1**-AgNP and SWCNT-**2**-AgNP) was used for TGA where the samples were heated to 800 °C (Figure 5). As

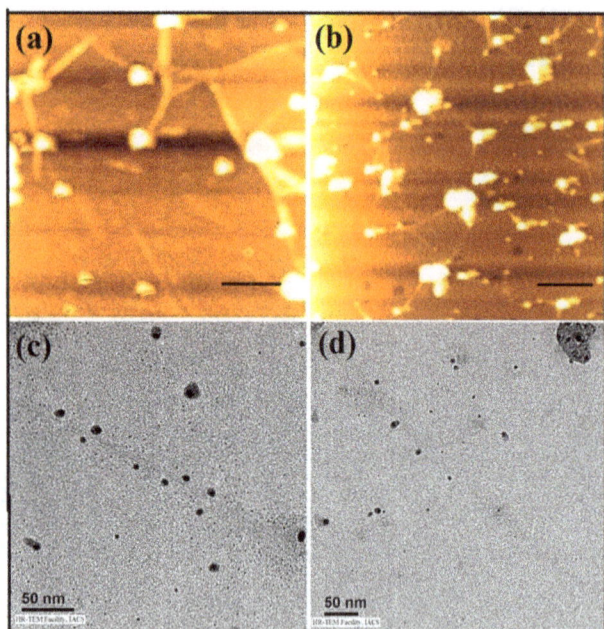

Figure 4. AFM images of (a) SWCNT-1-AgNP, (b) SWCNT-2-AgNP (the scale bars in the AFM images indicate 200 nm) and TEM images of (c) SWCNT-1-AgNP, (d) SWCNT-2-AgNP.

control 10 mg of amphiphiles **1** and **2** were also subjected to the same TGA process of heating. The thermal decomposition pattern of the amphiphiles clearly showed its complete decomposition within 400°C. The TGA plots SWCNT-**1**, SWCNT-**2** indicated the presence of 41 and 39% nanotubes in their respective dispersion. Importantly, the comparison of the TGA plots between the dispersed nanotubes and the AgNP decorated SWCNTs showed that SWCNT-**1**-AgNP contains 41% SWCNT and 14% AgNP while SWCNT-**2**-AgNP contains 39% SWCNT and 12% AgNP. Similarly, AgNP was prepared as control using only amphiphile in the absence of the nanotubes and quantified using

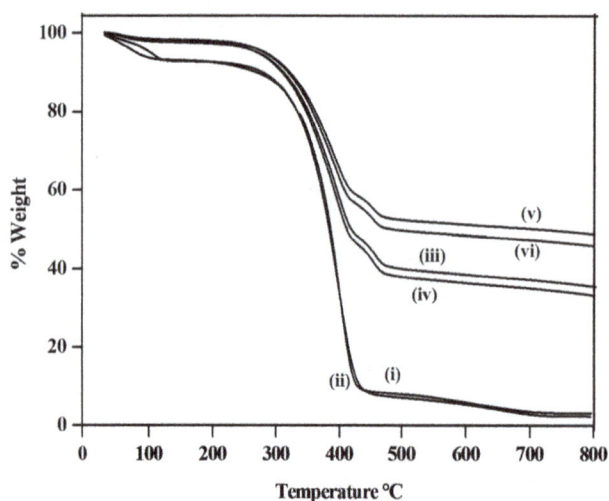

Figure 5. TGA analysis of (i) 1 (ii) 2 (iii) SWCNT-1 (iv) SWCNT-2 (v) SWCNT-1-AgNP and (vi) SWCNT-2-AgNP.

TGA (Figure S5). The nanoconjugates were found to contain 30% nanoparticle and 70% amphiphile in both the nanohybrids.

Antibacterial activity

Having ensured the formation of the nanotube-nanoparticle hybrids, the conjugates were taken for antibacterial studies. The antibacterial activity was tested against two Gram-positive bacteria *Bacillus subtilis* (*B. subtilis*), *Micrococcus luteus* (*M. luteus*) and two Gram-negative bacteria *Escherichia coli* (*E. coli*) and *Klebsiella aerogenes* (*K. aerogenes*). The study was carried out using colony count method. At first the antibacterial activity of the amphiphilic dispersing agents (**1, 2**) alone was tested against all the said bacterial strains. Both dispersing agents **1** and **2** were inefficient in killing all the four bacterial strains up to 100 μg/mL after 3 h of incubation under shaking condition followed by spread plating for 24 h [36]. Next the antibacterial activity of the SWCNT-dispersing agent was studied against the above mentioned bacterial strains under similar experimental conditions. SWCNT-**1** exhibited moderate bacterial killing efficacy having 35% and 45% killing of *B. subtilis* and *M. luteus*, respectively at 10 μg/mL of the dispersion (Figure 6a,b). This percent killing increased to 48% and 60% when the concentration of the nanotube dispersion was increased to 25 μg/mL. Similarly in case of SWCNT-**2** almost 40% and 45% killing of *B. subtilis* and *M. luteus*, respectively was observed at 25 μg/mL of the dispersion (Figure 6a,b). However when these nanotube dispersions (SWCNT-**1** and SWCNT-**2**) were incubated with Gram-negative bacterial strains *E. coli* and *K. aerogenes* under identical conditions, very poor killing efficacy (<10%) was observed up to 25 μg/mL concentration of both dispersions (Figure 6c,d). The outer membrane of Gram-positive bacteria is composed of peptidoglycan layer comprising of polymeric sugar, and amino acid and phosphoryl substituted teichoic and techuronic acid residues as well as carboxylate groups. However, in case of Gram-negative bacteria the cell membrane is composed of an extra layer of lipopolysaccharide (LPS) and phospholipids in addition to the peptidoglycan layer [36,64–65] This added shielding of the peptidoglycan layer by the presence of LPS might have inhibited the bacteria killing ability of dispersed SWCNTs against Gram-negative bacteria.

To widen the antibacterial spectrum of the nanotube dispersion, antibacterial activity of SWCNT-AgNP conjugates was tested against the above mentioned bacterial strains. AgNPs were synthesized under sunlight as mentioned above and the multi-modal bacteria killing efficacy of the nanohybrids was tested against both types of bacterial strains [56–58]. Encouragingly, more than 90% of Gram-positive *B. subtilis* and *M. luteus* were killed at 10 μg/mL concentration of SWCNT-**1**-AgNP and SWCNT-**2**-AgNP under similar experimental conditions (Figure 6a,b). Notable improvement in the Gram-positive bacteria killing efficiency was observed for nanotube-nanoparticle hybrids in contrast to that of SWCNT-**1** and SWCNT-**2** dispersions (35–45%) devoid of any AgNPs. Most promisingly, the Gram-negative bacteria which were almost resistant to nanotube dispersions, get efficiently killed by SWCNT-AgNP nanohybrids. SWCNT-**1**-AgNP and SWCNT-**2**-AgNP at 10 μg/mL killed 85–88% *E. coli* and more than 90% *K. aerogenes* under the experimental condition as mentioned above (Figure 6c,d). Thus these newly developed SWCNT-AgNP nanohybrids were equally effective in killing both Gram-positive and Gram-negative bacteria. The killing efficiency of AgNPs synthesized using **1** and **2** in absence of SWCNT was also studied. However, up to 50 μg/mL of AgNP-**1** and AgNP-**2**, negligible killing of (~10%) was observed against all the four bacterial strains after 3 h of incubation and spread plating

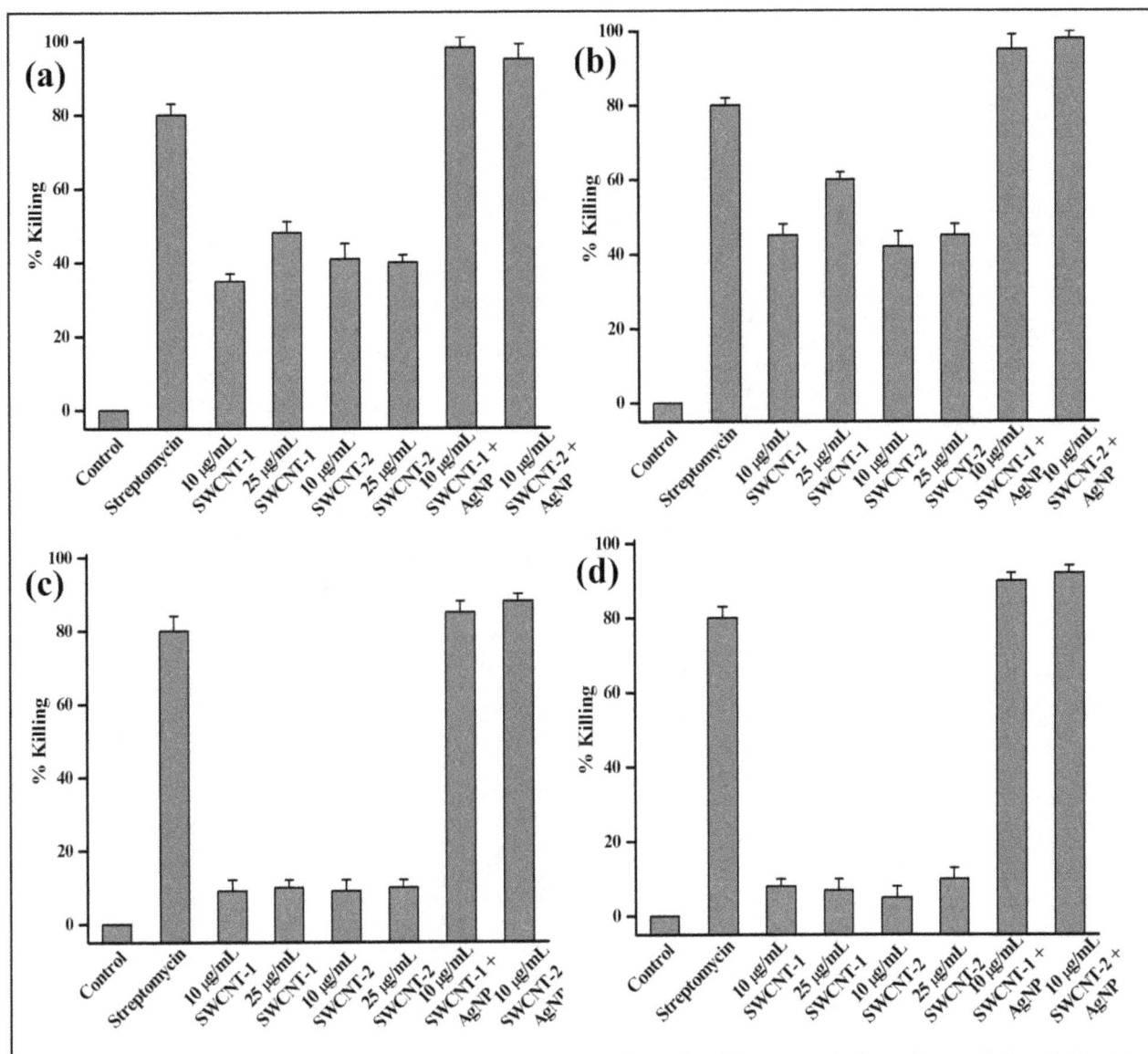

Figure 6. Percentage killing of (a) *B. subtilis,* **(b)** *M. luteus,* **(c)** *E. coli* **and (d)** *K. aragneosa* **after 3 h of incubation spread plating for 24 h with the hybrids.** Percent killing was determined using colony count method.

for 24 h (Figure S6–S9). Although in previous instances the antibacterial activity was achieved at a comparable concentration of AgNPs, which was probably due to the cumulative effect of the nature of capping agent and the nanoparticle [56]. In the present case, the neutral capping agents do not contribute to the antibacterial activity and the nanoparticles fail to exhibit notable antibacterial activity up to 50 μg/mL. These results clearly delineate the role of the dispersed SWCNTs in increasing the local concentration of AgNPs in vicinity of the bacteria membrane and ultimately facilitating the bacteria killing by the nanoparticles. The intrinsic cell permeability of the nanotubes possibly further aided the interactions of AgNPs along with the nanotube with the bacteria for disintegrating the cell membrane. The presence of nanotube in the newly fabricated SWCNT-AgNP hybrid indeed bolsters the bacteria killing efficiency of the nanoparticles presumably by acting as a cargo transporter. Hence, the antibacterial activity against both Gram-positive and Gram-

negative bacteria is achieved by these nanohybrids using the inherent multimodal killing mechanism of AgNPs along with the killing ability of dispersed SWCNTs. Although exact bacteria killing mechanism by AgNPs is not yet understood, AgNPs may act by attacking the phosphorus-containing DNA or interacting with the mitochondria leading to cell death [16]. It may also interact with the sulfur-containing proteins that would hinder the regular cell function. In addition, the release silver ion from AgNPs inside the bacterial cell may lead to bacterial death due to oxidative stress. However due to the lack of stability of the SWCNT in the absence of **1** or **2** it was not possible to study the effect of SWCNT. Similarly in the absence of the dispersing agents no AgNP synthesis takes place hence the effect of SWCNT and AgNP could not be studied separately.

Having studied the antibacterial activity by colony count method, the bacteria killing process was further investigated using live/dead bacterial viability kit. This staining kit is composed of two nucleic acid binding stains, known as SYTO 9 and propidium iodide (PI). Cell membrane permeable SYTO 9 binds to the nucleic acid of both living and dead cells, while propidium iodide can only bind to the nucleic acid of dead cells. Consequently, SYTO 9 labels live bacteria with green fluorescence and PI labels membrane-compromised bacteria with red fluorescence. Gram positive *B. subtilis* and Gram-negative *E. coli* were suspended in 0.9% NaCl and incubated for 3 h with 50 μg/mL of SWCNT-**1** and 10 μg/mL SWCNT-**1**-AgNP. The suspensions were then centrifuged and the supernatant was removed. Subsequently the bacteria were re-suspended in saline containing the live/dead kit and incubated for 30 min. The suspension was then cast onto a slide and visualised under fluorescence microscope. The green fluorescence for untreated *B. subtilis* and *E. coli* indicated that they were alive (Figure S10a,d). Incubation of these bacteria with SWCNT-**1** dispersion exhibited red fluorescence for dead *B. subtilis* (Figure S10b) and green fluorescence for live *E. coli* (Figure S10e). However, predominant presence of dead bacteria (red fluorescence) was observed upon incubation of both *B. subtilis* and *E. coli* with SWCNT-**1**-AgNP (Figure S10c,f). The observed fluorescence microscopic results corroborated well with the data obtained using colony count method.

To get an insight into the morphology of the bacteria upon incubation with the developed nanohybrids, field emission scanning electron microscopic (FESEM) images of untreated and nanohybrid treated *B. subtilis* and *E. coli* were taken (Figure 7). Both untreated cells showed the normal shape and size of prokaryotic cells (Figure 7a,d). However, upon incubation with 50 μg/mL of SWCNT-**1** dispersion, the cell membrane of Gram-positive *B. subtilis* gets disintegrated (Figure 7b) while cell membrane integrity was mostly unaffected for *E. coli* (Figure 7e). This compromised membrane structure with irregular shape is more predominant in the presence of SWCNT-**1**-AgNP (10 μg/ mL) for both *B. subtilis* as well as *E. coli* (Figure 7c,f). The disrupted membrane resulted in the loss of cytoplasmic constituents leading to cell death. The lethal effect of the SWCNT-AgNP nanohybrids was evident in comparison to only SWCNT-amphiphile dispersion. The synergic influence of inherent multimodal killing mechanism of AgNPs along with cell penetrating ability of SWCNTs made these newly developed nanohybrids very efficient in killing both Gram-positive and Gram-negative bacteria.

Eukaryotic cell viability

The biomedicinal application of these newly developed antibacterial nanohybrids will become pertinent only if they exhibit compatibility with eukaryotic cells. The antimicrobial nanohybrids should be lethal to microbes and safe to mammalian cells. The biocompatibility of these nanohybrids to normal eukaryotic cells was studied using MTT based cell viability assay. Initially nanotube dispersions of SWCNT-**1** and SWCNT-**2** at a concentration range of 0-50 μg/mL were incubated with Chinese Hamster Ovarian cells (CHO cells) for 24 h. Encouragingly, up to 50 μg/mL of the nanotube dispersion, 87% and 88% cells remained alive upon incubation with SWCNT-**1** and SWCNT-**2** were used, respectively (Figure 8a). Next the percentage cell viability of CHO cells in the presence of SWCNT-amphiphile-AgNP was checked within a concentration range of 0–25 μg/mL. With varying concentrations of the nanohybrids, promisingly, 80% cells were viable after 24 h of incubation with SWCNT-**1**-AgNP and SWCNT-**2**-AgNP (Figure 8b). Interestingly, within this range of concentration of the nanohybrids ~90% bacteria undergoes death while >80% eukaryotic cells remain unaffected. This selective killing of bacterial cells in contrast to normal cells probably takes place due to the difference in the constitution of the cell membrane. The bacterial membranes are rich in the thiol (–SH) containing protein present that facilitates its interaction with AgNP of the nanohybrid due to the affinity of silver toward sulphur leading to its death [56–58].

Antibacterial biocompatible film

The antibacterial materials which are also viable to eukaryotic cells are finding surging significance in biomedicine and tissue engineering. In fact, because of the materials induced nosocomial infections in living systems, designing antimicrobial biomaterials for tissue engineering is on the rise. To this end it would be interesting if the present antibacterial nanohybrids could be integrated with known tissue engineering scaffolds to develop superior soft nanocomposites. Agar-gelatin (2:1) hydrogel cross-

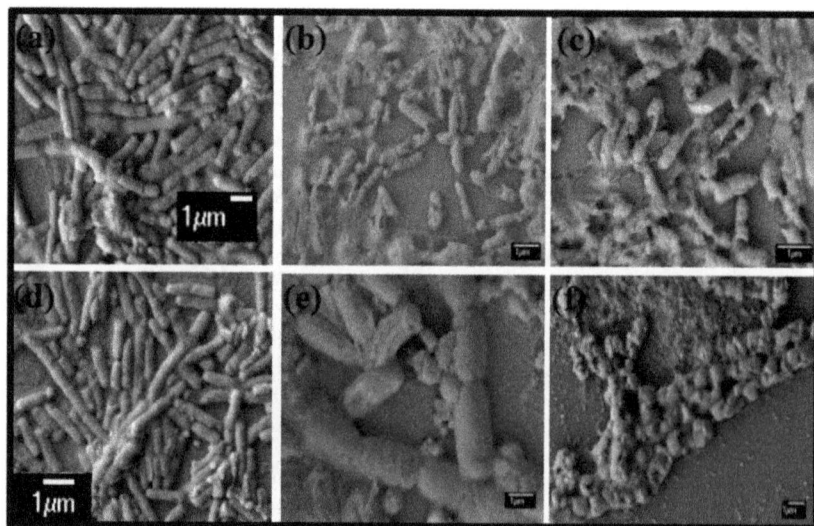

Figure 7. FESEM images of *B. subtilis* **incubated with (a) control (b) SWCNT-1 (c) SWCNT-1-AgNP and** *E. coli* **incubated with (d) control (e) SWCNT-1 and (f) SWCNT-1-AgNP.**

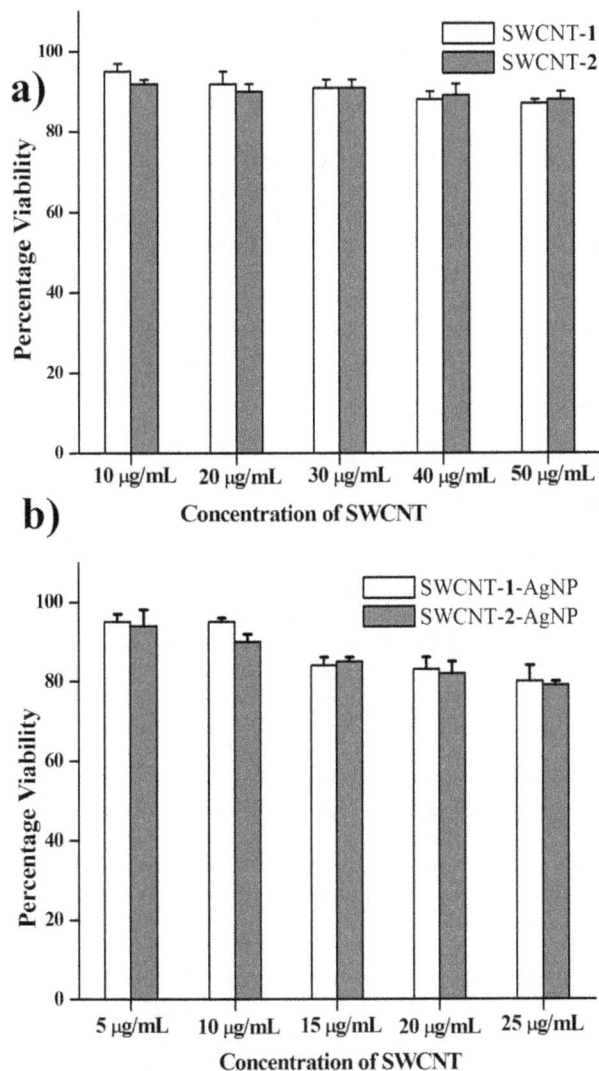

Figure 8. Percentage viability of CHO cells treated with (a) SWCNT-1 and SWCNT-2, (b) SWCNT-1-AgNP and SWCNT-2-AgNP for 24 h.

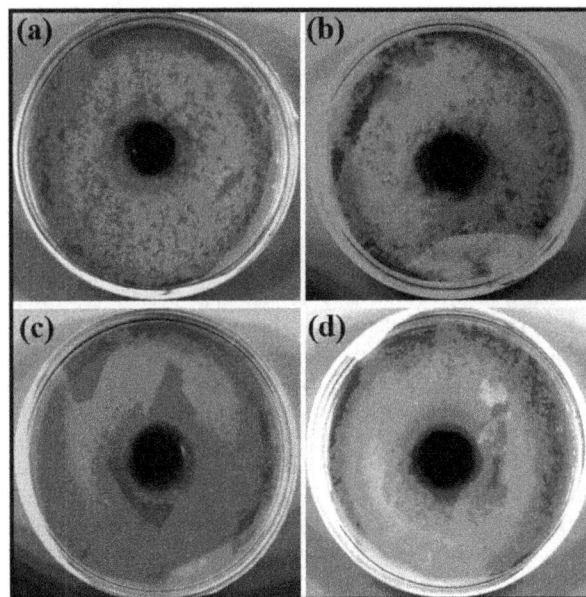

Figure 9. Antibacterial activity of agar-gelatin films against *B. subtilis* (a) SWCNT-1 (b) SWCNT-1-AgNP and *E. coli* (c) SWCNT-1 (d) SWCNT-1-AgNP composites.

affected the cell growth feature on its surface. In this regard we studied the growth and proliferation of normal CHO cells on the surface of these antibacterial polymeric films containing SWCNT-amphiphile (50 µg/mL) and SWCNT-amphiphile-AgNP (10 µg/mL). Cells grown in DMEM were seeded into 24 well plates containing the films and the viability of the cells was studied 24 h post incubation using live/dead viability assay kit (Figure S11). Encouragingly in all the instances bright green spindle shaped cells were observed and almost no red cell was seen, which indicates the healthy nature and growth eukaryotic cells on the agar-gelatin film.

Conclusion

In summary, biocompatible SWCNT dispersing was prepared and characterized. The SWCNT-dispersion was only lethal towards Gram-positive bacteria. However no bactericidal activity was observed against the Gram-negative bacteria. The dispersions were subsequently used for the synthesis and capping of AgNPs under sunlight. The newly developed AgNP based SWCNT included self assemblies had potent (~90%) antibacterial activity against both Gram-positive and Gram-negative bacteria. Interestingly the new hybrids showed substantial cell viability towards normal eukaryotic CHO cells after 24 h of incubation. Finally, inclusion of these nanotube-nanoparticle hybrids into well known tissue engineering scaffold agar-gelatin films, made the tissue engineering scaffold intrinsically antibacterial to both Gram-positive and Gram-negative bacteria despite being non-toxic toward mammalian cells. Therefore, the designed soft nanocomposite promises to have immense implications in biomedicine including tissue engineering.

Materials and Methods

Materials

Silica gel of 60–120 mesh, L-tryptophan, L-tyrosine, cetyl alcohol, *N*, *N*'-dicyclohexylcarbodiimide (DCC), 4-*N*, *N*-(dimethy-

linked with glutaraldehyde is well-known matrix that has been used for the proliferation of mice fibroblast cells (NIH3T3) [56,66]. However, such materials need to have the antibacterial property to be utilized as ideal tissue engineering scaffold. To this objective, the newly developed SWCNT-amphiphile dispersion (50 µg/mL) and SWCNT-amphiphile-AgNP nanohybrids (10 µg/mL) were infused into the agar-gelatin matrices (2:1). As expected in the absence of either of the dispersion or nanohybrids, normal growth of both *B. subtilis* and *E. coli* was observed. However in the presence of nanohybrid doped material, formation of a clear zone of inhibition was noted (Figure 9, Table S1). This zone of inhibition was less in case of SWCNT-**1** doped polymeric films in both bacterial strains (Figure 9a,c). In fact in case of the Gram-negative strain almost no zone of inhibition could be determined. Importantly, the zone of inhibition increased in the presence of SWCNT-**1**-AgNP nanohybrids (Figure 9b,d, Table S1). Hence, the inclusion of these soft nanohybrids made the tissue engineering scaffold agar-gelatin intrinsically antibacterial.

Now it has become necessary to investigate whether the infusion of the antibacterial soft nanohybrids with the agar-gelatin film lamino) pyridine (DMAP), 1-hydroxybenzotriazole (HOBT), succinic anhydride, solvents and all other reagents were procured from SRL, India. Milli-Q water was used throughout the study. Thin layer chromatography was performed on precoated silica gel 60-F_{254} plates of Merck. $CDCl_3$, Amberlite Ira-400 chloride ion exchange resins were obtained from Aldrich Chemical Company. Ethylene diaminetetraacetic acid (EDTA) and reagents required to prepare the nutrient broth culture medium like peptone, yeast extract, and agar powder were purchased from Himedia Chemical Company, India. The live/dead baclight bacterial viability kit, live/dead viability/cytotoxicity kit for mammalian cells were purchased from Molecular Probes, Invitrogen Chemical Company. All materials used in the cell culture study such as gelatin, DMEM, heat inactivated FBS, trypsin from porcine pancreas and 3-(4,5-dimethyl-2-thiazolyl)-2,5-diphenyl-2H-tetrazolium bromide (MTT), PEG ($M_n = 550$) were obtained from Sigma Aldrich Chemical Company. ^1H NMR spectra were recorded on AVANCE 300 MHz (BRUKER) spectrometer. The UV-visible absorption spectra were recorded on a Perkin Elmer Lambda 25 spectrophotometer. Mass Spectrometric (MS) data were acquired by Electron Spray Ionization (ESI) technique on a Q-tof-Micro Quadruple mass spectrophotometer, Micromass. Raman spectra was recorded using laser light (514.5 nm, scattering angle: 908, integration time: 10 s, 20 scans, 75 mW) on a Horiba Jobin Yvon instrument (Model T64000). TEM experiments were performed on a JEOL JEM 2010 high-resolution microscope operated at an accelerating voltage 200 kV. AFM was performed on Veeco, model AP0100 microscope in non-contact mode. Field emission scanning electron microscopy (FESEM) was performed on JEOL-6700F microscope.

Synthesis of amphiphiles 1 and 2

Briefly, Boc-protected L-amino acids were coupled with n-hexadecanol using DCC (1 equivalent) and catalytic amount of DMAP in presence of 1 equivalent of HOBT in dry dichloromethane (DCM). Boc-protected ester was then purified through column chromatography using 60–120 mesh silica gel and acetone/hexane as the eluent. Column purified materials were then subjected to deprotection by trifluoroacetic acid in dry DCM. A drop of anisole was added during the preparation of compound 1 and this was not required for compound 2. After 2 h of stirring, solvents were removed on a rotary evaporator and the mixture was taken in ethyl acetate. The organic part was washed with 10% aqueous sodium carbonate solution followed by brine to neutrality. The organic part was concentrated to get the corresponding amines. The obtained amine was then refluxed with succinic anhydride in dry DCM for 6 h. It was then washed with brine to remove the unreacted acid anhydride. The acid thus obtained was coupled with PEG unit (having $M_n = 550$) using DCC coupling reaction as mentioned. The coupled product was then purified through column chromatography using 60–120 mesh silica gel and $CHCl_3$/MeOH as the eluent. Overall yield was ~70–80%.

Characterization of 1 and 2

^1H NMR of 1 (500 MHz, $CDCl_3$, Me_4Si, 25 °C) $\delta = 0.82$–0.85 (m, 3H), 1.22–1.73 (m, 28H), 2.49–2.70 (m, 4H), 3.25–3.29 (m, 2H), 3.34 (s, 3H), 3.50–3.69 (m, 46H), 3.70–4.10 (m, 2H), 4.23–4.24 (m, 2H), 4.29–4.31 (m, 1H), 7.02–7.52 (m, 5H). MS (ESI): m/z calculated for $C_{56}H_{98}O_{17}N_2$: 1070.69; found: 1071.3874 [M + H]$^+$.

^1H NMR of 2 (500 MHz, $CDCl_3$, Me_4Si, 25 °C) $\delta = 0.85$–0.88 (m, 3H), 1.25–1.63 (m, 28H), 2.43–2.64 (m, 4H), 2.93–3.18 (m, 2H), 3.37 (s, 3H), 3.45–3.81 (m, 46H), 4.09–4.24 (m, 4H), 4.76–

4.78 (m, 1H), 6.29–7.26 (d, 4H). MS (ESI): m/z calculated for $C_{54}H_{97}O_{18}N$: 1047.67; found: 1048.3489 [M + H]$^+$.

Preparation of SWCNT-amphiphile dispersion

To an aqueous solution (4 mL) of the amphiphiles 1 and 2 (2.5 mg/mL) SWCNT (1 mg) was added. The solution was tip-sonicated and bath sonicated for 30 min. The suspension was centrifuged at 2500 g for 90 min to remove the heavy bundles. The amount of the dispersed SWCNTs in the supernatant was calculated from the observed absorbance value at 550 nm that was derived from the previously reported calibration plot of absorbance versus concentration using sodium dodecylbenzene sulfonate (SDBS). The percentage dispersion was calculated as the ratio of the amount of SWCNTs in the dispersion to the amount of SWCNTs initially added.

Sample preparation for TEM, AFM, Zeta (ζ)-potential and Raman spectroscopy

The SWCNT suspension obtained after centrifugation at 2500 g was ultracentrifuged at 375000 g to remove excess amphiphile and the pellet was re-dispersed in water (SWCNT-1 and SWCNT-2). The aqueous dispersion obtained was used for ζ-potential, TEM, AFM and Raman spectroscopy experiments. A drop of the SWCNT suspension was placed on a 300-mesh Cu-coated TEM grid and dried under vacuum for 4 h before taking the image. Similarly in case of AFM studies, a drop of the dispersion was cast on a freshly cleaved mica surface and the samples were air-dried overnight before imaging. The bundle diameter was calculated from the height profile of the nanotubes. From 20 AFM images of a statistical analysis of the bundle diameter and nanotue length were calculated by plotting histogram [49]. Similarly Raman was recorded by excitation of the sample using 514.5 nm laser.

In situ synthesis of silver nanoparticle (AgNP)

SWCNT (1 mg) was dispersed by sonication in 2.5 mg/mL amphiphile solution as mentioned above. The solution was centrifuged to remove the heavier bundles. To the supernatant $AgNO_3$ was added in the ratio 1:10 ($AgNO_3$: amphiphile). The solution was exposed to sunlight for 15 min and the spectra of the samples were taken. A clear peak generation was observed having maxima at 425 nm and 410 nm in case of SWCNT-1 and SWCNT-2, respectively. The peaks intensified upon exposure of the hybrid for 30 min. The prepared samples were centrifuged at 375000 g twice and the supernatant was discarded to remove excess unbound amphiphile and unconverted $AgNO_3$.

TEM and AFM sample preparation of the SWCNT-amphiphile-AgNP nanohybrids

SWCNT-1-AgNP (5 μL) of and SWCNT-2-AgNP (5 μL) composites were placed on 300-mesh carbon coated copper grid and freshly cleaved mica and dried under vacuum for 4 h before taking TEM and AFM images.

Quantification of the SWCNT-amphiphile-AgNP nanohybrids

The samples were then subjected to thermo gravimetric analysis (TGA). SWCNT (30 mg) was dispersed using 1 and 2 as described above. The supernatant was collected and half of the obtained supernatant was taken for the synthesis of AgNP and $AgNO_3$ was added in the ratio as mentioned above. Next, all suspensions were centrifuged twice at 375000 g to remove excess surfactant and unconverted $AgNO_3$. The obtained pellet was then lyophilized.

TGA of each conjugate (SWCNT-**1**, SWCNT-**2**, SWCNT-**1**-AgNP and SWCNT-**2**-AgNP) was done where the samples were heated to 800 °C. The thermal decomposition pattern was monitored. Amphiphiles **1** and **2** as well as AgNP-**1** and AgNP-**2** were also subjected to the same process of heating as control experiments.

Microorganisms and culture conditions

The antibacterial activity of the nanohybrids was tested against Gram-positive and Gram-negative bacteria. The nutrient broth medium was prepared using peptone (5 g), yeast extract (3 g) in 1 L sterile water. Solid medium for all antibacterial experiments was done by addition of 15 g agar was added in 1 L of the above prepared nutrient broth medium. All the bacteria were purchased from Institute of Microbial Technology, Chandigarh, India. For the experiments, a representative single colony of bacteria was picked up with a wire loop and that loopful was spread on nutrient agar slant to give single colonies and incubated at 37 °C for 24 h. These cultures were diluted as per requirement to give a working concentration in the range of 10^6–10^9 colony forming units (cfu)/mL.

Measurements of antibacterial activity

Antimicrobial activities of SWCNT samples were studied against *Bacillus subtilis* (*B. subtilis*), *Micrococcus luteus* (*M. luteus*), *Escherichia coli* (*E. coli*) and *Klebsiella aerogenes* (*K. aerogenes*). Purified SWCNT solid samples were first dispersed in solution of **1** and **2** (2.5 µg/mL) by sonication and centrifugation. The AgNP was synthesized as described above and the excess AgNO$_3$ and surfactant was removed by ultra centrifugation at 37500 g. A 10 mL portion of dispersion was incubated with 1 mL of bacterial suspensions (10^6–10^7 cfu/mL) for 2 h under shaking at 37 °C. The antimicrobial evaluations were carried out by a colony forming count method. For the colony forming count method, 100 µL serial 10-fold dilutions with saline solution was spread onto agar plates and left to grow overnight at 37 °C. Colonies were counted and compared with control plates to calculate percentage killing. All treatments were prepared in duplicate and repeated on at least three separate occasions.

FESEM of nanocomposite treated bacteria

The morphological changes of B. subtilis and E. coli were investigated by field emission scanning electron microscopy (FESEM). Treated bacterial suspensions were concentrated by centrifugation at 5000 rpm and then the cells were dropped on a glass slide to dry at room temperature. The dried samples were sputter coated with gold for FESEM imaging.

Fluorescence microscopic study for bacteria

To examine bacterial cell viability live/dead bacterial kit was used. The kit consists of a mixture of SYTO 9 and propidium iodide which are two nucleic-acid binding stains. *B. subtilis* (5×10^6–7.5×10^6 cfu/mL) and *E. coli* (3.75×10^7–7.5×10^7 cfu/mL) cells (1 mL) were treated with SWCNT-**1** (50 µg/mL) and SWCNT-**1**-AgNP (10 µg/mL) and also untreated cells were taken in centrifuge tube as control. The mixtures were centrifuged at 5,000 rpm for 5 min. Then, media was removed and the cells were re-dispersed in 0.9 wt% saline. Finally, the dye mixture was added and incubated in dark at room temperature for 15–20 min. After incubation, 5 µL of the solution mixture was mounted over microscope slides, air-dried and viewed under the microscope (BX61, Olympus) (ex/em ~495 nm/515 nm for SYTO 9 and ex/em ~ 495 nm/635 nm for propidium iodide).

Cell cultures

Mouse fibroblast NIH3T3 cells were obtained from National Center for Cell Science (NCCS), Pune (India), and cultured in DMEM medium containing 10% FBS, 100 mg/L streptomycin and 100 IU/mL penicillin. Cells were grown in 25 mL cell culture flask and incubated at 37 °C in a humidified atmosphere of 5% CO_2 to approximately 70–80% confluence. Fresh media was added after every 2–3 days and subculture was performed every 7 days. Next, the adherent cells were detached from the surface of the culture flask by trypsinization and seeded into plates for cytotoxicity assay and imaging experiments.

Cytotoxicity assay

Cytotoxicity of nanocomposites SWCNT-**1**, SWCNT-**2**, SWCNT-**1**-AgNP and SWCNT-**2**-AgNP were assessed by the microculture MTT reduction assay [67]. This assay is based on the reduction of a soluble tetrazolium salt by mitochondrial dehydrogenase of the viable cells to water insoluble coloured product, formazan. The amount of formazan formed can be measured spectrophotometrically after dissolution of the dye in DMSO. The activity of the enzyme and the amount of the formazan produced is proportional to the number of cells alive. Cells were seeded at a density 15,000 cells per well in a 96-well microtiter plate for 18–24 h before the assay. Stock solutions of all the composites were prepared in water. Sequential dilutions of these stock solutions were done during the experiment to vary the concentrations of SWCNT-amphiphile (5–50 µg/mL) and SWCNT-amphiphile-AgNP (5–25 µg/mL) in the microtiter plate. The cells were incubated for 4 h at 37 °C under 5% CO_2. Then, 15 µL MTT stock solution (5 mg/mL) in phosphate buffer saline was added to the above mixture and further incubated for 4 h. The precipitated formazan was dissolved thoroughly in DMSO and absorbance at 570 nm was measured using BioTek Elisa Reader. The number of surviving cells were expressed as percent viability = (A_{570}(treated cells)-background/A_{570}(untreated cells)-background) $\times100$.

Preparation of agar-gelatin gel film

The mixture (1 wt%) of agar and gelatin in the weight ratio 2:1 was taken in PBS (pH 7.4) and dissolved by heating. In case of SWCNT and SWCNT-amphiphile-AgNP containing film preparation, nanocomposite solution was added to this homogeneous mixture of agar-gelatin so that 50 µg/mL of SWCNT-**1** and 10 µg/mL of SWCNT-**1**-AgNP could be attained. The components were crosslinked by the addition of glutaraldehyde (0.15 wt%) to the hot solution. Each solution (1 mL) was poured into 24-wells tissue culture flask and left at room temperature. Then the plates containing gels were dried in the oven for 12 h at 50 °C to form thin films. These prepared films were treated with 0.1 mM glycine for 1 h to block the remaining aldehyde group. The films were then washed several times with Milli-Q water to remove excess glycine and then with PBS to neutralize the surface. These films were then again dried and UV sterilized overnight and kept in sterile vacuum desiccators for further experiment.

Antibacterial activity of agar-gelatin films

The antibacterial activity of agar-gelatin films against *B. Subtilis* and *E. coli* was followed in nutrient agar plates. The agar-gelatin films with and without the nanohybrids was placed on the middle of the freshly prepared agar plates. For this, each bacterium was cultured on nutrient agar slant at 37 °C for 24 h. These overnight cultures of bacteria were diluted as required to get a concentration of ~10^5 cfu/mL. This bacteria containing solution (1 mL) was added to nutrient agar plate. The plates were then incubated for 24 hours at 37 °C. The bacteria killing ability of the films was followed by measuring the zone of inhibition in the agar plate.

Biocompatibility of agar-gelatin films

The NIH3T3 cell attachment studies were done on control well (24 well plate) without any film and on agar-gelatin films with SWCNT-1, SWCNT-2, SWCNT-1-AgNP and SWCNT-2-AgNP. Films were soaked with cell culture media for 3 h. After 3 h the media was removed and 5×10^5 cells were seeded in each well with 1 mL media. The cells were incubated for 24 h in a 5% CO_2 atmosphere. After 24 h the adherent cells were washed with PBS and cell viability was examined under a fluorescence microscope using the live/dead viability/cytotoxicity kit for mammalian cells. The kit contains Calcein AM and Ethidium homodimer-1(EthD-1). The supplied 2 mM EthD-1 stock solution (4 μL) and 4 mM calcein AM stock solution (1 μL) was added to 2 mL of sterile, tissue culture-grade PBS and the mixture was vortexed to ensure thorough mixing. The final stock solution (500 μL) of calcein AM (2 μM) and EthD-1 (4 μM) was then added directly to each well containing NIH3T3 cells. After 30 min incubation in darkness, the cells were viewed under Olympus IX51 inverted microscope (ex/em ~495 nm/515 nm for calcein AM and ex/em ~ 495 nm/635 nm) for EthD-1.

Supporting Information

Figure S1   Histogram for the determination of average bundle diameter of the nanotubes.

Figure S2   Histogram for the determination of average length of the nanotubes.

Figure S3   Raman spectra of pristine SWCNT.

Figure S4   Time dependent UV-vis spectra of synthesized AgNP by (a) SWCNT-1 and (b) SWCNT-2 after (i) 15 min and (ii) 30 min.

Figure S5   TGA analysis of AgNP-1 and AgNP-2.

Figure S6   Percentage killing of _B. subtillis_ after 3 h of incubation and spread plating for 24 h with the varying concentration of AgNP capped with 1 and 2. Percent killing was determined using colony count method.

Figure S7   Percentage killing of _M. leuteus_ after 3 h of incubation and spread plating for 24 h with the varying concentration of AgNP capped with 1 and 2. Percent killing was determined using colony count method.

Figure S8   Percentage killing of _E. coli_ after 3 h of incubation and spread plating for 24 h with the varying concentration of AgNP capped with 1 and 2. Percent killing was determined using colony count method.

Figure S9   Percentage killing of _K. aragneosa_ after 3 h of incubation and spread plating for 24 h with the varying concentration of AgNP capped with 1 and 2. Percent killing was determined using colony count method.

Figure S10   Fluorescence micrographs of _B. subtilis_ incubated with (a) control (b) SWCNT-1 (c) SWCNT-1-AgNP and _E. coli_ incubated with (d) control (e) SWCNT-1 and (f) SWCNT-1-AgNP followed by incubation with live/dead kit.

Figure S11   Live dead images of CHO cells grown on agar gelatin films containing (a,b) SWCNT-1 (c,d) SWCNT-1-AgNP (e,f) SWCNT-2 and (g,h) SWCNT-2-AgNP.

Table S1   Zone of Inhibition (mm) for Agar-Gelatin Films Containing Soft Nanohybrids.

Author Contributions

Conceived and designed the experiments: SB SKM PKD. Performed the experiments: SB SKM. Analyzed the data: SB SKM PKD. Contributed to the writing of the manuscript: SB PKD.

References

1. Makovitzki A, Avrahami D, Shai Y (2006) Ultrashort antibacterial and antifungal lipopeptides. Proc Natl Acad Sci U S A 103: 15997–16002.
2. Pritz S, Pätzel M, Szeimies G, Dathea M, Bienert M (2007) Synthesis of a chiral amino acid with bicyclo[1.1.1]pentane moiety and its incorporation into linear and cyclic antimicrobial peptides. Org Biomol Chem 5: 1789–1794.
3. Zumbuhel A, Ferreira L, Kuhn D, Astashkina A, Long L, et al. (2007) Antifungal hydrogels. Proc Natl Acad Sci U S A 104: 12994–12998.
4. Xing B, Yu CW, Chow KH, Ho PL, Fu D, et al. (2002) Hydrophobic interaction and hydrogen bonding cooperatively confer a vancomycin hydrogel: a potential candidate for biomaterials. J Am Chem Soc 124: 14846–14847.
5. Yang Z, Xu K, Guo Z, Guo Z, Xu B (2007) Intracellular enzymatic formation of nanofibers results in hydrogelation and regulated cell death. Adv Mater 19: 3152–3156.
6. Tiller JC (2003) Increasing the local concentration of drugs by hydrogel formation. Angew Chem Int Ed Engl 42: 3072–3075.
7. Tiller JC, Lee SB, Lewis K, Klibanov AM (2002) Polymer surfaces derivatized with poly[vinyl-n-hexylpyridinium] kill airborne and waterborne bacteria. Biotechnol Bioeng 79: 465–471.
8. Sieber SA, Marahiel MA (2005) Molecular mechanisms underlying nonribosomal peptide synthesis: approaches to new antibiotics. Chem Rev 105: 715–738.
9. Brown ED, Wright GD (2005) New targets and screening approaches in antimicrobial drug discovery. Chem Rev 105: 759–774.
10. Haldar J, Kondaiah P, Bhattacharya S. (2005) Synthesis and antibacterial properties of novel hydrolyzable cationic amphiphiles: incorporation of multiple head groups leads to impressive antibacterial activity. J Med Chem 48: 3823–3831.
11. Waschinski CJ, Zimmermann J, Salz U, Hutzler R, Sadowski G, et al. (2008) Design of contact-active antimicrobial acrylate based materials using biocidal macromers. Adv Mater 20: 104–108.
12. Waschinski CJ, Barnert S, Theobald A, Schubert R, Kleinschmidt F, et al. (2008) Insights in the antibacterial action of poly(methyloxazoline)s with a biocidal end group and varying satellite groups. Biomacromolecules 9: 1764–1771.
13. Panáček A, Kvítek L, Prucek R, Kolář M, Večeřová R, et al. (2006) Silver colloid nanoparticles: synthesis, characterization and their antibacterial activity. J Phys Chem B 110: 16248–16253.
14. Morones JR, Elechiguerra JL, Camacho A, Holt K, Kouri JB, et al. (2005) The bactericidal effect of silver nanoparticles. Nanotechnology 16: 2346–2353.
15. Rai M, Yadav A, Gade A (2009) Silver nanoparticles as a new generation of antimicrobials. Biotechnol Adv 27: 76–83.
16. Jain J, Arora S, Rajwade JM, Omray P, Khandelwal S, et al. (2009) Silver nanoparticles in therapeutics: development of an antimicrobial gel formulation for topical use. Mol Pharmaceutics 6: 1388–1401.
17. Wigginton NS, Titta AD, Piccapietra F, Dobias J, Nesatyy VJ, et al. (2010) Binding of silver nanoparticles to bacterial proteins depends on surface modifications and inhibits enzymatic activity. Environ Sci Technol 44: 2163–2168.

18. Lara HH, Ayala-Núñez NV, Turrent LdCI, Padilla CR (2010) Bactericidal effect of silver nanoparticles against multidrug resistant bacteria. World J Microbiol Biotechnol 26: 615–621.

19. Kvítek L, Panáček A, Soukupová J, Kolář M, Večeřová R, et al. (2008) Effect of surfactants and polymers on stability and antibacterial activity of silver nanoparticles. J Phys Chem C 112: 5825–5834.

20. Kim JS, Kuk E, Yu KN, Kim JH, Park SJ, et al. (2007) Antimicrobial effects of silver nanoparticles. Nanomedicine 3: 95–101.

21. Sondi I, Salopek-Sondi B (2004) Silver nanoparticles as antimicrobial agent: a case study on E. coli as a model for gram-negative bacteria. J Colloid Interface Sci 275: 177–182.

22. Kumar A, Vemula PK, Ajayan PM, John G (2008) Silver-nanoparticle-embedded antimicrobial paints based on vegetable oil. Nat Mater 7: 236–241.

23. Gupta A, Matsui K, Lo JF, Silver S. (1999) Molecular basis for resistance to silver cations in Salmonella. Nat Med 5: 183–188.

24. Li XZ, Nikaido H, Williams K (1997) Silver-resistant mutants of *Escherichia coli* display active efflux of Ag⁺ and are deficient in porins. J Bacteriol 179: 6127–6132.

25. Daima HK, Selvakannan PR, Shukla R, Bhargava SK, Bansal V (2013) Fine-tuning the antimicrobial profile of biocompatible gold nanoparticles by sequential surface functionalization using polyoxometalates and lysine. PLOS ONE 8: e79676

26. Daima HK, Selvakannan PR, Kandjani AE, Shukla R, Bhargava SK, et al. (2014) Synergistic influence of polyoxometalate surface corona towards enhancing the antibacterial performance of tyrosine-capped Ag nanoparticles. Nanoscale 6: 758–765.

27. Xu X, Zhou M (2008) Antimicrobial gelatin nanofibers containing silver nanoparticles. Fiber Polym 9: 685–690.

28. Ajayan PM (1999) Nanotubes from carbon. Chem Rev 99: 1787–1800.

29. Shao L, Tobias G, Salzmann CG, Ballesteros B, Hong SY, et al. (2007) Removal of amorphous carbon for the efficient sidewall functionalisation of single-walled carbon nanotubes. Chem Commun 5090–5092.

30. Goodwin AP, Tabakman SM, Welsher K, Sherlock SP, Prencipe G, et al. (2009) Phospholipid-dextran with a single coupling point: a useful amphiphile for functionalization of nanomaterials. J Am Chem Soc 131: 289–296.

31. Prato M, Kostarelos K, Bianco A (2008) Functionalized carbon nanotubes in drug design and discovery. Acc Chem Res 41: 60–68.

32. Yuan W, Jiang G, Che J, Qi X, Xu R, et al. (2008) Deposition of silver nanoparticles on multiwalled carbon nanotubes grafted with hyperbranched poly[amidoamine] and their antimicrobial effects. J Phys Chem C 112: 18754–18759.

33. Kang S, Pinault M, Pfefferle LD, Elimelech M (2007) Single-walled carbon nanotubes exhibit strong antimicrobial activity. Langmuir 23: 8670–8673.

34. Kang S, Herzberg M, Rodrigues DF, Elimelech M (2008) Antibacterial effects of carbon nanotubes: size does matter! Langmuir 24: 6409–6413.

35. Arias LR, Yang L (2009) Inactivation of bacterial pathogens by carbon nanotubes in suspensions. Langmuir 25: 3003–3012.

36. Liu S, Wei L, Hao L, Fang N, Chang MW, et al. (2009) Sharper and faster "nano darts" kill more bacteria: a study of antibacterial activity of individually dispersed pristine single-walled carbon nanotube. ACS Nano 3: 3891–3902.

37. Niu A, Han Y, Wu J, Yu N, Xu Q (2010) Synthesis of one-dimensional carbon nanomaterials wrapped by silver nanoparticles and their antibacterial behavior. J Phys Chem C 114: 12728–12735.

38. Deokar AR, Lin LY, Chang CC, Ling YC (2013) Single-walled carbon nanotube coated antibacterial paper: preparation and mechanistic study. J Mater Chem B 1: 2639–2646.

39. Pramanik S, Konwarh R, Barua N, Buragohain AK, Karak N (2014) Bio-based hyperbranched poly(ester amide)–mwcnt nanocomposites: multimodalities at the biointerface. Biomater Sci 2: 192–202.

40. Jung JH, Hwang GB, Lee JE, Bae GN (2011) Preparation of airborne ag/cnt hybrid nanoparticles using an aerosol process and their application to antimicrobial air filtration. Langmuir 27: 10256–10264.

41. Vargas JH, Campos JBG, Romero JL, Prokhorov E, Barcenas GL, et al. (2014) Chitosan/MWCNTs-decorated with silver nanoparticle composites: dielectric and antibacterial characterization. J Appl Polym Sci 2014, DOI: 10.1002/APP.40214.

42. Witus LS, Rocha JDR, Yuwono VM, Paramonov SE, Weisman RB, et al. (2007) Peptides that non-covalently functionalize single-walled carbon nanotubes to give controlled solubility characteristics. J Mater Chem 17: 1909–1915.

43. Backes C, Schmidt CD, Rosenlehner K, Hauke F, Coleman JN, et al. (2010) Nanotube surfactant design: the versatility of water-soluble perylene bisimides. Adv Mater 22: 788–802.

44. Usrey ML, Strano MS (2009) Controlling single-walled carbon nanotube surface adsorption with covalent and noncovalent functionalization. J Phys Chem C 113: 12443–12453.

45. Li Y, Cousins BG, Ulijn RV, Kinloch IA (2009) A study of the dynamic interaction of surfactants with graphite and carbon nanotubes using Fmoc-amino acids as a model system. Langmuir 25: 11760–11767.

46. Das D, Das PK (2009) Superior activity of structurally deprived enzyme−carbon nanotube hybrids in cationic reverse micelles. Langmuir 25: 4421–4428.

47. Cousins BG, Das AK, Sharma R, Li Y, McNamara JP, et al. (2009) Enzyme-activated surfactants for dispersion of carbon nanotubes. Small 5: 587–590.

48. Backes C, Schmidt CD, Hauke F, Bottcher C, Hirsch A (2009) High population of individualized swcnts through the adsorption of water-soluble perylenes. J Am Chem Soc 131: 2172–2184.

49. Pal A, Chhikara BS, Govindaraj A, Bhattacharya S, Rao CNR (2008) Synthesis and properties of novel nanocomposites made of single-walled carbon nanotubes and low molecular mass organogels and their thermo-responsive behavior triggered by near IR radiation. J Mater Chem 18: 2593–2600.

50. Xie H, O-Acevedo A, Zorbas V, Baughman RH, Draper RK, et al. (2005) Peptide cross-linking modulated stability and assembly of peptide-wrapped single-walled carbon nanotubes. J Mater Chem 15: 1734–1742.

51. Becraft EJ, Klimenko AS, Dieckmann GR (2009) Influence of alternating L-/D-amino acid chiralities and disulfide bond geometry on the capacity of cysteine-containing reversible cyclic peptides to disperse carbon nanotubes. Pept Sci 92: 212–221.

52. Dutta S, Kar T, Brahmachari S, Das PK (2012) pH-responsive reversible dispersion of biocompatible swcnt/graphene-amphiphile hybrids. J Mater Chem 22: 6623–6631.

53. Brahmachari S, Das D, Das PK (2010) Superior SWNT dispersion by amino acid based amphiphiles: designing biocompatible cationic nanohybrids. Chem Commun 46: 8386–8388.

54. Brahmachari S, Das D, Shome A, Das PK (2011) Single-walled nanotube/amphiphile hybrids for efficacious protein delivery: rational modification of dispersing agents. Angew Chem Int Ed 50: 11243–11247.

55. Brahmachari S, Ghosh M, Dutta S, Das PK (2014) Biotinylated amphiphile-single walled carbon nanotube conjugate for target-specific delivery to cancer cells. J Mater Chem B 2: 1160–1173.

56. Shome, A.; Dutta, S.; Maiti, S.; Das, PK. (2011) In situ synthesized Ag nanoparticle in self-assemblies of amino acid based amphiphilic hydrogelators: development of antibacterial soft nanocomposites. Soft Matter 7: 3011–3022.

57. Dutta S, Shome A, Kar T, Das PK (2011) Counterion-induced modulation in the antimicrobial activity and biocompatibility of amphiphilic hydrogelators: influence of in-situ-synthesized Ag-nanoparticle on the bactericidal property. Langmuir 27: 5000–5008.

58. Dutta S, Kar T, Mandal D, Das PK (2013) Structure and properties of cholesterol-based hydrogelators with varying hydrophilic terminals: biocompatibility and development of antibacterial soft nanocomposites. Langmuir 29: 316–327.

59. Adhikari B, Banerjee A (2010) A short-peptide-based hydrogel: a template for the *in situ* synthesis of fluorescent silver nanoclusters by using sunlight. Chem Eur J 16: 13698–13705.

60. Mandal D, Ghosh M, Maiti S, Das K, Das PK (2014) Water-in-oil microemulsion doped with gold nanoparticle decorated single walled carbon nanotube: scaffold for enhancing lipase activity. Colloids and Surfaces B: Biointerfaces 113: 442–449.

61. Zhao B, Hu H, Yu A, Perea D, Haddon RC (2005) Synthesis and characterization of water soluble single-walled carbon nanotube graft copolymers. J Am Chem Soc 127: 8197–8203.

62. Salzmann CG, Lee GKC, Ward MAH, Chu BTT, Green MLH (2008) Highly hydrophilic and stable polypeptide/single-wall carbon nanotube conjugates. J Mater Chem 18: 1977–1983.

63. Salzmann CG, Chu BTT, Tobias G, Llewellyn SA, Green MLH (2007) Quantitative assessment of carbon nanotube dispersions by raman spectroscopy. Carbon 45: 907–912.

64. Roy S, Das PK (2008) Antibacterial hydrogels of amino acid-based cationic amphiphiles. Biotechnol Bioeng 100: 756–764.

65. Mitra RN, Shome A, Paul P, Das PK (2009) Antimicrobial activity, biocompatibility and hydrogelation ability of dipeptide-based amphiphiles. Org Biomol Chem 7: 94–102.

66. Verma V, Verma P, Kar S, Ray P, Ray AR (2007) Fabrication of agar-gelatin hybrid scaffolds using a novel entrapment method for in vitro tissue engineering applications. Biotechnol Bioeng 96: 392–400.

67. Hansen MB, Nielsen SE, Berg K (1989) Re-examination and further development of a precise and rapid dye method for measuring cell growth/cell kill. J Immunol Methods 119: 203–210.

## 9

The Functional Potential of Microbial Communities in Hydraulic Fracturing Source Water and Produced Water from Natural Gas Extraction Characterized by Metagenomic Sequencing

Arvind Murali Mohan[1,2], Kyle J. Bibby[1,3,4], Daniel Lipus[1,3], Richard W. Hammack[1], Kelvin B. Gregory[1,2]*

1 National Energy Technology Laboratory, Pittsburgh, Pennsylvania, United States of America, 2 Department of Civil and Environmental Engineering, Carnegie Mellon University, Pittsburgh, Pennsylvania, United States of America, 3 Department of Civil and Environmental Engineering, University of Pittsburgh, Pittsburgh, Pennsylvania, United States of America, 4 Department of Computational and Systems Biology, University of Pittsburgh Medical School, Pittsburgh, Pennsylvania, United States of America

Abstract

Microbial activity in produced water from hydraulic fracturing operations can lead to undesired environmental impacts and increase gas production costs. However, the metabolic profile of these microbial communities is not well understood. Here, for the first time, we present results from a shotgun metagenome of microbial communities in both hydraulic fracturing source water and wastewater produced by hydraulic fracturing. Taxonomic analyses showed an increase in anaerobic/ facultative anaerobic classes related to *Clostridia*, *Gammaproteobacteria*, *Bacteroidia* and *Epsilonproteobacteria* in produced water as compared to predominantly aerobic *Alphaproteobacteria* in the fracturing source water. The metabolic profile revealed a relative increase in genes responsible for carbohydrate metabolism, respiration, sporulation and dormancy, iron acquisition and metabolism, stress response and sulfur metabolism in the produced water samples. These results suggest that microbial communities in produced water have an increased genetic ability to handle stress, which has significant implications for produced water management, such as disinfection.

Editor: Robert J. Forster, Agriculture and Agri-Food Canada, Canada

Funding: Work was supported by the National Energy Technology Laboratory's Regional University Alliance (NETL-RUA), a collaborative initiative of the NETL, this technical effort was performed under the RES contract DE-FE0004000 (http://www.netl.doe.gov/). The funders had no role in study design, data collection and analysis, decision to publish, or preparation of the manuscript.

Competing Interests: The authors have declared that no competing interests exist.

* Email: kelvin@cmu.edu

Introduction

High-volume hydraulic fracturing operations for natural gas development from deep shale produce millions of gallons of wastewater over the lifetime of the well [1], [2], [3], commonly termed as 'produced water'. This produced water contains elevated concentrations of salts, metals, hydrocarbons and radioactive elements [3], [4], [5], [6], [7]. Microbial communities in produced water can utilize hydrocarbons as sources of carbon and energy [8] and transform redox labile salts and metals. This can give rise to significant water management challenges [9] and increased production costs [10], [11]. For instance, sulfidogenic and acid producing bacteria can cause corrosion of metal infrastructure, souring of natural gas, and reduced formation permeability [10], [11], [12], [13].

Deleterious microbial activity is commonly controlled with biocides at significant cost to the driller. However, despite biocide use, microbial activity is prevalent in produced water. Previous studies have shown that biocide effectiveness may be limited by high salt concentrations, organic compounds, and long residence times in the subsurface [14], [15], [16]. Other studies have shown that microbial communities in produced water are distinct from those in the injected fracturing fluid, and correlate well with changes in geochemical and environmental conditions [5], [15], [17]. This implies that the common practice of recycling produced water for subsequent hydraulic fracturing may introduce adapted populations into the formation [5].

Over the past decade molecular ecology surveys based on the 16S rRNA gene have increased our knowledge about the taxonomic composition of microbial communities in reservoir environments [5], [15], [17], [18], [19], [20], [21], [22]. However, these studies offer limited insights on the metabolic capabilities of the microbial community, as they rely on taxonomic inference based on 16S rRNA gene similarity to previously isolated microorganisms. As an example of the limitations of using previously isolated microorganisms to infer metabolic capability, the 'core genome' of the well-studied *Escherichia coli* is typically less than 50% of the genes in the genome, and <30% of the *E. coli* pan-genome [23]. On the other hand, shotgun metagenomic surveys enable access to complete genetic information within microbial genomes from uncultured, mixed consortia [24], [25], [26]. These surveys have provided significant insights on the functional potential of microorganisms in diverse environments such as marine samples [25], corals [27], activated sludge [28], permafrost [29], hydrocarbon and sandstone reservoirs [30], [31], and swine gut [32]. Despite the importance of microbial activity in produced water brines from hydraulic fracturing operations, the functional potential of associated microbial communities has not yet been studied. In this study, the metagenome of fracturing source water and produced water at two different time points from a Marcellus Shale natural gas well in Westmoreland County, PA was generated using Illumina MiSeq technology. The microbial ecology from 16S rRNA surveys and chemical composition of these samples has been described in a previous publication [5]. Sequences from each sample were assembled into contiguous sequences (contigs) and analyzed for taxonomic affiliations and functional potential of the microbial communities.

Materials and Methods

Sampling

Samples of hydraulic fracturing source water, and produced water on days 1 and 9 were collected from a horizontally drilled Marcellus Shale natural gas well in Westmoreland County, Pennsylvania, U.S.A in October 2011. The source water used for fracturing was a mix of fresh reservoir water (~80%) and produced water (~20%) from previous fracturing operations. Fracturing additives amended to the source water included proppant (silica sand), scale inhibitor (ammonium chloride), biocide (mixture of tributyl tetradecyl phosphonium chloride, methanol and proprietary chemicals), hydrochloric acid, gel (paraffinic solvent), breaker (sodium persulfate) and friction reducer (hydrotreated petroleum distillate). Details regarding the sampling procedure and chemical additives used in the fracturing process are described elsewhere [5]. The aqueous geochemical characteristics of these samples were described previously [5] (Table S1).

DNA extraction, library preparation and Illumina sequencing

Unfiltered water samples were centrifuged at 6,000 *g* for 30 min in an Avanti J-E centrifuge (Beckman Coulter, Brea, CA) to pellet cells. DNA was extracted from 0.25 g of cell pellet using MO BIO power soil DNA isolation kit (MO BIO, Carlsbad, CA) according to the manufacturer's instructions. DNA was prepared using Nextera XT DNA sample preparation kit (Illumina, San Diego, CA) according to manifacturer's instructions at Genewiz (South Plainfield, NJ). DNA for sequencing was quantified using qPCR prior to clustering, and sequenced using the Illumina MiSeq (Illumina, San Diego, CA) with a 2×250 PE configuration at Genewiz, NJ. Sequencing demultiplexing was performed on the Illumina MiSeq instrument using sample-specific barcodes.

Bioinformatic analyses

The raw unpaired sequences were checked for sequencing tags and adapters using the predict function implemented within the TagCleaner program [33]. No sequencing tags or adapters were identified. Sequences were then subjected to quality control using the FastX toolkit within the Galaxy platform [34] with a minimum length 100 and minimum quality score 20. The velvet assembler [35] was used to assemble sequences that passed quality control into contiguous sequences. The assembly parameters were empirically optimized for the dataset prior to assembly (Table S2); the dataset was processed using a kmer length of 77. Generated contigs >500 bp in length were uploaded to the MG-RAST server [36] with associated metadata files for taxonomic affiliations and functional annotations. Sequence similarity searches in MG-RAST was performed using the BLAT tool [37]. The metagenomes from fracturing source water, day 1 produced water, and day 9 produced water are available in the MG-RAST server [36] under accession nos. 4525703.3, 4525704.3 and 4525705.3, respectively. Taxonmic assignments of selected funcional categories from MG-RAST were execcuted in MGTAXA [38], [39], on the Galaxy bioinformatics workbench [40], [34], using default parameters and taxonomy as defined by the NCBI taxonomic tree. Data is for contig abundance and does not reflect read mapping.

As an additional assembly-independent analysis, sequence data was mapped against reference genomes downloaded from NCBI (Table S3) with CLC Genomics Workbench (Version 6.5.1, CLC Bio, Aarhus, Denmark) [41] using default parameters and no masking. Reference genomes were selected based upon taxonomic observations in MG-RAST annotation and a previous microbial ecology investigation [5]. Prior to mapping, sequencing data was trimmed to a minimum length of 100 bp and minimum quality score of 20. Furthremore, sequences for the sulfite reductase subunits A and B (dsrA/dsrB) (Table S4a) and the suflur metabolism gene adenylyl sulfate reductase subunit A (apsA) (Table S4b) were downloaded from NCBI and mapped against the trimmed sequencing data using CLC Genomics Workbench (Version 6.5.1, CLC Bio, Arhus, Denmark).

Results and Discussion

A total of 10 002, 17 055 and 16 661 contigs from the fracturing source water, produced water day 1 and day 9 samples, respectively, were uploaded to MG-RAST for downstream analyses. All uploaded contigs passed MG-RAST quality control and de-replication filters. The metagenomics sequence statistics are summarized in Table 1.

Taxonomic composition

Taxonomic affiliations were assigned to contigs with predicted proteins and rRNA genes based on comparison with the M5NR database. Alpha diversity (predicted phylotypes) for the fracturing source water, produced water day 1 and day 9 samples were 90, 79 and 88, respectively (Figure S1). Rarefaction curves for each of the samples were asymptotic suggesting that the majority of taxonomic diversity was recovered from the samples (Figure S1). Alpha diversity values and rarefaction curves were obtained using the MG-RAST tool.

Bacteria constituted the dominant domain (97–99% of the total community) in all samples. However, a shift in bacterial community composition was detected between the samples at the class and order levels (Figure 1, 2). Contigs affiliated to the class *Alphaproteobacteria* constituted the majority of the community in the fracturing source water (81%) and produced water day 1 (67%) samples (Figure 1). Within *Alphaproteobacteria*, the dominant order detected was *Rhodobacterales* (68–88% of the *Alphaproteobacteria*; 55–59% of the total community) in both the source water and produced water day 1 samples (Figure 2). The relative abundance of *Alphaproteobacteria* decreased to <2% of the community in the produced water day 9 sample. Previous

Table 1. Metagenomic sequence statistics of fracturing source water (SW), produced water day 1 (PW day 1) and produced water day 9 (PW day 9).

	SW	Pw day 1	PW day 9
Total base pair (bp) count	7,939,565 bp	18,254,354 bp	15,253,129 bp
No. of Contigs	10,002	17,055	16,661
Mean length of Contigs	793±809 bp	1,070±1,195 bp	915±651 bp
% GC content in Contigs	59±8%	55±13%	43±9%
% Contigs containing predicted proteins with known functions	83%	93.1%	80.8%
% Contigs containing predicted proteins with unknown functions	16.6%	6.6%	18.9%
% Contigs containing rRNA genes	0.4%	0.3%	0.3%
Identified protein features	9,919	20,687	16,982
Identified functional categories	8,041	16,948	13, 570

qPCR analysis of these samples suggests that that the total bacterial population remained constant at 10^6–10^7 copies of 16S RNA gene/ml [5].

An increase in the number of contigs associated with the class *Clostridia* was observed in the produced water day 1 sample (17%) as compared to the fracturing source water (1%). However, the relative abundance of *Clostridia* decreased to 3% in the produced water day 9 sample. The majority of the *Clostridia* in the produced water day 1 sample were affiliated to the order *Thermoanaerobacterales* (94% of *Clostridia*; 16% of the total community) (Figure 2). *Gammaproteobacteria* sequences constituted a minor fraction (6%) of the total community in the fracturing source water and produced water day 1 samples but increased in relative abundance to constitute the dominant class (52%) in the produced water day 9 sample. Within the *Gammaproteobacteria* of the produced water day 9 sample, dominant orders included *Vibrionales* (67% of *Gammaproteobacteria*) and *Alteromonadales* (23% of *Gammaproteobacteria*) (Figure 2). The day 9 samples also showed an increase in relative abundance of *Epsilonproteobacteria* (16%) and *Bacteroidia* (10%) classes as compared to the other samples (<2% of the total community). The major bacterial phyla, classes and orders identified in this study were consistent with previous 16S rRNA gene based clone library and pyrosequencing surveys of these samples (Figure S2) [5]. These results indicate a shift towards facultative anaerobic/anaerobic and halophilic communities in the produced water samples as compared to a predominantly aerobic community in the fracturing source water. At the class level, in each of the samples less that 3% of the total sequences did not affiliate to any taxonomic group.

A minor fraction of the total community was represented by contigs affiliated to *Archaea* (0.1–0.4%), *Viruses* (0.3–1%) and *Eukaryota* (0.4–1.4%) domains. These domains were not analyzed for in the previous 16S rRNA gene survey of these samples [5], and were not considered in more detailed functional classification of the metagenomes.

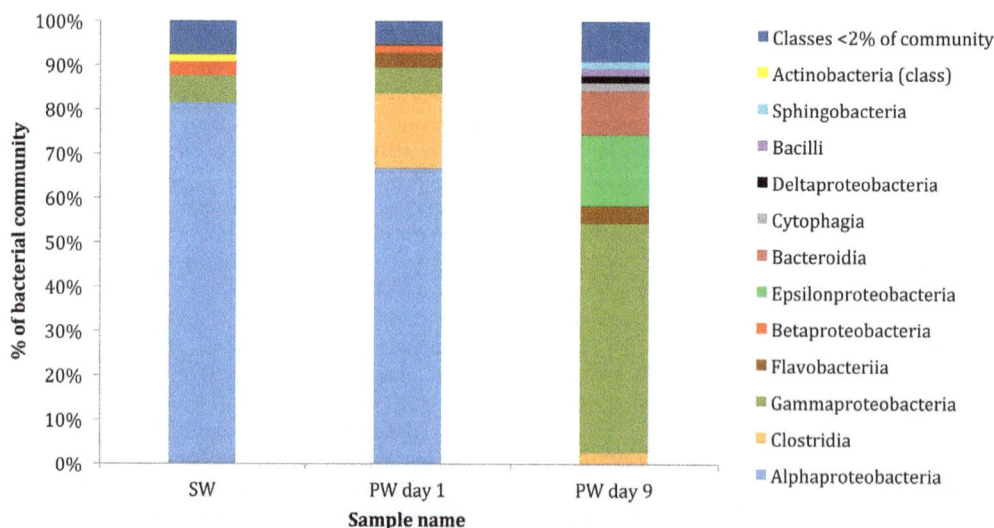

Figure 1. Class level affiliations assigned to contigs with predicted proteins and rRNA genes in source water (SW), produced water day 1 (PW day 1) and produced water day 9 (PW day 9). Total community includes *Bacteria, Archaea, Viruses* and *Eukaryota.*

Order (Class)	SW	PW day 1	PW day 9
Rhodobacterales (Alphaproteobacteria)	yellow	yellow	
Caulobacterales (Alphaproteobacteria)			
Rhizobiales (Alphaproteobacteria)			
Sphingomonadales (Alphaproteobacteria)			
Burkholderiales (Betaproteobacteria)			
Bacteroidales (Bacteroidia)			
Thermoanaerobacterales (Clostridia)			
Clostridiales (Clostridia)			
Flavobacteriales (Flavobacteria)			
Alteromonadales (Gammaproteobacteria)			
Vibrionales (Gammaproteobacteria)			
Enterobacteriales (Gammaproteobacteria)			
Cytophagales (Cytophagia)			
Campylobacterales (Epsilonproteobacteria)			

Color code

2-5%	
>5-10%	
>10-20%	
>20-30%	
>30-40%	
>40-60%	

Figure 2. Order level affiliations assigned to contigs with predicted proteins and rRNA genes in source water (SW), produced water day 1 (PW day 1) and produced water day 9 (PW day 9). Total community includes *Bacteria, Archaea, Viruses* and *Eukaryota*. Only orders representing >2% of the total community are shown in the figure.

Mapping results

Metagenomic reads were mapped against a diverse set of reference genomes to confirm MG-RAST taxonomic results and only reference genomes with good mapping results are discussed in this section. Reference genome mapping results confirmed taxonomic MG-RAST contig analysis. The best mapping results

Coverage of Reference Genomes	SW	PW day 1	PW day 9
Dinoroseobacter shibae	0.36	0.39	0.20
Thermoanaerobacter sp.	0.01	0.86	0.05
Thermoanaerobacter pseudethanolicus	0.02	0.87	0.05
Ruegeria pomeroyi	0.40	0.43	0.23
Thermoanaerobacter tengcongensis	0.01	0.45	0.03
Roseobacter denitrificans	0.30	0.33	0.15
Flavobacterium psychorophilum	0.18	0.27	0.05
Arcobacter butzleri	0.31	0.52	0.47
Arcobacter nitrofigilis	0.23	0.50	0.39
Marinobacter hydrocarbonoclasticus	0.14	0.19	0.83
Bacteroides fragilis	0.02	0.06	0.13
Sulfospirllium deleyianum	0.05	0.07	0.08
Sulfurimonas denitrificans	0.15	0.13	0.12
Parabacteroides distasonis	0.02	0.07	0.10
Phenylobacterium zucineum	0.33	0.27	0.08
Jannaschia sp.	0.25	0.27	0.12
Rhodobacter sphaeroides	0.38	0.42	0.21
Hyphomonas neptunium	0.17	0.22	0.06
Flavobacterium johnsoniae	0.09	0.14	0.03
Rhodobacter capsulatus	0.34	0.37	0.18
Marinobacter adhaerens	0.14	0.17	0.68
Clostridium difficile	0.02	0.06	0.06
Roseovarius sp.	0.79	0.79	0.51
Roseovarius nubinhibens	0.47	0.50	0.28
Vibrio campbelli	0.06	0.25	0.51
Thermoanaerobacter mathranii	0.01	0.88	0.05

Color code

>0-0.1	
>0.1-0.2	
>0.2-0.3	
>0.3-0.4	
>0.4-0.5	
>0.5-0.6	
>0.6-0.7	
>0.7-0.8	
>0.8-0.9	

Figure 3. Fraction of genome coverage for source water (SW), produced water day 1 (PW day 1) and produced water day 9 (PW day 9) samples. Reads were mapped against reference genomes using CLC Genomic workbench version 6.5.1 using default parameters. Shown are fractions of reads mapped against each reference genome included in the analysis for all three samples.

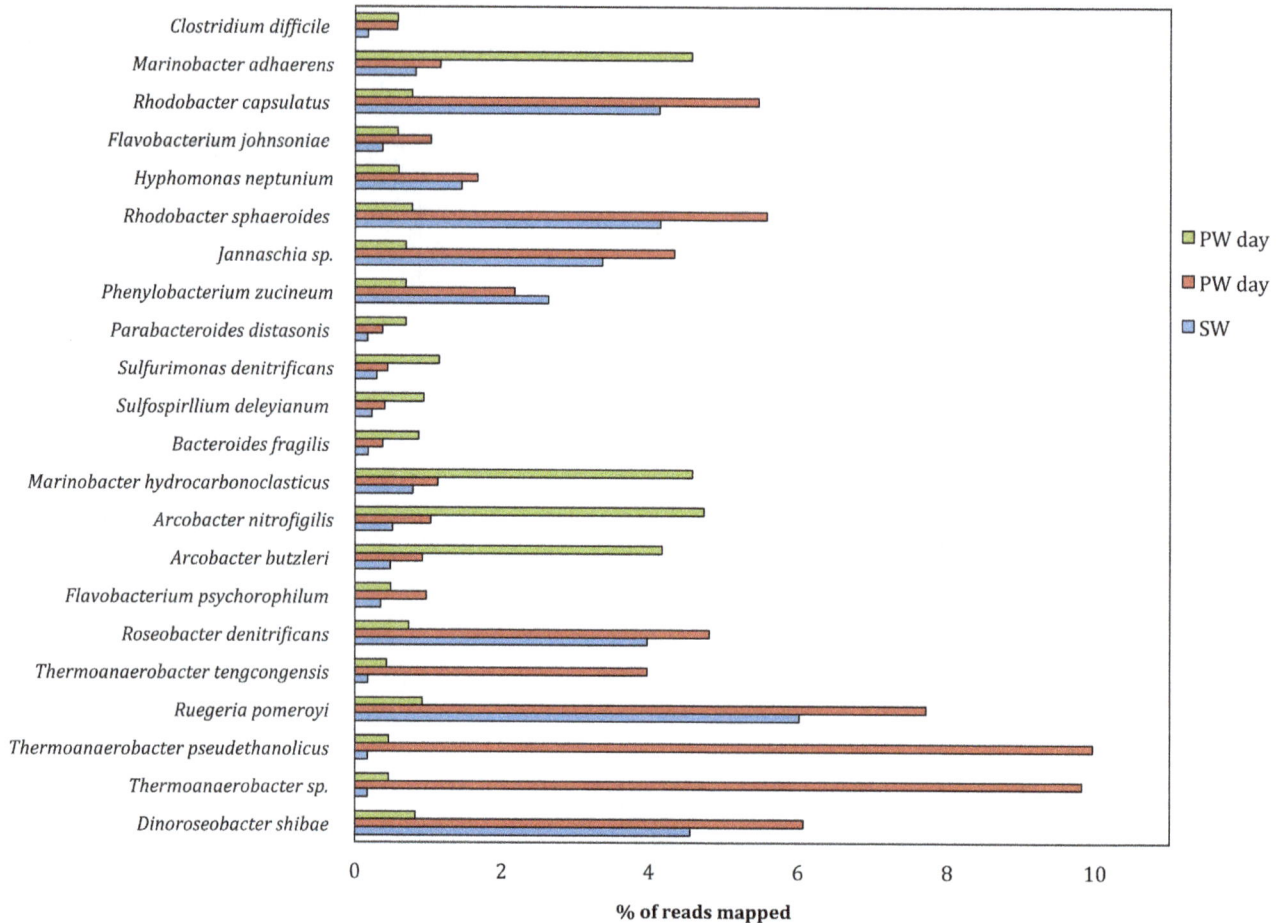

Figure 4. Read distribution for source water (SW), produced water day 1 (PW day 1) and produced water day 9 (PW day 9) samples. Reads were mapped against reference genomes using CLC Genomic workbench version 6.5.1 using default parameters. Shown are percentages of reads mapped against each reference genome included in the analysis for all three samples.

for source water were obtained when sequences were mapped against reference genomes of *Alphaproteabacteria*, specifically of the order *Rhodobacterales* (Figures 3, 4). Similarly, produced water day 1 sample mapping results suggest that it was dominated by bacteria of the orders *Rhodobacterales* and *Thermoanaerobacterales* (Figures 3, 4). A distinct shift in bacterial community was observed between produced water day 1 samples and produced water day 9 samples based on mapping results. Best mapping results for produced water day 9 samples were obtained for reference genomes in the order *Campylobacterales* and *Alteromondales* further supporting the MG-RAST results (Figures 3, 4). Produced water samples demonstrated a distinctive signature with reads mapping best to few select reference genomes, while source water sample reads were distributed more evenly throughout all included reference genomes. For four reference genomes (*Thermoanaerobacter* sp. X514, *Thermoanaerobacter pseudethanolicus*, *Thermoanaerobacter mathranii* in produced water day 1 samples and *Marinobacter hydrocarbonoclasticus* DSM 7299 in produced water day 9 sample) more than 80% coverage was achieved suggesting that these species could play important roles in the microbial community of the representative sample (Figure 3). Highest observed reference genome coverage for source water sample sequences were 79% for *Roseovarius* sp. 217, 40% for *Ruegeria pomeroyi* and 38% for *Rhodobacter sphaeroides* (Fig-

ure 3). For produced water day 1 samples, about 10% of all trimmed sequencing reads mapped against the three *Thermanaerobacter* genomes included in the analysis and 8–13% of reads mapped successfully against *Roseovarius* sp. 217 and *Roseovarius nubinhibens* genomes (Figure 4). 7.7% of produced water day 1 reads mapped against the *Ruegeria pomeroyi* genome (Figure 4). 4–6% of reads for produced water day 9 samples mapped against two different *Marinobacter* and *Arcobacter* reference genomes and one *Vibrio* reference genome (Figure 4). Almost 16% of all reads from source water samples mapped against *Roseovarius* sp. 217 and approximately 4–6% of reads for source water sample mapped against each *Dinoroseobacter shibae*, *Ruegeria pomeroyi*, *Rhodobacter sphaeroides* and *Rhodobacter capsulatus* genomes (Figure 4). All mapping results are summarized in Table S3. The high number of reads form source water and produced water day 1 samples mapping against *Roseovarius* species is in agreement with previous 16S rRNA gene sequencing [5], implying the *Roseovarius* species might be of importance in these waters. *Roseovarius sp.* was previously identified in natural gas brines from the Marcellus shale and its potential implications are discussed elsewhere [9].

The goal of this analysis was to provide an independent confirmation of MG-RAST results. Mapping results depend on the reference genomes selected and these reference genomes might

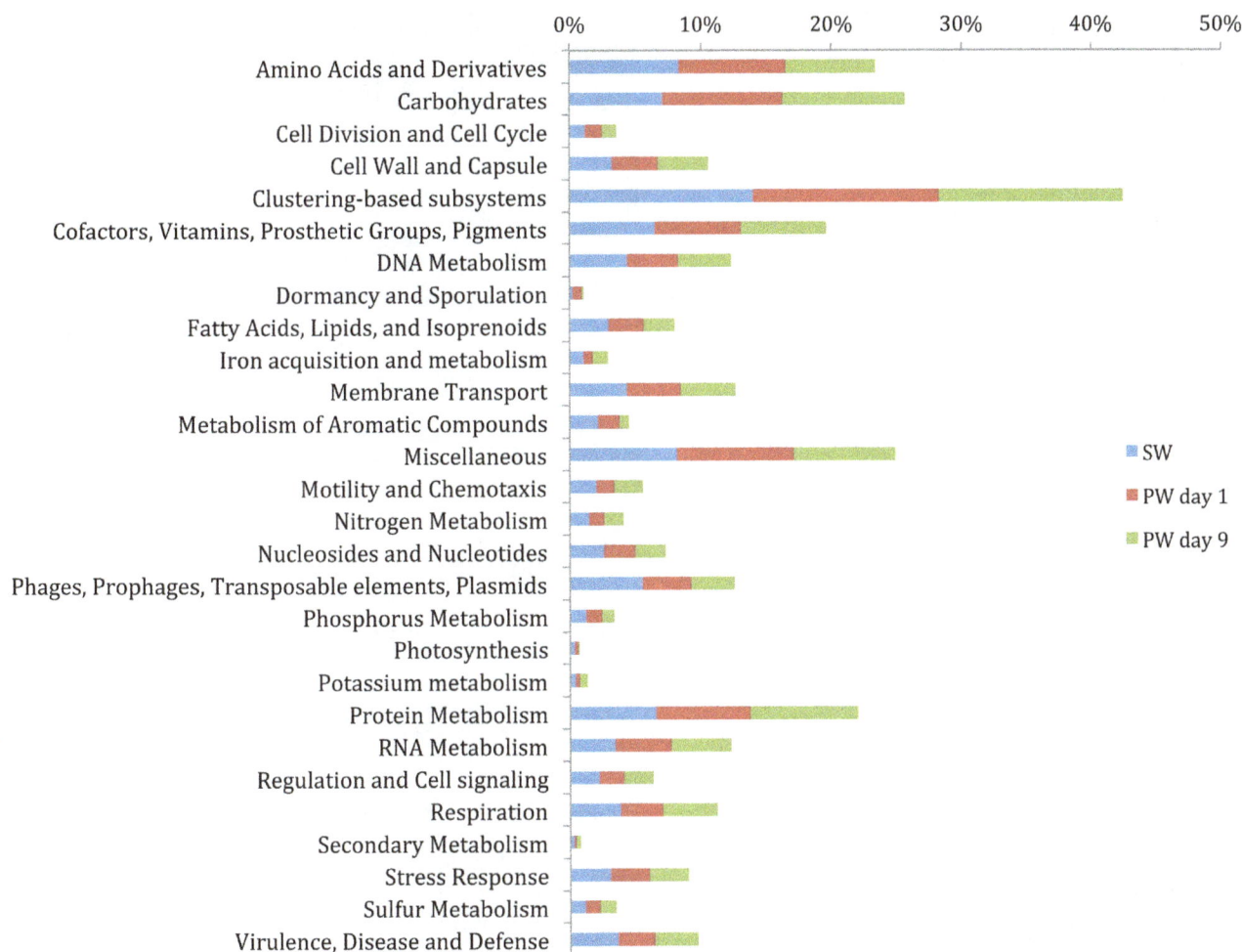

Figure 5. Actual abundance of contigs belonging to Level 1 functional categories in source water (SW), produced water day 1 (PW day 1) and produced water day 9 (PW day 9). Functional annotations were assigned based on the Subsystems database.

not be the same isolates found in the environment. While reference genomes for uncultured microorganisms from oil/gas environments are limited, the positive results achieved by this mapping analysis confirm the initial taxonomic assessment.

Sulfur metabolism gene mapping results

Very few reads in all three samples were successfully mapped against the sulfur metabolism genes dsrA and dsrB. 7 reads of produced water day 1 sample and 55 reads of produced water day 9 sample were successfully mapped against the dsrA/dsrB gene of *Desulfovibrio desulfuricans* with a coverage of 28% and 78% respectively (Table S4a). In addition 10 reads of produced water day 9 sample were successfully mapped against the dsrA/dsrB gene of *Desulfotignum balticum* with a coverage of 19% (Table S4a). For aspA genes, the produced water day 9 sample showed best results with 16, 11, 9 and 6 reads successfully mapped against aspA genes of *Desulfovirbio alaskensis*, *Desulfococcus mulitvorans*, *Desulfotignum balcticum* and *Desulfobacterium autotorphicum* with a coverage of 94%, 46%, 33% and 31% respectively (Table S4b). Very few source water and produced water day 1 reads were mapped successfully against the aspA genes included in the analysis (Table S4b). These results suggest that sulfur metabolism could play a more important role in produced water day 9 sample due the higher abundance of genes associated with sulfur

metabolism. Organisms that can metabolize sulfur compounds to sulfide are of interest in oil and gas environments because of their potential role in infrastructure corrosion, gas souring, worker safety as well as environmental health concerns.

Functional classification of metagenomes

The SEED subsystems database [42], was used to predict the metabolic potential of fracturing source water and produced water samples. Level 1 indicates the broadest set of functional categories to which sequences are assigned, and Level 2 refers to more specific functional assignments within Level 1 categories. The abundance of contigs designated to Level 1 functional categories is illustrated in Figure 5. The metabolic potential (based on Level 1 and Level 2 functional categories) between the samples was compared in a normalized manner (Figure 6, 7) to account for differences in community structure, size of the library, gene content between samples and to effectively compare low abundance functional categories [43]. Read normalization was performed within the MG-RAST analysis pipeline, in accordance with standards for metagenomic analysis.

The five most abundant Level 1 functional categories in all three samples were found to be clustering-based subsystems (e.g. genes where functional coupling is evident but function is unknown;~14%), carbohydrate metabolism (7–9%), amino acids

Level 1 functional categories	SW	PW day 1	PW day 9
Amino Acids and Derivatives			
Carbohydrates			
Cell Division and Cell Cycle			
Cell Wall and Capsule			
Clustering-based subsystems			
Cofactors, Vitamins, Prosthetic Groups, Pigments			
DNA Metabolism			
Dormancy and Sporulation			
Fatty Acids, Lipids, and Isoprenoids			
Iron acquisition and metabolism			
Membrane Transport			
Metabolism of Aromatic Compounds			
Miscellaneous			
Motility and Chemotaxis			
Nitrogen Metabolism			
Nucleosides and Nucleotides			
Phages, Prophages, Transposable elements, Plasmids			
Phosphorus Metabolism			
Photosynthesis			
Potassium metabolism			
Protein Metabolism			
RNA Metabolism			
Regulation and Cell signaling			
Respiration			
Secondary Metabolism			
Stress Response			
Sulfur Metabolism			
Virulence, Disease and Defense			

Color code:	
>0-0.1	
>0.1-0.2	
>0.2-0.3	
>0.3-0.4	
>0.4-0.5	
>0.5-0.6	
>0.6-0.7	
>0.7-0.8	
>0.8-0.9	
>0.9-1	

Figure 6. Normalized abundance (values of 0–1) of contigs belonging to Level 1 functional categories in source water (SW), produced water day 1 (PW day 1) and produced water day 9 (PW day 9). Functional annotations were assigned based on the Subsystems database.

and derivatives (7–8%), miscellaneous (eg: genes associated with iron sulfur cluster assembly and Niacine-Choline transport and metabolism; 8–9%), protein metabolism (6–8%), suggesting the dominant role of these functional categories in all samples (Figure 5). These functional categories were similarly identified as dominant in previous studies of soil [44], [45], marine samples [24],[46], activated sludge [24], freshwater [24] and hypersaline environments [24]. Normalization of gene abundance data shows a relative increase in each of the above functional categories in the produced water samples as compared to the fracturing source water (Figure 6) implying that core systems necessary for survival are enriched in the produced water community.

While comparison of gene abundance affiliated with the dominant broad Level 1 categories suggests similar functional profiles across samples, analysis of more specific Level 2 functional categories shows sample specific differences in metabolic capabilities (Figure 7). Differences in metabolic potential indicate a selective pressure exerted in the subsurface for microbes with particular metabolic capabilities. For instance, within the Level 1 carbohydrate metabolism category, sequences related to Level 2 functional categories such as mono-, di-, oligo- and polysaccharides, and aminosugar metabolism were present in higher relative abundance in the produced water samples (Figure 7). This finding correlates well with the expected higher content of carbohydrates

in produced water samples [5]. Carbohydrates and polysaccharide compounds added during hydraulic fracturing can serve as carbon and energy sources for microbial activity [8]. Within the Level 1 protein metabolism category, sequences affiliated with the Level 2 selenoprotein category were detected only in the produced water samples (Figure 7). One possible explanation is the role of selenoproteins in combating oxidative stress [47], which may arise from elevated concentrations of organic or inorganic dissolved constituents in produced water [48]. Results showed that *Rhodobacterales* were the dominant population involved in oxidative stress response in source water and produced water day 1 samples (Figure 8). However, *Alteromonadales* and *Vibrionales* were the dominant orders involved in oxidative stress response in produced water day 9 sample (Figure 8). Within the Level 1 clustering subsystem, genes affiliated with the Level 2 carbohydrate metabolism show a relative increase in the produced water samples as compared to fracturing source water (Figure 7). An increase in the relative abundance of genes related to carbohydrate metabolism in produced water compared to fracturing source water suggests the potential for utilization of hydrocarbons added either as fracturing fluid amendments or those derived from the shale formation and an overall shift to a more heterotrophic microbial community.

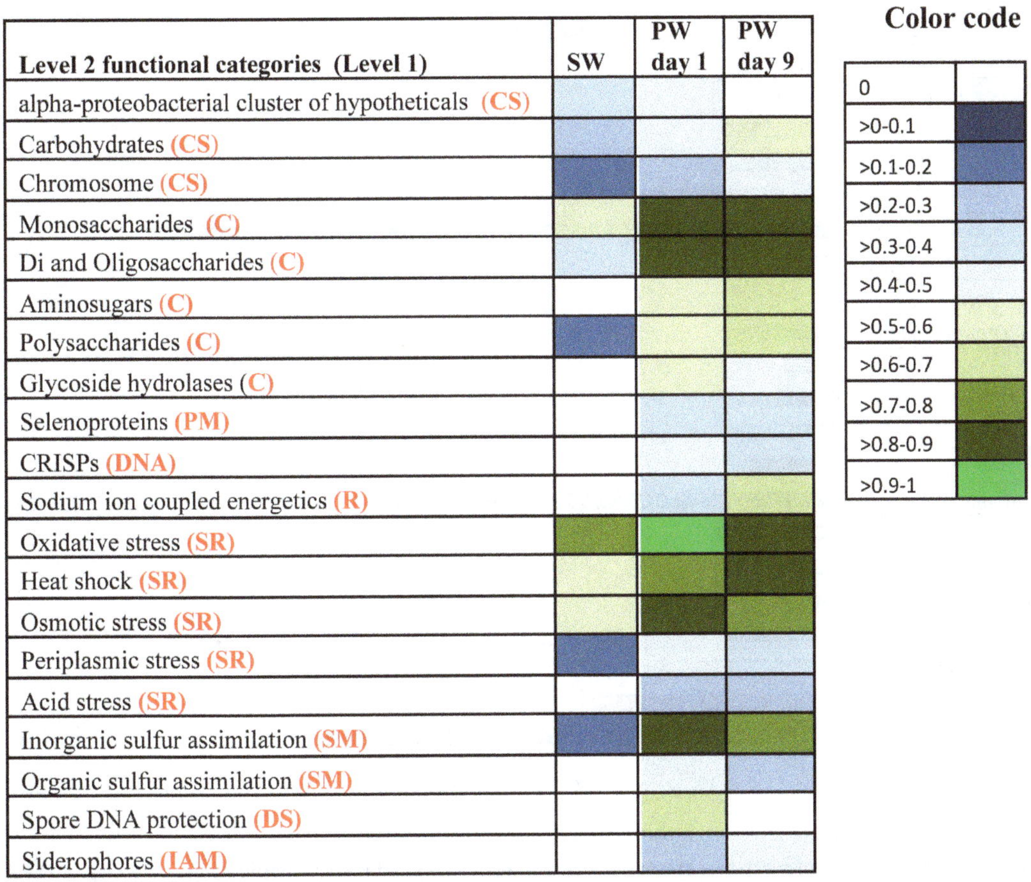

Level 2 functional categories (Level 1)	SW	PW day 1	PW day 9
alpha-proteobacterial cluster of hypotheticals (CS)			
Carbohydrates (CS)			
Chromosome (CS)			
Monosaccharides (C)			
Di and Oligosaccharides (C)			
Aminosugars (C)			
Polysaccharides (C)			
Glycoside hydrolases (C)			
Selenoproteins (PM)			
CRISPs (DNA)			
Sodium ion coupled energetics (R)			
Oxidative stress (SR)			
Heat shock (SR)			
Osmotic stress (SR)			
Periplasmic stress (SR)			
Acid stress (SR)			
Inorganic sulfur assimilation (SM)			
Organic sulfur assimilation (SM)			
Spore DNA protection (DS)			
Siderophores (IAM)			

Color code

0	
>0-0.1	
>0.1-0.2	
>0.2-0.3	
>0.3-0.4	
>0.4-0.5	
>0.5-0.6	
>0.6-0.7	
>0.7-0.8	
>0.8-0.9	
>0.9-1	

Figure 7. Normalized abundance (values of 0–1) of contigs belonging to selected Level 2 functional categories within associated Level 1 categories in source water (SW), produced water day 1 (PW day 1) and produced water day 9 (PW day 9). Functional annotations were assigned based on the Subsystems database. The affiliations of Level 2 categories to Level 1 categories are coded as follows CS- Clustering based subsystems; C- Carbohydrates; PM- Protein metabolism; DNA- DNA metabolism; R- Respiration; SR- Stress response; SM- Sulfur metabolism; DS- Dormancy and sporulation; IAM- Iron acquisition and metabolism.

Less abundant Level 1 functional categories showing an increase in normalized abundance in produced water samples (Figure 6) included genes affiliated with stress response (3%), respiration (3–4%), iron acquisition and metabolism (1%), sulfur metabolism (1%), and dormancy and sporulation (0.2–1%). Analysis of Level 2 functional categories within these Level 1 domains identified differences in metabolic potential between these samples (Figure 7). Within the Level 1 stress response domain, produced water samples showed a greater relative abundance of sequences affiliated with Level 2 categories such as acid stress, heat shock, periplasmic stress and osmotic stress (Figure 7). The increase in the relative abundance of these genes suggests a response to external stress experienced by the produced water microbial community. Results suggest that produced water day 1 population involved in osmotic stress response was dominated by the order *Rhodobacterales* and produced water day 9 population involved osmotic stress response was dominated by the orders *Vibrionales* and *Alteromonadales* (Figure 9). Subsurface stresses can include increased subsurface temperatures (>40°C) [49], addition of HCl and biocides to fracturing fluid, and higher concentrations of dissolved salts (Table S1) [5]. Within the Level 1 respiration category, sequences affiliated to the Level 2 category of sodium ion coupled energetics were undetected in fracturing source water (Na$^+$ 2.9 g/ L) but increased in relative abundance with time in produced water samples (Na$^+$ concentrations in PW day 1 and day 9 were 13.9 and 43 g/L) (Figure 7). This suggests that the produced water microbial community could use sodium ion coupled energetics for their energy needs, consistent with previous observations in saline environments [50]. In the Level 1 domain of sulfur metabolism, the relative abundance of genes affiliated with Level 2 functional categories of inorganic and organic sulfur assimilation increased in produced water samples as compared to fracturing source water (Figure 7). Genes recovered from produced water day 1 show that populations involved in sulfur metabolism were dominated by the orders *Rhodobacterales* and *Thermoanaerobacterales* (Figure 10). However, sulfur metabolism in produced water day 9 samples was dominated by the orders *Vibrionales* and *Bacteroidales* (Figure 10). Within the Level 1 domain of iron metabolism, sequences affiliated with siderophores, undetected in the fracturing source water, increased with time in produced water samples (Figure 7). Siderophores are strong chelators of ferric iron secreted and are utilized by bacteria for iron metabolism [51]. Relative increase in siderophore affiliated genes correlates with an increase in total iron concentrations with time in produced water (4.2–81.6 mg/L) (Table S1). Within the Level 1 dormancy and sporulation category, high relative abundance of Level 2 spore DNA protection related sequences in produced water day 1 sample (Figure 7) suggests the potential for long term dormancy of cells through DNA protection [52]. BLAT analysis [37] showed that these genes were similar to those present in *Thermoanaerobacter*, a

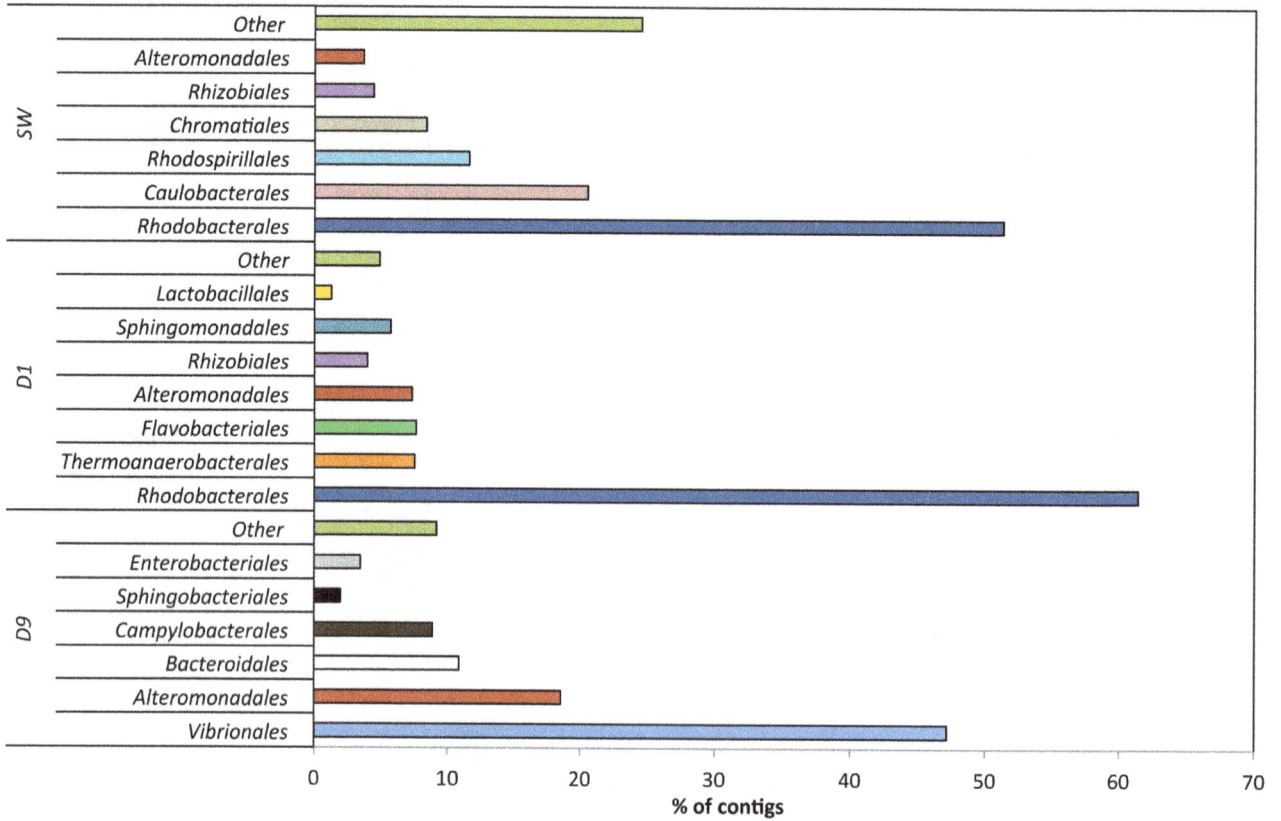

Figure 8. Taxonomic classification of oxidative stress contigs for each analyzed water sample as assigned by MGTAXA. SW- Source water; D1- Produced water day 1; D9- Produced water day 9. Only the top six bacterial orders to which most contigs were assigned to are shown in the figure. The less abundant bacterial orders are grouped as "other".

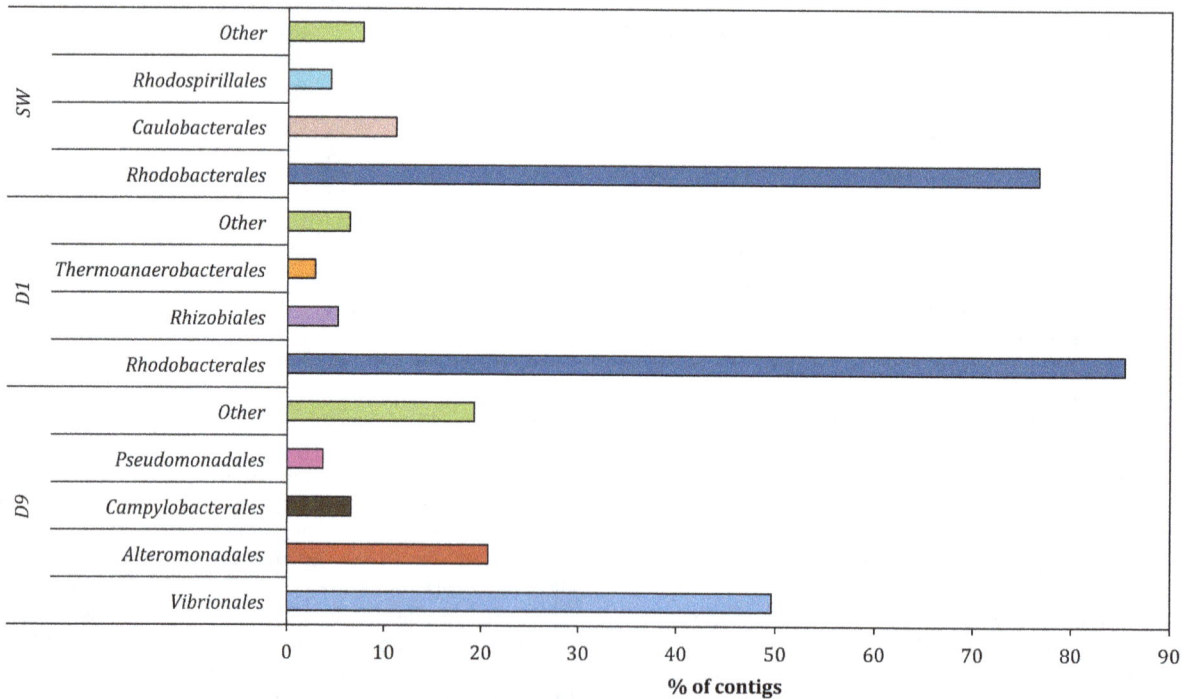

Figure 9. Taxonomic classification of osmotic stress contigs for each analyzed water sample as assigned by MGTAXA. SW- Source water; D1- Produced water day 1; D9- Produced water day 9. Only the top four bacterial orders to which most contigs were assigned to are shown in the figure. The less abundant bacterial orders are grouped as "other".

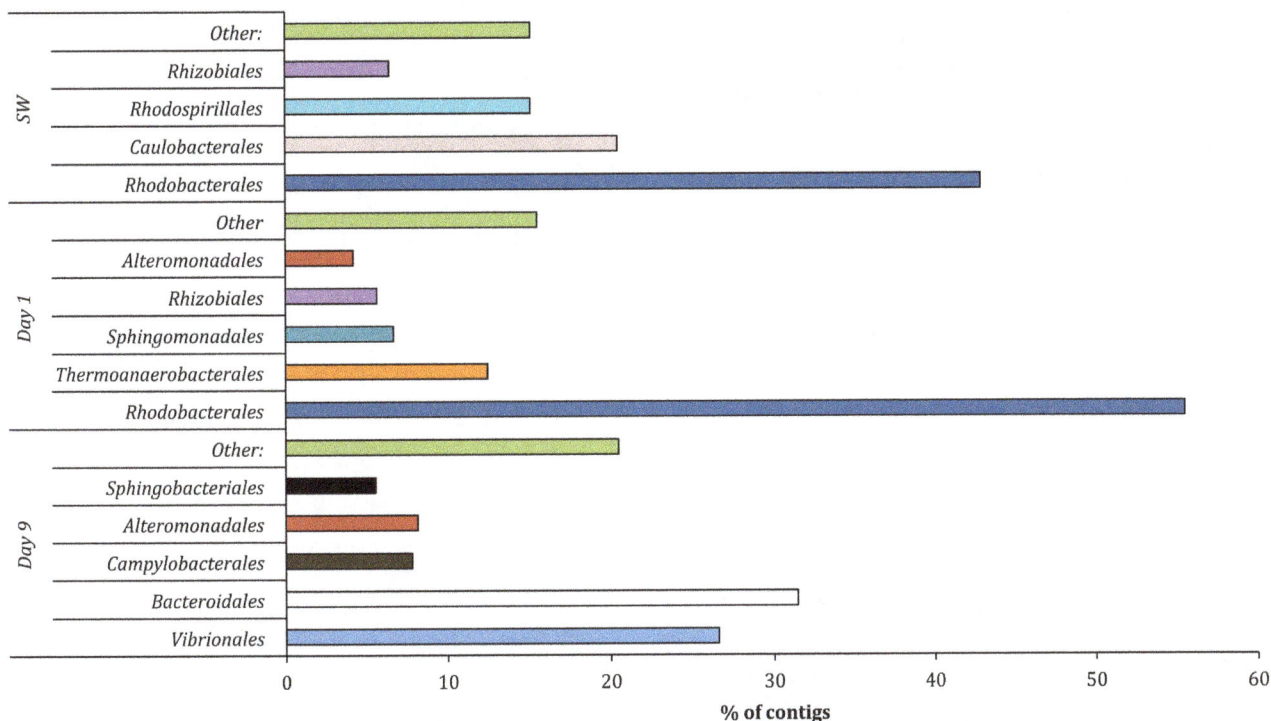

Figure 10. Taxonomic classification of sulfur metabolism contigs for each analyzed water sample as assigned by MGTAXA. SW-Source water; D1- Produced water day 1; D9- Produced water day 9. Only the top five bacterial orders to which most contigs were assigned to are shown in the figure. The less abundant bacterial orders are grouped as "other".

bacterial order that constituted 16% of the total community in this sample (Figure 2). An increase in the relative abundance of spore forming bacteria and genes affiliated with sporulation and dormancy is an important consideration in biocide application, and may provide an explanation for the previously observed limited efficacy of biocides [5].

Concluding Remarks. This study is the first shotgun metagenomic analysis of produced water from hydraulic fracturing for natural gas production and provides novel insights on taxonomic and functional potential of this pertinent yet unexplored environment. Taxonomic analysis showed that *Bacteria* constituted the dominant (>98%) domain in both fracturing source water and produced water samples. Results demonstrated the emergence of distinct bacterial classes and orders in the produced water samples and fracturing source water samples. These bacterial taxa were consistent with results from a previous 16S rRNA gene based survey of these samples [5]. The metabolic profile showed both a relative increase and functional changes in genes responsible for carbohydrate metabolism, respiration, sporulation and dormancy, iron acquisition and metabolism, stress response and sulfur metabolism in the produced water samples as compared to the fracturing source water sample. These results suggest that the microbial community is responsive to changes in hydrocarbon content, induced stresses such as increase in temperature, addition of biocides, and an increase in concentration of dissolved salts such as iron and sulfur. The detection of genes affiliated with sodium ion coupled energetics exclusively in the produced water samples suggests the use of sodium ion based energetics by microorganisms in these sodium rich environments. Understanding the evolving metabolic capabilities of microbial communities in produced water will help the

industry and its regulators improve environmental and economic sustainability of oil and gas extraction through more informed water management decisions.

Supporting Information

Figure S1 Plot of refraction curves with associated Alpha diversity in fracturing source water (SW), produced water day 1 (PW day 1) and produced water day 9 (PW day 9).

Figure S2 Sequences affiliated to major bacterial phyla in source water, Produced water day 1 and Produced water day 9 using 16S rRNA gene pyrosequencing and metagenomics.

Table S1 Chemical composition of source water and produced water (PW) samples days 1, 9 and 187.

Table S2 Assembly optimization statistics. Velvet 1.2.08 was used to optimize assembly of Source Water derived sequences.

Table S3 Mapping results for source water, produced water day 1 and produced water day 9 sequencing data against selected bacteria species reference genomes. Mapping analysis was performed using CLC Genomics Workbench version 6.5.1 with default parameters.

Table S4 Mapping results, (A), for source water, produced water day 1 and produced water day 9 sequencing data against the genome sequences of the dsrA/dsrB gene of selected microbial organisms. Mapping analysis was performed using CLC Genomics Workbench version 6.5.1 with default parameters. (B) Mapping results for source water, produced water day 1 and produced water day 9 sequencing data against the genome sequences of the apsA gene of selected microbial organisms. Mapping analysis was performed using CLC Genomics Workbench version 6.5.1 with default parameters.

Author Contributions

Conceived and designed the experiments: AMM KBG KJB RWH. Performed the experiments: AMM. Analyzed the data: AMM KJB DL. Contributed reagents/materials/analysis tools: KBG KJB. Wrote the paper: AMM KJB DL RWH KBG.

References

1. Veil JA (2010), Water Management Technologies Used by Marcellus Shale Gas Producers, ANL/EVS/R-10/3, prepared by Environmental Science Division, Argonne National Laboratory for the U.S. Department of Energy, Office of Fossil Energy, National Energy Technology Laboratory, July ANL/EVS/R-10/3.
2. Arthur JD, Bohm B, Coughlin B, Layne M, Cornue D (2009) Evaluating the Environmental Implications of Hydraulic Fracturing in Shale Gas Reservoirs, *SPE 121038*, In: SPE Americas Environmental and Safety Conference. San Antonio, TX, March 23–25.
3. Gregory KB, Vidic RD, Dzombak DA (2011) Water Management Challenges Associated with the Production of Shale Gas by Hydraulic Fracturing. Elements 7: 181–186.
4. Barbot E, Vidic N, Gregory KB, Vidic RD (2013) Spatial and Temporal Correlation of Water Quality Parameters of Produced Waters from Devonian-Age Shale following Hydraulic Fracturing. Environ Sci Technol 47: 2562–2569.
5. Murali Mohan A, Hartsock A, Bibby K, Hammack RW, Vidic RD, et al. (2013) Microbial Community Changes in Hydraulic Fracturing Fluids and Produced Water from Shale Gas Extraction. Environ Sci Technol 47(22): 13141–13150.
6. Soeder DJ, Kappel WM (2009) USGS Fact Sheet 2009–3032.
7. Hill D, Lombardi T, Martin J (2004) Fractured Shale Gas Potential In New York. Northeastern Geol. Environ Sci 26: 57–78.
8. Moore SL, Cripps CM (2010). Bacterial Survival in Fractured Shale Gas Wells of the Horn River Basin (CSUG/SPE 137010). CSUG pp. 1–14.
9. Murali Mohan A, Hartsock A, Hammack RW, Vidic RD, Gregory KB (2013) Microbial Communities in Flowback Water Impoundments from Hydraulic Fracturing for Recovery of Shale Gas. FEMS Microbiology Ecology 86(3): 567–580.
10. Kermani M, Harrop D (1996) The impact of corrosion on oil and gas industry. SPE Production Facilities 11: 186–190.
11. Little BJ, Lee JS (2007) Microbiologically influenced corrosion.Wiley and Sons Inc., Hoboken, NJ.
12. Fichter JK, Johnson K, French K, Oden R (2008) Use of Microbiocides in Barnett Shale Gas Well Fracturing Fluids to Control Bacteria Related Problems (Paper No. 08658). In NACE International Corrosion Conference and Expo pp. 1–14.
13. Roberge PR (2000) Handbook of Corrosion Engineering. McGraw-Hill, New York.
14. Struchtemeyer CG, Morrison MD, Elshahed MS (2012) A critical assessment of the efficacy of biocides used during the hydraulic fracturing process in shale natural gas wells. International Biodeterioration & Biodegradation 71: 15–21.
15. Struchtemeyer CG, Elshahed MS (2012) Bacterial communities associated with hydraulic fracturing fluids in thermogenic natural gas wells in North Central Texas, USA. FEMS Microbiology Ecology 81: 13–25.
16. Williams TM, Mcginley HR (2010) Deactivation of Industrial Water Treatment Biocides (Paper No. 10049). In NACE International Corrosion Conference and Expo pp. 1–15.
17. Davis JP, Struchtemeyer CG, Elshahed MS (2012) Bacterial communities associated with production facilities of two newly drilled thermogenic natural gas wells in the Barnett Shale (Texas, USA). Microbial Ecology 64: 942–954.
18. Dahle H, Garshol F, Madsen M, Birkeland NK (2008) Microbial community structure analysis of produced water from a high-temperature North Sea oil-field. Antonie van Leeuwenhoek 93: 37–49.
19. Pham VD, Hnatow LL, Zhang S, Fallon RD, Jackson SC, et al. (2009) Characterizing microbial diversity in production water from an Alaskan mesothermic petroleum reservoir with two independent molecular methods. Environmental Microbiology 11: 176–187.
20. Grabowski A, Nercessian O, Fayolle F, Blanchet D, Jeanthon C (2005) Microbial diversity in production waters of a low-temperature biodegraded oil reservoir. FEMS Microbiology Ecology 54: 427–443.
21. van der Kraan GM, Bruining J, van Loosdrecht MCM, Muyzer G (2010) Microbial diversityofan oil water processing site and its associated oil field: the possible role of microorganisms as information carriers from oil-associated environments. FEMS Microbiology Ecology 71: 428–443.
22. Gittel A, Sørensen KB, Skovhus TL, Ingvorsen K, Schramm A (2009) Prokaryotic community structure and sulfate reducer activity in water from high-temperature oil reservoirs with and without nitrate treatment. Appl Environ Microbiol 75: 7086–7096.
23. Hendrickson H (2009) Order and disorder during *Escherichia coli* divergence. PLoS genetics, 5, e1000335. Available: http://www.plosgenetics.org/article/info%3Adoi%2F10.1371%2Fjournal.pgen.1000335
24. Dinsdale EA, Edwards RA, Hall D, Angly F, Breitbart M et al. (2008) Functional metagenomic profiling of nine biomes. Nature 452: 629–632.
25. DeLong EF, Preston CM, Mincer T, Rich V, Hallam SJ, et al. (2006) Community genomics among stratified microbial assemblages in the ocean's interior. Science 311: 496–503.
26. Tringe SG, Rubin EM (2005) Metagenomics: DNA sequencing of environmental samples. Nature reviews. Genetics 6: 805–814.
27. Wegley L, Edwards R, Beltran Rodriguez-Brito1 HL, Rohwer F (2007) Metagenomic analysis of the microbial community associated with the coral *Porites astreoides*. Environmental Microbiology 9: 2707–2719.
28. Yu K., Zhang T (2012) Metagenomic and metatranscriptomic analysis of microbial community structure and gene expression of activated sludge. PloS one 7, e38183. Available: http://www.plosone.org/article/info%3Adoi%2F10.1371%2Fjournal.pone.0038183
29. Yergeau E, Hogues H, Whyte LG, Greer CW (2010) The functional potential of high Arctic permafrost revealed by metagenomic sequencing, qPCR and microarray analyses. The ISME journal 4: 1206–1214.
30. Dongshan AN, Caffrey SM, Soh J, Agrawal A, Brown D, et al. (2013) Metagenomics of Hydrocarbon Resource Environments Indicates Aerobic Taxa and Genes to be Unexpectedly Common. Environ Sci Technol 47: 10708–10717.
31. Dong Y, Kumar CG, Chia N, Kim PJ, Miller P, et al. (2013) *Halomonas sulfidaeris*-dominated microbial community inhabits a 1.8 km-deep subsurface Cambrian Sandstone reservoir. Environmental Microbiology 16(6): 1695–1708.
32. Lamendella R, Domingo JWS, Ghosh S, Martinson J, Oerther DB (2011) Comparative fecal metagenomics unveils unique functional capacity of the swine gut. BMC Microbiology 11:103. Available: http://www.biomedcentral.com/1471-2180/11/103
33. Schmieder R, Lim YW, Rohwer F, Edwards R (2010) TagCleaner: Identification and removal of tag sequences from genomic and metagenomic datasets. BMC Bioinformatics 11:341. Available: http://www.ncbi.nlm.nih.gov/pmc/articles/PMC2910026/
34. Goecks J, Nekrutenko A, Taylor J, The Galaxy Team (2010) Galaxy: A comprehensive approach for supporting accessible, reproducible, and transparent computational research in the life sciences. Genome biology 11:R86. Available: http://genomebiology.com/2010/11/8/R86
35. Zerbino DR (2010) Using the Velvet de novo assembler for short-read sequencing technologies. Curr Protoc Bioinformatics 31:11.5.1–11.5.12. Available: http://www.ncbi.nlm.nih.gov/pmc/articles/PMC2952100/
36. Meyer F, Paarmann D, D'Souza M, Olson R, Glass EM, et al. (2008) The metagenomics RAST server - a public resource for the automatic phylogenetic and functional analysis of metagenomes. *BMC Bioinformatics* 2008, 9: 386. Available: http://www.biomedcentral.com/1471-2105/9/386
37. Kent WJ (2002) BLAT—The BLAST-Like Alignment Tool. Genome Research 12: 656–664. Available: http://www.ncbi.nlm.nih.gov/pmc/articles/PMC187518/
38. Tovchigrechko A, Sul SJ, MGTAXA- A free software for taxonomic classification of metagenomic sequences with machine learning techniques. Available: http://andreyto.github.io/mgtaxa/
39. Brady A, Salzberg SL (2009) Classification with interpolated markov models. Nature Methods 6 (9): 673–676.
40. Giardine B, Riemer C, Hardison C, Burhans R, Elnitski L (2005) Galaxy: A platform for interactive large scale genome analysis. Genome Research 15: 1451–1455. Available: http://www.ncbi.nlm.nih.gov/pmc/articles/PMC1240089/
41. CLC Genomics Workbench, "Version 6.5.1", CLC bio A/S Science Park Aarhus Finlandsgade, 10–12. Available: http://www.clcbio.com/products/clc-genomics-workbench/
42. Overbeek R, Begley T, Butler RM, Choudhuri JV, Chuang HY, et al. (2005) The subsystems approach to genome annotation and its use in the project to annotate 1000 genomes. Nucleic acids research 33: 5691–5702.
43. Shi Y, Tyson GW, Eppley JM, DeLong EF (2011) Integrated metatranscriptomic and metagenomic analyses of stratified microbial assemblages in the open ocean. The ISME journal 5: 999–1013.
44. Delmont TO, Prestat E, Keegan KP, Faubladier M, Robe P, et al. (2012) Structure, fluctuation and magnitude of a natural grassland soil metagenome. The ISME Journal 6 (9): 1677–1687.

45. Urich T, Lanzén A, Qi J, Huson DH, Schleper C, et al. (2008) Simultaneous assessment of soil microbial community structure and function through analysis of the meta-transcriptome. PloS one 3, e2527. Available: http://www.plosone.org/article/info%3Adoi%2F10.1371%2Fjournal.pone.0002527

46. Gilbert JA, Field D, Huang Y, Edwards R, Li W, et al. (2008) Detection of large numbers of novel sequences in the metatranscriptomes of complex marine microbial communities. PloS one 3, e3042. Available: http://www.plosone.org/article/info%3Adoi%2F10.1371%2Fjournal.pone.0003042

47. Lu J, Holmgren A (2009) Selenoproteins. The Journal of biological chemistry 284: 723–727.

48. Valavanidis A, Vlahogianni T, Dassenakis M, Scoullos M (2006) Molecular biomarkers of oxidative stress in aquatic organisms in relation to toxic environmental pollutants. Ecotoxicol Environ Saf 64: 178–189.

49. Driscoll FG (1986) Groundwater and Wells. Johnson Filtration Inc.: St Paul, MN.

50. Kogure K (1998) Bioenergetics of marine bacteria. Current Opinion in Biotechnology 9: 278–282.

51. Sandy M, Butler A (2010) Microbial Iron Acquisition: Marine and Terrestrial Siderophores. Chem Rev 109: 4580–4595.

52. Setlow P (1992) Mini Review: I Will Survive: Protecting and Repairing Spore DNA. Journal of Bacteriology 174: 2737–2741.

## 10

Pronounced Metabolic Changes in Adaptation to Biofilm Growth by *Streptococcus pneumoniae*

Raymond N. Allan[1,2]*, **Paul Skipp**[3,4], **Johanna Jefferies**[1,5], **Stuart C. Clarke**[1,5,6], **Saul N. Faust**[1,2,6], **Luanne Hall-Stoodley**[1,2,7☯], **Jeremy Webb**[3,6☯]

1 Academic Unit of Clinical and Experimental Sciences, Faculty of Medicine and Institute for Life Sciences, University of Southampton, Southampton, United Kingdom, 2 Southampton NIHR Wellcome Trust Clinical Research Facility, University Hospital Southampton NHS Foundation Trust, Southampton, United Kingdom, 3 Centre for Biological Sciences, University of Southampton, Southampton, United Kingdom, 4 Centre for Proteomic Research, Institute for Life Sciences, University of Southampton, Southampton, United Kingdom, 5 Public Health England, Southampton, United Kingdom, 6 Southampton NIHR Respiratory Biomedical Research Unit, University Hospital Southampton NHS Foundation Trust, Southampton, United Kingdom, 7 Microbial Infection and Immunity, Centre for Microbial Interface Biology, The Ohio State University, Columbus, Ohio, United States of America

Abstract

Streptococcus pneumoniae accounts for a significant global burden of morbidity and mortality and biofilm development is increasingly recognised as important for colonization and infection. Analysis of protein expression patterns during biofilm development may therefore provide valuable insights to the understanding of pneumococcal persistence strategies and to improve vaccines. iTRAQ (isobaric tagging for relative and absolute quantification), a high-throughput gel-free proteomic approach which allows high resolution quantitative comparisons of protein profiles between multiple phenotypes, was used to interrogate planktonic and biofilm growth in a clinical serotype 14 strain. Comparative analyses of protein expression between log-phase planktonic and 1-day and 7-day biofilm cultures representing nascent and late phase biofilm growth were carried out. Overall, 244 proteins were identified, of which >80% were differentially expressed during biofilm development. Quantitatively and qualitatively, metabolic regulation appeared to play a central role in the adaptation from the planktonic to biofilm phenotype. Pneumococci adapted to biofilm growth by decreasing enzymes involved in the glycolytic pathway, as well as proteins involved in translation, transcription, and virulence. In contrast, proteins with a role in pyruvate, carbohydrate, and arginine metabolism were significantly increased during biofilm development. Downregulation of glycolytic and translational proteins suggests that pneumococcus adopts a covert phenotype whilst adapting to an adherent lifestyle, while utilization of alternative metabolic pathways highlights the resourcefulness of pneumococcus to facilitate survival in diverse environmental conditions. These metabolic proteins, conserved across both the planktonic and biofilm phenotypes, may also represent target candidates for future vaccine development and treatment strategies. Data are available via ProteomeXchange with identifier PXD001182.

Editor: Jens Kreth, University of Oklahoma Health Sciences Center, United States of America

Funding: This work was funded by The Bill and Melinda Gates Foundation grant GCE 1006884 (http://www.gatesfoundation.org). Funding was received by JW, SCC, SNF, LHS, and JJ. The funders had no role in study design, data collection and analysis, decision to publish, or preparation of the manuscript.

Competing Interests: The authors have declared that no competing interests exist.

* Email: r.allan@soton.ac.uk

☯ These authors contributed equally to this work.

Introduction

Pneumococcal disease accounts for a significant global burden of morbidity and mortality, particularly in younger children and older adults. *Streptococcus pneumoniae* is a Gram-positive facultative anaerobe that colonises the majority of children in the first year of life and is associated with mostly asymptomatic carriage in the human respiratory tract. However, multifactorial and incompletely understood mechanisms allow *S. pneumoniae* to cause localised infection such as otitis media and sinusitis, and/or invasive disease (meningitis, bacteremia and pneumonia). The transition from asymptomatic colonisation of the respiratory mucosal epithelium without inflammation to the initiation of infection is thought to correlate with the expression of specific cell surface molecular properties such as adhesins and capsule

polysaccharides under different conditions [1]. Although pneumococcal conjugate vaccines have reduced carriage and disease caused by capsular strains included in these vaccines, there has been a rise in the prevalence of capsular strains not included in the vaccines (serotype replacement). With over 90 different pneumococcal serotypes, there is a need for more effective vaccines that are conserved across serotypes, immunogenic in all age groups, and effective against both colonisation and infection by *S. pneumoniae* [2].

Aggregated pneumococci adherent to human middle ear mucosae of children undergoing treatment for chronic otitis media (COM) indicate that *S. pneumoniae* is present in biofilms on the mucosal epithelium [3], an observation supported by animal models [4,5]. Pneumococcal biofilms were also observed on adenoid mucosal epithelium from children undergoing adenoid-ectomy for the treatment of obstructive sleep apnea, suggesting that pneumococcal biofilms are present in the nasopharynx in both the presence and absence of overt infection [6,7]. The ability of S. pneumoniae to develop biofilms has been shown to affect both colonisation and disease states [8–10], and understanding S. pneumoniae biofilm development is important for studying the mechanisms of persistence.

S. pneumoniae adapts to different environments in the human host by modulating proteins and carbohydrates that may function as virulence factors. In addition to different capsule serotypes, multiple proteins modulate cell attachment and contribute to both colonisation and invasion. In vitro studies with pneumococcal clinical strains have demonstrated that biofilm growth downreg-ulates or disrupts capsule expression, and induces the production a complex extracellular matrix consisting of carbohydrates, proteins and DNA [11–14] and differential expression of a broad spectrum of proteins [15,16]. We hypothesized that protein expression patterns during biofilm growth might highlight important pheno-typic changes facilitating pneumococcal persistence or virulence. On the other hand, conserved protein expression in both planktonic and biofilm growth, might inform better vaccine strategies since these proteins would be present regardless of the mode of growth or type of infection.

Serotype 14 is a common serotype in pneumococcal infection worldwide, causing both localised and invasive disease. It exhibits high levels of antibiotic resistance [17] and is prevalent in paediatric disease such as otitis media [18]. Serotype 14 was chosen in this study as a model strain because of its high biofilm-forming index based on multiparametric analysis of different clinical S. pneumoniae strains [13]. Previous proteomic analyses of S. pneumoniae growth have used conventional 1DE and 2DE gel-based methodologies [15,16,19–21]. iTRAQ (isobaric tag for relative and absolute quantitation) is a high-throughput gel-free proteomic profiling approach enabling quantitative high resolution comparison of protein profiles between multiple phenotypes or growth conditions within a single experiment. To investigate the pneumococcal serotype 14 biofilm proteome, we compared nascent 1-day biofilms lacking complex structure, highly struc-tured 7-day biofilms, and log-phase planktonic growth to determine: 1) distinct patterns of differential expression comparing planktonic, nascent and late-phase biofilm growth, and 2) conserved patterns of protein regulation during planktonic and biofilm growth.

Methods

Bacterial strain and growth conditions

A Serotype 14 (ST124) blood sample isolate provided by the UK Public Health England (Health Protection Agency) Microbi-ology Services laboratory, Southampton, was selected based on multi-parametric, statistically-based criteria for biofilm develop-ment [13] [personal observation RNA/LHS]. The strain was subcultured from a frozen stock onto Columbia blood agar (CBA) plates (Oxoid, U.K.) and incubated for 14 hours at $37°C/5\%$ CO_2. Colonies were re-suspended in 2 ml Brain Heart Infusion (BHI) broth (Oxoid, U.K.), centrifuged at $3,200 \times g$ for 5 min to remove cell debris and secreted lytic enzymes (personal commu-nication, Public Health England), and the supernatant used to inoculate fresh dilute BHI broth. For planktonic growth, cultures were grown to mid-exponential phase ($OD_{600} = 0.3$; $\sim 1 \times 10^8$ cells).

Biofilm Formation

For biofilm formation, mid-exponential planktonic cultures (2 ml$\times 10^8$ cells) were used to inoculate individual wells of polystyrene 6-well plates (Corning Incorporated, Costar, U.S.A.), and all wells supplemented with 2 ml BHI diluted 1:5 with non-pyrogenic sterile dH_2O. Cultures were incubated under static conditions at $37°C/5\%$ CO_2 with replacement of warm, fresh diluted BHI daily.

Colony Forming Unit (CFU) Enumeration and Total Cell Counts

All medium was removed and biofilms washed twice using diluted BHI. Biofilms were resuspended in 1 ml Hank's balanced salt solution (HBSS) as described [13]. Briefly, bacterial suspen-sions (n = 3) underwent 10-fold dilutions in HBSS and 5×20 µl of each dilution spot plated onto CBA plates and incubated at $37°C/$ 5% CO_2. For total cell counts (n = 3) bacterial suspensions were diluted 10-fold, stained with Syto 9 (Life Technologies, U.S.A.) according to manufacturer instructions, and individual cells counted using a haemocytometer (Marienfeld-Superior, Germany) and a Leica epifluorescent microscope using a 100x oil immersion lens.

Scanning Electron Microscopy (SEM)

Biofilms (n = 3) were grown in 6-well plates containing ethanol-sterilized 13 mm glass coverslips (V.W.R., U.K.) under static conditions at $37°C/5\%$ CO_2. All medium was removed and biofilms washed twice using diluted BHI. Coverslips were removed using sterile forceps, placed in primary fixative solution (3% gluteraldehyde, 0.1 M sodium cacodylate (pH 7.2), 0.15% Alcian blue), and incubated at $4°C$ overnight. The primary fixative was replaced with 0.1 M sodium cacodylate (pH 7.2), followed by the secondary fixative solution (0.1 M osmium tetroxide, 0.1 M sodium cacodlyate; pH 7.2), and finally with 0.1 M sodium cacodylate (pH 7.2). Each treatment was incubated for 1 h at room temperature. Coverslips were processed through an ethanol series to 100%, critical point dried, and sputter coated and biofilms imaged using FEI Quanta 200 scanning electron microscope.

Confocal Laser Scanning Microscopy (CLSM)

Biofilms (n = 6) were grown in 35 mm glass bottom microwell dishes (MatTek Corporation, U.S.A.) under static conditions at $37°C$ with 5% CO_2, washed twice with HBSS and stained with LIVE/DEAD BacLight Bacterial Viability Kit (Life Technologies, U.S.A.) according to manufacturer instructions. Biofilms were examined immediately using a Leica SP5 LSCM with inverted stand using a 63x oil immersion lens, performing sequential scanning using 0.5 µm sections. Three random fields of view were captured and images were analysed using Leica LCS Software.

Protein Extraction

For planktonic protein samples, cultures were grown to mid-exponential phase and centrifuged. The supernatant was discarded and the pellet washed twice in 1 ml Hanks' Balanced Salt Solution (HBSS; Gibco, Life Technologies, U.K.) at $9,500 \times g$. For biofilm protein samples 1 and 7 day cultured biofilms were washed twice with HBSS and resuspended in 1 ml HBSS using sterile cell scrapers. The bacterial suspension was centrifuged ($9,500 \times g$), the supernatant removed and the pellet resuspended in 0.1 M triethylammonium bicarbonate (TEAB) buffer (Sigma-Aldrich, U.K.) with 0.1% Rapigest SF surfactant (Waters, U.K.). Both the planktonic and biofilm samples were lysed in lysing matrix B (MP

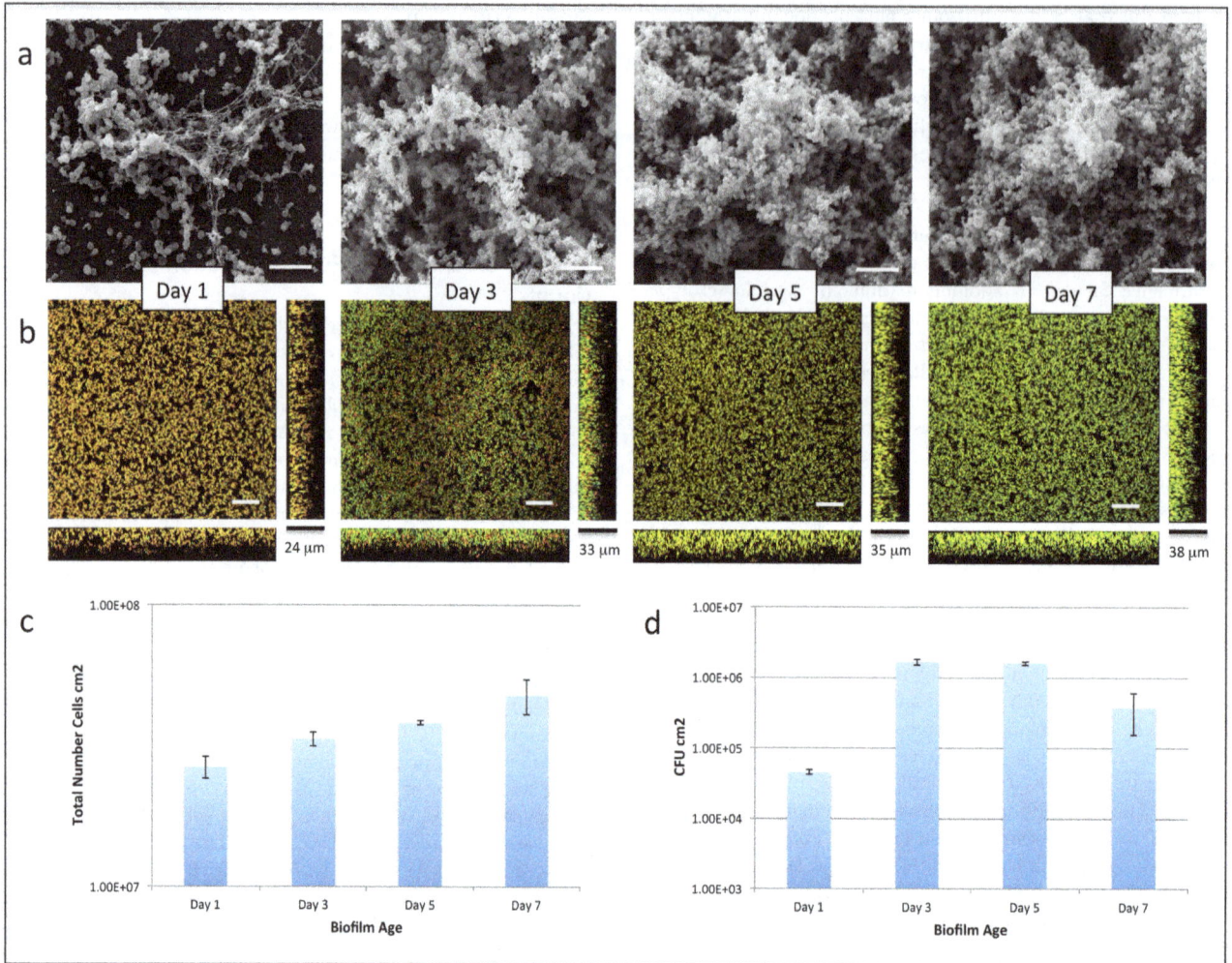

Figure 1. *S. pneumoniae* **serotype 14 day 1, 3, 5 and 7 biofilms imaged using (a) scanning electron microscopy with Alcian Blue staining (8,000x magnification; scale bar: 5 μm), and (b) confocal microscopy using Live/Dead staining, with sideviews (YZ – right) and (XZ – bottom) sagittal sections of the biofilm.** Scale bar beside YZ sagittal sections represents average biofilm thickness. Scale bar in xy pane: 30 μm. (c) Bar chart comparing the total number of cells between day 1, 3, 5 and 7 biofilms. (d) Bar chart comparing number of viable cells between day 1, 3, 5 and 7 biofilms through CFU cm^{-2} measurement.

Bioscience, U.K.) using a Hybaid Ribolyser Homogenizer (Hybaid, U.K.) in six 30 second sessions with 30 second storage on ice between sessions. The lysates were centrifuged at 855×g/ 5 min and the supernatant retained (soluble protein fraction). Protein quantification of the soluble protein fraction was performed using Coomassie Protein Assay Reagent (Thermo Fisher Scientific, U.K.), measuring optical density (595 nm) and quantified against albumin standards (Thermo Fisher Scientific, U.K.).

iTRAQ labelling

Comparative analyses of protein expression between planktonic log-phase culture and 1-day and 7-day biofilms were performed on 3 technical replicates of 3 biological replicates. iTRAQ labelling of samples was performed according to the manufacturer's instructions using an iTRAQ Reagent-8 plex Multiplex Kit (AB Sciex U.K. Limited). Protein samples were normalized to ~60 μg and 2 μl of the reducing agent and 50 mM (tris-(2-carboxyethyl) phosphine) were added and mixed. Samples were incubated at 60°C for 1 h. Methyl methane-thiosulfonate in isopropanol (1 μl

of 200 mM) was then added and incubated for a further 10 min at room temperature. Proteins were digested by adding 10 μl of 1 mg/ml trypsin in 80 mM CaCl$_2$. Samples were incubated for 16 h at 37°C. Labelling of tryptic peptides with iTRAQ 8-plex tags was achieved by mixing the appropriate iTRAQ labeling reagent with the relevant sample, and incubation at room temperature for 2 h.

Mass Spectrometry

Two dimensional separations were performed using a nanoAcquity 2D UPLC system (Waters). For the first dimension separation, 2.0 μl of the prepared protein lysates containing 3.2 μg of iTRAQ labeled peptides were injected onto a 5 μm Xbridge BEH130 C18, 300 μm ID×50 mm (Waters) column equilibrated in 20 mM ammonium formate, pH 10 (buffer A). The first dimension separation was achieved by increasing the concentration of acetonitrile (buffer B) in 7 steps consisting of 11, 14, 16, 20, 25, 50, 65%. At each step the programmed percentage composition was held for 1 min at a low rate of 1000 nl/min and the eluant diluted by buffer C (H$_2$0+0.1% formic acid)

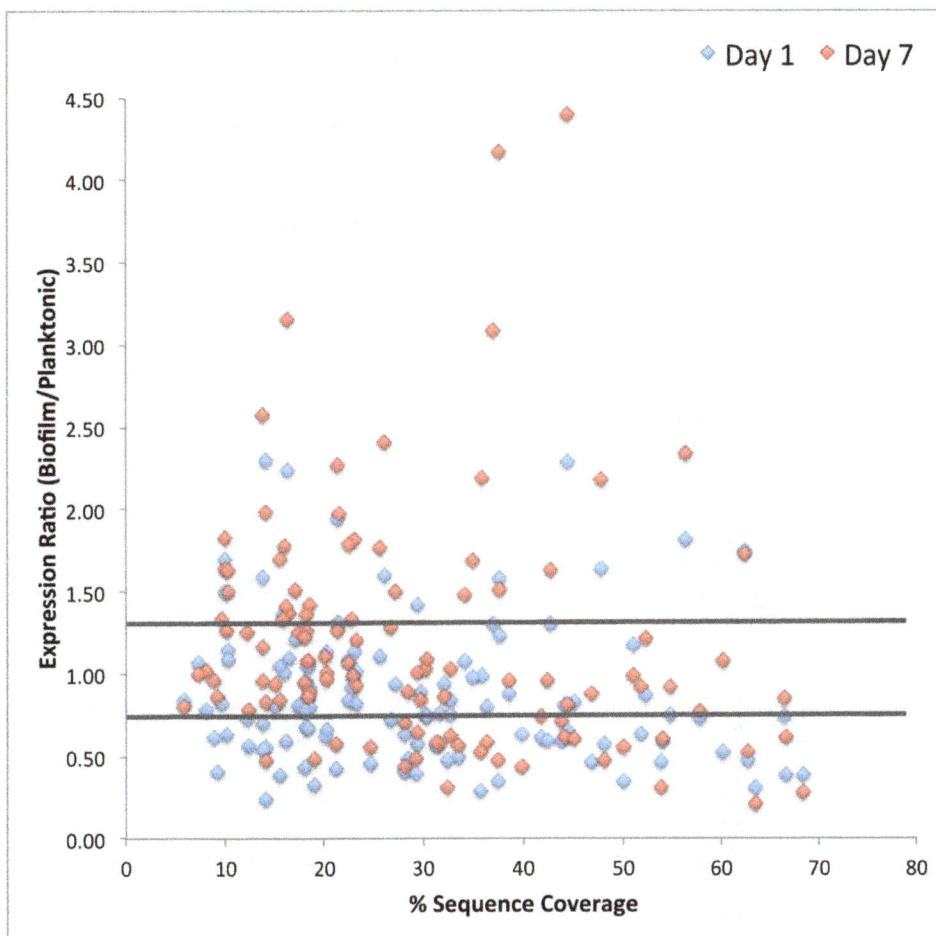

Figure 2. Volcano plot of the complete iTRAQ proteomic dataset comparing the fold change in protein expression between mid-exponential planktonic *S. pneumoniae* and day 1 and day 7 biofilms. Proteins represented met inclusion criteria of ≥3 peptide matches, ≥50 protein score, and ≥5% sequence coverage (p<0.05). Comparative protein data with >1.3 and <0.77 ratios (marked by horizontal lines) were identified as having differential expression.

from the second dimension pump at a flow rate of 20 μl/min, effectively diluting the ammonium formate and acetonitrile, allowing trapping of the eluting peptides onto a Symmetry C18, 180 μm×20 mm trapping cartridge (Waters). After 20 min washing of the trap column, peptides were separated using an in-line second dimension analytical separation performed on a 75 μm ID×200 mm, 1.7 μm BEH130 C18, column (Waters) using a linear gradient of 5 to 40% B (buffer A = water containing 0.1% formic acid, buffer B = 95% acetonitrile containing 0.1% formic acid) over 120 min with a wash to 85% B at a flow rate of 300 nl/min. All separations were automated and performed on-line to the mass spectrometer.

All data were acquired using a Q-tof Global Ultima (Waters Ltd, Manchester, UK) fitted with a nanoLockSpray source. A survey scan was acquired from *m/z* 350 to 1700 with the switching criteria for MS to MS/MS including ion intensity and charge state. The collision energy used to perform MS/MS was varied according to the mass and charge state of the eluting peptide. Peak lists were generated using ProteinLynx Global Server 2.4 (Waters Ltd, Manchester, UK). The following parameters were used for processing the MS/MS spectra; normal background subtraction with a 25% background threshold and medium de-isotoping with a threshold of 1%, no smoothing was performed.

Peak list generation and database searching

Peak lists from the MS/MS analysis of were submitted to the MASCOT search engine version 2.2.1 (Matrix Science, London, UK) and the data searched against a protein translation of the CGSP14 NCBI genome (http://www.ncbi.nlm.nih.gov/Taxonomy/Browser/wwwtax.cgi?id = 516950). The corresponding quantitative information using the iTRAQ reporter ions was also obtained via MASCOT.

A maximum of one missed cleavage for tryptic digestion and fixed modifications for methyl methane-thiosulfonation of cysteine and the N-terminus and lysine side chains using the 8-plex iTRAQ label were allowed (Applied Biosystems, Warrington, UK). Variable modification for the oxidation of methionine and iTRAQ modification of tyrosine were also allowed. Precursor ion and sequence ion mass tolerances were set at 100 ppm and 0.3 Da respectively. Protein identifications required the assignment of ≥2 different peptides with a significance threshold for accepting a match of p<0.05. The protein ratios were calculated using MASCOT version 2.2.1, where the peptide ratios were weighted and median normalisation was performed, automatic outlier removal was chosen and the peptide threshold was set to 'at least homology'. False discovery rates were calculated by running all spectra against a decoy database using the MASCOT software.

Category	Protein	Accession No.	Protein Expression Ratios 1-day Biofilm/ Planktonic	7-day Biofilm/ Planktonic	Peptide Matches	% Sequence Coverage
Translation	elongation factor Tu	YP_001836191	0.76	0.92	14	54.7
	elongation factor EF-2	YP_001835001	1.07	1.49	15	34.1
	elongation factor Ts	YP_001836897	0.82	0.62	10	44.2
	30S ribosomal protein S1	YP_001835523	0.61	0.74	13	41.8
	30S ribosomal protein S7	YP_001835000	0.39	0.62	4	66.7
	50S ribosomal protein L4	YP_001834936	0.63	0.92	8	51.7
	30S ribosomal protein S5	YP_001834952	0.83	0.60	5	45.1
	50S ribosomal protein L1	YP_001835309	0.66	0.81	8	44.5
	50S ribosomal protein L3	YP_001834935	0.43	0.57	3	21.2
	50S ribosomal protein L23	YP_001834937	0.64	0.44	4	39.8
	50S ribosomal protein L5	YP_001834947	0.59	0.61	7	53.9
	30S ribosomal protein S9	YP_001835026	0.46	0.56	3	24.6
	30S ribosomal protein S3	YP_001834941	0.41	0.71	4	28.1
	30S ribosomal protein S2	YP_001836898	0.35	0.48	4	37.5
	50S ribosomal protein L22	YP_001834940	0.38	0.28	4	68.4
	30S ribosomal protein S8	YP_001834949	0.80	0.59	3	36.4
	50S ribosomal protein L10	YP_001836061	0.77	0.58	4	31.6
	50S ribosomal protein L17	YP_001834961	0.49	0.57	5	33.6
	50S ribosomal protein L13	YP_001835025	0.25	0.52	3	35.8
	translation initiation factor IF-2	YP_001835236	0.84	0.81	5	5.9
	elongation factor Tu family protein	YP_001835355	0.68	1.37	5	18.2
	50S ribosomal protein L14	YP_001834945	0.35	0.56	4	50
	30S ribosomal protein S18	YP_001836242	0.57	0.48	4	48.1
	50S ribosomal protein L18	YP_001834951	0.31	0.22	3	63.4
	50S ribosomal protein L15	YP_001834954	0.63	0.44	3	28.1
	30S ribosomal protein S10	YP_001834934	0.73	0.77	4	57.8
	30S ribosomal protein S4	YP_001834800	0.59	0.96	4	42.4
	50S ribosomal protein L6	YP_001834950	0.60	0.71	4	43.8
	50S ribosomal protein L2	YP_001834938	0.94	0.86	4	32.1
	30S ribosomal protein S13	YP_001834958	0.57	0.58	3	31.4
	50S ribosomal protein L29	YP_001834943	0.48	0.31	3	32.4
	30S ribosomal protein S15	YP_001836318	0.40	0.49	3	29.2
	ribosomal subunit interface protein	YP_001836890	1.94	2.27	3	21.4
Transcription	DNA-directed RNA polymerase alpha subunit	YP_001834960	0.75	1.03	4	30.2
	DNA-binding protein HU	YP_001835885	0.47	0.31	3	53.8
	transcription elongation factor NusA	YP_001835233	0.55	0.84	4	14
	DNA-directed RNA polymerase beta~ subunit	YP_001836642	0.82	1.33	3	9.6
	ribose-phosphate pyrophosphokinase	YP_001835901	1.31	1.27	3	21.3
Metabolism — Glycolysis	enolase (phosphopyruvate hydratase)	YP_001835867	0.47	0.52	18	62.7
	pyruvate kinase	YP_001835590	0.87	1.21	16	52.3
	phosphoglycerate kinase	YP_001835194	1.17	0.99	14	51
	glyceraldehyde-3-phosphate dehydrogenase	YP_001836696	0.58	0.65	8	29.5
	phosphoglyceromutase	YP_001836345	0.74	0.85	8	66.5
	glucose-6-phosphate isomerase	YP_001836754	0.33	0.48	8	18.9
	6-phosphofructokinase	YP_001835589	0.84	1.02	5	32.8
	fructose-bisphosphate aldolase	YP_001835283	0.56	0.78	3	12.3
	phosphoglucomutase	YP_001836198	0.64	1.26	3	10.1
Metabolism — Pyruvate Metabolism	L-lactate dehydrogenase	YP_001835798	1.23	1.51	10	37.7
	formate acetyltransferase	YP_001835173	1.30	3.09	14	37
	alcohol dehydrogenase	YP_001835014	1.30	1.63	9	42.8
	alcohol dehydrogenase, iron-containing	YP_001836708	1.51	1.63	6	9.9
	alcohol dehydrogenase, zinc-containing	YP_001836739	1.70	1.82	3	9.9
Metabolism — Amino-acid Metabolism	arginine deiminase	YP_001836835	1.58	4.17	8	37.7
	ornithine carbamoyltransferase	YP_001836836	2.24	3.16	4	16.3
	aspartate aminotransferase	YP_001836677	0.62	0.96	3	8.9
	asparagine synthetase AsnA	YP_001836652	1.31	1.97	4	21.5
	adenylosuccinate synthetase	YP_001834736	0.24	0.48	3	14
	glutamate dehydrogenase	YP_001835989	0.78	1.02	3	8
Metabolism — Carbohydrate Metabolism	tagatose 1,6-diphosphate aldolase	YP_001835825	2.29	4.39	11	44.5
	6-phosphogluconate dehydrogenase	YP_001835088	0.94	1.50	6	27.2
	sugar ABC transporter, ATP-binding protein	YP_001836282	1.60	2.41	5	26.1
	PTS system, IIB component, putative	YP_001835318	1.82	2.34	4	56.4
	UDP-glucose 4-epimerase	YP_001836305	0.81	1.25	5	17.4
	phosphoenolpyruvate-protein phosphotransferase	YP_001835838	0.96	1.26	7	18.2
	PTS system, mannose-specific IIAB components	YP_001835012	1.42	1.01	7	29.5
	sugar ABC transporter, sugar-binding protein	YP_001836373	1.59	2.57	5	13.8
	PTS system, mannose-specific IID component	YP_001835010	1.13	1.82	6	23.1
	maltose/maltodextrin ABC transporter, maltose/maltodextrin-binding protein	YP_001836789	1.21	1.51	4	17.1
	PTS system, fructose specific IIABC components	YP_001835543	0.55	0.96	4	13.8
Metabolism — Other Metabolic	glycogen phosphorylase family protein	YP_001836787	1.07	1.00	4	7.2
	uracil phosphoribosyltransferase	YP_001835412	0.52	1.08	7	60.2
	pyridine nucleotide-disulfide oxidoreductase	YP_001836287	0.62	1.11	4	20.1
	ATP synthase subunit B	YP_001836209	0.75	1.09	9	30.3
	ATP synthase subunit A	YP_001836211	0.73	1.26	4	12.2
	3-ketoacyl-(acyl-carrier-protein) reductase	YP_001835136	0.48	0.89	4	28.4
	3-oxoacyl-(acyl carrier protein) synthase	YP_001835137	0.81	0.94	4	15
Virulence / Competence / Stress	pyruvate oxidase	YP_001835396	0.99	2.19	17	35.9
	NADH oxidase	YP_001836175	0.39	0.84	4	15.5
	general stress protein 24, putative	YP_001836498	1.74	1.73	10	62.4
	molecular chaperone DnaK	YP_001835207	0.75	0.63	16	32.8
	chaperonin GroEL	YP_001836598	1.14	1.01	7	20.2
	superoxide dismutase, manganese-dependent	YP_001835432	0.84	1.07	3	22.4
Protein Biosynthesis / Alteration / Degradation	ATP-dependent Clp protease, ATP-binding subunit, putative	YP_001835059	2.30	1.99	3	14.1
	endopeptidase O	YP_001836337	0.81	0.89	6	18.5
	glutamyl-tRNA amidotransferase subunit A	YP_001835148	1.02	1.21	3	23.2
	signal recognition particle protein	YP_001835967	1.01	1.78	3	15.9
	glycyl-tRNA synthetase beta subunit	YP_001836180	1.04	1.70	4	15.5
	arginyl-tRNA synthetase	YP_001836762	1.15	1.50	3	10.3
	phenylalanyl-tRNA synthetase beta subunit	YP_001835264	1.09	1.63	3	10.2
	lysyl-tRNA synthetase	YP_001835379	0.68	1.08	3	18.3
	aspartyl-tRNA synthetase	YP_001836796	0.78	1.22	3	18.1
	trigger factor	YP_001835116	0.88	0.96	9	38.6
Cell Division	cell division protein FtsZ	YP_001836355	0.66	0.97	3	20.3
	cell division protein DivIVA	YP_001836350	0.82	0.93	3	23.3
	septation ring formation regulator EzrA	YP_001835473	1.09	1.37	4	16.4
	cell division protein FtsH	YP_001834729	1.09	1.78	3	22.5
Transport	ABC transporter, substrate-binding protein	YP_001834869	1.11	1.77	4	25.7
	bacteriocin transport accessory protein	YP_001836201	1.64	2.18	4	47.8
	translocase	YP_001836391	0.41	0.86	3	9.1
Miscellaneous	polyribonucleotide nucleotidyltransferase	YP_001835270	0.71	1.17	6	13.7
	thioredoxin	YP_001836471	0.89	0.84	3	29.8
	lipoate-protein ligase, putative	YP_001835854	0.92	1.42	4	18.5
	putative manganese-dependent inorganic pyrophosphatase	YP_001836237	0.87	0.86	3	18.3
	flavodoxin	YP_001835981	0.47	0.88	3	46.9
	lipoprotein	YP_001835504	0.99	0.99	5	23
	SPFH domain-containing protein/band 7 family protein	YP_001836841	0.44	0.95	3	18.1
Unknown Function	hypothetical protein SPCG_1532	YP_001836249	0.73	1.29	5	26.8
	hypothetical protein SPCG_0438	YP_001835155	1.05	1.07	7	18.4
	hypothetical protein SPCG_0428	YP_001835145	0.98	1.69	4	34.9
	hypothetical protein SPCG_1503	YP_001836220	1.36	1.34	4	15.8
	hypothetical protein SPCG_2153	YP_001836870	0.59	1.41	4	16.1
	hypothetical protein SPCG_1901	YP_001836618	0.92	1.33	3	22.8

> 2.0	> 2 fold upregulation
1.3 - 1.99	< 2 fold upregulation
0.77 - 1.29	no change in expression
0.5 - 0.76	< 2 fold downregulation
< 0.5	> 2 fold downregulation

Figure 3. Comparative iTRAQ analyses of *S. pneumoniae* **serotype 14 1-day and 7-day old** *in vitro* **biofilms to mid-exponential planktonic population protein expression.** Inclusion criteria: ≥3 peptide matches, ≥50 protein score, ≥5% sequence coverage (p<0.05). Comparative protein data with >1.3 and <0.77 ratios identified as having differential expression.

	Protein	Accession No.	Protein Expression Ratios		Score	% Sequence Coverage
			1-day Biofilm/ Planktonic	7-day Biofilm/ Planktonic		
Translation	30S ribosomal protein S6	YP_001836244	0.19	0.24	126	26
	50S ribosomal protein L19	YP_001835977	0.50	1.05	107	40
	50S ribosomal protein L11	YP_001835308	0.27	0.14	51	37.6
	30S ribosomal protein S17	YP_001834944	0.55	0.74	762	34.9
	50S ribosomal protein L9	YP_001836888	0.72	0.51	197	15.3
	peptide chain release factor 1	YP_001835715	0.89	1.19	225	10
	30S ribosomal protein S12	YP_001834999	1.44	0.92	46	24.1
Transcription	transcription antitermination protein NusG	YP_001836690	0.78	0.96	62	14.2
	RNA polymerase sigma factor	YP_001835924	0.67	0.94	77	18.4
Metabolism — **Pyruvate Metabolism**	acetoin dehydrogenase complex, E3 component, dihydrolipoamide dehydrogenase, putative	YP_001835853	0.49	0.72	162	14.3
	lactate oxidase	YP_001835380	0.70	1.40	85	8.2
Amino-acid Metabolism	methionine--tRNA ligase	YP_001835454	0.25	0.50	38	7.1
	2,3,4,5-tetrahydropyridine-2-carboxylate N-succinyltransferase, putative	YP_001836779	0.60	0.73	159	13.8
	3-deoxy-7-phosphoheptulonate synthase	YP_001836389	1.55	1.46	198	12
	serine hydroxymethyltransferase	YP_001835719	1.39	1.39	124	11.2
	glutamine amidotransferase, class I	YP_001835906	0.97	1.39	87	13.1
	cysteine synthase	YP_001836894	1.92	2.57	159	9.5
Sugar Metabolism	glucose-6-phosphate 1-dehydrogenase	YP_001835777	0.58	0.83	200	7.5
	galactose-6-phosphate isomerase	YP_001835822	1.23	1.87	263	22.7
	D-fructose-6-phosphate amidotransferase	YP_001834993	0.90	1.65	263	9.8
	transketolase	YP_001836712	1.30	1.67	240	6.5
	UTP-glucose-1-phosphate uridylyltransferase	YP_001836775	1.34	1.14	221	17.4
	galactose-6-phosphate isomerase	YP_001835823	1.60	1.77	137	26.3
	PTS system, IIC component	YP_001834780	0.90	2.38	58	10.3
Other Metabolic	acyl carrier protein	YP_001835133	0.54	0.54	1620	40.5
	acetyl-CoA carboxylase alpha subunit	YP_001835142	0.76	1.22	201	15.7
	aminopeptidase N	YP_001835463	0.84	1.08	269	8.7
	UDP-N-acetylglucosamine 1-carboxyvinyltransferase	YP_001836648	1.17	0.77	143	15.5
	UDP-N-acetylmuramoyl-L-alanyl-D-glutamate synthetase	YP_001835360	1.14	1.26	98	9.6
	purine nucleoside phosphorylase	YP_001835495	1.16	1.19	77	14.8
	ATP synthase subunit B	YP_001836213	1.45	1.20	120	7.9
	enoyl-(acyl-carrier-protein) reductase	YP_001835134	1.60	1.23	78	11.7
	bifunctional GMP synthase/glutamine amidotransferase protein	YP_001836150	1.25	1.94	43	11.5
	catabolite control protein A	YP_001836682	0.78	1.37	22	5.2
	glycerol kinase	YP_001836871	0.90	3.54	195	5.6
	acetyl-CoA carboxylase beta subunit	YP_001835141	1.44	2.27	53	14.6
Virulence / Competence / Stress	universal stress protein	YP_001836679	0.84	1.86	200	31
	pneumolysin	YP_001836615	1.10	1.36	170	15.5
	DNA-binding response regulator CiaR	YP_001835464	1.51	1.75	59	19.1
Protein Biosynthesis / Alteration / Degradation	glutamyl-tRNA synthetase	YP_001836753	0.76	1.03	210	14.8
	ATP-dependent protease ATP-binding subunit	YP_001836271	0.54	0.89	195	9.5
	asparaginyl-tRNA synthetase	YP_001836245	0.58	1.15	50	11.4
	aminopeptidase C	YP_001835009	0.79	1.21	164	10.6
	seryl-tRNA synthetase	YP_001835127	1.11	1.14	135	8.8
	prolyl-tRNA synthetase	YP_001834991	0.83	1.34	151	13.8
	peptidase T	YP_001835701	1.07	1.51	63	10.6
	peptidylprolyl isomerase	YP_001835673	1.22	1.60	46	8
Transport	oligopeptide ABC transporter, ATP-binding protein AmiE	YP_001836579	0.47	0.62	72	10.1
	oligopeptide ABC transporter, ATP-binding protein AmiF	YP_001836578	0.87	1.56	54	10
	ABC transporter, ATP-binding protein	YP_001835529	1.69	2.26	164	16.3
Miscellaneous	FeS assembly protein SufB	YP_001835533	0.60	1.13	47	11.5
	lipoprotein	YP_001834870	0.81	1.10	95	16.2
	cell wall surface anchor family protein	YP_001834797	1.12	0.88	46	2.2
	thiol peroxidase	YP_001836341	0.82	1.43	129	31.4
	nitroreductase family protein	YP_001835300	1.06	1.32	94	12.4
	inositol-5-monophosphate dehydrogenase	YP_001836911	1.15	1.33	72	11.8
	GTP-binding protein EngA	YP_001836398	1.48	3.28	154	15.8
	glucose-inhibited division protein A	YP_001834840	1.37	4.01	64	8
	Gfo/Idh/MocA family oxidoreductase	YP_001836376	1.28	2.86	42	7.6
Unknown	hypothetical protein SPCG_1897	YP_001836614	0.79	0.66	401	11.8
	hypothetical protein SPCG_0966	YP_001835683	0.83	0.94	343	20.9
	hypothetical protein SPCG_0759	YP_001835476	1.06	1.26	61	22.2

> 2.0	> 2 fold upregulation
1.3 - 1.99	< 2 fold upregulation
0.77 - 1.29	no change in expression
0.5 - 0.76	< 2 fold downregulation
< 0.5	> 2 fold downregulation

Figure 4. Qualitative data listing proteins identified with 2 peptide matches using comparative iTRAQ analysis of *S. pneumoniae* **serotype 14 1-day and 7-day old** *in vitro* **biofilms to planktonic population protein expression.** Comparative protein data with >1.3 and <0.77 ratios identified as having differential expression.

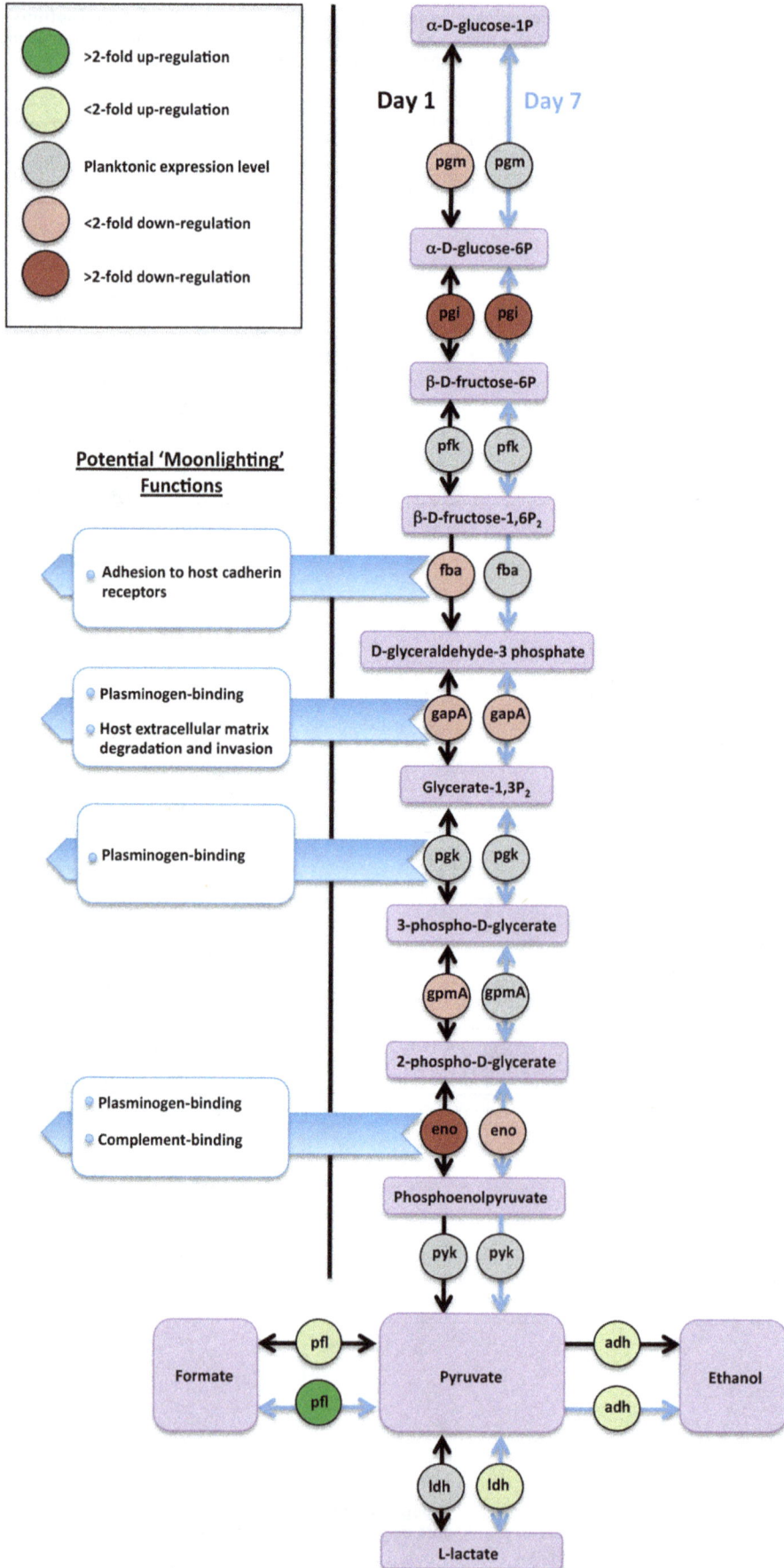

Figure 5. Change in expression of enzymes involved in the glycolysis/gluconeogenesis pathway and pyruvate metabolism of *S. pneumoniae* serotype 14 during biofilm formation. Based on iTRAQ expression data comparing mid-exponential planktonic cultures to day 1 and 7 biofilms. Proteins with potential moonlighting capability highlighted with their putative alternative functions. Pgm - phosphoglucomutase; pgi – glucose-6-phosphate isomerase; pfk–6-phosphofructokinase; fba – fructose biphosphate aldolase; gapA – glyceraldehyde-3-phosphate dehydrogenase; pgk – phosphoglycerate kinase; gpmA – phosphoglyceromutase; eno – enolase; pyk – pyruvate kinase, ldh – lactate dehydrogenase; adh – alcohol dehydrogenase, pfl – formate acetyltransferase.

Inclusion criteria for quantitative analysis were set at ≥ 3 peptide matches, ≥ 50 protein score, $\geq 5\%$ sequence coverage (p< 0.05). Comparative protein data with >1.3 and <0.77 ratios were identified as having differential expression. For qualitative identification the inclusion criteria were ≥ 2 peptide matches, 50 protein score and 10% sequence coverage.

Results

Pneumococcus grew over 7 days from unstructured cell clusters to a structurally complex biofilm comprised of metabolically active, viable cells

SEM demonstrated that day 1 biofilms were characterized by the presence of small clusters of cells interspersed amongst individual cells and diplococci (Fig. 1a) surrounded by extracellular material

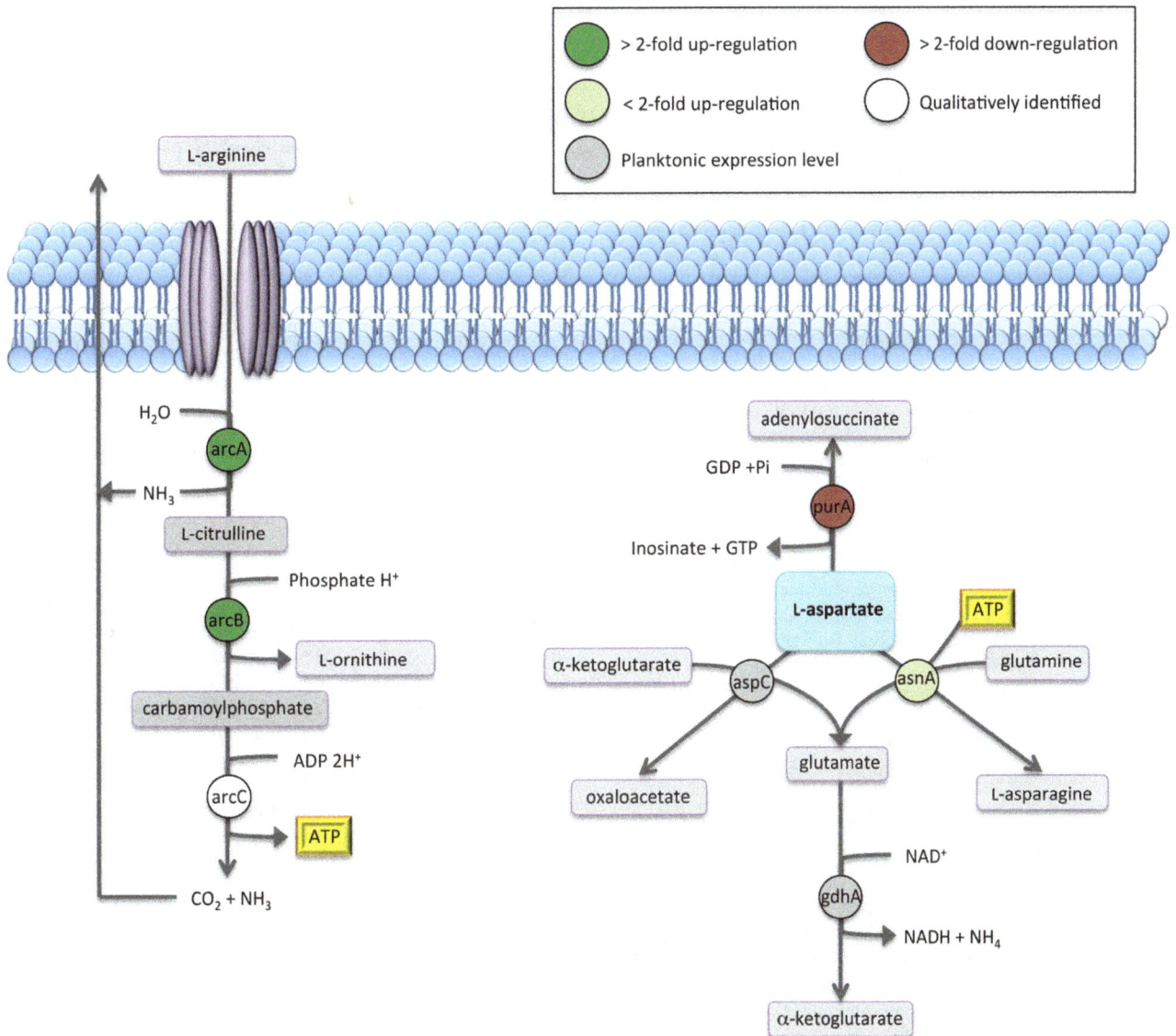

Figure 6. Change in expression of enzymes involved in amino-acid metabolism of day 7 *S. pneumoniae* serotype 14 biofilms. Based on iTRAQ expression data comparing mid-exponential planktonic cultures to day 7 biofilms. Highlighting the arginine deiminase (ADI) pathway consisting of arginine deiminase (arcA), ornithine carbamoyltransferase (arcB) and carbamate kinase (arcC), and components of L-aspartate metabolism consisting of adenylosuccinate synthetase (purA), aspartate aminotransferase (aspC), asparagine synthetase (asnA), and glutamate dehydrogenase (gdhA).

linking numerous clusters together. Adherent growth after 7 days resulted in structurally complex biofilms characterized by large towers of cells connected by chains of cells. CLSM following Live/Dead staining demonstrated few live cells in day 1 biofilms, whereas day 7 biofilms were comprised of live cells adjacent to copious amounts of extracellular matrix (Fig. 1b). Day 7 biofilms also demonstrated increased numbers of metabolically active cells following Calcein violet staining (data not shown), and a ten-fold increase in CFU/cm^2 compared with nascent (1-day) biofilms (Fig. 1c & 1d). These results agree with other reports showing that Live/Dead correlates with viability in pneumococcal biofilms [12–14].

Pneumococcus exhibited extensive proteomic modulation during biofilm development

Overall, 244 individual proteins were identified using iTRAQ profiling of which 112 met inclusion criteria for quantitative analysis (≥3 peptide matches; >5% sequence coverage and a 50+ score; P<0.05) (Fig. 2 & 3). Additionally, 62 proteins with 2 peptide matches were qualitatively identified (Fig. 4). The majority (>80%) of proteins were differentially expressed during biofilm development. Nascent (1-day) pneumococcal biofilms differed significantly from log-phase planktonic culture with 47% (53/112) of identified proteins down-regulated and only 16% (18/112) up-regulated at this time point (Fig. 2).

Adherent growth for 7 days resulted in further changes in protein expression compared with early biofilm growth with 44% of proteins (49/112) overall differentially expressed. Nearly a quarter (24%; 27/112) of the proteins differentially expressed in the nascent 1-day old biofilm returned to levels similar to log-phase planktonic growth in the late-phase 7-day old biofilm, and a further 16% (18/112) proteins were up-regulated; only 3/112 (2.7%) were down-regulated in late compared with early biofilm growth. When late biofilm growth was compared with log-phase planktonic protein expression 35/112 proteins (31%) were up-regulated, 48/112 (43%) were similarly expressed, and 29/112 (26%) were down-regulated. Nearly 20% (21/112) of proteins were quantitatively unchanged in all profiles. iTRAQ profiling suggests the transition to adherent growth was a dynamic adaptive process consistent with previous proteomic analyses demonstrating remarkable changes in expression profiles during pneumococcal biofilm development. Pneumococcal translational capacity was substantially reduced during biofilm development. Of the 33 identified proteins specific to translation, 29 (88%) were down-regulated with over half (15/29) more than two-fold, together with increased levels of a ribosomal subunit interface protein, which facilitates arrest of translation [22].

Glycolytic proteins were significantly decreased during nascent pneumococcal biofilm development

Previous proteomic analyses of S. pneumoniae indicate that biofilm and planktonic pneumococci substantially differ in a number of proteins involved in biosynthesis and metabolism [15,16,21], and iTRAQ profiling broadly supported these. Thirty-eight metabolic proteins were quantitatively identified and divided into 4 subsets specific to glycolysis, amino acid, carbohydrate and pyruvate metabolism, with a final group comprised of proteins with miscellaneous metabolic roles (Fig. 3). Of all metabolic proteins quantified, 37% (14/38) were decreased in the nascent biofilm, and 32% (12/38) increased. After 7 days of adherent growth however, all but 4 (11%) proteins initially down-regulated in the nascent biofilm, returned to levels similar to log-phase planktonic growth. Additionally, 40% (15/38) of metabolic

proteins were increased in late biofilm growth suggesting that pneumococcal biofilms adapted to adherent growth, exhibiting protein levels observed in log-phase growth or up-regulating specific proteins associated with pyruvate, sugar and amino acid metabolism (Fig. 3).

Compared with previous pneumococcal proteomic analyses, iTRAQ identified all 9 proteins mediating the glycolytic (Embden-Meyerhof-Parnas) pathway allowing us, for the first time, to fully and quantitatively characterize changes in the pathway during biofilm formation. The majority of these proteins were decreased in nascent biofilm development consistent with other reports demonstrating down-regulation of glycolytic proteins in early TIGR4 biofilms and following epithelial cell contact [23,24]. However, after 7 days of adherent growth, half of these proteins were quantitatively commensurate with levels observed in log-phase planktonic growth (Fig. 3 & 5). All proteins associated with pyruvate metabolism were up-regulated in adherent pneumococci (Fig. 3 & 5).

Proteins involved with amino-acid metabolism and carbohydrate utilization were up-regulated in adherent pneumococci

In contrast to the decrease in glycolytic proteins in adherent bacteria, there was a significant increase in proteins involved in amino acid and sugar metabolism. Analysis of these proteins revealed that 47% (8/17) were increased in nascent biofilms, and 3/17 (18%) were decreased. By day 7 59% (10/17) were significantly increased, and only one decreased, suggesting that adherent pneumococci utilized alternative metabolic pathways during biofilm development. Six proteins associated with amino acid metabolism were identified, 3 of which were up-regulated in both nascent and 7 day biofilms (Fig. 3 & 6). Arginine deiminase (ADI) and ornithine carbamoyltransferase, enzymes which facilitate the degradation of L-arginine to generate 1 mol ATP, were up-regulated >3-fold (Fig. 3 & 6). Aspartate aminotransferase and adenyolsuccinate synthetase were down-regulated in adherent pneumococci, while glutamate dehydrogenase corresponded with planktonic levels (Fig. 3 & 6).

Our data suggest S. pneumoniae utilised a range of carbohydrates during biofilm development, since 90% (10/11) of the proteins associated with sugar metabolism were either up-regulated or unchanged during nascent biofilm growth relative to log-phase planktonic growth. Furthermore, (64%; 7/11) proteins were significantly increased in day 7 biofilms with the remainder consistent with planktonic S. pneumoniae (Fig. 3). Five PTS system proteins were identified, with mannose-specific IIAB and IID components, and a putative IIB galactitol-specific component up-regulated. The phosphoenolpyruvate-protein (PEP) PTS, an active-transport system specific for carbohydrates, was unchanged between log-phase planktonic and adherent bacteria. A fructose II ABC component, was down-regulated in nascent biofilms, but commensurate with planktonic growth at day 7 biofilms (Fig. 3).

Additionally, three ABC transporter system proteins were significantly up-regulated in biofilm pneumococci (Fig. 3). An ATP-binding protein and a sugar-binding protein from a sugar ABC transporter were both increased >2 fold in day 7 adherent S. pneumoniae and correspond to msmK and the sialic acid ABC transporter protein (SatABC). A multi-substrate ABC transporter (maltose/maltodextrin-binding protein of the maltose/maltodextrin ABC transporter) corresponding to MalX was also up-regulated in the pneumococcal biofilm profile. Tagatose 1,6-diphosphate aldolase demonstrated the highest levels of expression in the biofilm phenotype (>4-fold at day 7). Several additional sugar metabolism proteins were identified qualitatively (Fig. 4), including a PTS system IIC component and another ABC transporter.

Proteins associated with pneumococcal virulence and stress response were quantitatively different in biofilms

There was a >2-fold decrease in the level of NADH oxidase (nox) in nascent *S. pneumoniae* biofilms similar to a previous report [9], however, in day 7 biofilms, levels were similar to planktonic levels (Fig. 3). Superoxide dismutase (SOD) was unchanged between log-phase planktonic and biofilm pneumococci, indicating a requirement for protection from molecular oxygen in both conditions. Pyruvate oxidase, required for most stages of infection was also quantitatively unchanged between nascent and planktonic culture, but was >2-fold higher during day 7 biofilms. Bacterocin transport accessory protein (Bta) was also significantly higher (>2-fold) at this timepoint, suggesting increased production of bacteriocins by biofilm pneumococci. Stress proteins GroEL and DnaK, chaperone proteins facilitating appropriate protein conformation, and trigger factor were either decreased or unchanged compared with planktonic bacteria, however the pneumococcal general stress protein 24 (gls24) was increased in both nascent and late phase biofilms.

Discussion

Biofilm growth by *S. pneumoniae* is incompletely understood, but mounting evidence suggests that it plays a role in colonisation and the ability to persist in the host [3,4,7–16,23,24]. We used a static biofilm system to model conditions present in chronic otitis media where there is an absence of flow [3,13] to interrogate the hard-wired, intrinsic ability of *S. pneumoniae* to develop biofilms and obtain sufficient quantities of protein for analysis, without the presence of proteins from cell lines or animal cells to interfere with quantitative proteomic analysis.

Our data (to our knowledge the first study of *S. pneumoniae* using a quantitative gel-free approach and the first proteomic analysis of biofilms using a serotype 14 clinical strain) identify proteins whose role in biofilm development has previously been uncharacterized. iTRAQ analyses suggest that the regulation of metabolism plays a pivotal role during pneumococcal adaptation to aggregated biofilm growth. The glycolytic pathway, for which a quantitative profile of all enzyme components has been characterized for the first time, was generally down-regulated. In contrast, there was significant up-regulation of proteins involved in carbohydrate, pyruvate and arginine metabolism. These findings suggest that targeting proteins essential for pneumococcal adaption to biofilm growth, and particularly those involved in sugar and arginine metabolism, may offer new targets for treating pneumococcal-associated infections. Several proteins were conserved in both the planktonic and biofilm phenotypes, and may represent candidates for protein-based vaccines.

Protein profiling highlighted pronounced changes in the pneumococcal proteome during adaptation to the biofilm phenotype consistent with previous studies [15,24]. Proteins related to translation, transcription, virulence and glycolysis were significantly decreased, particularly in nascent biofilms. Together with experiments showing many dead and metabolically inactive cells in nascent biofilms, pneumococcus appeared to adopt a quiescent phase while it adapted to biofilm growth. These results also suggest this time point may be characterized by increased pneumococcal lysis with lysed bacteria contributing substantially to the extracellular matrix, since *LytA* mutants failed to produce biofilms with an extracellular matrix [14]. These results are also similar to reports showing that pneumococcal biofilms are suggestive of a quiescent

phase that may not be strongly immunogenic [16,23]. After 7 days however, 2/3 of identified proteins quantitatively increased, with most of the remaining proteins statistically similar to log-phase planktonic growth. These data correlated with increased CFU/ cm^2, and viable, metabolically-active biofilm pneumococci *in situ* suggesting that biofilms were comprised of both viable and dead cells during adaption to biofilm growth (Fig. 1).

This nascent phase may be similar to the prioritization of carbon source utilization recently observed in planktonic lag phase growth in *E. coli* [25]. Moreover, the decrease in proteins associated with cell division and translation may be important in the decreased susceptibility of biofilm pneumococci to antibiotics that target the replication machinery and protein and cell wall synthesis [13,14,23,26]. Even a small population of metabolically active cells in structurally complex biofilms would comprise a reservoir of viable pneumococci with the potential for propagation to new sites within the host [1]. Notably, other reports indicate that biofilm *S. pneumoniae* are hyperadhesive [23,27]. However, biofilm development is multifactorial with disease potential related to clinical isolate, extracellular matrix production and previous cultivation with epithelial cells [12–14,28,29].

In our in vitro study, a key difference in planktonic and biofilm proteomic profiles was the expression of metabolic proteins. This is the first study to provide quantitative analysis of the complete set of glycolytic enzymes and indicated reduced levels of the majority of these proteins during the adaptation to biofilm growth, consistent with previous data showing decreased expression of 6 glycolytic proteins [16]. However, our analysis at a later time point in biofilm development suggests that further biofilm development resulted in half of these proteins returning to levels similar to planktonic pneumococci by day 7. Interestingly, 4 glycolytic proteins represent potential "moonlighting" proteins comprising a class of anchorless cell membrane proteins with dual protein binding and metabolic functions [30,31]. Phoshopyruvate hydratase (α-enolase), glyceraldehyde-3-phosphate dehydrogenase (GapDH) and fructose bisphosphate aldolase (Fba) were significantly decreased in pneumococcal biofilms, whereas phosphoglycerate kinase (Pgk) was unchanged. Decreased expression of these proteins in *in vitro* biofilms suggests that *S. pneumoniae* modulated these bifunctional proteins upon adapting to biofilm growth and may promote a non-invasive phenotype capable of reducing an inflammatory response that facilitates persistence [1]. Given the streamlined pneumococcal genome, the ability of proteins to perform different functions depending on their localization in the cell would be a resourceful adaptation. Consistent with this possibility an SPFH domain-containing protein/band 7 family protein was decreased in nascent biofilms, but increased >2-fold in day 7 biofilms. These lipid raft-associated proteins aggregate to form membrane micro-domains that function in the recruitment of multi-protein complexes.

Evidence of increased pyruvate metabolism, and particularly lactate dehydrogenase, suggest that pneumococcal biofilm cells respond to a changing oxygen environment over time. The marked increase in arginine deiminase and ornithine carbamoyltransferase, two of three enzymes comprising the ADI system in pneumococcal biofilms, warrant further study. The ADI system has been shown to play a role in pathogenesis, energy metabolism and the protection of streptococcal biofilms from acidic conditions [32], and is widely distributed among bacteria with conservation of structure, but diversification of operons and regulation [33]. In particular, ADI has been characterized in oral streptococci where it plays a role in protection from lethal acidic conditions, production of ATP and survival. Recently, extracellular *Streptococcus intermedius* ADI was shown to inhibit *Porphyromonas gingivalis* biofilm formation by an interspecies signaling mechanism [33]. ADI in other streptococci has been shown to inhibit PBMC proliferation to be up-regulated in response to human serum [34,35].

iTRAQ quantitative data indicate that carbohydrate metabolism is especially important during pneumococcal adaptation to biofilm growth. Proteins associated with carbohydrate utilization, including phosphotransferase system (PTS) and ABC transporter proteins that determine carbohydrate substrate selection and fermentation were significantly increased in biofilm bacteria [36–39] suggesting that *S. pneumoniae* was capable of metabolising a wide range of carbohydrates during biofilm development. Several PTS systems and ABC transporters, many of which have been shown to play a role in pathogenesis [36], were upregulated in the biofilm including a mannose-specific IIAB component up-regulated in 1 day adherent bacteria and mannose-specific IID significantly up-regulated at 7-days (Fig. 3). These membrane spanning proteins, corresponding to the ManL and ManN nomenclature in D39 and TIGR 4, are widely conserved and have been shown to transport glucose and galactose, GlcNAc and GlcN, in addition to mannose [37,40]. The phosphoenolpyruvate-protein (PEP) PTS, an active-transport system specific for carbohydrates phosphorylating incoming sugar substrates during translocation across the cell membrane, was unchanged between planktonic and biofilm pneumococci suggesting this system was active in both growth conditions consistent with PEP's role in pyruvate metabolism. Inhibition of PTS systems demonstrated attenuated virulence in *Staphylococcus aureus* and *Haemophilus influenzae* [41].

Several ABC transporter system proteins were significantly upregulated in pneumococcal biofilms. A multi-substrate ABC transporter (maltose/maltodextrin-binding protein of the maltose/maltodextrin ABC transporter) corresponding to *MalX* was identified along with several additional sugar metabolism proteins. Pneumococcal ABC transporters may utilise multiple carbohydrates as nutrient substrates [36,37,40], since free carbohydrates are scarce in the human nasopharynx whereas complex glycans are abundant. Specifically, an ATP-binding protein corresponding to *msmK* was increased over two-fold in day 7 biofilms. MsmK acts as a common ATPase for multiple carbohydrate transporters in pneumococcus and mutation of *msmK* reduced colonization in a mouse model [38]. Similarly, in a screen of pneumococcal biofilm mutants, several PTS and ABC transporters were shown to be defective in colonization [42].

Pneumococcal biofilm development also resulted in a quantitative increase in several proteins associated with a transparent phenotype, which plays an important role in colonization and adherence [5,43]. Tagatose-1, 6-diphosphate aldolase, ABC transporter sugar-binding protein, ADI and ornithine carbamoyl-transferase were similarly upregulated in a transcriptional profiling study of *S. pneumoniae* comparing transparent and opaque colony variants [43] and have been linked to pneumococcal colonisation of the airway in several studies [9,16,44].

Pneumococcal virulence proteins associated with infection, persistence and competitive fitness were mostly downregulated during biofilm development. NADH oxidase (*nox*) decreased over two-fold in nascent biofilms. NADH oxidase regulates competence, virulence, and pneumococcal persistence through its actions as an oxygen sensor, in detoxifying oxygen, and in improving the efficiency of glucose catabolism [45] and plays an important role in pneumococcal infection in animal models of both pneumonia and otitis media [46]. However, pyruvate oxidase (*spxB*) was significantly upregulated in day 7 biofilms. SpxB appears to be

important in carbohydrate selection and capsule production [47], and in reducing competition by other organisms, since pneumococcal extracellular H_2O_2 inhibits the growth of competing bacteria [48,49]. Recently pyruvate oxidase was shown to be essential for pneumococcal aggregate formation and biofilm growth [27] and mutation of *spxB* resulted in changes in colonization and transparent/opaque phenotype expression [47,50]. Pneumococcal biofilm growth resulted in increased levels of bacteriocin transport accessory protein (Bta). Since small antimicrobial peptides play a role in pneumococcal competition with closely related bacterial species [51] biofilm growth may enhance *S. pneumoniae* fitness [1].

Finally quantitative proteomic analyses of pneumococcal biofilms may be useful in identifying cross-protective protein antigen candidates that target colonization across serotypes. The iTRAQ dataset highlighted both proteins that were modulated between planktonic and biofilm modes of growth, and those stably expressed in both conditions and may be useful in screening potential antigens for mucosal or transcutaneously administered vaccines [52,53]. Several prominent proteins quantified in iTRAQ proteomic profiles appear to be immunogenic. For example, nearly two-thirds of the 17 immunogenic proteins identified by MALDI-TOF analysis are present in iTRAQ biofilm profiles, including 2 of 2 proteins that were highly immunogenic across all age groups: pyruvate oxidase and α-enolase [54]. Fba and GapDH were other strongly immunogenic glycolytic enzymes and 6PGD, a surface-associated protein with putative adhesin-like activity, was significantly increased in day 7 pneumococcal biofilms. Present in multiple pneumococcal strains and highly immunogenic in adults, 6PGD was also immunoprotective in a mouse model of infection [55]. Although NADH oxidase (nox) was recently evaluated as a potential vaccine candidate that resulted in decreased pneumococcal colonisation in the nasopharynx and lungs in mice [56] iTRAQ data suggest this protein was downregulated in nascent biofilms. The maltose/maltodextrin ABC transporter binding protein MalX, however, also identified as a vaccine candidate, was quantitatively unchanged in nascent biofilms and increased in day 7 biofilm pneumococci [40]. Notably, this protein has also been identified through proteomic analysis to be a widely conserved pneumococcal antigen capable of eliciting mucosal T_H17 responses that abrogated colonization [53].

S. pneumoniae metabolic regulation appears to be of key importance in biofilm growth. Although marked downregulation of glycolysis, was observed, alternative metabolic pathways likely enhance the ability of pneumococcus to adapt to biofilm growth and facilitate survival. Such metabolic proteins may provide novel cross protective target antigens underlying pneumococcal persistence in multiple serotypes.

Acknowledgments

We would like to thank David Johnston and the Southampton Biomedical Imaging Unit for their support in imaging the biofilms. We acknowledge the PRIDE team for the deposition of our mass spectrometry data to the ProteomeXchange Consortium (http://proteomecentral. proteomexchange.org) via the PRIDE partner repository with the dataset identifier PXD001182.

Author Contributions

Conceived and designed the experiments: RNA PS JJ SCC SNF LHS JW. Performed the experiments: RNA. Analyzed the data: RNA LHS PS. Contributed reagents/materials/analysis tools: PS JJ SCC SNF LHS JW. Contributed to the writing of the manuscript: RNA PS JJ SCC SNF LHS JW.

References

1. Weiser JN (2010) The pneumococcus: why a commensal misbehaves. J Mol Med (Berl) 88: 97–102.

2. Kadioglu A, Weiser JN, Paton JC, Andrew PW (2008) The role of Streptococcus pneumoniae virulence factors in host respiratory colonization and disease. Nat Rev Microbiol 6: 288–301.

3. Hall-Stoodley L, Hu FZ, Gieseke A, Nistico L, Nguyen D, et al. (2006) Direct detection of bacterial biofilms on the middle-ear mucosa of children with chronic otitis media. JAMA 296: 202–211.

4. Reid SD, Hong W, Dew KE, Winn DR, Pang B, et al. (2009) Streptococcus pneumoniae forms surface-attached communities in the middle ear of experimentally infected chinchillas. J Infect Dis 199: 786–794.

5. Weiser JN, Markiewicz Z, Tuomanen EI, Wani JH (1996) Relationship between phase variation in colony morphology, intrastrain variation in cell wall physiology, and nasopharyngeal colonization by Streptococcus pneumoniae. Infect Immun 64: 2240–2245.

6. Hoa M, Tomovic S, Nistico L, Hall-Stoodley L, Stoodley P, et al. (2009) Identification of adenoid biofilms with middle ear pathogens in otitis-prone children utilizing SEM and FISH. Int J Pediatr Otorhinolaryngol 73: 1242–1248.

7. Nistico L, Kreft R, Gieseke A, Coticchia JM, Burrows A, et al. (2011) Adenoid reservoir for pathogenic biofilm bacteria. J Clin Microbiol 49: 1411–1420.

8. Camilli R, Pantosti A, Baldassarri L (2011) Contribution of serotype and genetic background to biofilm formation by Streptococcus pneumoniae. Eur J Clin Microbiol Infect Dis 30: 97–102.

9. Oggioni MR, Trappetti C, Kadioglu A, Cassone M, Iannelli F, et al. (2006) Switch from planktonic to sessile life: a major event in pneumococcal pathogenesis. Mol Microbiol 61: 1196–1210.

10. Marks LR, Reddinger RM, Hakansson AP (2012) High levels of genetic recombination during nasopharyngeal carriage and biofilm formation in Streptococcus pneumoniae. MBio 3.

11. Moscoso M, Garcia E, Lopez R (2006) Biofilm formation by Streptococcus pneumoniae: role of choline, extracellular DNA, and capsular polysaccharide in microbial accretion. J Bacteriol 188: 7785–7795.

12. Allegrucci M, Sauer K (2007) Characterization of colony morphology variants isolated from Streptococcus pneumoniae biofilms. J Bacteriol 189: 2030–2038.

13. Hall-Stoodley L, Nistico L, Sambanthamoorthy K, Dice B, Nguyen D, et al. (2008) Characterization of biofilm matrix, degradation by DNase treatment and evidence of capsule downregulation in Streptococcus pneumoniae clinical isolates. BMC Microbiol 8: 173.

14. Marks LR, Parameswaran GI, Hakansson AP (2012) Pneumococcal interactions with epithelial cells are crucial for optimal biofilm formation and colonization in vitro and in vivo. Infect Immun 80: 2744–2760.

15. Allegrucci M, Hu FZ, Shen K, Hayes J, Ehrlich GD, et al. (2006) Phenotypic characterization of Streptococcus pneumoniae biofilm development. J Bacteriol 188: 2325–2335.

16. Sanchez CJ, Hurtgen BJ, Lizcano A, Shivshankar P, Cole GT, et al. (2011) Biofilm and planktonic pneumococci demonstrate disparate immunoreactivity to human convalescent sera. BMC Microbiol 11: 245.

17. Ding F, Tang P, Hsu MH, Cui P, Hu S, et al. (2009) Genome evolution driven by host adaptations results in a more virulent and antimicrobial-resistant Streptococcus pneumoniae serotype 14. BMC Genomics 10: 158.

18. Hiller NL, Janto B, Hogg JS, Boissy R, Yu S, et al. (2007) Comparative genomic analyses of seventeen Streptococcus pneumoniae strains: insights into the pneumococcal supragenome. J Bacteriol 189: 8186–8195.

19. Bae SM, Yeon SM, Kim TS, Lee KJ (2006) The effect of protein expression of Streptococcus pneumoniae by blood. J Biochem Mol Biol 39: 703–708.

20. Encheva V, Gharbia SE, Wait R, Begum S, Shah HN (2006) Comparison of extraction procedures for proteome analysis of Streptococcus pneumoniae and a basic reference map. Proteomics 6: 3306–3317.

21. Lee KJ, Bae SM, Lee MR, Yeon SM, Lee YH, et al. (2006) Proteomic analysis of growth phase-dependent proteins of Streptococcus pneumoniae. Proteomics 6: 1274–1282.

22. Agafonov DE, Kolb VA, Spirin AS (2001) Ribosome-associated protein that inhibits translation at the aminoacyl-tRNA binding stage. EMBO Rep 2: 399–402.

23. Sanchez CJ, Kumar N, Lizcano A, Shivshankar P, Dunning Hotopp JC, et al. (2011) Streptococcus pneumoniae in biofilms are unable to cause invasive disease due to altered virulence determinant production. PLoS One 6: e28738.

24. Orihuela CJ, Radin JN, Sublett JE, Gao G, Kaushal D, et al. (2004) Microarray analysis of pneumococcal gene expression during invasive disease. Infect Immun 72: 5582–5596.

25. Schultz D, Kishony R (2013) Optimization and control in bacterial Lag phase. BMC Biol 11: 120.

26. McCoy LS, Xie Y, Tor Y (2011) Antibiotics that target protein synthesis. Wiley Interdiscip Rev RNA 2: 209–232.

27. Blanchette-Cain K, Hinojosa CA, Akula Suresh Babu R, Lizcano A, Gonzalez-Juarbe N, et al. (2013) Streptococcus pneumoniae Biofilm Formation Is Strain Dependent, Multifactorial, and Associated with Reduced Invasiveness and Immunoreactivity during Colonization. MBio 4.

28. Trappetti C, van der Maten E, Amin Z, Potter AJ, Chen AY, et al. (2013) Site of isolation determines biofilm formation and virulence phenotypes of Streptococcus pneumoniae serotype 3 clinical isolates. Infect Immun 81: 505–513.

29. Hakansson AP (2014) Pneumococcal Adaptive Responses to Changing Host Environments. J Infect Dis.

30. Fulde M, Bernardo-Garcia N, Rohde M, Nachtigall N, Frank R, et al. (2013) Pneumococcal phosphoglycerate kinase interacts with plasminogen and its tissue activator. Thromb Haemost 111.

31. Chhatwal GS (2002) Anchorless adhesins and invasins of Gram-positive bacteria: a new class of virulence factors. Trends Microbiol 10: 205–208.

32. Casiano-Colon A, Marquis RE (1988) Role of the arginine deiminase system in protecting oral bacteria and an enzymatic basis for acid tolerance. Appl Environ Microbiol 54: 1318–1324.

33. Cugini C, Stephens DN, Nguyen D, Kantarci A, Davey ME (2013) Arginine deiminase inhibits Porphyromonas gingivalis surface attachment. Microbiology 159: 275–285.

34. Degnan BA, Palmer JM, Robson T, Jones CE, Fischer M, et al. (1998) Inhibition of human peripheral blood mononuclear cell proliferation by Streptococcus pyogenes cell extract is associated with arginine deiminase activity. Infect Immun 66: 3050–3058.

35. Yang Q, Zhang M, Harrington DJ, Black GW, Sutcliffe IC (2011) A proteomic investigation of Streptococcus agalactiae reveals that human serum induces the C protein beta antigen and arginine deiminase. Microbes Infect 13: 757–760.

36. Hartel T, Eylert E, Schulz C, Petruschka L, Gierok P, et al. (2012) Characterization of central carbon metabolism of Streptococcus pneumoniae by isotopologue profiling. J Biol Chem 287: 4260–4274.

37. Bidossi A, Mulas L, Decorosi F, Colomba L, Ricci S, et al. (2012) A functional genomics approach to establish the complement of carbohydrate transporters in Streptococcus pneumoniae. PLoS One 7: e33320.

38. Marion C, Aten AE, Woodiga SA, King SJ (2011) Identification of an ATPase, MsmK, which energizes multiple carbohydrate ABC transporters in Streptococcus pneumoniae. Infect Immun 79: 4193–4200.

39. Hardy GG, Magee AD, Ventura CL, Caimano MJ, Yother J (2001) Essential role for cellular phosphoglucomutase in virulence of type 3 Streptococcus pneumoniae. Infect Immun 69: 2309–2317.

40. Buckwalter CM, King SJ (2012) Pneumococcal carbohydrate transport: food for thought. Trends Microbiol 20: 517–522.

41. Kok M, Bron G, Erni B, Mukhija S (2003) Effect of enzyme I of the bacterial phosphoenolpyruvate: sugar phosphotransferase system (PTS) on virulence in a murine model. Microbiology 149: 2645–2652.

42. Munoz-Elias EJ, Marcano J, Camilli A (2008) Isolation of Streptococcus pneumoniae biofilm mutants and their characterization during nasopharyngeal colonization. Infect Immun 76: 5049–5061.

43. King SJ, Hippe KR, Gould JM, Bae D, Peterson S, et al. (2004) Phase variable desialylation of host proteins that bind to Streptococcus pneumoniae in vivo and protect the airway. Mol Microbiol 54: 159–171.

44. Marion C, Stewart JM, Tazi MF, Burnaugh AM, Linke CM, et al. (2012) Streptococcus pneumoniae can utilize multiple sources of hyaluronic acid for growth. Infect Immun 80: 1390–1398.

45. Auzat I, Chapuy-Regaud S, Le Bras G, Dos Santos D, Ogunniyi AD, et al. (1999) The NADH oxidase of Streptococcus pneumoniae: its involvement in competence and virulence. Mol Microbiol 34: 1018–1028.

46. Yu J, Bryant AP, Marra A, Lonetto MA, Ingraham KA, et al. (2001) Characterization of the Streptococcus pneumoniae NADH oxidase that is required for infection. Microbiology 147: 431–438.

47. Carvalho SM, Farshchi Andisi V, Gradstedt H, Neef J, Kuipers OP, et al. (2013) Pyruvate oxidase influences the sugar utilization pattern and capsule production in Streptococcus pneumoniae. PLoS One 8: e68277.

48. Regev-Yochay G, Trzcinski K, Thompson CM, Malley R, Lipsitch M (2006) Interference between Streptococcus pneumoniae and Staphylococcus aureus: In vitro hydrogen peroxide-mediated killing by Streptococcus pneumoniae. J Bacteriol 188: 4996–5001.

49. Regev-Yochay G, Trzcinski K, Thompson CM, Lipsitch M, Malley R (2007) SpxB is a suicide gene of Streptococcus pneumoniae and confers a selective advantage in an in vivo competitive colonization model. J Bacteriol 189: 6532–6539.

50. Syk A, Norman M, Fernebro J, Gallotta M, Farmand S, et al. (2014) Emergence of Hypervirulent Mutants Resistant to Early Clearance During Systemic Serotype 1 Pneumococcal Infection in Mice and Humans. J Infect Dis.

51. Dawid S, Roche AM, Weiser JN (2007) The blp bacteriocins of Streptococcus pneumoniae mediate intraspecies competition both in vitro and in vivo. Infect Immun 75: 443–451.

52. Novotny LA, Clements JD, Bakaletz LO (2013) Kinetic analysis and evaluation of the mechanisms involved in the resolution of experimental nontypeable Haemophilus influenzae-induced otitis media after transcutaneous immunization. Vaccine 31: 3417–3426.

53. Moffitt KL, Gierahn TM, Lu YJ, Gouveia P, Alderson M, et al. (2011) T(H) 17-based vaccine design for prevention of Streptococcus pneumoniae colonization. Cell Host Microbe 9: 158–165.

54. Ling E, Feldman G, Portnoi M, Dagan R, Overweg K, et al. (2004) Glycolytic enzymes associated with the cell surface of Streptococcus pneumoniae are

antigenic in humans and elicit protective immune responses in the mouse. Clin Exp Immunol 138: 290–298.

55. Daniely D, Portnoi M, Shagan M, Porgador A, Givon-Lavi N, et al. (2006) Pneumococcal 6-phosphogluconate-dehydrogenase, a putative adhesin, induces protective immune response in mice. Clin Exp Immunol 144: 254–263.

56. Muchnik L, Adawi A, Ohayon A, Dotan S, Malka I, et al. (2013) NADH oxidase functions as an adhesin in Streptococcus pneumoniae and elicits a protective immune response in mice. PLoS One 8: e61128.

## 11

The Symbiotic Biofilm of *Sinorhizobium fredii* SMH12, Necessary for Successful Colonization and Symbiosis of *Glycine max* cv Osumi, Is Regulated by Quorum Sensing Systems and Inducing Flavonoids via NodD1

Francisco Pérez-Montaño[1]*, Irene Jiménez-Guerrero[1], Pablo Del Cerro[1], Irene Baena-Ropero[2], Francisco Javier López-Baena[1], Francisco Javier Ollero[1], Ramón Bellogín[1], Javier Lloret[2], Rosario Espuny[1]

1 Departamento de Microbiología, Facultad de Biología, Universidad de Sevilla, Sevilla, Spain, 2 Departamento de Biología, Facultad de Ciencias, Universidad Autónoma de Madrid, Madrid, Spain

Abstract

Bacterial surface components, especially exopolysaccharides, in combination with bacterial Quorum Sensing signals are crucial for the formation of biofilms in most species studied so far. Biofilm formation allows soil bacteria to colonize their surrounding habitat and survive common environmental stresses such as desiccation and nutrient limitation. This mode of life is often essential for survival in bacteria of the genera *Mesorhizobium*, *Sinorhizobium*, *Bradyrhizobium*, and *Rhizobium*. The role of biofilm formation in symbiosis has been investigated in detail for *Sinorhizobium meliloti* and *Bradyrhizobium japonicum*. However, for *S. fredii* this process has not been studied. In this work we have demonstrated that biofilm formation is crucial for an optimal root colonization and symbiosis between *S. fredii* SMH12 and *Glycine max* cv Osumi. In this bacterium, *nod*-gene inducing flavonoids and the NodD1 protein are required for the transition of the biofilm structure from monolayer to microcolony. Quorum Sensing systems are also required for the full development of both types of biofilms. In fact, both the *nodD1* mutant and the lactonase strain (the lactonase enzyme prevents AHL accumulation) are defective in soybean root colonization. The impairment of the lactonase strain in its colonization ability leads to a decrease in the symbiotic parameters. Interestingly, NodD1 together with flavonoids activates certain quorum sensing systems implicit in the development of the symbiotic biofilm. Thus, *S. fredii* SMH12 by means of a unique key molecule, the flavonoid, efficiently forms biofilm, colonizes the legume roots and activates the synthesis of Nod factors, required for successfully symbiosis.

Editor: Roy M. Roopll, East Carolina University School of Medicine, United States of America

Funding: This work was supported by the following grants: AGL2012-38831 from the Spanish Ministerio de Economía y Competitividad, http://www.idi.mineco.gob.es/portal/site/MICINN/ P10-AGR-5821 and P11-CVI-7050 from the Junta de Andalucía, Consejería de Innovación, Ciencia y Empresas. http://www.juntadeandalucia.es/organismos/economiainnovacioncienciayempleo.html. Dr. Pérez-Montaño's work was supported by an FPU fellowship from the Spanish Ministerio de Ciencia y Tecnología and a contract from the University of Seville. http://www.idi.mineco.gob.es/portal/site/MICINN/ http://www.us.es/. The funders had no role in study design, data collection and analysis, decision to publish, or preparation of the manuscript.

* Email: fperezm@us.es

Introduction

Rhizobia are bacteria that interact with the roots of leguminous plants to develop symbiotic nodules when environmental nitrogen is limited. In these symbiotic structures atmospheric nitrogen is reduced to ammonium making it available to the plant. This process requires the exchange of molecular signals between both members of the symbiosis. Legume roots exude flavonoids which induce, via a mechanism involving the NodD family of transcription activators, the biosynthesis and secretion of strain-specific lipo-chito-oligosaccharides, also known as nodulation factors (Nod factors), which trigger the initiation of nodule organogenesis [1]. Plant flavonoids, besides inducing Nod factor production, attract the bacteria to the legume root [2]; induce type III secretion

machinery via NodD1 by activation of the transcriptional regulator TtsI [3,4]; and activate the rhizobial Quorum Sensing (QS) systems [5].

QS is defined as a regulation mode of bacterial gene expression in response to changes in their population density, which is mediated by small diffusible signal molecules called autoinducers (AI) [6,7]. QS-regulated genes are involved in adaptive changes in the physiology of the bacterial population, which modify their behaviour when a certain cell density is reached. QS modulates a broad variety of phenotypes, such as biofilm formation, toxin and exopolysaccharide production, virulence, plasmid transfer and motility, which usually are essential for successful establishment of a symbiotic or a pathogenic relationship with the eukaryotic hosts [8,9,10,11,12]. In plant-associated bacteria QS coordinates the expression of genes involved in virulence, colonization and symbiosis [13].

In all species studied so far, bacterial surface components (flagella, lipopolysaccharides and especially exopolysaccharides) in combination with the presence of bacterial QS signals are crucial for the formation of biofilms [12]. Biofilms are defined as bacterial communities surrounded by a self-produced polymeric matrix, and attached to an inert or a biotic surface [14]. In the course of biofilm formation, the initial reversible attachment to the surface is followed by an irreversible attachment and multiplication of the bacteria forming microcolonies that develop mature communities with a three-dimensional structure, in some cases permeated by channels, which act as the biofilm circulatory system. All these processes are coordinated by bacterial QS systems [14,15]. Many rhizobia have been described as forming microcolonies or biofilms when they colonize legume roots. These biofilms are mainly composed of water and bacterial cells. However, the three-dimensional structure of the biofilm is due to an extracellular matrix, which is formed by exopolysaccharides (EPS) [16]. In *S. meliloti*, the biofilm formation is affected and/or regulated by nutritional and environmental conditions [17,18]; exopolysaccharides and flagella [19]; ExoR with the ExoS–ChvI two-component system [20]; Nod factors synthesized by *nod* genes [19]; and regulation of exopolysaccharide biosynthesis by means of QS systems [21]. Development of biofilms by rhizobia is crucial to overcoming environmental stresses and in certain species the biofilm formation is clearly an important feature of symbiotic ability [12,19,22,23,24]. *S. fredii* SMH12, a wide host-range rhizobium that nodulates soybean and dozens of other legumes, produces at least three AHLs, *N*-octanoyl homoserine lactone (C8-HSL), 3-oxo *N*-octanoyl homoserine lactone (3-oxo-C8-HSL) and *N*-tetradecanoyl homoserine lactone (C14-HSL). In the presence of the flavonoid genistein, an activator of its nodulation genes, the overall production of AHLs is enhanced, being detected the C14-HSL only in bacterial cultures supplemented with inducing flavonoids [5]. SMH12 possesses at least one gene, *traI*, which is responsible for the synthesis of the 3-oxo-C8-HSL [5]; and at least two non-identified *luxI*-type gene involved in the synthesis of the C8-HSL and the C14-HSL. Furthermore, in this bacterium the AHLs and the QS systems are involved in the biofilm formation on the abiotic surface [25].

In this work, the role and regulation of biofilm formation in *S. fredii* SMH12 have been studied during symbiosis with *Glycine max* cv. Osumi. For this purpose, the wild-type strain, a *nodD1* mutant and a lactonase overproducing strain of SMH12 were constructed. The lactonase enzyme hydrolyzes the ester bond of the homoserine lactone ring of acylated homoserine lactones preventing these signalling molecules from binding to their target transcriptional regulators. Both *nodD1* and lactonase strains showed an impaired ability for colonization of *Glycine max* roots with respect to the wild-strain, probably due to an abnormal symbiotic biofilm formation which determines a less effective symbiosis. Interestingly, QS-biofilm formation and effective nodulation are connected through flavonoids, since these molecules initiate the molecular dialogue between bacteria and plants and allow the symbiotic biofilm formation. In summary, this report unequivocally demonstrates that the development of biofilm is crucial for successful root colonization and optimal symbiosis in *S. fredii* SMH12.

Materials and Methods

Strains and media

Bacterial strains and plasmids used in this work are listed in Table 1. *S. fredii* SMH12 and derivative strains were grown at 28°C in tryptone-yeast extract (TY) medium [26], yeast extract mannitol (YM) medium [27] with a lower mannitol concentration (3 g l^{-1}) and low-phosphate minimal glutamate mannitol (MGM) medium [21], supplemented when necessary with genistein 3.7 μM as inducing flavonoid, umbelliferone 6.2 μM as non-inducing flavonoid, or with commercial AHLs [25]. *Agrobacterium tumefaciens* NT1 (pZLR4) was grown at 28°C in YM with carbenicillin (100 μg ml^{-1}) and gentamicin (30 μg ml^{-1}). *Escherichia coli* strains were cultured in Luria-Bertani (LB) medium (28) at 37°C. When required, the media were supplemented with the appropriate antibiotics as described by Lamrabet et al. [29]. Commercial AHLs were dissolved in methanol and used at different concentrations. Flavonoids and AHLs were purchased from Fluka (Sigma-Aldrich, USA).

Plasmids were transferred from *E. coli* to SMH12 by conjugation as described by Simon [30] using plasmid pRK2013 as helper. Recombinant DNA techniques were performed according to the general protocols of Sambrook et al. [28]. For hybridization, DNA was blotted to Hybond-N nylon membranes (Amersham, UK), and the DigDNA method of Roche was employed according to the manufacturer's instructions. PCR amplifications were performed as previously described [31]. Using this methodology the plasmid pMUS534 was employed for the homogenotization of the mutated version of the *nodD1* gene in *S. fredii* SMH12, generating the mutant strain SVQ648. The double recombination event was confirmed by Southern blotting (data not shown).

For the growth curves of the different strains, bacteria were grown in 5 ml of YM medium to early stationary phase and then diluted to OD$_{600}$ values around 0.03 in fresh YM medium with or without flavonoids. Growth was monitored by measuring the OD$_{600}$ for at least 46 h. Each experiment was performed two days, eight replicates each day.

RNA isolation, cDNA synthesis and Quantitative RT-PCR

Bacterial RNA was extracted from bacterial cultures from the microtiter plates in the same conditions employed for the biofilm assays and described below. For the RNA extraction the PowerBiofilm RNA Isolation Kit was employed following the manufacturer's instructions (MO BIO, USA). Three independent RNA extractions were performed.

To quantify the expression of the *S. fredii* SMH12 *traI* gene using quantitative RT-PCR, primers rt-traI-F and rt-traI-R described by Pérez-Montaño et al. [5] were used. Expression was calculated relative to bacteria grown without flavonoid. The *S. fredii* SMH12 RNA *16S* gene was used as internal control to normalize gene expression. RNA *16S* primers used are described in the same work [5]. The expression data shown are the mean (± standard deviation of the mean) for three biological replicates. The fold change in the target gene, normalized to RNA *16S*, and relative to gene expression in the culture without flavonoids was calculated. PCR was conducted on the LightCycler 480 (Roche, Switzerland) with the following conditions: 95°C, 10 min; 95°C, 30 s; 50°C, 30 s; 72°C, 20 s; forty cycles, followed by the melting curve profile from 60 to 95°C to verify the specificity of the reaction. The threshold cycles (Ct) were determined with the iCycler software and the individual values for each sample were generated by averaging three technical replicates that varied less than 0.5 per cycle.

Table 1. Resistance phenotypes: Gm^R, Km^R, Nx^R, Ap^R and Tc^R, gentamicin, kanamycin, nalidixic acid, ampicillin and tetracycline, respectively.

Strain or plasmid	Relevant properties	Reference
Agrobacterium tumefaciens NT1 (pZRL4)	A.tumefaciens without pTiC58; with pZRL4, which carries the fusion traG::lacZ and the gene traR, Gm^R	[13]
Escherichia coli DH5α	SupE44, ΔlacU169, 5hsdR17, recA1, endA1, gyrA96, thi-1, relA1, Nx^R	[25]
Sinorhizobium fredii SMH12	Wild-type strain, Ap^R	[40]
SMH12 (pME6000)	SMH12 with the plasmid pME6000, Tc^R	This work
SMH12 (pME6863)	SMH12 with the plasmid pME6863, which carries the lactonase gene, Tc^R	This work
SMH12 (pMP2463)	SMH12 with the plasmid pMP2463, which carries the gene of the GFP, Gm^R	This work
SMH12 (pMP6000) (pMP2463)	SMH12 with the plasmids pME6000 and pMP2463, which carry the empty plasmid and the green fluorescent protein, respectively, Tc^R Gm^R	This work
SMH12 (pMP6863) (pMP2463)	SMH12 with the plasmids pME6863 and pMP2463, which carry the gene that encode the lactonase and the green fluorescent protein, respectively, Tc^R Gm^R	This work
SVQ648	SMH12 nodD1::lacZ-Gm^R	This work
SVQ648 (pMP2463)	SVQ648 with the plasmid pMP2463, which carries the gene of the GFP, Gm^R	This work
pBBR1MCS-5	Broad-host-range cloning vector, Gm^R	[41]
pK18mobsacB	Cloning vector (suicide in rhizobia), Km^R	[42]
pME6000	Broad-host-range cloning vector, Tc^R	[43]
pME6863	pME6000::aiiA, plasmid carrying the lactonase gene, Tc^R	[44]
pMP2463	pBBR-MCS-5::egfp-1, plasmid carrying the green fluorescent protein gene, Gm^R	[45]
pMUS534	Plasmid pK18mob derivative containing nodD1::lacZ-Gm^R	[46]
pRK2013	Helper plasmid, Km^R	[47]

Confocal laser scanning microscopy

The different events of biofilm formation were visualized by confocal laser scanning microscopy using a method described by Russo et al. [32]. Bacterial cultures were placed in 8 well chambered cover glass slides containing a borosilicate glass base 1 μm thick (Thermo Fischer Scientific Inc., USA) for 4 days without shaking. To avoid desiccation, the chambers were incubated in a humid sterilized petri dish. Confocal microscope image capture was carried out with Leica TCS SP2 (Leica Microsystems, Germany). In silico 3D reconstruction analysis was executed using the computer program ImageJ (Java, USA).

Biofilm formation assay

The biofilm formation assay was based on the method described by O'Toole and Kolter [33] with modifications [34]. Cultures were grown in 5 ml of low-phosphate MGM medium, diluted to

Table 2. Plant responses to inoculation of Glycine max cv. Osumi with S. fredii SMH12 and derivatives.

Inoculant	Number of nodules	Fresh mass of nodules (g)	Plant-top dry mass (g)
None	0±0*	0±0*	0.48±0.11*
SMH12	124.30±12.06	1.56±0.20	1.34±0.18
SVQ648	0±0*	0±0*	0.46±0.10*
SMH12 (pME6863)	76.20±16.02*	1.04±0.29	1.10±0.30
SMH12 (pME6000)	131.00±22.68	1.60±0.38	1.35±0.25

Data represent means ± sd of six soybean jars. Each jar contained two soybean plants. Determinations were made six weeks after inoculation. Mutant nodD1 and the lactonase strain parameters were individually compared with the parental strain SMH12 parameters by using the Mann-Whitney non-parametric test. Values tagged by * are significantly different at the level α = 5%.

Table 3. *Glycine max* cv. Osumi root attachment.

Inoculant	CFU/cm of proximal root	CFU/cm of lateral root
None	0 ± 0*	0 ± 0*
SMH12	$1.68\pm0.18\times10^5$	$1.22\pm0.29\times10^4$
SVQ648	$1.10\pm0.42\times10^5$	$3.83\pm0.5\times10^3$*
SMH12 (pME6863)	$5.00\pm0.10\times10^4$*	$8.45\pm0.46\times10^2$*
SMH12 (pME6000)	$1.55\pm0.27\times10^5$	$1.20\pm0.39\times10^4$

Soybean plants were inoculated with the test strains, one centimetre of the proximal root and of the lateral roots of each plant were collected after 7 days, and the bacteria present were resuspended and plated. Data are the mean $\pm$ SD of at least three independent experiments performed in triplicate. Mutant *nodD1* and the lactonase strain attachment were individually compared with the parental strain SMH12 attachment by using the Mann-Whitney non-parametric test. Values tagged by * are significantly different at the level $\alpha=5\%$.

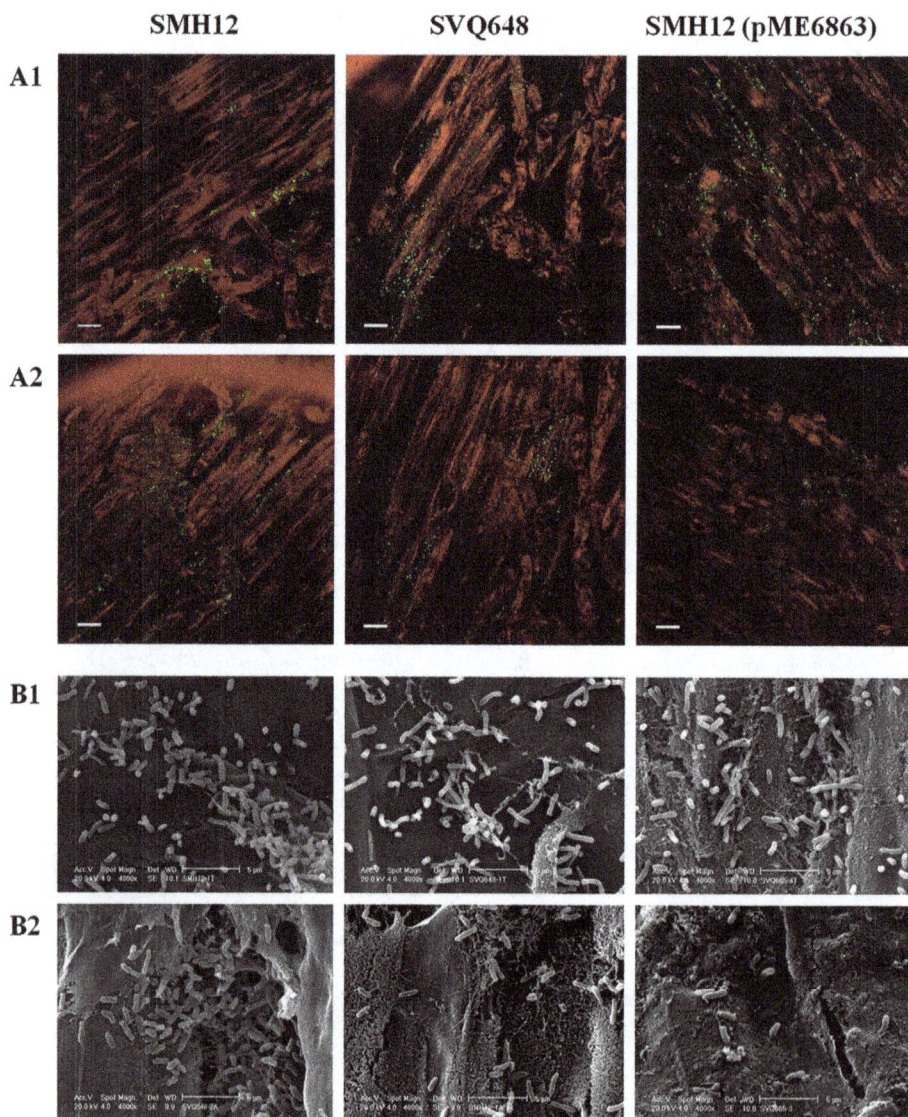

Figure 1. Visualization of the soybean root colonization. A. Epifluorescence microscopy analysis of the colonization of the soybean rhizosphere by gfp-tagged bacteria [SMH12, SVQ648, SMH12 (pME6863)]. Roots were visualized 7 days after inoculation. 1. Proximal root. 2. Lateral roots. Bar, 100 μm. B. Scanning microscopy analysis of the colonization of the soybean rhizosphere by SMH12, SVQ648 and SMH12 (pME6863). Roots were visualized 7 days after inoculation. 1. Proximal root. 2. Lateral roots. Bar, 5 μm. SMH12: wild-type, SVQ648: *nodD1* mutant, SMH12 (pME6863): lactonase strain.

an OD_{600} of 0.2, with or without flavonoids and AHLs, and inoculated with 100 µl aliquots and placed on polystyrene microtiter plates, U form (Deltalab S.L., Spain). The plates were inverted and incubated at 28°C for 7 days with gentle rocking. Cell growth was analyzed by measuring OD_{600} using a microtiter reader Synergy HT (Biotek, USA). The culture in each well was removed carefully; the wells were dried, washed three times with 0.9% NaCl and dried again. Biofilms in each well were stained with 100 µl of 0.1% crystal violet for 20 minutes, then washed with water three times and dried again. Finally, 100 µl of 96% ethanol were added to each well and the OD_{570} was measured. Every experiment was performed six times with eight replicates each time.

Quantification of overall AHLs

Supernatants of the rhizobial strains grown for 7 days in 96 wells on U shaped polystyrene microtiter plates (Deltalab S.L., Spain) were collected and sterilized by microfiltration. To determine the overall autoinducer production in each condition the method described by Pérez-Montaño et al. was used. [5]. Briefly, 25 µl, 2.5 µl or 0.25 µl of the supernatants were mixed with YM to obtain a final volume of 2.5 ml reaching supernatant concentrations of 1%, 0.1% and 0.01% (v/v). The mixtures were inoculated with approximately 10^7 cells ml^{-1} of the A. tumefaciens NT1 (pZLR4), incubated with shaking for 12 h at 28°C and assayed for β-galactosidase activity [35]. As controls, 125 µl of distilled water or 125 µl of N-(3-oxo-hexanoyl)-L-homoserine lactone (3-oxo-C6-HSL) 5.5 µM, were used. To obtain the standard curve with synthetic AHLs, 125 µl of 3-oxo-C8-HSL, C8-HSL or C14-HSL at different concentration were added. The experiments were repeated independently five times with 3 replicates.

Thin Layer Chromatography

For TLC analysis, cultures previously removed from each well of the microtiter plate were extracted with the same volume of dichloromethane, evaporated to dryness and analyzed by thin-layer chromatography as described by Pérez-Montaño et al. [25]. Briefly, 1 µl of each culture extract was loaded on TLC plates (HPTLC plates RP-18 $_{F254s}$ 1.13724 and 1.05559, Merck, Germany) using methanol:water (60:40 v/v) as eluent, dried and overlaid with a soft agar culture of the biosensor A. tumefaciens NT1 (pZLR4).

Plant assays

Nodulation assays on Glycine max (L.) Merrill cultivar Osumi were performed as described by de Lyra et al. [36]. Plants were inoculated with approximately 5×10^8 bacteria and were grown in Leonard jars with Fåhraeus nutrient solution [27] for 42 days with a 16 hour-photoperiod at 26°C in light and 18°C in the dark with 70% of humidity. Shoots were dried at 70°C for 48 h and weighed. Experiments were performed three times. For root colonization assays and microscopy, seeds were surface sterilized, germinated and inoculated with a modified method described in Pérez-Montaño et al. [25]. Briefly, Glycine max (L.) Merrill cultivar Osumi seeds were soaked for 30 seconds in 96% ethanol and for 8 minutes in commercial bleach. Then, seeds were washed repeatedly with sterilized distilled water, germinated and checked for sterility in LB medium. Each plant was inoculated with approximately 5×10^8 bacteria. Plants were grown under hydroponic controlled conditions in Fåhraeus solution for 7 days in the same conditions as those mentioned above.

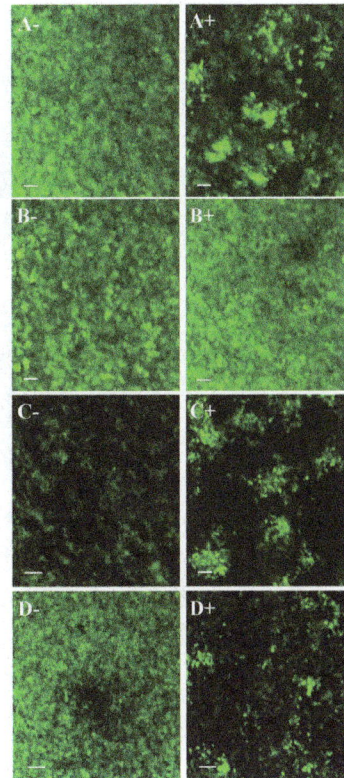

	Main fluorescence value
SMH12	100 ± 20.06
SMH12 + genistein (3.7 µM)	57.32 ± 5.76 *
SVQ648	80.16 ± 7.85
SVQ648 + genistein (3.7 µM)	105.86 ± 13.39
SMH12 (pME6863)	43.81 ± 2.51 *
SMH12 (pME6863) + genistein (3.7 µM)	51.37 ± 6.88 *
SMH12 (pME6000)	93.23 ± 17.15
SMH12 (pME6000) + genistein (3.7 µM)	56.37 ± 3.72*

Figure 2. Biofilm structure of *S. fredii* SMH12 and derivatives on glass surfaces: reconstruction of the Z-stacks and measure of the surface coverage. Main fluorescence value of the wild-type strain was arbitrarily given a value of 100. Averages and standard deviations of five randomized optical fields per strain corresponding to two independent experiments are shown. The asterisks indicate a significant different at the level α = 5% with respect to wild-type strain by using the Mann-Whitney non-parametrical test. Left side corresponds to cultures without flavonoids. Right side corresponds to cultures with inducing flavonoid. A. SMH12. B. SVQ648. C. SMH12 (pME6863). D. SMH12 (pME6000). Bar, 20 µm. SMH12: wild-type, SVQ648: *nodD1* mutant, SMH12 (pME6863): lactonase strain. SMH12 (pME6000): carrying the empty plasmid.

Epifluorescence and scanning electron microscopy

Epifluorescence microscopy assays were carried out with *S. fredii* SMH12 and derivatives carrying GFP expressing plasmid pME2463. After growth, roots were excised (one centimetre of the proximal area of the main root and the last centimetre of a lateral root) and thoroughly washed with water to eliminate any loosely

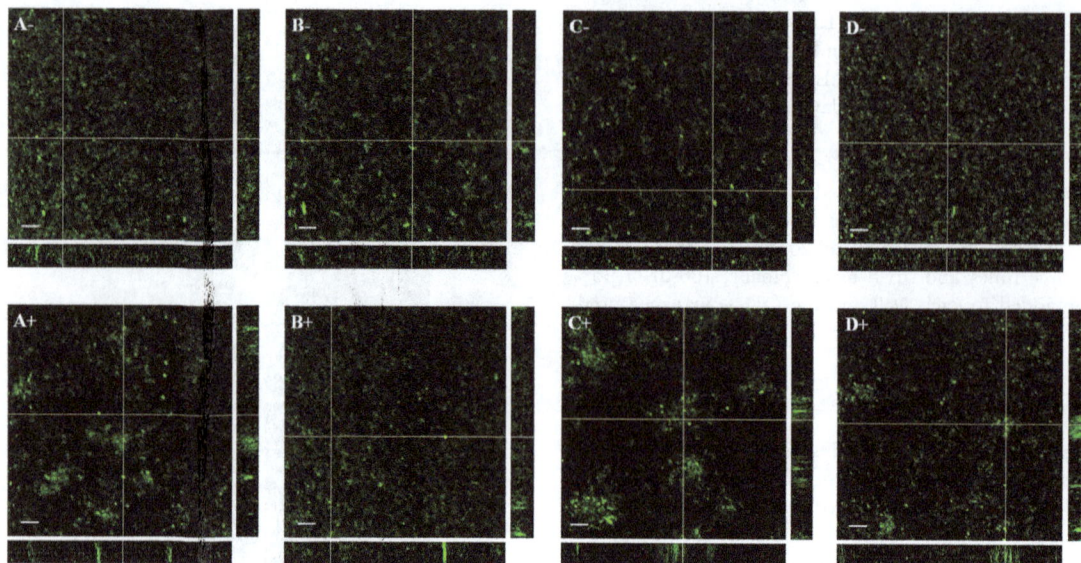

Figure 3. Biofilm structure of *S. fredii* SMH12 and derivatives on glass surfaces: reconstruction of the XY-axis, XZ-axis and YZ-axis. The top corresponds to cultures without flavonoids. The bottom corresponds to cultures with inducing flavonoid. A. SMH12. B. SVQ648. C. SMH12 (pME6863). D. SMH12 (pME6000). Bar, 20 μm. SMH12: wild-type, SVQ648: *nodD1* mutant, SMH12 (pME6863): lactonase strain. SMH12 (pME6000): carrying the empty plasmid.

attached bacteria. Afterwards, roots were soaked for 3 minutes in 0.5% crystal violet to avoid auto-fluorescence, following 3 washes with distilled water. Microscopy analysis was carried out using an Olympus BH2-FRCA microscope with 40x and 100x magnifications (Olympus, Japan). Images of GFP-labelled bacterial cells were obtained by using a filter set consisting of a 400 to 490 nm (BP490) bandpass exciter, a 505 nm dicroic filter and a 530 nm longpass emitter (EO530). In all cases, exposure length of red and green channels was 30 seconds.

For electron scanning microscopy, 7 day-soybean plants were collected and the roots were excised as described above. Sample preparation and visualization were done according to Barahona *et al.* [37].

Root attachment

Whole plants were carefully taken from the hydroponic solution and roots were excised and thoroughly washed with water to eliminate any loosely attached bacteria. Remaining attached bacteria were recovered from the root by vortexing (one centimetre of the proximal area of the main root or the last centimetre of a lateral root) for 2 min in a tube containing YM medium and plating the appropriate dilutions on YM medium plates. Experiments were performed three times with three plants per assay.

Table 4. β-galactosidase activity obtained using an adapted assay with *A. tumefaciens* NT1 (pZLR4) as bioreporter and grown in the presence of supernatants from biofilm cultures (1% v/v).

	Miller units	n (%)
Control (YM)	143.5±6.4	
Genistein (37 nM)	122.1±25.6	
Umbelliferone (62 nM)	131.2±7.2	
3-oxo-C6-HSL (5.5 μM)	824.7±21.2	
SMH12	717.6±23.6	100
SMH12+genistein (3.7 μM)	843.9±24.6	117*
SVQ648	702.7±28.4	97
SVQ648+genistein (3.7 μM)	745.8±39.2	104
SMH12 (pME6863)	150.2±53.8	21*
SMH12 (pME6863)+genistein (3.7 μM)	152.2±9.71	21*

Data are the mean ± SD of three independent experiments performed in triplicate.
n: percentage of induction of each supernatant with respect to SMH12 without flavonoids, defined as 100%.
Each β-galactosidase activity using biofilm supernatant was individually compared to that obtained in SMH12 without flavonoids by using the Mann-Whitney non-parametrical test. Numbers on the percentage of induction column followed by * are significantly different at the level α = 5%.

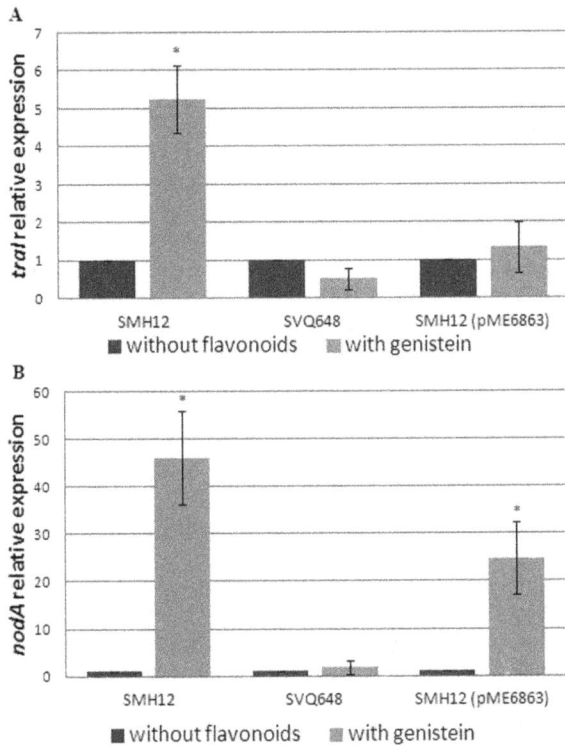

Figure 4. Quantitative RT-PCR analysis of the expression of *tral* and *nodA* from *S. fredii* SMH12 and derivatives from biofilm cultures. Expression data shown are the mean (± standard deviation of the mean) for three biological replicates. Expression was calculated relative to the expression without flavonoids of the wild-type strain by using the Mann-Whitney non-parametrical test. The asterisks indicate a significant different at the level $\alpha = 5\%$. White bars: biofilm cultures without flavonoids. Gray bars: biofilm cultures with genistein. A. *tral* relative expression. B. *nodA* relative expression. SMH12: wild-type, SVQ648: *nodD1* mutant, SMH12 (pME6863): lactonase strain.

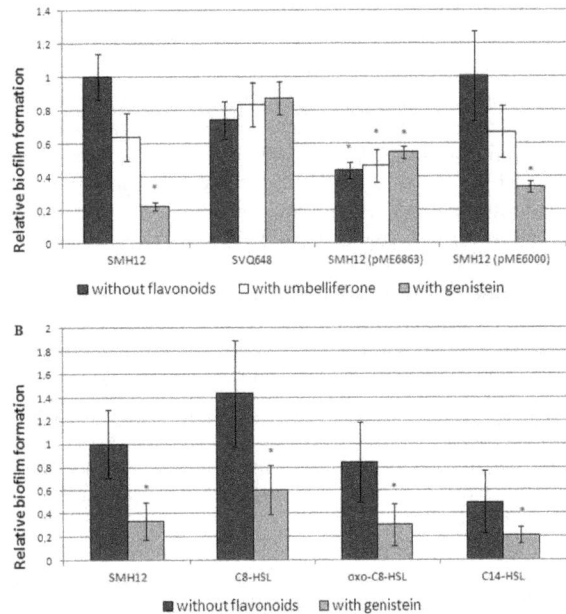

Figure 5. Adhesion of *S. fredii* SMH12 and derivatives on polystyrene surfaces. Biofilms were measured as the amount of crystal violet absorbed by the biofilm formed on multi-well plates and determined by absorbance at 570 nm after de-staining with ethanol (see methods). Absorbance of the wild-type strain was arbitrarily given a value of 1. Averages and standard deviations of eight replicas per strain corresponding to five independent experiments are shown. The asterisks indicate a significant different at the level $\alpha = 5\%$ with respect to wild-type strain by using the Mann-Whitney non-parametrical test. A. Dark gray bars correspond to experiments performed without flavonoids, white to experiments with umbelliferone and light grey bars to experiments with genistein. B. Dark gray bars correspond to experiments performed without flavonoids and light grey bars to experiments with genistein. 3-oxo-C8-HSL and C8-HSL are used at 5.5 μM. C14-HSL is used at 55 μM. SMH12: wild-type, SVQ648: *nodD1* mutant, SMH12 (pME6863): lactonase strain. SMH12 (pME6000): carrying the empty plasmid.

Results

Nodulation Test

In a previous work [5] we described that certain plant flavonoids, besides inducing synthesis of Nod factors, increase the overall production of QS signals in *S. fredii* SMH12. In many bacteria, these systems regulate a broad variety of phenotypes which are important for the successful establishment of a relationship with the eukaryotic hosts. For this reason, a nodulation test was performed to elucidate if SMH12 QS systems are regulating any important phenotype during the symbiosis with *Glycine max*. The symbiotic properties of the wild-type strain, SVQ648 (SMH12 *nodD1*::LacZ::Gm^R), hereafter *nodD1* mutant, SMH12 (pME6863), hereafter lactonase strain (this enzyme prevents AHL accumulation), and SMH12 (pME6000), which carries the empty plasmid, were determined in plant infection tests with soybean cultivar Osumi which are effectively nodulated by the wild-type strain *S. fredii* SMH12.

As expected, the *nodD1* mutant did not induce nodule development due to its inability to produce Nod factors, which are responsible for the formation of these structures in the soybean roots. Consequently, the plant top dry mass was significantly lower (a 60%, where * is the statistical significance of the differences observed using the Mann–Whitney non-parametric test) in plants inoculated with the *nodD1* mutant than in those inoculated with

the parental strain SMH12. In the cases of the plants inoculated

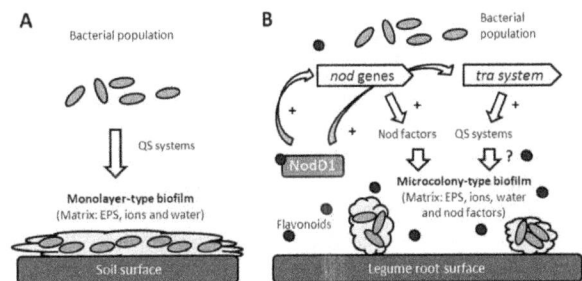

Figure 6. Model of biofilm formation in *S. fredii* SMH12. A. In soil. B. In rizosphere. *S. fredii* SMH12 forms monolayer-type biofilm when colonize soil surfaces. When the bacterium colonizes the legume root (in presence of inducing flavonoids), it forms microcolony-type biofilm, which is necessary for a successful root colonization and symbiosis between rhizobia and legume.

with the lactonase strain, the top dry mass, number of nodules and fresh mass of nodules formed were lower than in the wild-type strain (20%, 40% and 35% respectively). Only in the case of the number of nodules were these differences statistically significant (where * is the statistical significance of the differences observed using the Mann–Whitney non-parametric test) (Table 2). To confirm that differences in this strain are not due to the presence of the plasmid, the symbiotic parameters of SMH12 carrying the empty plasmid (pME6000) were studied. No changes were observed with respect to the wild-type strain. In summary, these results indicate that QS systems could regulate some important phenotypes during the symbiosis with *Glycine max*.

Plant root colonization

As commented in the introduction, in plant-associated bacteria one of the processes controlled by bacterial QS systems is the root colonization. For this reason three independent assays to study root colonization were carried out: root attachment quantification, observation studies of root by epifluorescence microscopy and electronic barrier microscopy. In these studies, proximal and lateral roots were analyzed separately to test differences in bacterial colonization in the whole root.

Root attachment assays showed a statistically significant reduction in the bacterial number per cm on the proximal area of the main root in the lactonase strain, the detected bacteria being around 10-fold less than in the other two strains. In lateral roots, colonization differences (statistically significant) were even more pronounced, these differences being 100-fold less than in the wild-type strain. A slight reduction in the number of bacteria per cm on the lateral root was also observed in the *nodD1* mutant (10-fold less than SMH12). No changes in root attachment were observed in plants inoculated with the SMH12 strain carrying the empty plasmid with respect to those inoculated with the wild-type strain (Table 3). No differences in the bacterial growth-curves were detected among the four strains (data not shown).

Epifluorescence microscopy and electronic barrier microscopy assays (Fig. 1) showed that in lateral roots of plants inoculated with *nodD1* mutant and lactonase strain, the soybean root surface presented lower bacterial density with respect to the wild-type strain. Interestingly, the wild-type strain seems to be also distributed more clustered than the other two strains along the main root and especially in the lateral roots, which could indicate different biofilm structures in these strains. All these results suggest that the *nodD1* mutant and especially the lactonase strain are defective in root colonization.

Influence of inducing flavonoids on biofilm structure

When rhizobia colonize the legume root surface microcolonies or biofilms are formed. Could the observed colonization differences be a consequence of abnormal biofilm formation in the lactonase strain and in the *nodD1* mutant? To study the role of the biofilm formation during symbiosis, the biofilm structures both in the presence/absence of nod-gen inducing flavonoid (genistein) were observed using confocal microscopy experiments. Biofilm images of the wild-type, the *nodD1* mutant, the lactonase and the control strain carrying the empty plasmid, all labelled with GFP, were obtained (Fig. 2 and Fig. 3). Results showed that after 4 days of growth without flavonoid, the formed biofilm consisted of a monolayer in all the studied strains being the surface coverage statistically lower (about 50%) in the lactonase strain biofilm (Fig. 2c). In the presence of genistein, the formed biofilm changed from monolayer to microcolony type in SMH12, being the coverage statistically lower (almost 45%) (Fig. 2a). In the case of the *nodD1* mutant, no change was observed in the biofilm three-dimensional structure or surface coverage with genistein (Fig. 2b). Moreover, the lactonase strain showed a less surface-coverage (43%) without genistein with respect to the wild-type, but

interestingly, in the presence of genistein the biofilm development underwent the transition to microcolony but no reduction in the coverage was obtained with respect to the lactonase strain without flavonoid (Fig. 2c). The control strain carrying the empty plasmid showed the same phenotype as the wild-type strain (Fig. 2d). These results suggest that, in *S. fredii* SMH12, the nodulation gene inducing flavonoids via NodD1 protein provoke a decrease in the bacterial attachment to an abiotic surface, since the monolayer biofilm structure (covering the whole surface) developed in the absence of flavonoid changed to microcolony-type (covering only some parts of the surface). Furthermore, these data indicate that QS signals must to be involved at least in the full formation of the monolayer-type biofilm on the glass surface.

Influence of inducing flavonoids on the QS systems

Our earlier results demonstrated that *nod* gene inducing flavonoids increase the overall production of QS signals in *S. fredii* SMH12 [5]. To investigate if the activation of the QS systems takes place during the biofilm formation a quantification of the overall AI production was carried out in each condition and strain during the biofilm formation experiments on microtiter plates using the *A. tumefaciens* NT1 (pZRL4) biosensor. Firstly, the optimal concentration of bacterial supernatant required to obtain the wider differences between each condition without reaching saturation were determined adding different concentrations of bacterial supernatants (1%, 0.1% and 0.01%). The optimal concentration for these assays was 1% of the total volume of the biosensor culture (data not shown). Under these experimental conditions, results showed that the presence of genistein provoked a statistically significant higher production of AIs only in SMH12 (17% more), indicating that *nod* gene inducing flavonoids enhanced the QS signals accumulation via NodD1. Furthermore, β-galactosidase activity only in the presence of these flavonoids was similar to that obtained in the negative control (Table 4). As expected, a strong decrease in AI accumulation was detected in the supernatants from the lactonase strain, with or without genistein (Table 4). A supplementary TLC with extracts of the supernatants of biofilm experiments in microtiter plates showed a complete degradation of all AHLs in the lactonase strain. In the SMH12 strain carrying the empty plasmid no changes in the AHL profile were observed with respect to the wild-type strain (data not shown).

Finally, to ascertain if the increase in AHL production with flavonoids in biofilm culture assays was correlated with an increase of gene transcription, a quantitative real time RT-PCR assays was carried out. Results showed that the expression of *traI*, an AHL-synthesis gene from SMH12 [5], significantly increased (5-fold) in the presence of genistein compared to the control without flavonoids. No changes in the *traI* expression were observed in the presence of flavonoid in the case of the *nodD1* mutant or in the lactonase strain, suggesting that the induction of the transcription takes place via NodD1 and that QS systems are necessary for the gene expression enhancement probably due to the typical positive feedback of the QS systems at high cellular density (Fig. 4a). As control, the expression of *nodA*, a gene positively regulated via NodD1 and flavonoids, was measured. As expected, a statistically significant increase was observed in *nodA* gene expression (46-fold and 24-fold, respectively) in both SMH12 and lactonase strains in the presence of genistein, but not in the *nodD1* mutant. Interestingly, the *nodA* gene expression in the lactonase strain in the presence of genistein was significantly lower than in the wild-type (Fig. 4b), which could explain both the reduced soybean nodulation and the less rizosphere attachment capacities of the lactonase strain, connecting the Nod factor production with these capacities (Table 2, Table 3, Fig. 1).

All these observations suggest that SMH12 *nod* gene inducing flavonoids enhance the transcription of certain AHL synthesis genes and the overall AHL production in biofilm experiments via NodD1 when the bacteria have reached the necessary cell density threshold.

Role of the QS signals on biofilm formation

Finally, for a more exhaustive analysis of the role of each QS signal in the biofilm formation, bacterial adhesion to polystyrene surface (biofilm formation experiments) was studied after 7 days of growing in low-phosphate MGM medium (Fig. 5). Firstly, it was confirmed that the glass coverage results obtained in confocal microscopy experiments are correlated with the bacterial adhesion values obtained in biofilm formation in polystyrene microtiter plate experiments. Thus, we could use this experimental approach to study the different biofilm structures. As shown in Figure 5, in the presence of genistein, SMH12 showed a significant reduction in the adhesion value to the abiotic surface (80%, where * is the statistical significance of the differences observed using the Mann–Whitney non-parametric test). Only a slight reduction was obtained using umbelliferone, a non-inducing flavonoid [5]. In the case of *nodD1* mutant, the adhesion values with or without flavonoids were similar to those obtained in the case of SMH12 without flavonoid. However, in the lactonase strain a lower adhesion was observed to polystyrene surfaces (50%) in all cases with respect to the wild-type strain without flavonoids, indicating that QS signals must be implied in the full differentiation of both types of biofilm on polystyrene surface since neither the high values without flavonoids nor the low values with inducing flavonoids are reached. To confirm that the differences observed in this strain were not due to the presence of the plasmid, the adhesion values were studied using SMH12 carrying the empty plasmid (pME6000). The experiment confirmed that this strain behaved like the wild-type strain (Fig. 5a). The absorbance at 600 nm after bacterial growth was similar in all strains and conditions (data not shown). Thus, the adhesion values to polystyrene surface in the wild type strain have a correlation with the values of surface coverage obtained on a glass surface with confocal microscopy experiments (Fig. 2).

Once this method was validated as a reporter of the formation of the different biofilm types (monolayer with high values of coverage/adhesion and microcolony with low values of coverage/adhesion), the influence of QS signals in the biofilm formation of SMH12 with or without inducing flavonoids was studied. For this purpose, the cognate AHLs of SMH12, 3-oxo-C8-HSL, C8-HSL and C14-HSL [5], were added to wild-type bacterial cultures at physiological concentrations (data not shown) in biofilm formation experiments on microtiter plates (Fig. 5b). A slight increase or decrease in the adhesion to the polystyrene surface was observed when the medium was supplemented with C8-HSL or C14-HSL, respectively. However, these differences were not statistically significant (Fig. 5b). The absorbance at 600 nm after bacterial growth was similar in all bacteria and conditions (data not shown).

Discussion

During the first stages of root colonization, leguminous plants exudate molecules which chemotactically attract rhizobia. Once in the rhizosphere, the rhizobial population colonizes the root surface, forming a bacterial community surrounded by a matrix produced by its own bacteria, the biofilm. In the family *Rhizobiaceae*, the bacterial intrinsic factors that are required for

the biofilm formation are mainly EPS and bacterial QS systems as well as in the case of *S. meliloti* the Nod factor production [12]. A fundamental question is whether the process of biofilm formation significantly affects nodulation in the symbiosis *S. fredii* SMH12-*G. max* cv. Osumi. To answer this question, the significance of QS systems and the *nod* gene inducing flavonoids on the biofilm formation, the colonization and the symbiosis with soybean were studied by means of both a *nodD1* mutant and a lactonase strain of *S. fredii* SMH12.

As expected, nodulation tests with soybean showed that the symbiotic phenotype of the *nodD1* mutant is similar to that obtained in non-inoculated plants, since this bacterium is unable to detect via NodD1 the flavonoids exuded by plants. Consequently the Nod factor production is blocked. Interestingly, results indicate that SMH12 QS systems could regulate some important process during the symbiosis, since the lactonase strain showed reductions in both the plant top dry mass and the fresh mass of nodules formed with respect to the wild-type strain but only in the number of nodules the reduction was statistically significant (Table 2). As described in the introduction, QS regulates a broad variety of phenotypes, including the biofilm formation. In the review of Rinaudi and Giordano [12] the mechanisms involved in both the rhizobial biofilm formation and the attachment to the plant roots are summarized. Taking into consideration the previous reports, they concluded that the biofilm lifestyle allows rhizobia to survive under unfavourable conditions and in certain species the biofilm formation is clearly an important feature of symbiotic ability [19,22,23,24]. However, the role and regulation of biofilm formation during the symbiosis *S. fredii* SMH12-soybean has not been reported. Our results suggest that there is a correlation between QS systems, biofilm formation, root colonization and symbiosis. Firstly, three independent experiments of bacterial root colonization showed a significant reduction in bacterial root colonization in the *nodD1* mutant and especially in the lactonase strain. These differences were more visible in the lateral root colonization (Table 3). Interestingly, in addition to the differences in the bacterial number, these two strains showed more of a spread surface distribution on roots compared to the wild-type, in which the bacteria appeared clustered (Fig. 1). These observations would indicate that the *nodD1* mutant and specially the lactonase strain do not colonize the root surface optimally due to an alteration in their biofilm formation ability. The impaired biofilm formation in the case of the lactonase strain would explain its less nodulation capacity in the soybean test (Table 2).

To corroborate this hypothesis, the biofilm structure was analyzed by confocal microscopy. Two types of biofilms were observed, the monolayer-type in the absence of flavonoid (bacterial population cover the entire surface homogeneously), and the microcolony-type in the presence of genistein (bacterial population is clustered) (Fig. 2a and Fig. 3a). These observations would indicate that on the surface of the legume root, with a high concentration of flavonoids, the bacterial biofilm could undergo a transition from monolayer to microcolony. Perhaps this biofilm structure allows the bacterial colonization of specific root areas (those with high flavonoid exudation), which would be the optimal for the symbiosis initiation (i.e. root hairs).

Furthermore, analysing these results we infer that QS systems is involved at least in the formation of the monolayer type, since the lactonase strain developed incomplete biofilm phenotypes without inducing flavonoids (Fig. 2c and Fig. 3c). In fact, most rhizobial biofilms investigated so far are regulated directly or indirectly by QS systems [12]. Interestingly, the transition to microcolony-type biofilm requires Nod factor production since this biofilm was not developed in the presence of genistein in the *nodD1* mutant

(Fig. 2b and Fig. 3b). Fujishige *et al.* [19] reported that *nodD1ABC* of *S. meliloti*, involved in the synthesis of Nod factors, are necessary for the three-dimensional architecture of the biofilm and, in fact, these molecules are present in the biofilm matrix. The mutation of any of these genes generates a monolayer-type biofilm. However, in contrast with our results, the presence of inducing flavonoids is not necessary for the development of three dimensional structures in the biofilm of *S. meliloti* [19].

Interestingly, our findings also suggest that flavonoids, beyond inducing the synthesis of Nod factors (which integrate the symbiotic biofilm matrix), are enhancing the QS systems in biofilm formation experiments (Table 4 and Fig. 4a). In the presence of inducing flavonoids, an increase in the *traI* expression and in the overall AI production was detected. These effects were not observed in the lactonase strain due to the positive feed-back regulation of the QS systems which only occurs when the threshold in cellular density is reached [6] (Fig. 4a and Table 4). Furthermore, it is clear that the enhancement of the *traI* gene expression takes place via NodD1, because in the *nodD1* mutant this over expression was not observed. He et al. [38] studied the QS systems of *Rhizobium* sp. NGR234, a *Sinorhizobium fredii* related bacterium, but they only unequivocally identified the *tra* system. This QS system is homologous to the one responsible for the plasmid Ti transfer in *A. tumefaciens*. This transference occurs only in the presence of opines, compounds that are produced by plants. Control of the Ti plasmid transference is modulated by TraM, a small protein that binds to and inactivates TraR, which senses the AI concentration. The induction in the presence of opines allows the synthesis of TraR at levels that overcome the inhibitory activity of TraM, activating the QS systems and the plasmid transfer [39]. In NGR234, the over expression of TraR activates not only the *tra* system but also other QS systems of the bacteria, since an increase in the overall AI production was detected [34]. However, so far, no natural compounds exuded by plants (similar to opines for *A. tumefaciens* QS systems) have been identified as inductor molecules for NGR234. Interestingly, results obtained in this paper and those reported in our previous work [5] indicate that the *nod* gene inducing flavonoids could be these inducing compounds for *S. fredii* SMH12.

In summary, experiments show that SMH12 QS systems must be involved at least in the full differentiation of monolayer-type biofilm and, moreover, in the presence of inducing flavonoids, these systems increase the AHL production and the biofilm undergoes a transition to microcolony-type (Fig. 4a, Table 4 and Fig. 2). Logically, this expression increase in the presence of flavonoids could be related to the developement of the microcolony-type biofilm, which would only occur in the rhizosphere. This

fact would allow, through a unique key molecule, the efficient colonization of the legume root and the activation of the synthesis of Nod factors, both required for successful symbiosis. Interestingly, the study of the role of QS signals on biofilm formation experiments (Fig. 5) showed no statistically differences in adhesion values with or without flavonoid after addition of the wild-type cognate AHLs (Fig. 5b). However, in the case of the lactonase strain, the decrease in the adhesion values in the presence of inducing flavonoids did not reach the values obtained in the wild-type strain. This result suggests that, despite the transition to microcolony-type biofilm is regulated directly by NodD1 protein and inducing flavonoids (Fig. 2), the QS signals could also be involved in the full differentiation of this biofilm on the polystyrene surface (Fig. 5a).

Thus, taking into consideration all the results, we propose the following model (Fig. 6): *S. fredii* SMH12 forms a QS-regulated monolayer-type biofilm (whose matrix would be mainly composed of EPS, water and ions) when bacterium colonizes soil. In the rhizosphere, the exudation of inducing flavonoids, besides inducing the synthesis of Nod factors which are necessary for the microcolony-type biofilm, potentiate via NodD1 the *tra* QS system, which leads to an overproduction of the QS signals. This accumulation and the synthesis of Nod factors are required for the full development of the microcolony-type biofilm, whose matrix should be composed of EPS, water, ions and Nod factors. The complete formation of this symbiotic biofilm is necessary for a successful root colonization and symbiosis between rhizobia and legume.

Despite demonstrating in this work that the rhizobial biofilm formation is important for colonization and symbiosis of *S. fredii* SMH12 with the soybean, there are still many aspects that must to be studied in the future to further clarify this process.

Acknowledgments

We would like to thank the Laboratorio de Microscopía de Barrido y Análisis por Energía Dispersiva de Rayos X from the Universidad Autónoma de Madrid for electronic barrier microscopy assays. Thanks to Servicio General de Biología of the CITIUS from the University of Seville for allowing us to use their laboratory equipment. Thanks to Diane Haun for English style supervision. Special thanks to Miguel Cámara, for his comments on an earlier version of the manuscript.

Author Contributions

Conceived and designed the experiments: FPM RB JL RE. Performed the experiments: FPM IJG PDC IBR. Analyzed the data: FPM RB JL RE. Contributed reagents/materials/analysis tools: FJO JL. Contributed to the writing of the manuscript: FPM FJL RE.

References

1. Murray JD. (2011) Invasion by invitation: rhizobial infection in legumes. Mol Plant-Microbe Interact 24: 631–639.
2. Somers E, Vanderleyden J, Srinivasan M. (2004). Rhizosphere bacterial signalling: a love parade beneath our feet. Crit Rev Microbiol 30: 205–240.
3. Krause A, Doerfel A, Göttfert M. (2002) Mutational and Transcriptional Analysis of the Type III Secretion System of *Bradyrhizobium japonicum*. Mol Plant Microbe Interact 12: 1228–1235.
4. López-Baena FJ, Vinardell JM, Pérez-Montano F, Crespo-Rivas JC, Bellogín RA, et al. (2008) Regulation and symbiotic significance of nodulation outer proteins secretion in *Sinorhizobium fredii* HH103. Microbiology 154: 1825–1836.
5. Pérez-Montaño F, Guasch-Vidal B, González-Barroso S, López-Baena FJ, Cubo T, et al. (2011) Nodulation-gene-inducing flavonoids increase overall production of autoinducers and expression of *N*-acyl homoserine lactone synthesis genes in rhizobia. Res Microbiol 162: 715–723.
6. Fuqua WC, Winans SC, Greenberg EP. (1994) Quorum sensing in bacteria: the LuxR-LuxI family of cell density-responsive transcriptional regulators. J Bacteriol 176: 269–275.

7. Miller MB, Bassler BL. (2001) Quorum sensing in bacteria. Annu Rev Microbiol 55: 165–199.
8. Marketon MM, Glenn SA, Eberhard A, Gonzalez JE. (2003) Quorum sensing controls exopolysaccharide production in *Sinorhizobium meliloti*. J Bacteriol 185: 25–331.
9. Ohtani K, Hayashi H, Shimizu T. (2002) The *luxS* gene is involved in cell-cell signalling for toxin production in *Clostridium perfringens*. Mol Microbiol 44: 171–179.
10. Quiñones B, Dulla G, Lindow SE. (2005) Quorum sensing regulates exopolysaccharide production, motility, and virulence in *Pseudomonas syringae*. Mol Plant-Microbe Interact 18: 682–693.
11. Rice SA, Koh KS, Queck SY, Labbate M, Lam KW, et al. (2005) Biofilm formation and sloughing in *Serratia marcescens* are controlled by quorum sensing and nutrient cues. J Bacteriol 187: 3477–3485.
12. Rinaudi LV, Giordano W. (2010) An integrated view of biofilm formation in rhizobia. FEMS Microbiol Lett 304: 1–11.
13. Cha C, Gao P, Chen YC, Shaw PD, Farrand SK. (1998) Production of acyl-homoserine lactone quorum-sensing signals by gram-negative plant-associated bacteria. Mol Plant-Microbe Interact 11: 1119–1129.

14. Costerton JW, Lewandowski Z, Caldwell DE, Korber DR, Lappin-Scott HM. (1995) Microbial biofilm. Annu Rev Microbiol 49: 711–745.

15. Stanley NR, Lazazzera BA. (2004) Environmental signals and regulatory pathways that influence biofilm formation. Mol Microbiol 52: 917–924.

16. Sutherland IW. (2001) Biofilm exopolysaccharides: a strong and sticky framework. Microbiology 147: 3–9.

17. Rinaudi L, Fujishige NA, Hirsch AM, Banchio E, Zorreguieta A, et al. (2006) Effects of nutritional and environmental conditions on Sinorhizobium meliloti biofilm formation. Res Microbiol 157: 867–875.

18. Fujishige NA, Kapadia NN, De Hoff PL, Hirsch AM. (2006) Investigations of Rhizobium biofilm formation. FEMS Microbiol Ecol 56: 195–206.

19. Fujishige NA, Lum MR, De Hoff PL, Whitelegge JP, Faull KF, et al. (2008) Rhizobium common nod genes are required for biofilm formation. Mol Microbiol 67: 504–515.

20. Wells DH, Chen EJ, Fisher RF, Long SR. (2007) ExoR is genetically coupled to the ExoS-ChvI two-component system and located in the periplasm of Sinorhizobium meliloti. Mol Microbiol 64: 647–664.

21. Rinaudi LV, González JE. (2009) The low-molecular-weight fraction of exopolysaccharide II from Sinorhizobium meliloti is a crucial determinant of biofilm formation. J Bacteriol 191: 7216–7224.

22. González JE, Marketon MM (2003) Quorum sensing in nitrogen-fixing rhizobia. Microbiol Mol Biol Rev 67: 574–592.

23. Loh JT, Yuen-Tsai JP, Stacey MG, Lohar D, Welborn A, et al. (2001) Population density-dependent regulation of the Bradyrhizobium japonicum nodulation genes. Mol Microbiol 42: 37–46.

24. Jitacksorn S1, Sadowsky MJ (2008) Nodulation gene regulation and quorum sensing control density-dependent suppression and restriction of nodulation in the Bradyrhizobium japonicum-soybean symbiosis. Appl Environ Microbiol 74: 3749–3756.

25. Pérez-Montaño F, Jiménez-Guerrero I, Sánchez-Matamoros RC, López-Baena FJ, Ollero FJ, et al. (2013) Rice and bean AHL-mimic quorum-sensing signals specifically interfere with the capacity to form biofilms by plant-associated bacteria. Res Microbiol 164: 749–760.

26. Beringer JE. (1974) R factor transfer in Rhizobium leguminosarum. J Gen Microbiol 84: 188–198.

27. Vincent JM (1970) The modified Fahraeus slide technique. In A manual for the practical study of root nodule bacteria, 144–145. Edited by J. M. Vincent. Oxford, UK: Blackwell Scientific Publications.

28. Sambrook J, Fritsch EF, Maniatis T(1989) Molecular cloning: a laboratory manual, 2nd edn. Cold Spring Harbor NY: Cold Spring Harbor Laboratory.

29. Lamrabet Y, Bellogín RA, Cubo T, Espuny R, Gil A, et al. (1999) Mutation in GDP-fucose synthesis genes of Sinorhizobium fredii alters Nod factors and significantly decreases competitiveness to nodulate soybeans. Mol Plant-Microbe Interact 12: 207–217.

30. Simon R. (1984) High frequency mobilization of gram-negative bacterial replicons by the in vitro constructed Tn5-Mob transposon. Mol Gen Genet 196: 413–420.

31. López-Baena FJ, Monreal JA, Pérez-Montano F, Guasch-Vidal B, Bellogín RA, et al. (2009) The absence of Nops secretion in Sinorhizobium fredii HH103 increases GmPR1 expression in Williams soybean. Mol Plant-Microbe Interact 22: 1445–1454.

32. Russo DM, Williams A, Edwards A, Posadas DM, Finnie C, et al. (2006) Proteins exported via the PrsD-PrsE type I secretion system and the acidic exopolysaccharide are involved in biofilm formation by Rhizobium leguminosarum. J Bacteriol 188: 4474–4486.

33. O'Toole GA, Kolter R. (1998) Initiation of biofilm formation in Pseudomonas fluorescens WCS365 proceeds via multiple, convergent signalling pathways: a genetic analysis. Mol Microbiol 28: 449–461.

34. Mueller K, González JE. (2011) Complex regulation of symbiotic functions is coordinated by MucR and quorum sensing in Sinorhizobium meliloti. J Bacteriol 193: 485–496.

35. Miller JH. (1972) Experiments in molecular genetics. Cold Spring Harbor Laboratory Press. Cold Spring Harbor, New York (USA).

36. de Lyra, MCCP, López-Baena FJ, Madinabeitia N, Vinardell JM, Espuny MR, et al. (2006) Inactivation of the Sinorhizobium fredii HH103 rhcJ gene abolishes nodulation outer proteins (Nops) secretion and decreases the symbiotic capacity with soybean. Int Microbiol 9: 125–133.

37. Barahona E, Navazo A, Yousef-Coronado F, Aguirre de Cárcer D, Martínez-Granero F, et al. (2010) Efficient rhizosphere colonization by Pseudomonas fluorescens f113 mutants unable to form biofilms on abiotic surfaces. Environ Microbiol 12: 3185–3195.

38. He X, Chang W, Pierce DL, Seib LO, Wagner J, et al. (2003) Quorum sensing in Rhizobium sp. strain NGR234 regulates conjugal transfer (tra) gene expression and influences growth rate. J Bacteriol 185: 809–822.

39. Piper KR, Farrand SK. (2000) Quorum sensing but not autoinduction of Ti plasmid conjugal transfer requires control by the opine regulon and the antiactivator TraM. J Bacteriol 182: 1080–1088.

40. Rodríguez-Navarro DN, Bellogín R, Camacho M, Daza A, Medina C, et al. (2003) Field assessment and genetic stability of Sinorhizobium fredii strain SMH12 for commercial soybean inoculants. Eur J Agron 19: 299–309.

41. Kovach ME, Elzer PH, Hill DS, Robertson GT, Farris MA, et al. (1995) Four new derivatives of the broad-host-range cloning vector pBBR1MCS, carrying different antibiotic-resistance cassettes. Gene 166: 175–176.

42. Schäfer A, Tauch A, Jager W, Kalinowski J, Thierbach G, et al. (1994) Small mobilizable multi-purpose cloning vectors derived from the Escherichia coli plasmids pK18 and pK19: selection of defined deletions in the chromosome of Corynebacterium glutamicum. Gene 145: 69–73.

43. Maurhofer M, Reimmann C, Schmidli-Sacherer P, Heeb S, Haas D, et al. (1998) Salicylic acid biosynthetic genes expressed in Pseudomonas fluorescens strain P3 improve the induction of systemic resistance in tobacco against tobacco necrosis virus. Phytopathology 88: 678–684.

44. Reimmann C, Ginet N, Michel L, Keel C, Michaux P, et al. (2002) Genetically programmed autoinducer destruction reduces virulence gene expression and swarming motility in Pseudomonas aeruginosa PAO1. Microbiology 148: 923–932.

45. Stuurman N, Pacios-Bras C, Schlaman HR, Wijfjes AH, Bloemberg G, et al. (2000) Use of green fluorescent protein color variants expressed on stable broad-host-range vectors to visualize rhizobia interacting with plants. Mol Plant Microbe-Interact 13: 1163–1169.

46. Vinardell JM, Ollero FJ, Hidalgo A, López-Baena FJ, Medina C, et al. (2004) NolR regulates diverse symbiotic signals of Sinorhizobium fredii HH103. Mol Plant Microbe-Interact 17: 676.

47. Figurski DH, Helinski DR. (1979) Replication of an origin-containing derivative of plasmid RK2 dependent on a plasmid function provided in trans. Proc Natl Acad Sci USA 76: 1648–1652.

## 12

Architecture and Assembly of the *Bacillus subtilis* Spore Coat

Marco Plomp[1], Alicia Monroe Carroll[2], Peter Setlow[2]*, Alexander J. Malkin[1]*

1 Biosciences and Biotechnology Division, Physical and Life Sciences Directorate, Lawrence Livermore National Laboratory, Livermore, California, United States of America, **2** Department of Molecular Biology and Biophysics, University of Connecticut Health Center, Farmington, Connecticut, United States of America

Abstract

Bacillus spores are encased in a multilayer, proteinaceous self-assembled coat structure that assists in protecting the bacterial genome from stresses and consists of at least 70 proteins. The elucidation of *Bacillus* spore coat assembly, architecture, and function is critical to determining mechanisms of spore pathogenesis, environmental resistance, immune response, and physicochemical properties. Recently, genetic, biochemical and microscopy methods have provided new insight into spore coat architecture, assembly, structure and function. However, detailed spore coat architecture and assembly, comprehensive understanding of the proteomic composition of coat layers, and specific roles of coat proteins in coat assembly and their precise localization within the coat remain in question. In this study, atomic force microscopy was used to probe the coat structure of *Bacillus subtilis* wild type and *cotA, cotB, safA, cotH, cotO, cotE, gerE,* and *cotE gerE* spores. This approach provided high-resolution visualization of the various spore coat structures, new insight into the function of specific coat proteins, and enabled the development of a detailed model of spore coat architecture. This model is consistent with a recently reported four-layer coat assembly and further adds several coat layers not reported previously. The coat is organized starting from the outside into an outermost amorphous (crust) layer, a rodlet layer, a honeycomb layer, a fibrous layer, a layer of "nanodot" particles, a multilayer assembly, and finally the undercoat/basement layer. We propose that the assembly of the previously unreported fibrous layer, which we link to the darkly stained outer coat seen by electron microscopy, and the nanodot layer are *cotH-* and *cotE-* dependent and *cotE-*specific respectively. We further propose that the inner coat multilayer structure is crystalline with its apparent two-dimensional (2D) nuclei being the first example of a non-mineral 2D nucleation crystallization pattern in a biological organism.

Editor: Etienne Dague, LAAS-CNRS, France

Funding: This work was supported by a grant from the National Institutes of Health (http://www.nih.gov/) (GM-19698) (PS), by a Department of Defense Multi-disciplinary University Research Initiative (http://www.arl.army.mil/www/default.cfm?page=472) through the United States Army Research Laboratory and the United States Army Research Office under contract number W911F-09-1-0286 (PS), and by the Lawrence Livermore National Laboratory (https://www.llnl.gov/) through Laboratory Directed Research and Development Grant 04-ERD-002 (AJM). Part of this work was performed under the auspices of the United States Department of Energy by the University of California, Lawrence Livermore National Laboratory under Contract W-7405-Eng-48. The funders had no role in study design, data collection and analysis, or preparation of the manuscript. This document was cleared by the Lawrence Livermore National Laboratory for publication.

* Email: setlow@nso2.uchc.edu (PS); malkin1@llnl.gov (AJM)

Introduction

Spores of bacteria of *Bacillus* species are formed in sporulation and are metabolically dormant and resistant to a large variety of environmental stress factors. While multiple factors contribute to spore resistance, one striking spore feature is the multilayer spore coat that provides protection against many toxic chemicals, as well as digestion by lytic enzymes and being eaten by several types of predatory eukaryotes [1–4]. The spore coat is assembled moderately late in sporulation from components synthesized in the mother cell compartment of the sporulating cell, and comprises the outer layers of spores of many *Bacillus* species, although spores of some species contain an outermost exosporium. Spore coat structure and assembly have been best studied in the model spore former *Bacillus subtilis* and ~70 spore specific proteins have been identified in the spore coat [2,3,5,6]. In addition, a number of these coat proteins undergo covalent

modifications including proteolytic cleavage, cross-linking, and tyrosine peroxidation.

The spore coat of *B. subtilis* has drawn attention not only because of its role in spore resistance but also because some coat proteins play significant roles in spore germination. However, much recent work on the spore coat has focused on determining overall spore coat structure as well as the mechanisms involved in the assembly of this large multi-molecular structure. Work to date has indicated that there are at least four coat layers that can be distinguished by electron microscopy (EM) as well as other means – undercoat, inner coat, outer coat, and an outermost glycoprotein layer called the crust [2,4,7,8]. Several of these individual layers also have sublayers, as the inner and outer coats have multiple lamellae. Most of the proteins in these various layers do not have specific roles in spore properties with the exception of a few coat enzymes, and most importantly, proteins that are essential for coat morphogenesis. The morphogenetic proteins include coat proteins such as CotE, CotH, CotO, SafA, and SpoVID, loss of any of which have drastic effects on overall coat architecture, as these proteins direct the assembly of different subsets of proteins into the coat [2,4,9–11]. In addition, the SpoIIID, GerE and GerR proteins have major effects on the expression of genes encoding coat proteins that are transcribed during sporulation, and this in turn has significant effects on coat properties and morphology [4–6].

A variety of studies of the functional repertoire of coat proteins have focused on the determination of the locations of these proteins in the spore coat and their specific roles in spore coat morphogenesis [5,7,12–15]. These studies have been extended and complemented by studies of direct interactions between various coat proteins, both *in vitro* and *in vivo* [4,16–20]. All of this work has given a picture of the molecular interactions in the spore coat, as well as the dependencies of the assembly of specific proteins into the coat. However, this type of analysis has not yet been complemented by detailed analysis of the structures of the various spore layers. Atomic force microscopy (AFM) has been used to unravel high-resolution structures of the coats of dormant and germinating spores of various *Bacillus* [14,21–30] and *Clostridium* [31] species. However, this analysis has generally been conducted on wild-type spores, with AFM data on only a few mutants lacking specific coat layers. Consequently, in this work we have used high-resolution AFM to analyze the surface structure of spores of wild-type *B. subtilis* spores as well as spores of a variety of mutant strains in order to reveal the surface morphology of various layers of the spore coat. The results from these analyses have provided high-resolution visualization of the various spore coat structures as well as several coat layers not reported previously. This information has allowed the formulation of a model for coat structure and provided further insight into the assembly of the spore coat.

Materials and Methods

Strains used in this study

The *B. subtilis* strains used in this study (Table 1) except one are isogenic with the wild-type strain PS832, a prototrophic derivative of strain 168. Preparation of strains by transformation with chromosomal DNA was as described [32].

Spore preparation

B. subtilis strains were grown at 37°C in Luria-Bertani (LB) [33] medium supplemented with the appropriate antibiotics when necessary. Chloramphenicol was used at a final concentration of 5 mg/liter, kanamycin at a final concentration of 10 mg/liter, and tetracycline at a final concentration of 10 mg/liter.

For spore preparation, *B. subtilis* strains were grown for 3 h in LB medium and then spread on 2× Schaeffer's-glucose medium agar plates without antibiotics [34]. Spores were harvested after incubation at 37°C for 5 d followed by incubation at room temperature for 2 d, and purified as described [34] by brief sonication and repeated washing with distilled water. All spore preparations, except for strain PS3735 (Δ*spoVID::kan*) (see below) were free (>98%) of vegetative and sporulating cells and germinated spores as determined by phase-contrast microscopy.

Spores of strain PS3735 (Δ*spoVID::kan*) were generally significantly contaminated with germinated spores and these germinated spores were removed by centrifugation in a one-step Histodenz™ (Sigma, St. Louis, MO) gradient. Four samples, each containing ~3 mg (dry weight) crude spores were suspended in 100 μl of 20% Histodenz™ that was layered on top of 2 ml of 50% Histodenz™ in four Ultra-Clear™ (11×34 mm) centrifuge tubes (Beckman Instruments, Palo Alto, CA) and then centrifuged at 14,000 rpm for 45 min at 20°C in a TLS 55 rotor. After centrifugation, the germinated spores in the supernatant fluid were removed, the pellets containing the dormant spores washed 5 times with 500 μl water and the final pellets were suspended in 500 μl water and combined. These purified spores were free (>98%) from vegetative and sporulating cells as well as germinated spores as determined by phase contrast microscopy.

Chemical decoating of spores

Spores (~6 mg dry weight) were decoated as described previously [35,36]. Briefly, spores were incubated for 90 min at 37°C in 1 ml of 50 mM Tris-HCl (pH 8.0)-8 M urea-10 mM EDTA-1% sodium dodecyl sulfate (SDS)-50 mM dithiothreitol (DTT). After incubation, the spores were centrifuged and the pellets were washed with 1 ml of water 6–10 times.

Atomic force microscopy

Droplets of ~2.0 μm of spore suspensions (~3×10⁹ spores/ml) were deposited on plastic cover slips and incubated for 10 min at room temperature and the sample substrate was carefully rinsed

Table 1. *B. subtilis* strains used in this study.

Strain	Genotype	Phenotype[a]	Source or reference[b]
PS832	wild-type		Laboratory stock
PS3394	Δ*cotE::tet*	Kan^r Tet^r	[1]
PS3735	Δ*spoVID::kan*	Kan^r	[1]
PS3736	Δ*cotH::cat*	Cm^r	[1]
PS3738	Δ*safA::tet*	Tet^r	[1]
PS4133	Δ*cotB::cat*	Cm^r	DL067→PS832
PS4134	Δ*cotO::tet*	Tet^r	PE250→PS832
DL063	Δ*cotA::cat*	Cm^r	[71]
DL067	Δ*cotB::cat*	Cm^r	[71]
PE250	Δ*cotO::tet*	Tet^r	[87]

[a]Abbreviations: Cm^r, chloramphenicol resistant; Kan^r, kanamycin resistant; Tet^r, tetracycline resistant.
[b]DNA from the strain to the left of the arrow was used to transform the strain to the right of the arrow.

with double-distilled water and allowed to dry. Our prior work with spores of other *Bacillus spp.* [22–25] demonstrated that spore morphological and structural attributes were reproduced both for spores analyzed within the same spore preparation and when multiple spore preparations were analyzed. Thus, in this study for each spore strain a single spore batch was analyzed by AFM with ~50–75 spores being imaged for each spore strain. Detailed experimental procedures for AFM imaging of spores were as described previously [22,24]. Images were collected using a Nanoscope IV atomic force microscope (Bruker Corporation, Santa Barbara, CA) operated in tapping mode. For rapid low-resolution analysis of spore samples, fast scanning AFM probes (DMASP Micro-Actuated, Bruker Corporation, Santa Barbara, CA) with resonance frequencies of ~210 kHz were utilized. For high-resolution imaging, SuperSharpSilicon (SSS) AFM probes (NanoWorld Inc, Neuchâtel, Switzerland) with tip radii <2 nm and resonance frequencies of ~300 kHz were used. Nanoscope software 5.30r3sr3 was used for acquisition and subsequent processing of AFM images. In order to successfully assess both overall low-resolution and high-resolution spore features, raw AFM images typically need to be modified. In particular, the *contrast enhancement* command, which runs a statistical differencing filter on the current image, was typically utilized. This filter can bring all the features of an image to the same height and equalize the contrast among them. This allows all features of an image to be seen simultaneously, and thus a single spore or a group of spores can be imaged at relatively low resolution while visualizing spore coat attributes at high resolution. Heights of spore surface features (i.e. folds, coat layers, etc.) were measured from *height* images using the *section* command, which allows measurements of vertical distance (height), horizontal distance, and the angle between two or more points on the surface. Tapping amplitude, phase and height images were collected simultaneously. Height images allow quantitative height determinations, providing precise measurements of spore surface topography. Amplitude and phase images do not provide height information. While amplitude and phase images provide similar morphological and structural information as do height images, they can often display a greater amount of structural detail and contrast compared with height images, often making them a preferred choice for presentation purposes. The surface roughness of spore surfaces for wild type and *cotE gerE* spores was evaluated as the root mean square (RMS) value R_q using AFM height images. R_q is the standard deviation of the Z values (height) within the given area and is calculated as: $\sqrt{\Sigma(Z_i\text{-}Z_{ave})^2/N}$, where Z_{ave} is the average of Z values within the given area, Z_i is the current Z value, and N is the number of points within the given area. R_q was determined for each spore from 4 μm^2 height images (pixel number -512^2) of multiple spores using a 400 nm^2 zoomed in area in the center of the spore. In order to eliminate tilt on the spore surface, prior to the measurement of the roughness, the image was flattened using the third flatten order in the *flatten* command. Step roughness levels were determined by manually digitizing steps' contour from AFM capture images with a plot digitizer (http://plotdigitizer. sourceforge.net/). Once the x and y coordinates of the step contours were obtained, the step perimeter length, S, was estimated from the sum of all segment lengths given by $S = \Sigma\sqrt{(\Delta x^2+\Delta y^2)}$, where the sum is carried over all digitized contour segments. The sinuosity index, which is a measure of step meandering/roughness, is then calculated by taking the ratio of the contour length S over the shortest path length between the two end points of the step (straight line). Note, that the value of the sinuosity index ranges from 1 (case of straight line) to infinity (case of a closed loop, where the shortest path length is zero).

Results

Surface architecture of wild-type and decoated spore surfaces

As seen previously by AFM [21,22], the prominent surface features of air-dried wild-type *B. subtilis* spores are surface ridges extending along the long axis of the spore (Fig. 1a,b; light blue arrows). The height of these surface ridges was generally 15–30 nm, occasionally exceeding 40 nm. Similar surface ridges have been observed on spores of *Bacillus anthracis* [29], *Bacillus cereus* [22,23], *Bacillus atrophaeus* [22,24], *Bacillus thuringiensis* [22,23], and *Clostridium novyi* NT [31]. This ridge formation appears to be due to coat folding caused by changes in spore size upon dehydration [22,24,30,37,38]. RMS roughness R_q of wild-type coat surfaces measured as described in the Methods section for 20 spores varied between 3.49 nm to 8.71 nm with an average R_q value of ~5.26 nm.

AFM studies of protozoal-digested coat-defective *B. subtilis* spores [27] showed that the *B. subtilis* spore's outer surface exhibits a thin layer without prominent structural features, which was defined as an amorphous layer (Fig. 1c,e; green arrows). EM of ruthenium red stained *B. subtilis* spores demonstrated the presence of an outermost glycoprotein layer, and it was suggested that this layer is an exosporium that is tightly attached to the coat layer [8]. Later, a combination of EM, fluorescence microscopy, and genetic analysis also demonstrated the existence of this outermost glycoprotein layer that was named the spore crust [7]. Thus the outermost layer revealed by AFM and the crust layer correspond to the same spore layer. The thickness of the outermost amorphous layer in *B. subtilis* spores as measured from AFM images (Fig. 1e) was not uniform and varied between 4–15 nm with an RMS roughness R_q of ~3 nm. Typically, the coverage of surfaces of *B. subtilis* spores with the amorphous layer was not complete, revealing an underlying rodlet layer, seen on all visualized wild-type spores, with a periodicity of ~7–8.5 nm (Fig. 1c–e; red arrows); note that these rodlets are also seen on the surfaces of the surface ridges. Rodlet structures similar to ones seen in Fig. 1 were previously described in freeze-etching EM [39–41] and AFM studies of both fungal [42,43] and bacterial (*B. atrophaeus*, *B. cereus* and *B. thuringiensis*) [22–25] spores, with rodlet structures on *B. atrophaeus*, and *B. cereus* spores exhibiting ~8 nm periodicity. Note, that depending on sporulation conditions for *B. thuringiensis*, rodlet structures were found either on the spore coat or as extrasporal structures that were present in spore preparations [30].

In order to remove spores' outer coat, *B. subtilis* spores were chemically decoated with urea-SDS at slightly alkaline pH as described in Methods. The great majority of the proteins removed by this type of treatment have been well characterized in work from a number of laboratories [44]. This treatment partially or completely removed the amorphous layer, and the outer surface of the decoated spores was now comprised primarily of the intact rodlet layer (Fig. 2b–f; red arrows), which was covered in some cases with remnants of the amorphous layer (Fig. 2a–c; green arrows). The 15–30 nm surface ridges were also seen on the air-dried decoated spores, similar to what was seen on intact spores, and again these ridges appear to contain rodlets (Fig. 2a–d; light blue arrows).

Surface architecture of spores lacking CotA, CotB and SafA

CotA and CotB are two outer coat proteins that are likely localized on or very near the spore's outer surface [5,13,14]. Loss of either of these proteins has no notable effect on spore resistance

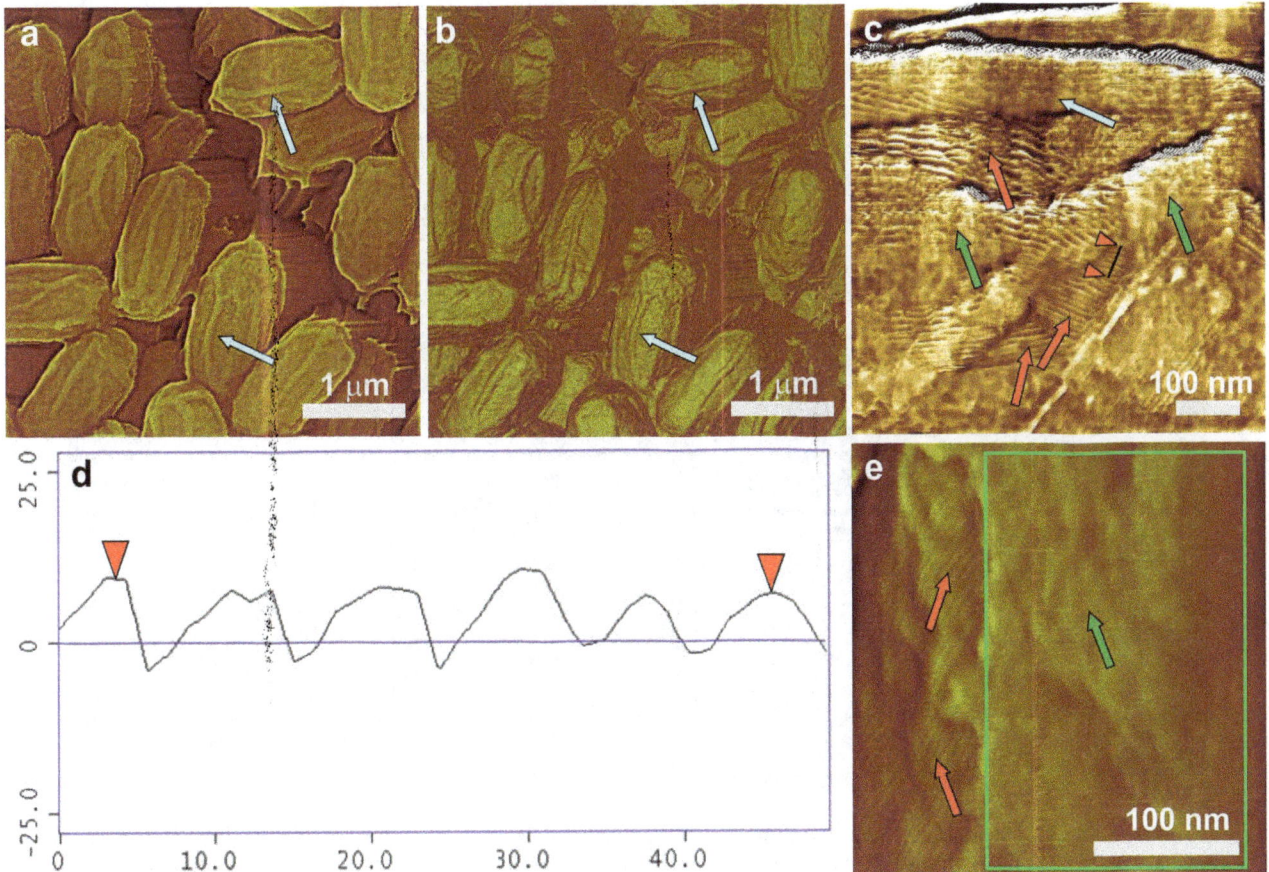

Figure 1. AFM images of *B. subtilis* wild-type spores. (a) Height and (b) phase images of spores with surface ridges (coincidental in both images) extending along the entire length of spores (several surface ridges noted by light blue arrows). (c) High-resolution height image of an area on a surface of a single spore showing surface ridges (light blue arrow), patches of an amorphous outermost layer (green arrows), and a rodlet layer (red arrows) seen beneath the amorphous layer. (d) A cross section line drawn perpendicular to rodlets (indicated with red arrows in (c)) showing a periodicity of ~8.2 nm. (e) High-resolution height image of an area on the surface of a single spore showing patches of an amorphous outermost layer (green arrow and green rectangle), and a rodlet layer (red arrows) seen beneath the amorphous layer.

properties or gross spore coat structure. We found that both *cotA* and *cotB* spore morphologies were indistinguishable from wild-type spores by AFM (Fig. 3), as all *cotA* and *cotB* spores were encased in the outermost amorphous and rodlet layers (Fig. 3c,d; green and red arrows, respectively) and exhibited 20–40 nm thick surface ridges (Fig. 3a–d; light blue arrows). The *cotA* and *cotB* spores also had an undulating surface topography from a subsurface layer (Fig. 3c,d; red circles) that was also seen in wild-type spores (data not shown).

In contrast to CotA and CotB, SafA plays a significant role in the assembly of at least some components of the spore's outer coat, and much of the coat in *safA* spores does not adhere tightly and can peel off [4,45]. We observed that the general surface morphology of *safA* spores as seen by AFM (Fig. 4) appears similar to that of wild-type, *cotA* and *cotB* spores, with amorphous and rodlet layers (Fig. 4c; green and red arrows, respectively) forming the outermost *safA* spores' coat layer. However, the degree of *safA* spore coat folding was different from that in wild-type spores. This resulted in the formation of surface ridges in *safA* spores (Fig. 4a–c; light blue arrows) that appeared shorter (e.g. not running along the whole spore surface as in Fig. 1a) and smaller (ridge heights of 10–20 nm) than in wild-type spores. Furthermore, some *safA* spores had no or minimal surface ridges (Fig. 4a; dark blue arrow), and ~25% of *safA* spores had an oversized spore coat

sacculus that appeared not to be firmly attached to the body of the spore itself (Fig. 4a,b; spores with adjacent green stars, and data not shown), consistent with previous work [44].

Surface architecture of spores lacking CotO and CotH

In addition to SafA, CotO and CotH also play significant roles in outer coat assembly, with perhaps some role in inner coat assembly as well [26,46]. As seen by AFM (Fig. 5), the outer surface of *cotO* spores was covered either completely or partially by a layer with a grainy appearance (Fig. 5b; brown arrow) and exhibited 15–40 nm thick ridges (Fig. 5a; light blue arrows). The thickness of the grainy layer was 8–20 nm as measured from the AFM images. High-resolution imaging of areas where the grainy structure density was low revealed that this layer actually has a fibrous structure, with the thickness of the thinnest fibers being ~2–4 nm (Fig. 5c; several fibers indicated with light yellow arrows). Thus, high densities of these fibrous structures appear to have assembled on the inner coat to form a layer that has a granular structure (Fig. 5b,c). Underneath the granular structure, multiple structural layers were observed (Fig. 5c; terraces of 3 consecutive layers numbered 1–3, and the edge of one terrace indicated by a purple arrow), and these terraces were decorated with "nanodot" particles (Fig. 5b,c; groups of nanodots indicated with black arrows, and a circle in 5c). While the heights of some

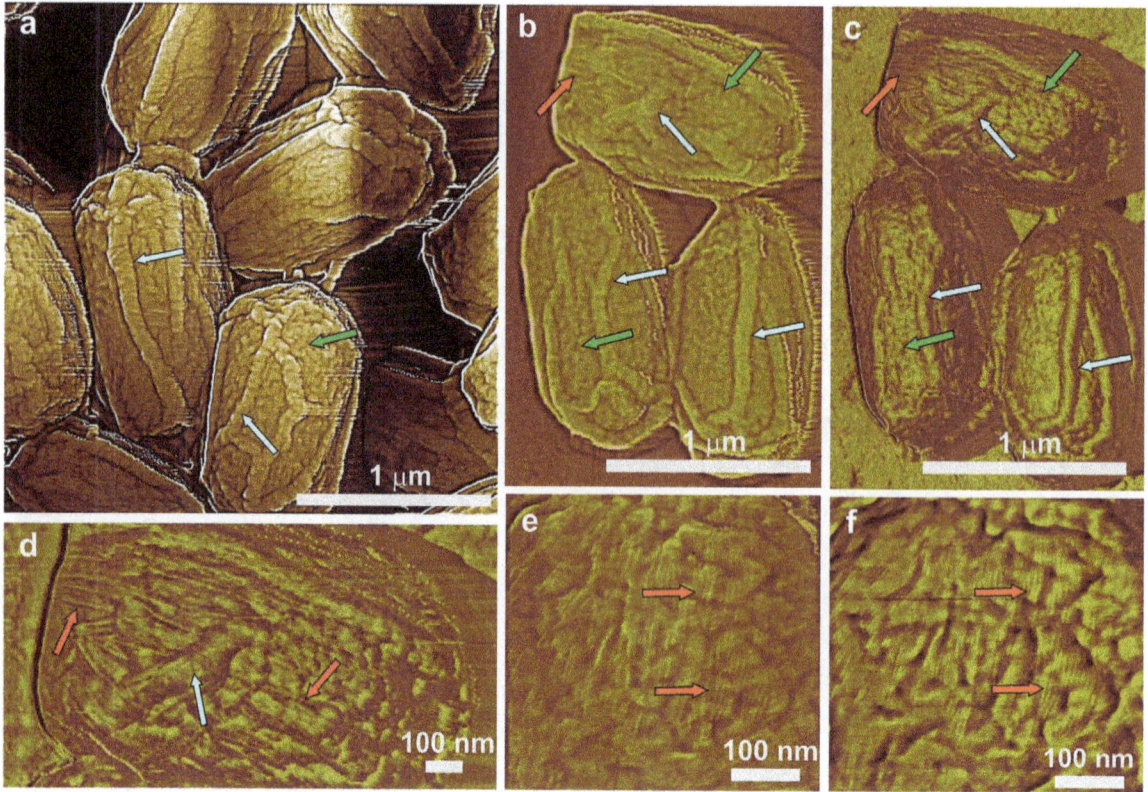

Figure 2. AFM images of decoated *B. subtilis* **wild-type spores.** Surface ridges extending along the entire length of spores are indicated with light blue arrows in height (a, b) and phase (c, d) images. Patches of rodlet structures are indicated with red arrows in (b–d). The green arrows in (a–c) indicate remnants of the amorphous outermost layer. High resolution height (e) and phase (f) images showing coincidental patches of rodlet structures denoted with red arrows.

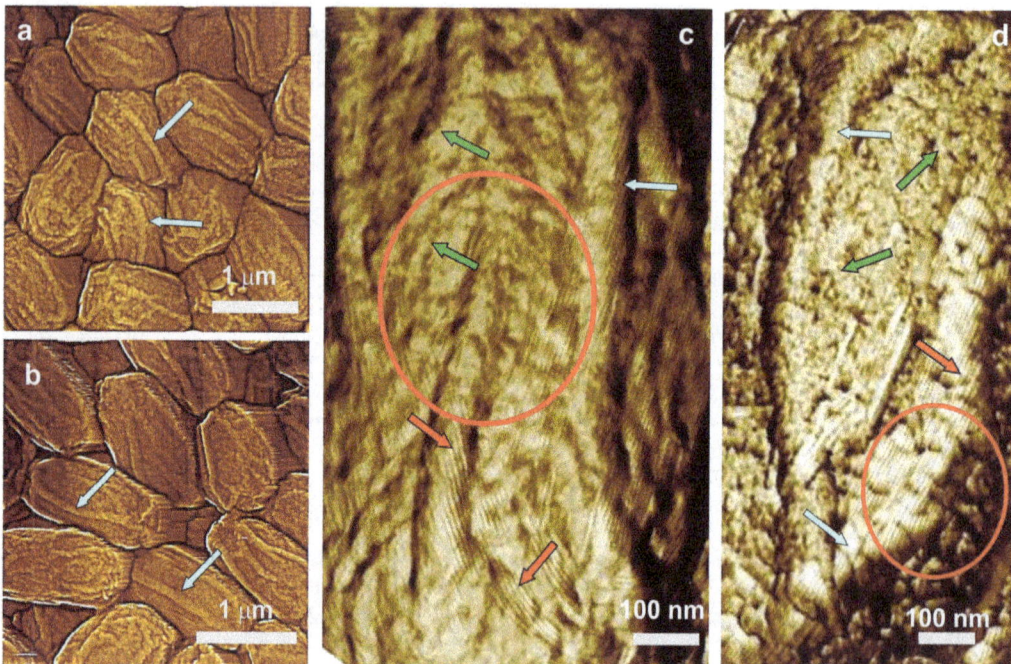

Figure 3. AFM images of *cotA* **and** *cotB* **spores.** Height images of *cotA* (a) and *cotB* (b) spores exhibit surface ridges similar to those in wild-type spores (light blue arrows). High-resolution phase images of single *cotA* (c) and *cotB* (d) spores show an irregular outermost amorphous layer (green arrows) as well as underlying rodlets (red arrows). In addition to the amorphous layer and rodlets seen on these spores' outermost surface, a strong undulating topography from a sub-surface layer is also present (red circles). Surface ridges in (c,d) are indicated with light blue arrows.

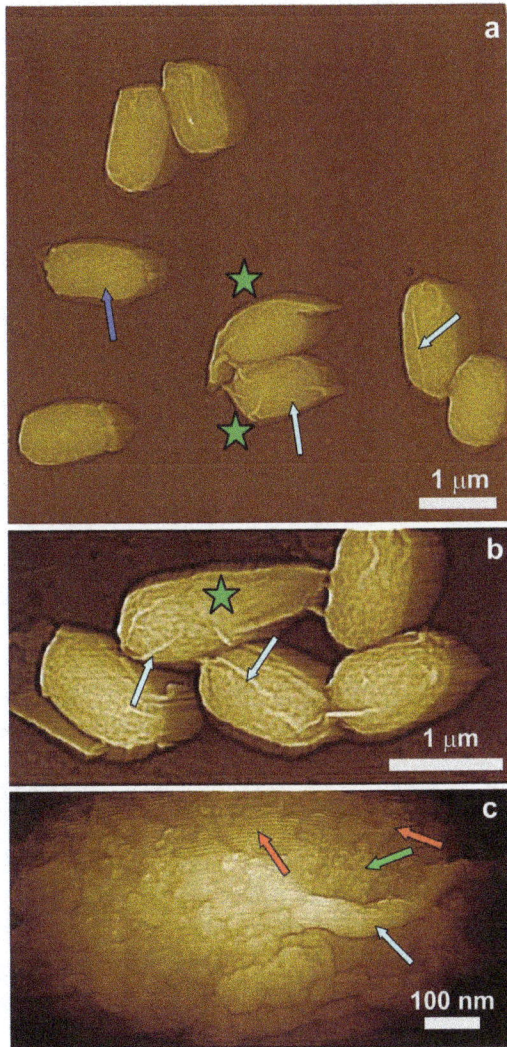

Figure 4. AFM height images of *safA* spores. (a–c) Several surface ridges are indicated with light blue arrows, and in (a) and (b) spores with an oversized sacculus are marked with adjacent green stars. A spore with at most minimal ridges is indicated with a dark blue arrow in (a). In panel (c), two patches of rodlet structure are indicated with red arrows, and a patch of an amorphous layer is indicated with a green arrow.

nanodots were as small as 3–4 nm, their typical height was 10–22 nm.

Significant numbers of *cotH* spores (Fig. 6) were also encased in the outermost amorphous and rodlet layers (Fig. 6b; green arrow and red arrows, respectively) and exhibited 15–40 nm thick surface ridges (Fig. 6a,b; light blue arrows). However, 10–15% of *cotH* spores were partially (Fig. 6b) or completely (Fig. 6c) devoid of outer spore coat layers. These *cotH* spores with a defective outer coat exhibited multilayer structures (Fig. 6c; two layers indicated with purple arrows) similar to ones observed on *cotO* spores (Fig. 5b,c). As seen in Fig. 6b,c, these layers again exhibited high densities of nanodots (Fig. 6b; one group of nanodots indicated with a black arrow) similar to ones seen on *cotO* spores (Fig. 5). Nanodot heights appeared smaller and more uniform on *cotH* spores compared with those on *cotO* spores, varying between 2.5–3.5 nm, and none of the *cotH* spores exhibited the fibrous/granular structural layer observed on *cotO* spores.

Surface architecture of *cotE*, *gerE* and *cotE gerE* spores

CotE is one of the major morphogenetic proteins in spore coat assembly, and *cotE* spores lack an outer coat and also have alterations in the inner coat layer [2]. AFM images (Fig. 7) revealed that the outermost surface of *cotE* spores is also a multilayer structure, composed of ~6 nm thick smooth layers (Fig. 7c; three consecutive layers marked as 1–3). These structures are identical to ones observed for *cotO* and *cotH* spores (Fig. 5,6). Note also that: i) the surface of *cotE* spores was devoid of nanodots; ii) the vast majority of *cotE* spores appeared to lack the outermost amorphous and rodlet layers; and iii) *cotE* spores exhibited no granular/fibrous surface structures.

In contrast to wild-type spores, 20–25% of *cotE* spores had no surfaces ridges (Fig. 7a; dark blue arrow) or shorter, thinner ridges (Fig. 7a; light blue arrows) that did not extend across the entire spore surface. The thickness of surface ridges that were seen were only 5–15 nm, less than for surface ridges on intact and decoated wild-type spores (Fig. 1,2). The other interesting morphological feature observed on many *cotE* spores was an oversized spore coat sacculus (Fig. 7a,b; green stars). This was also seen on some *safA* spores (Fig. 4a,b; adjacent green stars), while the wild-type spore coat was always tightly fitted (Fig. 1).

The multilayer outer structure of *cotE* spores (Fig. 7b,c) exhibited step growth patterns similar to those observed on surfaces of inorganic [47,48] and macromolecular [49–52] crystals. An example of similar structures observed on the surface of a growing trypsin crystal [53] is shown in Fig. 7d. Similar patterns were also observed for the inner coat of *C. novyi* NT [31] and *B. anthracis* spores (Plomp and Malkin, unpublished data). As seen in Fig. 7c, layers of structure forming the inner coat of *B. subtilis* spores are similar in morphology to the surface of trypsin crystals (Fig. 7d), with both showing rough steps with many kinks and a number of 5–10 nm (Fig. 7c) and 70–90 nm (Fig. 7d) wide holes (Fig. 7c,d; purple arrows and circles, respectively). The sinuosity index, which is a further measure of the step roughness (see Methods), was estimated for steps on surfaces of *cotE* spores (Fig. 7c) and trypsin crystals (Fig. 7d) as 3.84 and 1.49 respectively. Note that high-resolution AFM observations, which allow at least 1 nm resolution for macromolecular crystalline layers [50,54], do not result in molecular scale visualization of the molecular packing within the spore coat layers.

While most *cotE* spores are encased only in the multilayer coat structure, <5% spores were completely covered by a rodlet layer (Fig. 8a,b; red arrow), with a periodicity of ~7.2 nm (Fig. 8c; insert). Occasionally, as seen in Fig. 8a,b 4–10 nm thick patches of the outermost amorphous layer were observed atop the rodlet layer of <10% of *cotE* spores (Fig. 8a,b; green arrows). In addition, on <5% of *cotE* spores a honeycomb-like coat layer with a periodicity of ~8–9 nm was observed atop the inner coat layer (Fig. 9a,b, orange arrow). Note, that loose honeycomb layers with remnants of rodlet structures on top of a honeycomb layer (Fig. 9c, orange and red arrows respectively), were occasionally observed in spore preparations. While >75% of *cotE* spores lacked a complete rodlet layer, these spores still exhibited patches of rodlet structure of different sizes assembled atop the inner spore coat layer (Fig. 9a; red arrows).

In contrast to CotE and other proteins noted above, GerE is a transcription factor that modulates the expression of some coat protein genes late in sporulation, including genes that encode proteins in the insoluble fraction of the spore coat [6]. A *gerE* mutation has drastic effects on overall spore coat structure, as: i) much of the *gerE* spores' coat adheres poorly [55]; and ii) some coat component(s) responsible for the strong X-ray scattering by the spore coat is either absent or misassembled on *gerE* spores,

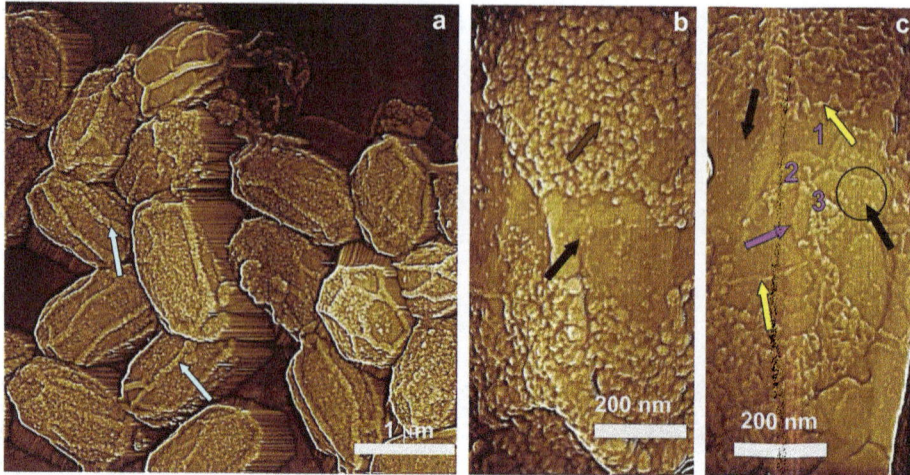

Figure 5. AFM images of *cotO* spores. (a) Height image of spores with surface ridges extending along the entire length of spores (light blue arrows). (b,c) High-resolution height images of areas on surfaces of single spores showing a dense fiber structure forming a granular structure (b; brown arrow) and individual fibers (c; light yellow arrows). In panel (c), three layers (terraces) of inner coat structure are numbered 1, 2, and 3 in purple. Step edges representing boundaries of each layer (one marked with a purple arrow) are visible. In panels (b) and (c), nanodots are marked with black arrows and one area with a high density of nanodots is circled in panel (c).

while this X-ray scattering is observed from *cotE* spores [56]. As seen by AFM (Fig. 10), *gerE* spores were devoid of the outer amorphous and rodlet layers, and fibrous structures. Most of these spores were only partially or completely covered by patches of irregular material (Fig. 10a; black stars; Fig. 10b), and 40–45% of *gerE* spores had only patches of this material (Fig. 10a; grey stars; Fig. 10c; grey arrow), with a thin layer of material covering the spore surface (Fig. 10b,c). The thickness of these patches of coat material was ~6 nm, a value similar to the thickness of the inner coat layers forming the multilayered coat structure (Fig. 5b, 7d).

The combination of *cotE* and *gerE* mutations has an even more drastic effect on spore coat structure than either mutation alone, as *cotE gerE* spores are almost completely devoid of a coat (Fig. 11),

except for a thin rind of insoluble material [28]. As reported previously [28], with the exception of small numbers of spores which have remnants of coat material (Fig. 11b; grey arrow), >90% of *cotE gerE* spores had none of the spore coat structures described above and their outer surface appeared rather smooth (Fig. 11a,b), although high-resolution imaging revealed a slightly bumpy textured outermost surface (Fig. 11c). The RMS roughness R_q of *cotE gerE* spore surfaces measured for 20 spores as described in the Methods section varied between 0.25 nm to 0.49 nm with average R_q value of ~0.38 nm. These severely coat-defective spores also appeared less rigid than intact spores, as *cotE gerE* spores within a closely packed monolayer were more deformed compared to ones with fewer near neighbors (Fig. 11a). Approx-

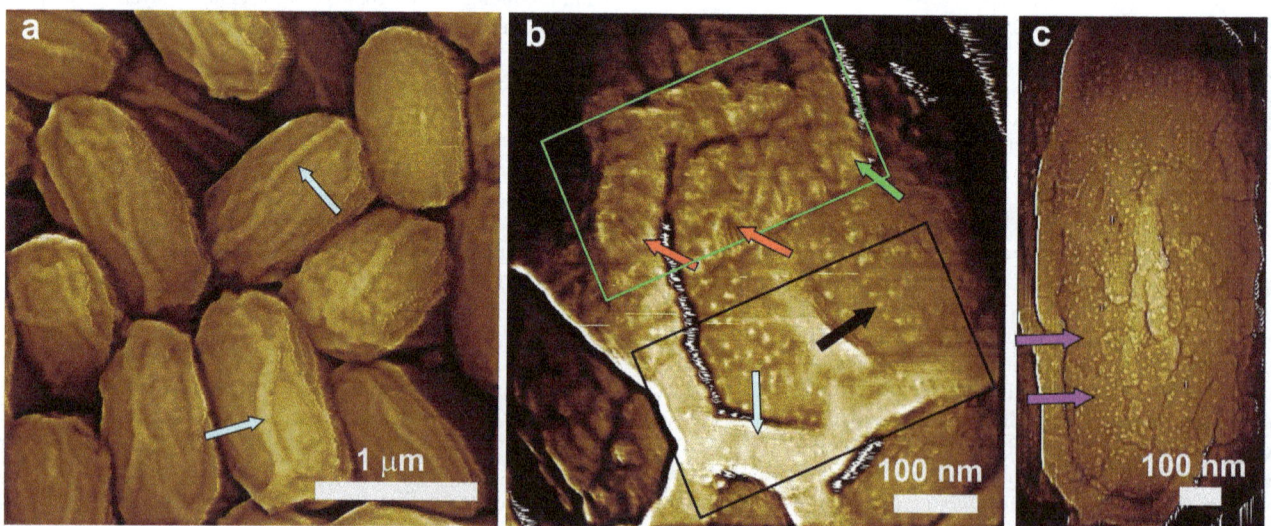

Figure 6. AFM images of *cotH* spores. (a) Height image of spores with surface ridges extending along the entire length of spores (light blue arrows). (b) High-resolution height image of a spore surface area showing the upper surface area (green rectangle) covered with an amorphous layer (green arrow) and rodlets (red arrows). The lower part of the outermost layer-free area (black rectangle) is covered with nanodots (black arrow). One of the surface ridges in (b) is indicated with a light blue arrow. In panel (c), a two–layer inner coat structure (two purple arrows noting the two layers) decorated with nanodots can be seen.

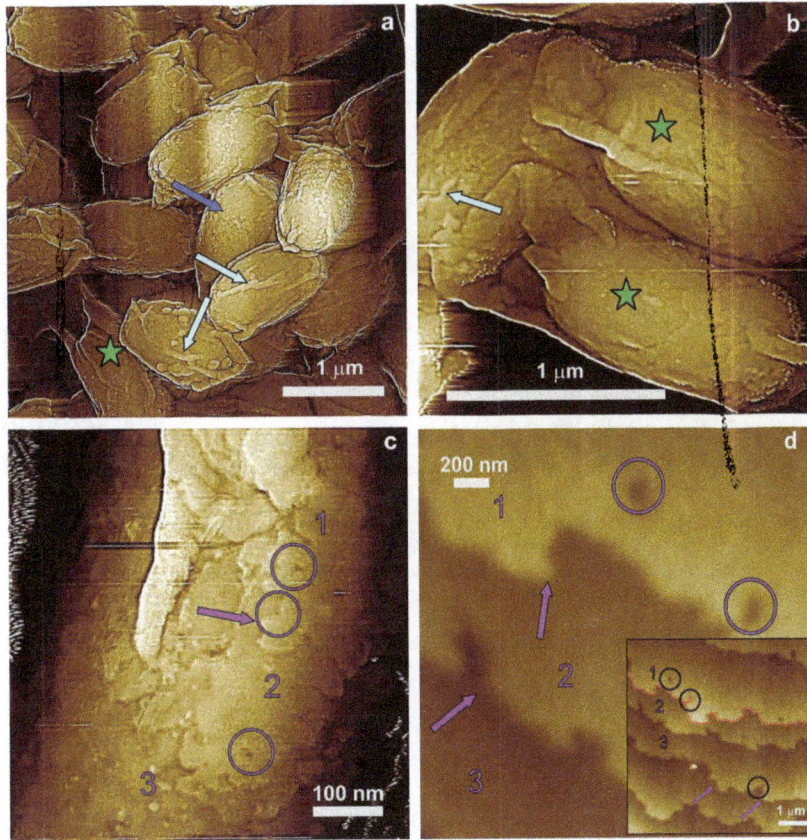

Figure 7. AFM images of *cotE* spores. (a,b) Height images of spores that exhibit surface ridges (light blue arrows), and several spores with an oversize sacculus are labeled with green stars. In (a) a spore with no apparent ridges is indicated with a dark blue arrow. (c) Height image of a multilayer inner coat structure. Three layers are indicated with numbers, and a kink on a step edge is marked with a purple arrow. Several holes in the layered structure are also indicated with purple circles. The hole in the middle circle corresponds to a pinning point on the step. (d) Height image of a multilayer layer structure similar to ones seen in Figs. 5b,c, 6c, and 7c, as seen on the surface of a trypsin crystal. Similar to the spore coat layers in (c), three layers, kinks and several holes are indicated with purple numbers, arrows and circles, respectively. The insert in (d) is a larger area of the crystal surface seen in (d). The same holes and three layers seen in (d) are indicated in the insert. The red line in (d) denotes the step contour, which was utilized for the measurement of the sinuosity index. Panel (d) is reprinted with permission from Plomp M, McPherson A, Larson SB, Malkin AJ (2001). Growth mechanisms and kinetics of trypsin crystallization. J Phys Chem B 105: 542–551. [52]. © (2001) American Chemical Society.

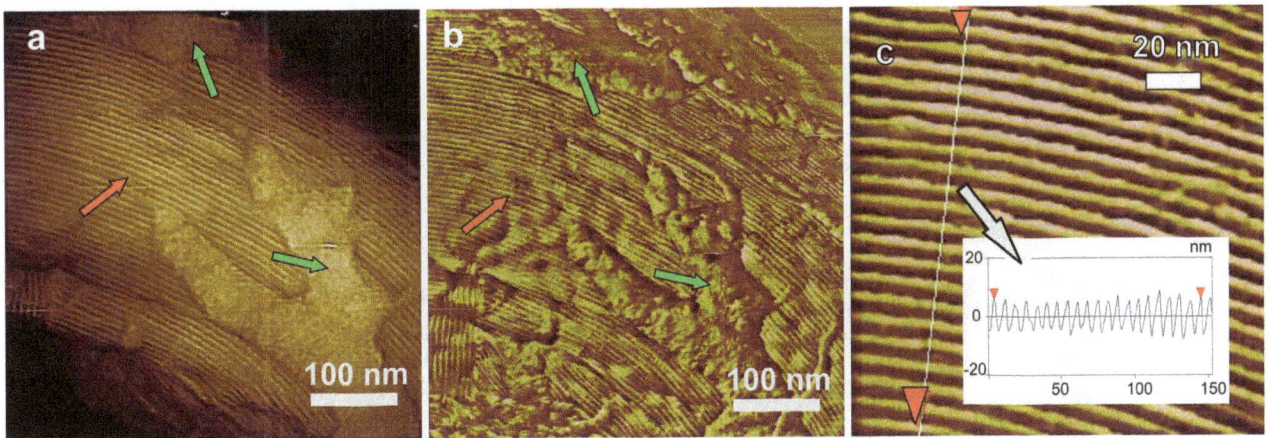

Figure 8. AFM images of *cotE* spores. (a) High-resolution height (a) and phase (b) images of the spore surface showing (coincidental in both images) a rodlet structure (red arrows) covered with patches of an amorphous layer (green arrows). (c) High-resolution height image of the spore surface with an insert with a cross section line drawn perpendicular to rodlets showing the periodicity of ~7.2 nm.

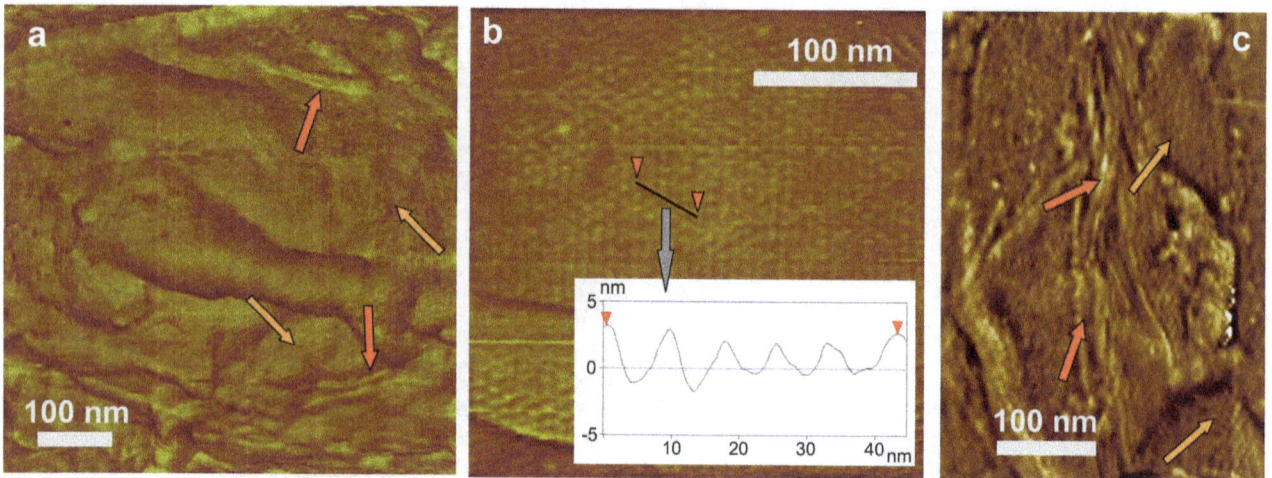

Figure 9. AFM images of *cotE* spores. (a,b) High-resolution height images of the spore surface showing a honeycomb structure (orange arrows in (a)) and patches of rodlets on top seen in (a) (red arrows). The insert in (b) is a cross section line along a honeycomb structure (indicated with a black line and red arrows in (b)) showing a periodicity of ~8.5 nm. (c) A portion of a loose honeycomb layer (orange arrows) with remnants of rodlet structures (red arrows), which were seen in *cotE* spore preparations.

imately 7% of *cotE gerE* spores also exhibited 80–100 nm wide and 30–40 nm deep depressions (Fig. 11a; black circles), which were also observed on some *gerE* spores (data not shown). Note, that neither *gerE* nor *cotE gerE* spores exhibited surface ridges (Fig. 10,11).

Surface architecture of *spoVID* spores

SpoVID is another major morphogenetic protein in spore coat assembly. This protein is essential for the adherence and assembly of the coat, and while the peptidoglycan cortex forms relatively normally in *spoVID* spores, the coat largely assembles as swirls in the cytoplasm, giving rise to spores with little coat material [2,10]. Consequently, the surface architecture of *spoVID* spores is drastically different from that of wild-type spores, as a number of *spoVID* spores were again encased in only loosely fitted coat sacculi (Fig. 12a; green stars). Indeed, for a number of *spoVID* spores, the coat sacculi were partially (Fig. 12 c,d; grey stars) or completely (Fig. 12b,d; white stars) sloughed off, releasing empty

sacculi (Fig. 12a, insert; dark blue star) and leaving spores encased in what appeared at lower resolution to be a rather smooth structure (Fig. 12a). Note that the shape of a number of the coatless *spoVID* spores was altered significantly compared either to other mutant spores described above or to *spoVID* spores still encased in coat sacculi. The shape of the coatless *spoVID* spores also varied significantly (Fig. 12b; spores with white stars), sometimes having a shape resembling a bowling pin.

The outer and internal surface structures of the coat sacculi released from *spoVID* spores were similar to the outermost surface structure of wild-type spores, as seen in a high-resolution image of a *spoVID* spore sacculus (Fig. 13a), and consisted of rodlet layers (Fig. 13a; red arrows) covered with amorphous material (Fig. 13a; green arrows). As illustrated in Fig. 13b, high-resolution images of the surfaces of the coatless *spoVID* spores revealed a 2–6 nm thick amorphous layer (Fig. 13b; grey arrow) and an underlying pitted surface structure (pink arrow).

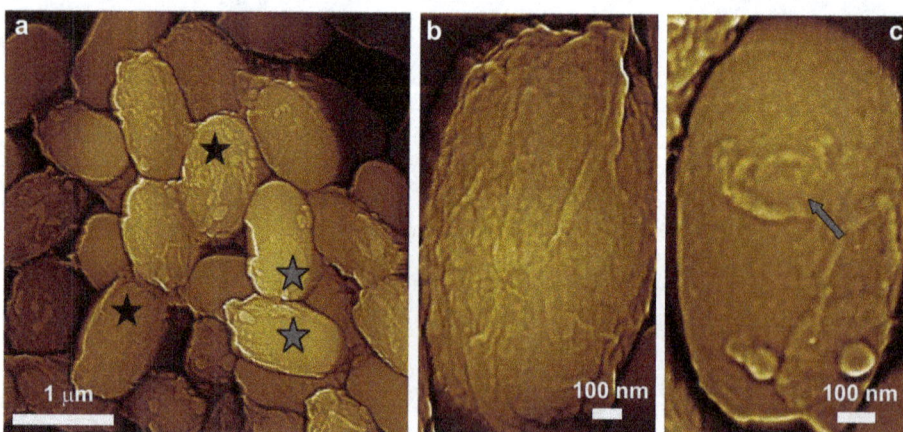

Figure 10. AFM height images of *gerE* spores. (a) *gerE* spores are either completely (black stars) or partially (grey stars) covered with coat material. (b) A spore that is completely encased in the coat material, and (c) a spore with patches of coat material (grey arrow).

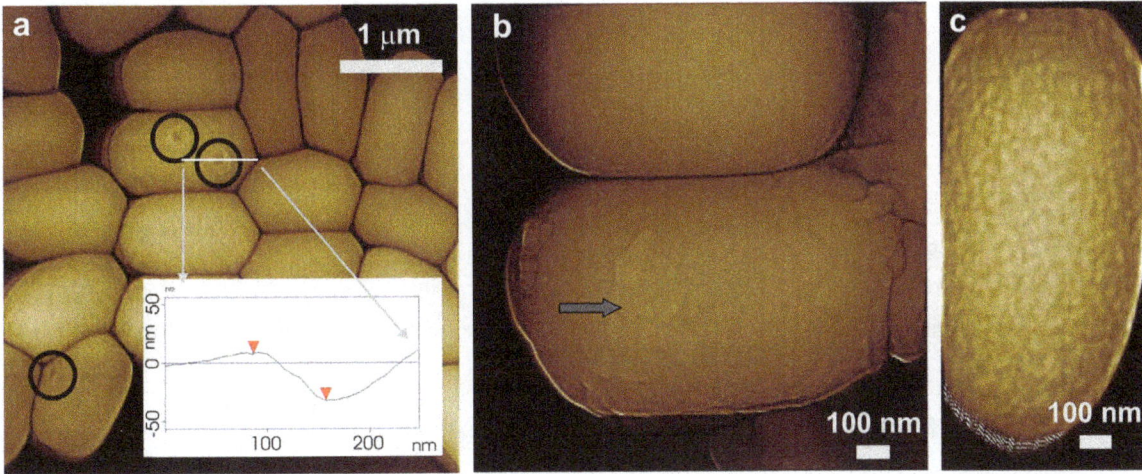

Figure 11. AFM height images of *cotE gerE* spores. (a) Spores which appeared to be devoid of spore coat material. Closely packed spores are more deformed than ones that are not surrounded by other spores. Some spores exhibit 80–100 nm wide and 30–40 nm deep depressions (black circles). The insert in (a) is a cross section line (indicated with a white line) drawn across the ~100 nm wide depression showing a depth of ~40 nm. (b) Image showing small patches of coat material (grey arrow) on the spore surface. (c) High-resolution image of a spore devoid of any obvious coat material, and showing a textured outermost surface.

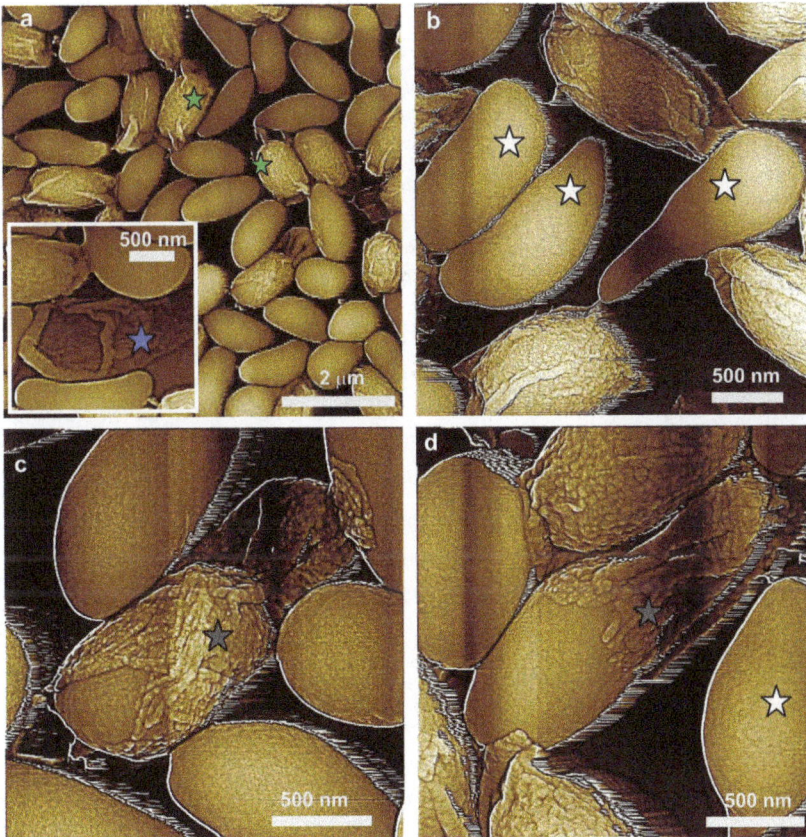

Figure 12. AFM height images of *sspoVID* spores. (a) Many of the *spoVID* spores are devoid of obvious spore coat material, although some *spoVID* spores are encased in loosely fitting coat sacculi (green stars); insert: an empty intact sacculus (blue star) present in a spore preparation. (b,d) Severely deformed spores without any visible coat material are indicated with white stars. (c, d) Spores with partially sloughed off coat sacculii are indicated with grey stars.

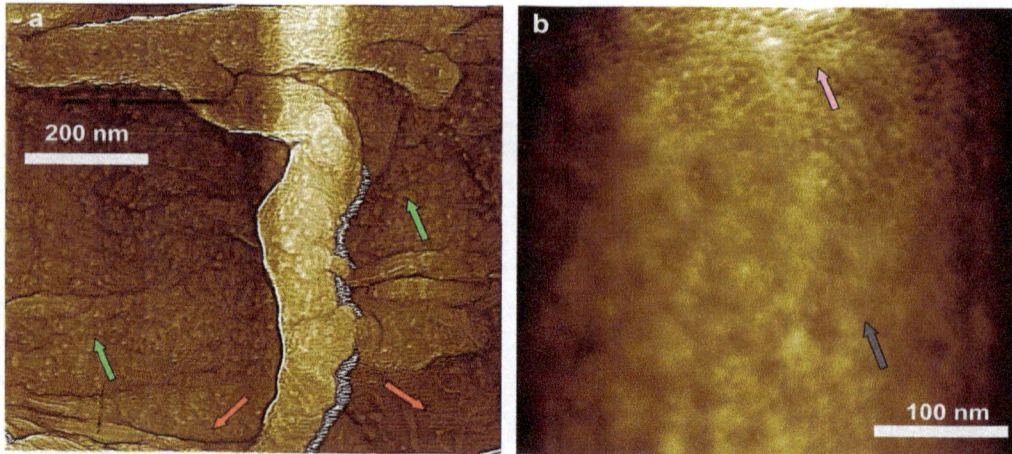

Figure 13. High-resolution AFM height images of *spoVID* **spores.** (a) External and internal surfaces of the empty coat sacculus in Fig. 12a, insert exhibit morphology similar to that of the outermost wild-type spore layer seen in Fig. 1b. The surface is comprised of rodlets (red arrows) and patches of amorphous material (green arrows). In (b) a pitted layer (pink arrow) is seen beneath a layer of coat material (grey arrow).

Discussion

Topography of the outer spore surface

The ridges on the surfaces of air-dried *B. subtilis* spores (Fig. 1a) have been seen previously in EM [39,40] and AFM [22–24,29,30] studies of spores of various *Bacillus* species. These ridges have been proposed to form due to the folding of the coat in response to dehydration, likely as a consequence of decreases in spores' internal volume [22,37,38]. Indeed, AFM measurements of the morphology of fully hydrated and air-dried spores demonstrate that surface ridges on dehydrated spores mostly disappear or decrease in size upon hydration [22]. Thus, spore coat flexibility can compensate for decreases in spore surface area upon drying by surface folding and ridge formation [22]. Current work demonstrated that this surface folding takes place in the spore coat, since dry *gerE*, *cotE gerE* and *spoVID* spores lacking much of the spore coat exhibit no surface ridges (Fig. 10–12). Note, that while *gerE* and *cotE gerE* spores as well as coat rinds produced by protozoal

digestion of spores exhibit some coat material [1,28] (Fig. 9,10), the amount of this material is either not sufficient or its proteomic composition is not appropriate to form surface ridges. In contrast, the presence of surface ridges on *cotO*, *cotH* and *cotE* spores lacking the amorphous and rodlet layers (Fig. 5–7) indicates again that surface ridge formation takes place within the multilayer spore coat structure.

The spore coats of *B. subtilis* mutants lacking morphogenetic coat proteins as well as chemically decoated wild-type spores have different thickness and composition, and these variations could affect the coat's elastic properties and thus change its folding and surface ridges. However, affecting the outer spore coat architecture by mutation or chemical decoating gave no large changes in spore surface ridge parameters or patterns. These results suggest that formation of spore surface ridges originates within multilayer coat structures, which are relatively unaffected by loss of some coat proteins, with the amorphous, rodlet and fibrous layers only following the ridge-associated topography. Our data showing pronounced changes in the surface folding of *safA* spores (Fig. 4) may indicate that these spores' multilayer coat structure is either thinner or more flexible than in wild-type, decoated wild-type, *cotA*, *cotB*, *cotH*, and *cotO* spores (Fig. 1–3,5,6). However, *cotE* spore coats with lower levels of surface folding typically have the same number of layers as do *cotH* and *cotO* spore coats. Perhaps the decreased surface folding of *cotE* spore coats is due to changes in the elastic properties of inner coat layers because one or more inner coat proteins are not assembled in *cotE* spores. Note, that the wide range of surface ridge parameters and folding patterns observed with spores of different species [22–24] and isogenic strains (this work) makes it problematic to assign these parameters as spore species-specific structural attributes.

Spore coat architecture

In addition to providing information on spore surface topography, AFM images allowed construction of a detailed model of coat architecture (Fig. 14). In this model starting from the outside, the coat consists of an amorphous layer/crust, a rodlet layer, a honeycomb layer, a fibrous layer, a nanodot particle layer, the multilayer assembly, and the undercoat/basement layer just above the pitted surface, which we tentatively assign as the spore cortex.

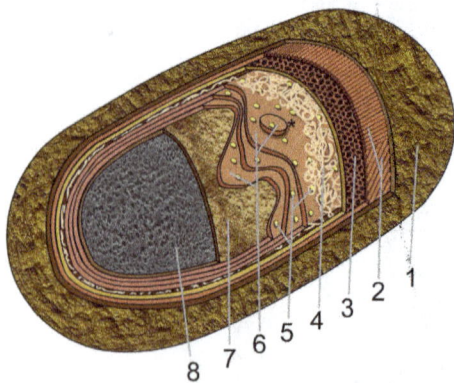

Figure 14. Model of the spore coat architecture of a single *B. subtilis* spore. The layers of the spore coat and the cortex are depicted as: (1) an outermost amorphous layer (the crust); (2) the rodlet layer; (3) the honeycomb layer; (4) the fibrous/granular layer, (5) the nanodot layer on top of a multilayer structure (6) ((with a 2D nucleus (indicated with *) seen on the upper layer)); and the basement layer (7), which is on the top of the cortex's outer pitted surface (8). Structural features of spore coat layers are not shown to scale.

The existence of an outermost tightly fitting spore layer was initially reported in thin section EM images of *B. subtilis* spores treated with a reducing agent [57]. It was suggested that this layer is an exosporium-like structure, which in EM images of untreated spores is usually indistinguishable from the darkly stained outer spore coat. While the amorphous spore coat layer reported here (Fig. 1) could correspond to this outer coat structure, it does not resemble an exosporium, as this outer layer has no paracrystalline basal layer typical of the exosporium of spores of the *B. cereus* group [41,58]. Rather the amorphous layer likely corresponds to the outer crust layer of *B. subtilis* spores that stains with ruthenium red and is glycoprotein-rich [7,8]. Patches of an outermost amorphous layer are also observed in AFM studies of *B. atrophaeus* spores (Plomp and Malkin, unpublished data).

Rodlet structures, similar to ones seen in Fig. 1b,c, were previously described on the outer surface of a diverse set of microorganisms (see [59]), including Gram-negative bacteria and various fungi. The fungal rodlet layers were resistant to treatment by detergents, organic solvents, enzymes, alkali and mild acids [60,61], and the structural proteins hydrophobins [62,63] and chaplins [64] were integral components of fungal rodlet structures. In several cases, these rodlets had a cross β-structure similar to that in the amyloid fibrils [64] associated with several neurodegenerative diseases [65]. The amyloid-like rodlet fibrils forming microbial outer surface layers appear to play important roles in attachment, dispersal and pathogenesis [59].

Rodlet structures were also reported in EM [38–40] and AFM studies [22–24,30] of spores of several *Bacillus* species, although the proteins that form these rodlet structures are not known. The structural similarities between *B. atrophaeus* rodlets seen during their germination-induced disassembly [25] and amyloid rodlets found on surfaces of fungi and bacteria suggest that *B. subtilis* coat rodlets may also be amyloids. However, full understanding of the function of rodlet structures in spores awaits elucidation. Interestingly, micro-etch pits form in the rodlet layer early in spore germination [25], and these could facilitate access of degradative enzymes to their targets in an otherwise tightly packed coat. Characterization of the strength and mechanical stiffness of individual amyloid fibrils of insulin reveals that these parameters are similar to those of steel and silk [66]. Thus, the spore coat' rodlet layer could play a role in protecting spores from mechanical stress, and a combination of rodlet and amorphous structures could provide spores with a wide range of physicochemical properties. Indeed, the existence of both hydrophobic (rodlets) and hydrophilic (glycoproteins) structures on the outermost layer might enable spores' successful dissemination as both as air-born and fully hydrated particles.

Current results indicate that the assembly of the outermost coat layer does not require CotA and CotB, two of the most abundant outer coat proteins [2,14,67–71]. While interactions between CotB and CotG are critical in guiding assembly of the outer coat layer, no coat assembly defect has been observed in *cotA* or *cotB* mutants [69,70]. In addition, *cotA* and *cotB* mutations have no effects on spore lysozyme resistance, germination [2] or surface appearance (Fig. 3), and CotB is absent from *cotH* spores [72] yet the outermost coat structure of CotH spores is similar to that of wild-type spores (Fig. 6), including both the rodlet and amorphous layers. Thus, neither CotA nor CotB appear to play important roles in directing the assembly of spores' outer layers. The similar surface ridges on *cotA* and *cotB* spores further suggests that loss of these proteins does not significantly alter the elastic properties of the spore coat.

Loss of SafA also does not affect the high-resolution architecture of spores' amorphous and rodlet layers (Fig. 4), consistent with

safA spores' lysozyme resistance [2,45]. However, SafA plays an important role in coat assembly, as in many *safA* spores the coat is loosely attached to the spore (Fig. 4). SafA is localized in the spore cortex near the inner coat, and SafA may help associate the spore cortex and coat [2,73]. The absence of surface ridges on a large portion of *safA* spores, along with relatively thinner existing surface ridges, also suggest that *safA* spores' coat is thinner and/or more flexible. This is consistent with EM analyses that indicate the *safA* spore coat often has 1–2 layers instead of the typical 3–5 layers [45].

Inner coat structures

In *B. cereus* [22] and *B. atrophaeus* spores [25] the coat's rodlet layer is underlain by a honeycomb structure also observed in *B. subtilis* spores (Fig. 9). Since disordered microporous inorganic substrates can effectively initiate three-dimensional protein crystallization [74], perhaps the spore coat's honeycomb structure represents a biological example of a microporous matrix that facilitates the ordered self-assembly of the coat's rodlet structure. Note that the 8–9 nm periodicity of the *B. subtilis* coat honeycomb layer is similar to periodicities of honeycomb structures for *B. cereus*, *B. thuringiensis* [22] and *C. novyi* NT [31] spores. This indicates that molecular dimensions of proteins forming these honeycomb structures are similar for different bacterial species, and the molecular composition of the honeycomb layer in different bacterial species may thus be similar.

Studies of *cotO*, *cotH* and *cotE* spores revealed consecutive structural layers of granular/fibrous material (*cotO* spores; Fig. 5), nanodots (*cotO* and *cotH* spores; Fig. 5, 6), and multilayer structures (*cotO*, *cotH* and *cotE* spores; Fig. 5–7). While spores of these mutants had the multilayer structure, only *cotO* spores retained the granular/fibrous structure and *cotE* spores lacked the nanodot layer. We propose that the granular/fibrous layer represents an outer spore coat layer that appears as a darkly stained irregular layer in EM images [69]. The thickness of this outer coat layer varies significantly in EM images on both the same spore and between spores, consistent with the range of granular/fibrous layer thickness observed on *cotO* spores.

On wild-type spores and spores of some mutants lacking specific coat proteins (i.e. *cotA* and *cotB*), the grainy/fibrous outer coat layer was largely obscured by the rodlet and amorphous layers. However, the force exerted by the AFM probe tip on the outermost spore layer allows visualization of underlying structures, as in the AFM visualization of a cytoskeleton beneath a cellular plasma membrane [75]. AFM phase imaging can probe micromechanical properties of sample materials (e.g. viscoelasticity) [76] and map surface inhomogeneity of these properties. Furthermore, when mechanical properties of two layers are significantly different, phase imaging can provide structural information on layers beneath the topmost layer [77]. Thus an irregular grainy layer can often be seen in AFM phase images beneath the outer rodlet structure (Fig. 3c,d), and we suggest that this underlying layer corresponds to a grainy/fibrous outer coat layer (Fig. 5, 414). Note, that an undulating surface morphology similar to that seen on *cotA* and *cotB* spores (Fig. 3c,d) was also observed on the surface of wild-type spores (data not shown).

Typically, multilayer structures on *cotO*, *cotH* and *cotE* spores contained 3–5 layers, consistent with the appearance of the lightly staining lamellar inner coat of *B. subtilis* spores seen by EM [68], and thus these multilayer structures may correspond to the *B. subtilis* spores' inner coat. We further suggest that the nanodots between the outer and inner coat layers but absent on *cotE* spores, might be CotE molecules that facilitate the assembly of the grainy/granular outer coat layer. The height of the smallest nanodots seen on *cotH* spores was ~3 nm, consistent with CotE's mol wt of 20.9 kDa [2], and this suggestion is consistent with current models of *B. subtilis* spore coat assembly that have CotE positioned between the inner and outer coat layers [2,26]. However, these nanodots could also be small coat protein aggregates, and further experiments, perhaps using AFM-based immunolabeling techniques (29), will be needed to identify the protein(s) forming the nanodots.

The *cotO* spores have no amorphous or rodlet layers, which could explain the partial lysozyme sensitivity of *cotO* spores [26]. However, the presence of these outer layers on the majority of *cotH* spores (Fig. 6) is consistent with their relatively normal lysozyme resistance [1]. The outer coat of *cotO* spores often appears disorganized and missing in EM thin sections [26] and is generally indistinguishable from that of *cotH* spores. CotO and CotH are suggested to be localized below the coat surface [13,26] and to participate in a late phase of coat assembly. However, our AFM analyses showed pronounced differences between *cotO* and *cotH* spore coats. In particular, CotO plays a critical role in the assembly of the amorphous and rodlet layers, while assembly of the fibrous outer coat requires CotH and CotE. AFM studies also indicated that these proteins play a role in assembly of the coat's amorphous and rodlet layers, consistent with biochemical, genetic and EM studies [26,46,72]. It has been suggested [26] that CotO and CotH also play an important role in inhibiting the tendency of outer coat protein layers to stack up resulting in the polymerization of the coat layers into closed shells. However, AFM demonstrates that *cotO* and *cotH* spore coats self-assemble to form contiguous shells rather than disorganized coats. At the same time, many *cotE* spores exhibited only a loose coat sacculus (Fig. 7), indicating that CotE plays an important role in the assembly of the inner coat and/or its attachment to the cortex as noted above.

The crucial role for CotE and GerE in proper coat assembly was further highlighted by the AFM of *gerE* and *cotE gerE* spores. First, loss of *gerE* prevented formation of the outer coat, rodlet, and amorphous layers. Second, while most *gerE* spores are encased in a loose structure formed by what appeared to be patches of the inner coat (Fig. 10), these structures do not resemble the inner coat multilayer structures described above. The *cotE gerE* spores were devoid of the amorphous and rodlet layers, and both complete inner and outer coats, and these spores' surface exhibited some roughness (Fig. 11). This surface likely corresponds to the basement/undercoat layer [4]. Thus, both CotE and GerE are crucial in proper assembly of the inner coat. Note, also that *cotE gerE* spores are less rigid than *cotE* or *gerE* spores. This increased deformability is due either to the loss of the inner coat or a role for CotE in the assembly and elastic properties of the basement layer. The nature of the 80–100 nm wide and 30–40 nm deep depressions seen in Fig. 11a is unclear, but we speculate that these holes may facilitate germinant access to the spore inner membrane, and are perhaps associated with the GerP proteins important in germinant movement through spores' outer layers [78].

Another coat protein important for proper spore coat assembly and attachment to the cortex is SpoVID, as a large percentage of *spoVID* spores lacked obvious coat structures, with some encased in a misassembled sacculus composed of amorphous and rodlet structures (Fig. 12). The thickness of the sacculi walls varied between 15–30 nm, indicating that the sacculi could contain coat material in addition to the rodlet and amorphous layers (Fig. 13a). Note that none of the *spoVID* spores visualized in this study exhibited the multilayer inner coat structures seen on *cotO*, *cotH* and *cotE* spores indicating that the inner coat is absent on *spoVID* spores. Most *spoVID* sacculi were only loosely attached to the spore body and were partially sloughed off, exposing a relatively smooth spore surface (Fig. 12). These AFM data are consistent with observations of swirls of spore coat in *spoVID* mother cells [10] (Fig. 12a, insert) and that SpoVID is required for the stable attachment of the coat. High-resolution imaging of the surface of *spoVID* spores indicated the existence of two prominent layers (Fig. 13b). One layer (13b; square) could correspond to the basement layer [4] and a pitted layer (Fig. 13b; black arrow) could correspond to either a subbasement coat layer or the cortex. Note, that *spoVID* spores lacking sacculi exhibit very high deformability (Fig. 12).

During wild-type *B. subtilis* sporulation, proteins forming honeycomb and rodlet coat layers self-assemble on the outer spore coat layer. Based on AFM results with *cotE* spores, the complete outer coat layer is not essential for formation of patches of the honeycomb and rodlet coat layers. Thus, the underlying integument is not crucial for assembly of the rodlet and honeycomb layers. Proteins that form honeycomb and rodlet spore coat structures must therefore be present during *cotE* spore formation, and self-assemble on the outer spore surface producing amorphous and rodlet layers (Fig. 8). Indeed, during *B. thuringiensis* sporulation rodlet proteins can self-assemble on the underlying spore coat, or in either the mother cell cytoplasm or the sporulation medium [23]. Hydrophobins, which form fungal rodlet layers, also self-assemble into rodlet fibrils *in vitro* (for review see [59]).

The multilayer structure forming the inner coat of *B. subtilis* spores exhibits patterns similar to ones described for the inner coats of spores of *C. novyi* NT [31] and *B. anthracis* (Plomp and Malkin, unpublished data). These patterns are also similar to those observed on surfaces of inorganic and macromolecular crystals. In addition to growth steps, these patterns include two-dimensional (2D) nuclei and screw dislocations that are major growth sources of inorganic, organic, and macromolecular crystals [79]. The presence of these growth patterns plus the smooth appearances of coat layers strongly point to a crystalline nature [79] of *B. subtilis* inner coat layers. While no screw dislocation sources similar to ones observed on the *C. novyi* NT inner spore coat [31] were seen on *B. subtilis* spores, on some spores with a low density of the grainy/fibrous outer layer, circular 2D nuclei were observed on the inner coat (Fig. 15a; dark blue arrows). This indicates that *B. subtilis* spores could represent the first case of non-mineral 2D nucleation growth patterns in a biological organism.

The observations above strongly suggest that assembly of inner spore coat layers proceeds by formation of 2D nuclei and their subsequent growth, similar to the birth-and-spread growth mechanism of conventional and macromolecular crystals [51,79]. In this model, 2D crystal growth takes place by generation and subsequent spread of 2D nuclei that provide a new crystalline layer on crystalline surfaces. Subsequent formation and growth of new 2D nuclei on this layer result in the formation of a new crystalline layer. An example of such growth, showing 2D nuclei on the surface of a crystal of satellite tobacco mosaic virus that are similar to ones seen in Fig. 15a, is presented in Fig. 15b (dark blue arrows). Typically, 2D growth takes place at high supersaturation (e.g. protein and precipitant concentrations used in macromolecular crystallization) [51,52,79], suggesting that relatively high concentrations of inner coat protein(s) are present during *B. subtilis* sporulation.

Step edges seen on the inner coat of *B. subtilis* spores showed significant roughness with many kinks (Fig. 7b), suggesting that formation of the inner coat was strongly affected by impurities. Similar patterns have been described for a wide range of crystalline surfaces (illustrated in Fig. 7d), where adsorption of

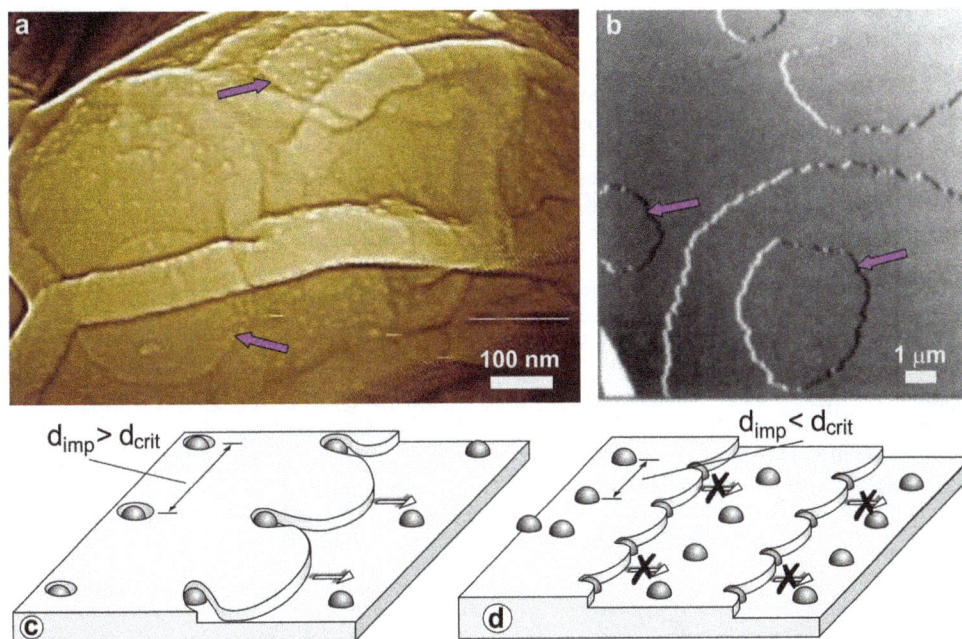

Figure 15. 2D nucleation and growth of inner spore coat layers. Panel (a) shows two putative 2D nuclei (purple arrows) on the inner coat surface of a *cotO* spore. Panel (b) shows 2D nuclei (purple arrows) on the surface of a satellite tobacco mosaic virus (STMV) crystal. This illustration is reproduced with permission from Malkin AJ, Kuznetsov YuG, Land TA, DeYoreo JJ, McPherson A (1995) Mechanisms of growth for protein and virus crystals. Nature Struct Biol. 2: 956–959 [48]. © (1995) Nature Publishing Group. (c) At a relatively small impurity (indicated as small balls) density, the average impurity distance d_{imp} is larger than d_{crit} and steps are able to advance. (d) At higher impurity densities, $d_{imp} < d_{crit}$, the curvature of step segments between impurities increases and steps are halted. Panels (c) and (d) are reproduced, with permission from Plomp M, McPherson A, Malkin AJ (2003). Repair of impurity-poisoned protein crystal surfaces. Proteins: Struct, Function, Bioinform 50: 486–495 [82]. © (2003) John Wiley and Sons.

impurities (ones present in solution), but not forming a layer at the step terraces and edges results in step roughening and cessation of growth [80–82]. Indeed the roughness and sinuosity of step edges on the inner coat of *B. subtilis* spores (Fig. 7c) are higher than observed for step edges on the surface of the trypsin crystal (Fig. 7d). This may indicate [80–83] that higher levels of impurities are adsorbed on the inner coat surface of *B. subtilis* spores compared to ones on the surface of trypsin crystals. Growth steps stop at sites of contact with impurity particles (indicated as small balls in Fig. 15c,d) that are adsorbed to the surface. However, portions of steps between neighboring impurity particles continue to grow, resulting in pinning of growth steps (Fig. 15c) as seen in Fig. 7c,d. Step advancement ceases (Fig. 15d) when at increased impurities' concentration, the distance between impurities/pinning points d_{imp} becomes smaller than the diameter of critical nuclei d_c necessary for step advancement [80]. One interesting feature of the inner coat is a number of ~5–10 nm holes (Fig. 7c), which may indicate locations of clusters of impurities [53,83]. Note, that in general the size of such holes is a function of the size of impurities or their clusters adsorbed on the surface. As described for a number of systems [80–83], these clusters of impurities may be responsible for pinning the advancement and cessation of growth of spore coat layers observed in Fig. 7c. Alternatively, such holes that were also observed on inner coat layers of *C. novyi* NT spores [31] and *B. anthracis* spores (Plomp and Malkin, unpublished data) could be an intrinsic feature of spore inner coat layers having a particular function. These results, combined with prior observation of screw dislocations on the inner coat of *C. novyi* NT spores [31], strongly suggest that inner spore coat assembly is governed by two crystallization mechanisms – growth on dislocations and 2D nucleation. These observations suggest that while spore coat proteins are produced

enzymatically [84], the assembly of these proteins into coat layers may be a self-assembly process similar to crystallization, and may be influenced by the sporulation conditions (protein and salt concentrations, pH, temperature, impurities) when these proteins assemble.

The lack of high-resolution crystalline lattice structures of the *B. subtilis* inner coat layers is similar to prior observations of *C. novyi* NT [31] and *B. anthracis* inner spore coat layers (Plomp and Malkin, unpublished data). It was suggested that proteins forming the *C. novyi* NT inner coat layers [31] are not globular, but rather peptides 'standing upright' in the layers, similar to peptide arrangements found in several organic crystals [85,86]. This hypothesis was based on the fact that for globular proteins, the ~6 nm height of the inner spore coat layers would not be considerably different in either perpendicular or lateral unit cell parameters, with the latter being amenable for AFM visualization [54]. Based on the lack of molecular scale AFM resolution of the crystalline lattice forming the *B. subtilis* inner coat layer, it is reasonable to suggest that proteins forming the inner coat might be also "standing upright" peptides [31,85,86].

In conclusion, the results presented in this communication provide further understanding of the structure and assembly of the *B. subtilis* spore coat. Furthermore, morphological and structural attributes of *B. subtilis* spores described here could thus serve as a baseline for future studies of effects of sporulation conditions on these structures. In addition, the similarities of some of the new findings with *B. subtilis* spores to findings with spores of *C. novyi* NT and other *Bacillus* species, suggest that the coat structure proposed in this work may generally be similar for spores of all of these species. While there is extensive knowledge of the individual proteins in the spore coat, as well as their location and assembly, there is much less knowledge of precise coat structure. In particular, the new high-resolution AFM studies have identified a number of new coat structural features, including the nanodots, the fibrous layer, and the terraced multilayer inner spore coat. Based on these results, we propose that the amorphous/crust layer and rodlets form the outermost spore structure, the fibrous layer and multilayer structure correspond to the outer coat and the inner coat respectively, with honeycomb and nanodot structures sandwiched between the outermost layer and the inner coat and the inner and the outer coats respectively.

Note, that high-resolution studies of fully hydrated *B. atrophaeus* [22,25] and *Clostridium novyi* NT spores [31] demonstrated that rodlet, honeycomb, and inner coat layer structures, similar to ones described here for *B. subtilis*, maintained the same patterns, lattice periodicities, and step heights as seen on air-dried spores.

Finally, the striking similarity between the appearance of the terraces and likely 2D nuclei in the multilayer inner coat and in inorganic and macromolecular crystals suggest that at least this part of the coat may assemble by crystallization mechanisms. A consequence of a crystallization spore coat assembly mechanism is that coat structure will be influenced by conditions during which these proteins self-assemble. In particular, variations in rates of 2D nucleation on spores could change the growth rate and hence the thickness of the spore coat, and this could influence spore properties such as their resistance and germination. The challenge now will be to correlate spore coat features identified in this work with specific coat proteins, and to understand how individual proteins contribute to these coat features, in particular, by using AFM-based immunolabeling techniques [29].

Acknowledgments

The authors are grateful to Patrick Eichenberger and Selim Elhadj for strain PE250 (ΔcotO::tet) and assisting in data analysis, respectively.

Author Contributions

Conceived and designed the experiments: AJM PS. Performed the experiments: MP AMC PS AJM. Analyzed the data: MP AJM AMC PS. Contributed reagents/materials/analysis tools: MP AMC PS AJM. Wrote the paper: MP PS AJM.

References

1. Klobutcher LA, Ragkousi K, Setlow P (2006) The *Bacillus subtilis* spore coat provides "eat resistance" during phagocytic predation by the protozoan *Tetrahymena thermophila*. Proc Natl Acad Sci USA 103: 165–70.

2. Henriques AO, Moran CP Jr (2007) Structure, assembly and function of the spore surface layers. Annu Rev Microbiol 61: 555–588.

3. Laaberki MH, Dworkin J (2008) Role of spore coat proteins in the resistance of *Bacillus subtilis* spores to *Caenorhabditis elegans* predation. J Bacteriol 190: 6197–6203.

4. McKenney PT, Driks A, Eichenberger P (2013) The *Bacillus subtilis* endospore: assembly and functions of the multilayered coat. Nature Rev Microbiol 11: 33–44.

5. McKenney PT, Eichenberger P (2012) Dynamics of spore coat morphogenesis in *Bacillus subtilis*. Mol Microbiol 83: 245–260.

6. de Hoon MJL, Eichenberger P, Vitkup D (2010) Hierarchical evolution of the bacterial sporulation network. Curr Biol 20: 735–745

7. McKenney PT, Driks A, Eskandarian HA, Grabowski P, Guberman J, et al. (2010) A distance-weighted interaction map reveals a previously uncharacterized layer of the *Bacillus subtilis* spore coat. Curr Biol 20: 934–938.

8. Waller LN, Fox N, Fox KF, Fox A, Price RL (2004) Ruthenium red staining for ultrastructural visualization of a glycoprotein layer surrounding the spore of *Bacillus anthracis* and *Bacillus subtilis*. J Microbiol Meth 58: 23–30.

9. Wang KH, Isidro AL, Domingues L, Eskandarian HA, McKenney PT, et al. (2009) The coat morphogenetic protein SpoIVD is necessary for spore encasement in *Bacillus subtilis*. Mol Microbiol 74: 634–649.

10. Beall BA, Driks A, Losick R, Moran CP Jr (1993) Cloning and characterization of a gene required for assembly of the *Bacillus subtilis* spore coat. J Bacteriol 175: 1705–1716.

11. Isticato R, Sirec T, Giglio R, Baccigalupi L, Pesce G, et al. (2013) Flexibility of the programme of spore coat formation in *Bacillus subtilis*: bypass of CotE requirement by overproduction of CotH. PLoS One 8: e74949.

12. Imamura D, Kuwana R, Takamatsu H, Watabe K (2010) Localization of proteins to different layers and regions of *Bacillus subtilis* spore coats. J Bacteriol 192: 518–524.

13. Imamura D, Kuwana R, Takamatsu H, Watabe K (2011) Proteins involved in formation of the outermost layer of *Bacillus subtilis* spores. J Bacteriol 193: 4075–4080.

14. Tang J, Krajcikova D, Zhu R, Ebner A, Cutting S, et al. (2007) Atomic force microscopy imaging and single molecule recognition force spectroscopy of coat proteins on the surface of *Bacillus subtilis* spore. J Mol Recog 20: 483–489.

15. Abhyankar W, Ter Beek A, Dekker H, Kort R, Brul S, et al. (2011) Gel-free proteomic identification of the *Bacillus subtilis* insoluble coat protein fraction. Proteomics 11: 4541–4550.

16. De Francesco M, Jacobs JZ, Nunes F, Serrano M, McKenney PT, et al. (2012) Physical interactions between coat morphogenetic proteins SpoIVD and CotE is necessary for spore encasement in *Bacillus subtilis*. J Bacteriol 194: 4941–4950

17. Kim H, Hahn M, Grabowski P, McPherson DC, Otte MM, et al. (2006) The *Bacillus subtilis* spore coat protein interaction network. Mol Microbiol 59: 487–502.

18. Krajcikova D, Lukacova M, Mullerova D, Cutting SM, Barak I (2009) Searching for protein-protein interactions within the *Bacillus subtilis* spore coat. J Bacteriol 191: 3212–3219.

19. Mullerova D, Krajcikova D, Barak I (2009) Interactions between *Bacillus subtilis* early spore coat morphogenetic proteins. FEMS Microbiol Lett 299: 74–85.

20. Qiao H, Krajcikova D, Xing C, Lu B, Hao J, et al. (2013) Study of the interactions between the key spore coat morphogenetic proteins CotE and SpoIVD. J Struct Biol 181: 128–135.

21. Chada VGR, Sanstad EA, Wang R, Driks A (2003) Morphogenesis of *Bacillus* spore surfaces. J Bacteriol 185: 6255–6261.

22. Plomp M, Leighton TJ, Wheeler KE, Malkin AJ (2005) The high-resolution architecture and structural dynamics of *Bacillus* spores. Biophys J 88: 603–608.

23. Plomp M, Leighton TJ, Wheeler KE, Malkin AJ (2005) Architecture and high-resolution structure of *Bacillus thuringiensis* and *Bacillus cereus* spore coat surfaces. Langmuir 21: 7892–7898.

24. Plomp M., Leighton TJ, Wheeler KE, Pitesky ME, Malkin AJ (2005) *Bacillus atrophaeus* outer spore coat assembly and ultrastructure. Langmuir 21: 10710–10716.

25. Plomp M., Leighton TJ, Wheeler KE, Hill HD, Malkin AJ (2007) *In vitro* high-resolution structural dynamics of single germinating bacterial spores. Proc Nac Acad Sci USA 104: 9644–9649.

26. McPherson DC, Kim H, Hahn M, Wang R, Grabowski P, et al. (2005) Characterization of the *Bacillus subtilis* spore morphogenetic coat protein CotO. J Bacteriol 187: 8278–8290.

27. Carroll AM, Plomp M, Malkin AJ, Setlow P (2008) Protozoal digestion of coat-defective *Bacillus subtilis* spores produces "rinds" composed of insoluble coat protein. Appl Environ Microbiol 74: 5875–5881.

28. Ghosh S, Setlow B, Wahome PG, Cowan AE, Plomp M, et al. (2008) Characterization of spores of *Bacillus subtilis* that lack most coat layers. J Bacteriol 190: 6741–6748.

29. Plomp M, Malkin AJ (2009) Mapping of proteomic composition on the surfaces of *Bacillus* spores by atomic force microscopy. Langmuir 25: 403–409.

30. Malkin AJ, Plomp M (2010) High-resolution architecture and structural dynamics of microbial and cellular system: Insights from high-resolution *in vitro* atomic force microscopy. In: Kalinin SV, Gruverman A, editors. Scanning probe microscopy of functional materials: nanoscale imaging and spectroscopy. New York: Springer. pp. 39–68.

31. Plomp M, McCaffery JM, Cheong I, Huang X, Bettegowda C, et al. (2007) Spore coat architecture of *Clostridium novyi* NT spores. J Bacteriol 189: 6457–6468.

32. Anagnostopoulos C, Spizizen J (1961) Requirements for transformation in *Bacillus subtilis*. J Bacteriol 81: 741–746.

33. Maniatis T, Fritsch EF, Sambrook J (1982) Molecular cloning: A laboratory manual. Cold Spring Harbor, NY: Cold Spring Harbor Laboratory. 545 p.

34. Nicholson WL, Setlow P (1990) Sporulation, germination and outgrowth. In: Harwood CR, Cutting SM, editors. Molecular biological methods for *Bacillus*. Chichester, UK: JohnWiley & Sons Ltd. pp. 391–450.

35. Ragkousi K, Setlow P (2004) Transglutaminase-mediated cross-linking of GerQ in the coats of *Bacillus subtilis* spores. J Bacteriol 186: 5567–75.

36. Monroe A, Setlow P (2006) Localization of the transglutaminase cross-linking sites in the *Bacillus subtilis* spore coat protein GerQ. J Bacteriol 188: 7609–16.

37. Westphal AJ, Price PB, Leighton TJ, Wheeler KE (2003) Kinetics of size changes of individual *Bacillus thuringiensis* spores in response to changes in relative humidity. Proc Natl Acad Sci USA 100: 3461–3466.

38. Driks A (2003) The dynamic spore. Proc Natl Acad Sci USA 100: 3007–3009.

39. Aronson AI, Fitz-James P (1976) Structure and morphogenesis of the bacterial spore coat. Bact Rev 40: 360–402.

40. Holt SC, Leadbetter ER (1969) Comparative ultrastructure of selected aerobic spore-forming bacteria: a freeze etching study. Bacteriol Rev 33: 346–378.

41. Wehrli E, Scherrer P, Kübler O (1980) The crystalline layers in spores of *Bacillus cereus* and *Bacillus thuringiensis* studied by freeze-etching and high resolution electron microscopy. Eur J Cell Biol 20: 283–289.

42. Dufrêne YF, Boonaert CJP, Gerin PA, Asther M, Rouxhet PG (1999) Direct probing of the surface ultrastructure and molecular interactions of dormant and germinating spores of *Phanerochaete chrysosporium*. J Bacteriol 181: 5350–5354.

43. Dufrêne YF (2004) Using nanotechnologies to explore microbial surfaces. Nature Rev Microbiol 2: 451–458.

44. Henriques AO, Moran CP Jr (2000) Structure and assembly of the bacterial endospore coat. Methods 20: 95–110.

45. Takamatsu H, Kodama T, Nakayama T, Watabe K (1999) Characterization of the *yrbA* gene of *Bacillus subtilis*, involved in resistance and germination of spores. J Bacteriol 181: 4986–4994.

46. Zilhão R, Naclerio G, Baccigalupi L, Henriques AO, Moran CP Jr, et al. (1999) Assembly requirements and role of CotH during spore coat formation in *Bacillus subtilis*. J Bacteriol 181: 2631–2633.

47. Maiwa K, Plomp M, van Enckevort WJP, Bennema P (1998) AFM observation of barium nitrate {111} and {100} faces: spiral growth and two-dimensional nucleation growth. J Cryst Growth 186: 214–223.

48. Rogilo DI, Fedina LI, Kosolobov SS, Ranguelov BS, Latyshev AV (2013) Critical terrace width for two-dimensional nucleation during Si growth on Si (111)-(7×7) surface. Phys Rev Lett 111: 036105.

49. Malkin AJ, Kuznetsov YuG, Land TA, DeYoreo JJ, McPherson A (1995) Mechanisms of growth for protein and virus crystals. Nature Struct Biol 2: 956–959.

50. Malkin AJ, Kuznetsov YuG, McPherson A (1999) *In situ* atomic force microscopy studies of surface morphology, growth kinetics, defect structure and dissolution in macromolecular crystallization. J Cryst Growth 196: 471–488.

51. Malkin AJ, McPherson A (2004) Probing of crystal interfaces and the structures and dynamic properties of large macromolecular ensembles with in situ atomic force microscopy. In: Lin XY, DeYoreo JJ, editors. From solid-liquid interface to nanostructure engineering, vol. 2. New York: Plenum/Kluwer Academic Publisher. pp. 201–208.

52. DeYoreo JJ, Vekilov PG (2003) Principles of crystal nucleation and growth. In: Dove PM, DeYoreo JJ, Weiner S, editors. Biomineralization. Washington, DC: Mineral Society of America. pp. 57–93.

53. Plomp M, McPherson A, Larson SB, Malkin AJ (2001). Growth mechanisms and kinetics of trypsin crystallization. J Phys Chem B 105: 542–551.

54. Kuznetsov YuG, Malkin AJ, Land TA, DeYoreo JJ, Barba AP, et al. (1997) Molecular resolution imaging of macromolecular crystals by atomic force microscopy. Biophys J 72: 2357–2364.

55. Moir A (1981) Germination properties of a spore coat-defective mutant of *Bacillus subtilis*. J. Bacteriol 146: 1106–1116.

56. Qiu X, Setlow P (2009) Structural and genetic analysis of X-ray scattering by spores of *Bacillus subtilis*. J Bacteriol 191: 7620–7622.

57. Sousa JCF, Silva MT, Balassa G (1976) Exosporium-like outer layer in *Bacillus subtilis* spores. Nature 263: 53–54.

58. Kaillas L, Terry C, Abbott N, Taylor R, Mullin N, et al. (2011) Surface architecture of endospores of the *Bacillus cereus/anthracis/thuringiensis* family at the subnanometer scale. Proc Natl Acad Sci USA 108: 16014–16019.

59. Gebbink MF, Claessen D, Bouma B, Dijkhuizen L, Wösten HA (2005) Amyloids – a functional coat for microorganisms. Nature Rev Microbiol 3: 333–341.

60. Hashimoto T, Wu-Yuan CD, Blumenthal HJ (1976) Isolation and characterization of the rodlet layer of *Trichophyton mentagrophytes* microconidial wall. J Bacteriol 127: 1543–1549.

61. Beever RE, Redgewell RJ, Dempsey G (1979) Purification and chemical characterization of the rodlet layer of *Neurospora crassa* conidia. J Bacteriol 140: 1063–1070.

62. Wessels JGH (1998) Hydrophobins: Proteins that change the nature of the fungal surface. Adv Microbial Physiol 38: 1–45.

63. Wösten HAB, de Vocht ML (2000) Hydrophobins, the fungal coat unravelled. Biochim Biophys Acta 1469: 79–86.

64. Claessen D, Stokroos I, Deelstra HJ, Penninga, NA Bormann C, et al. (2004) The formation of the rodlet layer of streptomycetes is the result of the interplay between rodlins and chaplins. Mol Microbiol 53: 433–443

65. Dobson CM (2003) Protein folding and misfolding. Nature 426: 884–890.

66. Smith JF, Knowles TPJ, Dobson CM, MacPhee CE, Welland ME (2006) Characterization of the nanoscale properties of individual amyloid fibrils. Proc Natl Acad Sci USA 103: 15806–15811.

67. Isticato R, Cangiano G, Tran HT, Ciabattini A, Medaglini D, et al. (2001) Surface display of recombinant proteins on *Bacillus subtilis* spores. J Bacteriol 183: 6294–6301.

68. Driks A (1999) *Bacillus subtilis* spore coat. Microbiol Mol Bio Rev 63: 1–20.

69. Zheng LB, Donovan WP, Fitz-James PC, Losick R (1988) Gene encoding a morphogenic protein required in the assembly of the outer coat of the *Bacillus subtilis* endospore. Genes Dev 2: 1047–1054.

70. Zilhão R, Serrano M, Isticato R, Ricca E, Moran CP Jr, et al. (2004) Interactions among CotB, CotG, and CotH during assembly of the *Bacillus subtilis* spore coat. J Bacteriol 186: 1110–1119.

71. Donovan W, Zheng LB, Sandman K, Losick R (1987) Genes encoding spore coat polypeptides from *Bacillus subtilis*. J Mol Biol 196: 1–10.

72. Naclerio G, Baccigalupi L, Zilhao R, de Felice M, Ricca E (1996) *Bacillus subtilis* spore coat assembly requires *cotH* gene expression. J Bacteriol 178: 4375–4380

73. Ozin AJ, Henriques AO, Yi H, Moran CP Jr (2000) Morphogenetic proteins SpoVID and SafA form a complex during assembly of the *Bacillus subtilis* spore coat. J Bacteriol: 1828–1833.

74. Frenkel D (2006) Physical chemistry - Seeds of phase change. Nature 443: 641–641.

75. Kuznetsov YuG, Malkin AJ, McPherson A (1997). Atomic force microscopy studies of living cells: Visualization of motility, division, aggregation, transformation and apoptosis. J Struct Biol 120: 180–191.

76. Magonov SN, Elings V, Whangbo MH (1997) Phase imaging and stiffness in tapping-mode atomic force microscopy. Surf Sci 375: L385–L391.

77. Magonov SN, Cleveland J, Elings V, Denley D, Whangbo M-H (1997) Tapping-mode atomic force microscopy study of the near-surface composition of a styrene-butadiene-styrene triblock copolymer film. Surf Sci 389: 201–211.

78. Butzin XY, Troiano AJ, Coleman WH, Griffiths KK, Doona CJ, et al. (2012) Analysis of the effects of a *gerP* mutation on the germination of spores of *Bacillus subtilis*. J Bacteriol 194: 5749–5758.

79. Chernov AA (1984) Modern crystallography. III. Crystal growth. Berlin: Springer-Verlag. 517 p.

80. Cabrera N, Vermilyea DA (1958) The growth of crystals from solution. In: Doremus RH, Roberts BW, Turnbul D, editors. Growth and perfection of crystals. New York: Wiley. pp. 393–410.

81. van Enckevort WJP, van der Berg ACJF, Kreuwel KBG, Derksen AJ, Couto MS (1996) Impurity blocking of growth steps: experiments and theory. J Cryst Growth 166: 156–161.

82. Land TA, Martin TL, Potapenko S, Palmore GT, DeYoreo JJ. (1999) Recovery of surfaces from impurity poisoning during crystal growth. Nature 399: 442–445.

83. Plomp M, McPherson A, Malkin AJ (2003). Repair of impurity-poisoned protein crystal surfaces. Proteins: Struct, Function, Bioinform 50: 486–495

84. Driks A (2002) Maximum shields: the assembly and function of the bacterial spore coat. Trends Microbiol 10: 251–254.

85. Hollander FFA, Plomp M, van de Streek CJ, van Enckevort WJP (2001) A two-dimensional Hartman-Perdok analysis of polymorphic fat surfaces observed with atomic force microscopy. Surf Sci 471: 101–113.

86. Plomp M, van Enckevort MJV, van Hoof PJCM, van de Streek CJ (2003) Morphology and dislocation movement in n-$C_{40}H_{82}$ paraffin crystals grown from solution. J Cryst Growth 249: 600–613.

87. Eichenberger P, Jensen ST, Conlon EM, van Ooij C, Silvaggi J, et al. (2003) The σ^E regulon and the identification of additional sporulation genes in *Bacillus subtilis*. J Mol Biol 327: 945–972.

## 13

A Thermostable *Salmonella* Phage Endolysin, Lys68, with Broad Bactericidal Properties against Gram-Negative Pathogens in Presence of Weak Acids

Hugo Oliveira[1], **Viruthachalam Thiagarajan**[2], **Maarten Walmagh**[3], **Sanna Sillankorva**[1], **Rob Lavigne**[3], **Maria Teresa Neves-Petersen**[4,5], **Leon D. Kluskens**[1], **Joana Azeredo**[1]*

1 Centre of Biological Engineering, University of Minho, Braga, Portugal, 2 School of Chemistry, Bharathidasan University, Tiruchirappalli, India, 3 Laboratory of Gene Technology, Katholieke Universiteit Leuven, Leuven, Belgium, 4 Nanomedicine Department, International Iberian Nanotechnology Laboratory, Braga, Portugal, 5 Medical Faculty, Aalborg University, Aalborg, Denmark

Abstract

Resistance rates are increasing among several problematic Gram-negative pathogens, a fact that has encouraged the development of new antimicrobial agents. This paper characterizes a *Salmonella* phage endolysin (Lys68) and demonstrates its potential antimicrobial effectiveness when combined with organic acids towards Gram-negative pathogens. Biochemical characterization reveals that Lys68 is more active at pH 7.0, maintaining 76.7% of its activity when stored at 4°C for two months. Thermostability tests showed that Lys68 is only completely inactivated upon exposure to 100°C for 30 min, and circular dichroism analysis demonstrated the ability to refold into its original conformation upon thermal denaturation. It was shown that Lys68 is able to lyse a wide panel of Gram-negative bacteria (13 different species) in combination with the outer membrane permeabilizers EDTA, citric and malic acid. While the EDTA/Lys68 combination only inactivated *Pseudomonas* strains, the use of citric or malic acid broadened Lys68 antibacterial effect to other Gram-negative pathogens (lytic activity against 9 and 11 species, respectively). Particularly against *Salmonella* Typhimurium LT2, the combinatory effect of malic or citric acid with Lys68 led to approximately 3 to 5 log reductions in bacterial load/CFUs after 2 hours, respectively, and was also able to reduce stationary-phase cells and bacterial biofilms by approximately 1 log. The broad killing capacity of malic/citric acid-Lys68 is explained by the destabilization and major disruptions of the cell outer membrane integrity due to the acidity caused by the organic acids and a relatively high muralytic activity of Lys68 at low pH. Lys68 demonstrates good (thermo)stability properties that combined with different outer membrane permeabilizers, could become useful to combat Gram-negative pathogens in agricultural, food and medical industry.

Editor: Kornelius Zeth, University of the Basque Country, Spain

Funding: This work was supported by the projects FCOMP-01-0124-FEDER-019446, FCOMP-01-0124-FEDER-027462 and PEst-OE/EQB/LA0023/2013 from "Fundação para a Ciência e Tecnologia" (FCT), Portugal. The authors thank the Project "BioHealth - Biotechnology and Bioengineering approaches to improve health quality", Ref. NORTE-07-0124-FEDER-000027, co-funded by the Programa Operacional Regional do Norte (ON.2 – O Novo Norte), QREN, FEDER. Hugo Oliveira acknowledges the FCT grant SFRH/BD/63734/2009. Maarten Walmagh held a PhD scholarship of the IWT Vlaanderen. The funders had no role in study design, data collection and analysis, decision to publish, or preparation of the manuscript.

Competing Interests: The authors have declared that no competing interests exist.

* Email: jazeredo@deb.uminho.pt

Introduction

Gram-negative bacterial pathogens are common causes of food-borne (e.g. *Salmonella*, *Escherichia coli* O157:H7, *Shigella*) and hospital-acquired (e.g. *Pseudomonas*, *Acinetobacter*) infectious diseases [1,2]. In an era in which the threat of antibiotic and multi-resistant bacteria is increasing and solutions are becoming scarce, it is important to search for alternative antimicrobials.

One promising alternative approach to prevent or destroy pathogenic bacteria is the use of bacterial cell wall hydrolases. These enzymes cause bacteriolysis by degrading the peptidoglycan (PG) layer, also known as murein, the major component of the bacterial cell wall and responsible for the mechanical integrity. Among these bacterial cell wall hydrolases, an increased interest has been given to bacteriophage (phage) endolysins. Endolysins are specialized PG-degrading enzymes, part of a universal lytic cassette system encoded by all double stranded DNA phages and expressed during the terminal stage of the reproduction cycle [3].

In contrast to their extensively knowledge and successful use in fighting Gram-positive pathogens (e.g. *Streptococcus*, *Staphylococcus* and *Bacillus*) [4], the application of endolysins against Gram-negative pathogens has been impaired by the presence of a protecting outer membrane (OM), preventing their entry into the cell and reach the PG [5]. EDTA is a well-known outer membrane permeabilizer (OMP) that acts as a chelator by removing stabilizing cations from the OM, notably Ca^{2+} and Mg^{2+} [10]. EDTA (at 0.5 mM) is the only OMP agent used so far, to enhance the activity of these bacterial cell wall hydrolases through OM permeabilization, although with a moderate and narrow bacterial host effect limited to *P. aeruginosa* species [6,7,8,9]. Organic acids are also reported to be membrane-active agents and hence potential permeabilizers [10]. Although some organic acids have, to a lesser extent, chelating properties, additional acidity can contribute to OM disruption [11]. Examples are citric acid (at 2 mM) and malic acid (at 5 mM) that were recently shown to weaken the OM of Gram-negative bacteria [12]. They are natural occurring compounds, versatile and widely used in food, cosmetic and pharmaceutical industries [10].

In this work it was intended to (*i*) biochemically characterize a new *Salmonella* phage endolysin (further abbreviated as Lys68), to enlarge the knowledge of a group of endolysins from Gram-negative affecting phages that remain scarcely explored and (*ii*) to develop an efficient anti-Gram-negative pathogen strategy by combining the endolysin with citric or malic acid and compared with EDTA, as it is the only OMP described so far to act synergistically with endolysins. We further present an explanatory hypothesis of the mechanism involved in the high and broad lytic ability of this endolysin in the presence of organic acids.

Materials and Methods

Bacteria, phage and chemicals

Bacterial strains used in this work are listed in **Table S1** and belong to the Centre of Biological Engineering bacterial collection (Braga, Portugal). All strains were grown in Lysogeny broth (LB) (Liofilchem, Italy) at 37°C and 120 rpm, with exception of *P. fluorescence* at 25°C. The *Salmonella* phage phi68 was isolated from faeces from a poultry farm (Braga, Portugal), and was recently characterized [13]. Hen egg white lysozyme (HEWL) (Fisher Scientific, USA), EDTA (Pronalab, Mexico) and HEPES, citric and malic acid (Sigma-Aldrich, USA) were purchased from the specified suppliers.

Cloning, protein overexpression and purification

The isolated *Salmonella* phage phi68 was partially sequenced showing resemblance to the sequenced *Salmonella enterica* serovar Enteritidis typing phage SETP3 (data not shown). Based on phage SETP3 genomic sequence available at NCBI database (reference sequence: NC_009232.2), primers (Invitrogen) were designed to amplify the putative endolysin gene by PCR from the isolated phage phi68 genomic DNA template, using Phusion High-Fidelity DNA Polymerase (NEB, UK). Forward primer (AGATAT<u>CA</u><u>TATG</u>TCAAACCGAAACATTAGC) and reverse primer (GTGGT<u>GCTCGAG</u>CTACTTAG) contained *Nde*I and *Xho*I restriction sites, respectively (underlined) [13]. The PCR amplification product was purified (DNA Clean & Concentrator-5k, Zymo Research, USA) and digested using *Nde*I and *Xho*I enzymes (NEB, UK), and cloned in the pET-28a expression vector (Novagen) with an N-terminal His$_6$ tag. The presence of the insert in the plasmid was confirmed by DNA sequencing (Macrogen, Amsterdam) (GenBank accession number for Lys68 nucleotide sequence, KJ475444).

E. coli BL21(DE3) harboring the endolysin vector was grown in 600 mL Luria Broth (LB) supplemented with 50 μg/mL kanamycin to an optical density (OD$_{600 \text{ nm}}$) of 0.6 (4 h, 120 rpm at 37°C). Recombinant protein expression was induced for 18 h at 16°C by the addition of isopropyl-β-D-thiogalactopyranoside (IPTG) to a final concentration of 0.5 mM. The culture was then centrifuged (9500×*g*, 30 min) and cells were disrupted by resuspending the pellet in 25 mL of lysis buffer (20 mM NaH$_2$PO$_4$, 0.5 M NaCl/ NaOH, pH 7.4), followed by three cycles of freeze-thawing (−80°C to room temperature). Maintaining the sample on ice, cells were further disrupted by sonication (Cole-Parmer, Ultra-sonic Processors) for 8–10 cycles (30 s pulse, 30 s pause). Insoluble

cell debris were removed by centrifugation (9500×*g*, 30 min, 4°C). The supernatant was collected and filtered (0.22 μm filters), and applied to Ni^{2+}-NTA resin stacked in 1 mL HisTrap HP columns (GE Healthcare, Waukesha, WI, USA) for purification, using protein-dependent imidazole concentrations in the washing and elution buffer (20 mM NaH$_2$PO$_4$, 0.5 M NaCl/NaOH, pH 7.4, 25–300 mM imidazole). Eluted protein was then dialysed against 10 mM PBS at pH 7.2 (using Maxi GeBAflex-tube Dialysis Kit - Gene Bio-Application L.T.D) and the protein concentration was determined using the BCA Protein Assay Kit with bovine serum albumin (BSA) as standard (Thermo Scientific).

Site-directed mutagenesis

To establish the identification of the putative catalytic residues within the endolysin sequence, two active site mutations (Glu18Ala and Thr35Ala) were introduced. Two sets of overlapping mutagenic primers (5'-CGCGGCATTC<u>GC</u>GGGGTTCCGGG, forward, 5'-CCCGGAACCCC<u>GC</u>GAA<u>T</u>GCCGCG, reverse, for Glu18Ala, and 5'-AGAATGAGAAGTACCTT<u>GCT</u>ATTGGC-TACGGCCAC, forward 5'-GTGGCCGTAGCCAAT<u>AG</u>-<u>C</u>AAGGTACTTCTCATTCT, reverse, for Thr35Ala, with mutation basepairs underlined), were applied using the Quick-Change II XL Site-Directed Mutagenesis Kit (Agilent Technologies). Plasmids containing mutated endolysin genes were then introduced in competent *E. coli* BL21(DE3) cells for heterologous production as described above.

Quantification and characterization of endolysin muralytic activity

The PG lytic (or muralytic) activity of the purified Lys68 was quantified using *P. aeruginosa* PAO1 cells permeabilized by chloroform/Tris treatments, resuspended in 80 mM phosphate buffer pH 7.2. This muralytic test has been extensively described for endolysins elsewhere [14]. Briefly, 30 μL of serial dilutions of enzyme were added to 270 μL of OM permeabilized cells. The activity of Lys68 was measured at room temperature through the decrease in OD$_{600 \text{ nm}}$ using a BIO-TEK Synergy HT Microplate Reader. The muralytic activity is based on the linear relation between lysis and concentration, according to a standardized calculation method described elsewhere [14]. Enzymatic activity is expressed as follows: activity (units/μM) = ((slope (OD$_{600 \text{ nm}}$/ min)/μM))/0.001), where the activity of 1 unit is defined as the concentration of enzyme (in μM) necessary to create a drop in OD$_{600 \text{ nm}}$ of 0.001 per minute. Obtained values for the negative control (30 μL of PBS pH 7.2) were subtracted from the sample values.

The pH dependence of the muralytic activity was assessed using the same 30 μL of serial dilutions of enzyme on 270 μL of OM permeabilized cells, but now resuspended in a universal pH buffer (10 mM KH$_2$PO$_4$, 10 mM Na-citrate and 10 mM H$_3$BO$_4$) adjusted to different pHs within the range of 3 and 10.

Thermostability was evaluated by incubating Lys68 first at several temperatures (40°C, 60°C, 80°C and 100°C) for 30 min and later at 100°C during different time intervals (5, 10, 20, 30 and 45 min) in a MJ Mini BIO-RAD Thermocycler. This was followed by a 20-min cooling step on ice, after which the activity was measured using 30 μL of 2 μM Lys68 on 270 μL of OM permeabilized cells resuspended in 80 mM phosphate buffer pH 7.2. The residual muralytic activity of each sample relative to the activity of unheated reference sample at time 0 (= 100% activity) was determined.

The determination of the endolysin lytic spectrum was conducted using the different strains listed in **Table S1**. Gram-negative cultures were grown overnight and diluted 1:100 the following day in fresh 5 mL LB and allowed to grow until reaching the mid-exponential phase (OD$_{600\ nm}$ of 0.6). Cells were then plated (100 μL) onto LB agar Petri dishes and allowed to grow for 8 h to form bacterial lawns. After, a step involving treatment with chloroform vapours was included to permeabilize the OM [15], prior to adding a 30-μL drop of 2 μM of Lys68. Lysis halos were visualized after 30 min incubation period.

Endolysin secondary structure characterization by circular dichroism

Circular dichroism (CD) experiments in the far- and near-UV region were performed using a Jasco J-815 CD spectrometer equipped with a water-cooled Peltier unit. The spectra were recorded in a cell width of 0.1-mm path length (110.QS, Hellma) from 185 to 360 nm for all proteins with 1 nm steps, scan speed of 20 nm/min, high sensitivity and a 16 s response time. Three consecutive scans for each sample and its respective buffer baseline were obtained. The averaged baseline spectrum was subtracted from the averaged sample spectrum measured under the same conditions. Secondary structure estimates were derived from the spectra using the CDSSTR [16] and CONTINLL [17,18] routine of the DICHROWEB [19,20] server run on the Set 4 optimised for a wavelength of 190–240 nm. Thermal denaturation/renaturation of the proteins was measured by monitoring the change in ellipticity in a cell width of 0.1 cm path-length at 222 nm over the range of 20°C to 75°C, in increments of 1°C/min. The experimental denaturation/renaturation profiles were analysed by a nonlinear least squares fit assuming a two-state transition and used to calculate the melting temperature (T_m). For CD measurements, a concentration of 8 μM of Lys68 was prepared in presence of an universal buffer (10 mM KH$_2$PO$_4$, 10 mM Na-citrate and 10 mM H$_3$BO$_4$) adjusted to different pH values within the range of 3 and 10 (for pH dependence tests) and in 80 mM phosphate buffer pH 7.2 (for thermostability tests). The same conditions were used for the analysis of the pH and temperature on the muralytic activity described above.

Endolysin antibacterial activity assay with OMPs

Gram-negative cells (listed in **Table S1**), grown overnight at 37°C, were diluted 1:100 in fresh LB and allowed to grow to an OD$_{600\ nm}$ of 0.6. Cells were then resuspended in 10 mM HEPES/NaOH (pH 7.2) to a final density of 10^8 colony forming units (CFU)/mL. Each culture (50 μL) was incubated for 30 min at room temperature with 25 μL of Lys68 (2 μM final concentration) together with 25 μL of water or 25 μL of OMPs. The final concentrations used were 0.5 mM EDTA, 2 mM of citric acid and 5 mM of malic acid (all prepared in water), concentrations described in literature as potential permeabilizers. Under the same conditions, the HEWL with similar catalytic activity (glycoside hydrolase) was assessed at a final concentration of 2 μM to compare with the Lys68 results.

Additionally and following the same protocol, the influence of different bacterial physiological states on the combinatorial effect of Lys68 and OMPs was evaluated after a 2 h incubation period against *Salmonella enterica* serovar Typhimurium LT2 cells grown for 3 h (planktonic exponential phase), 12 h (stationary phase) and 24 h (biofilm cells - in this case, 200 μL volume used in 96-microplate wells). In case of biofilm assays, 2 μL of overnight normalized cultures (OD adjusted to 1.0 in LB medium) were transferred to 96-well polystyrene plates (Orange Scientific) containing 198 μL of LB. To establish mature biofilms the plate was incubated for 24 h at 37°C and 120 rpms. Afterwards, the culture medium was removed and non-adherent cells removed by washing the biofilms twice with 200 μL of 10 mM HEPES/

NaOH (pH 7.2). Cells were then incubated with 100 μL of HEPES together with 50 μL Lys68 (2 μM final concentration) and 50 μL of citric and malic acid (2 mM and 5 mM final concentrations, respectively).

For all experiments, water (instead of EDTA/citric/malic acid) or PBS (instead of Lys68 or HEWL) was used as a negative control. The effect of the Lys68 and outer membrane permeabilizer mixtures on Gram-negative cells (planktonic and biofilms) was assessed by serial dilution plating in 10 mM PBS pH 7.2 buffer to quantify the number of CFUs. The antibacterial activity was expressed as the relative inactivation in logarithmic units (= log$_{10}$ (N$_0$/N$_i$) with N$_0$ = number of untreated cells (in the negative control) and N$_i$ = number of treated cells counted after incubation). Averages ± standard deviations for all experiments are given for n = 4 repeats.

OM permeabilization disruption mechanism

In order to assess the citric and malic acid permeabilization mechanism, three independent assays with *Salmonella* Typhimurium LT2 cells were made, as previously described with some modifications. Firstly, the reaction was supplemented with 5 mM of MgCl$_2$, an extra source of divalent cations. Secondly, the cells were incubated with 3.5 and 4 mM HCl instead of the OMP agents, to mimic the pH drop achieved by 2 mM citric acid (pH 4.2) and 5 mM malic acid (pH 3.8), respectively. All reactions followed an incubation period of 30 min, after which cell suspensions were diluted, plated on LB agar plates and the CFUs were counted as previously described.

Finally, a third complementary reaction was carried out, to visualize the OMPs effect. Here, *S.* Typhimurium LT2 cells were incubated with 2 mM citric acid and 2 μM KZ-EGFP protein (instead of Lys68). KZ-EGFP is a previously described PG binding domain from a *Pseudomonas* phage endolysin KZ144 coupled to a green fluorescent protein that has strong affinity for Gram-negative PG [21]. A negative wild-type EGFP control was also included for comparison. After a 30-min incubation period, the sample was washed twice in 10 mM HEPES/NaOH (pH 7.2). Then, a 30-μl drop was spotted onto a microscope slide and images were recorded using confocal laser scanning microscopy (FluoView FV1000 microscope, Olympus) in search for fluorescent *Salmonella*. *Salmonella* cells treated with water (instead of citric acid) and EFGP were used as a negative control.

Results

In silico analysis

The 489-bps endolysin gene (*lys68*) from *Salmonella* phage phi68, encoding a 162-amino acid protein with a deduced molecular mass of 19.6 kDa, was amplified and sequenced. Afterwards, a bioinformatic analysis was conducted to analyse the primary and secondary structures of Lys68 as well as a possible identification of its putative catalytic residues.

Primary structure comparisons using BlastP, revealed high homology (>90%) to six other endolysin proteins from *Salmonella* phages, namely SETP3, ST4, SE2, vB_SenS-Ent1, SS3e and wksl3, that have not yet been characterized *in vitro*. Lys68 is predicted to be a globular protein with a conserved domain between amino acids 1–151 and belongs to the glycoside hydrolase family GH24 (**Figure S1**) [22]. This class of enzymes degrades the cell wall by catalyzing the hydrolysis of 1, 4-linkages between *N*-acetylmuramic acid and *N*-acetyl-D-glucosamine residues.

Secondary structure analysis using HHpred output with PDB as a database, which contains all publicly available protein structures, identified endolysins from *Enterobacteria* phages P1, P22 and T4 as relevant homologs that have high α-helical content. This is in agreement with the 9 α-helices (54.9%) and 7 β-sheets (4.3%) regions predicted using the PSIPRED server (**Figure S1a**).

T4, P1 and P22 endolysins are well-known muramidases (GH24) with different catalytic residues identified. While in T4 endolysin experimental evidence shows a catalytic triad (Glu-11-Asp-20-Thr-26) being responsible for glycosidase reaction [23], the P1 and P22 active sites have only been predicted by sequence similarity as a catalytic diad (Glu-42-Thr-57) and triad (Glu-35-Asp-44-Thr-50) respectively. In case of Lys68, BlastP analysis identified the residues Glu-18 and Thr-35 as highly conserved, and indicates them as probable catalytic residues of Lys68 (**Figure S1b**). No aspartic acids were found between the Glu and Thr residues.

Lys68 spectra and pH dependence on the activity/stability

Overexpression of Lys68 in *E. coli* BL21(DE3) yielded a soluble protein of 14.3 mg/L with >95% purity, as judged by SDS-PAGE analysis (data not shown). To confirm the predicted catalytic activity, Lys68 endolysin 30 μL drops (2 μM) were spotted on several OM-permeabilized Gram-negative cells, and all were efficiently lysed.

Regarding pH dependence of the muralytic activity, Lys68 remains active over a pH range from 4.0 to 10.0 (maintaining 41.17% ±4.81 and 46.31% ±0.96 of its activity, respectively) (**Figure 1a**), with a pH optimum around 7.0 (pH of the bacterial cytoplasm). The muralytic activity of Lys68 at optimal pH is 400 Units/μM and decreases by 23.3% (non-significant) and 74.9% (significant) when the protein is stored for two months at 4°C and −20°C, respectively (**Figure S2**). Therefore, even when preparing single-use aliquots of Lys68 at −20°C, in the absence of glycerol, protein functionality is significantly influenced by either an increase the protein concentration ionic strength or a pH shift after formation of ice crystals during storage.

To elucidate the major structural features of Lys68, we also carried out CD studies using a universal buffer at different pH values (**Figure 1b**). The Lys68 CD profile exhibited two negative dichroic minima at 222 nm and 208 nm and a positive dichroic maximum at 192 nm, which is characteristic of a protein with high α-helix content. Deconvolution of the CD spectra, using a dataset as specified in Materials and Methods, allowed us to determine that 37% of Lys68 folds as α-helices, 17% as β-sheet, 18% as turns, while 27% was unordered, which matches with the α-helix content predicted in the PSIPRED. Interestingly, when the pH was between 4.0 and 10.0, the obtained spectra were almost identical (insignificant ellipticity values changes of approximately 1%) and the predicted secondary structures content remained essentially unchanged. These data show that the secondary structure content of Lys68, at various pH conditions, is highly stable, even at pH 4.0. Conformational changes occurred around pH 3.0 with loss of secondary structure. The ellipticity at 222 nm at pH 3.0 decreased compared to values observed between pH 4.0–10.0.

Thermostability

Thermostability tests were also evaluated on the endolysin muralytic activity and stability. Activity assays showed non-significant reductions on the enzymatic activity after 30 min incubation of the enzyme at 40°C and at 60°C, maintaining 54.7% of its residual activity at 80°C. When heated to 100°C, the muralytic activity significantly dropped to 25.8%, 10.2% and 0.27% of the initial activity after 20, 30 and 45 min incubation, respectively (**Figure 2a, 2b**).

CD spectra of Lys68 (protein ellipticity) was also measured under the same temperatures (40, 60, 80 and 100°C) (**Figure 2c, 2d**). It was observed that from 4°C to 40°C, Lys68 was very stable not changing its secondary structure. As the incubation temperature was increasing from 40°C to 60°C, 80°C and 100°C, the changes in secondary structures were evident with a continuous decrease in the ellipticity of the 208 nm and 220 nm bands. In accordance, the CD thermal stability shows that the loss of muralytic activity upon thermal stress is accompanied with gradual loss of endolysin secondary structure.

To confirm a possible refolding mechanism after thermal stress, the loss of Lys68 CD signal (ellipticity) at 222 nm (dichroic band characteristic for α-helical proteins) was monitored, when increasing the temperature of the sample from 20°C to 75°C at

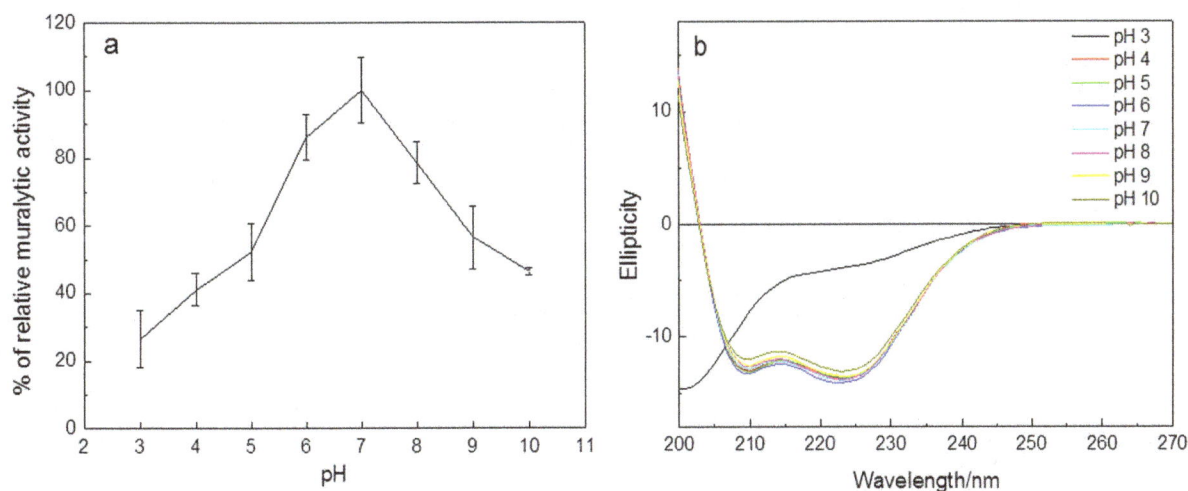

Figure 1. Influence of pH on the muralytic activity and secondary structure stability of Lys68. a) pH optimization curve for enzymatic activity of Lys68. Relative muralytic activity is measured as the slope of the OD_{600nm}/min curve, given in percentage by comparing to the activity at pH 7.0 (the highest measured value) (Y-axis) on OM-permeabilized *P. aeruginosa* PAO1 substrate and is shown for a pH range between 3 and 10 (X-axis). Averages and standard deviations of three repeated experiments are given. b) CD spectra of Lys68 as a function of pH using a universal buffer with adjusted pH (3.0-10.0).

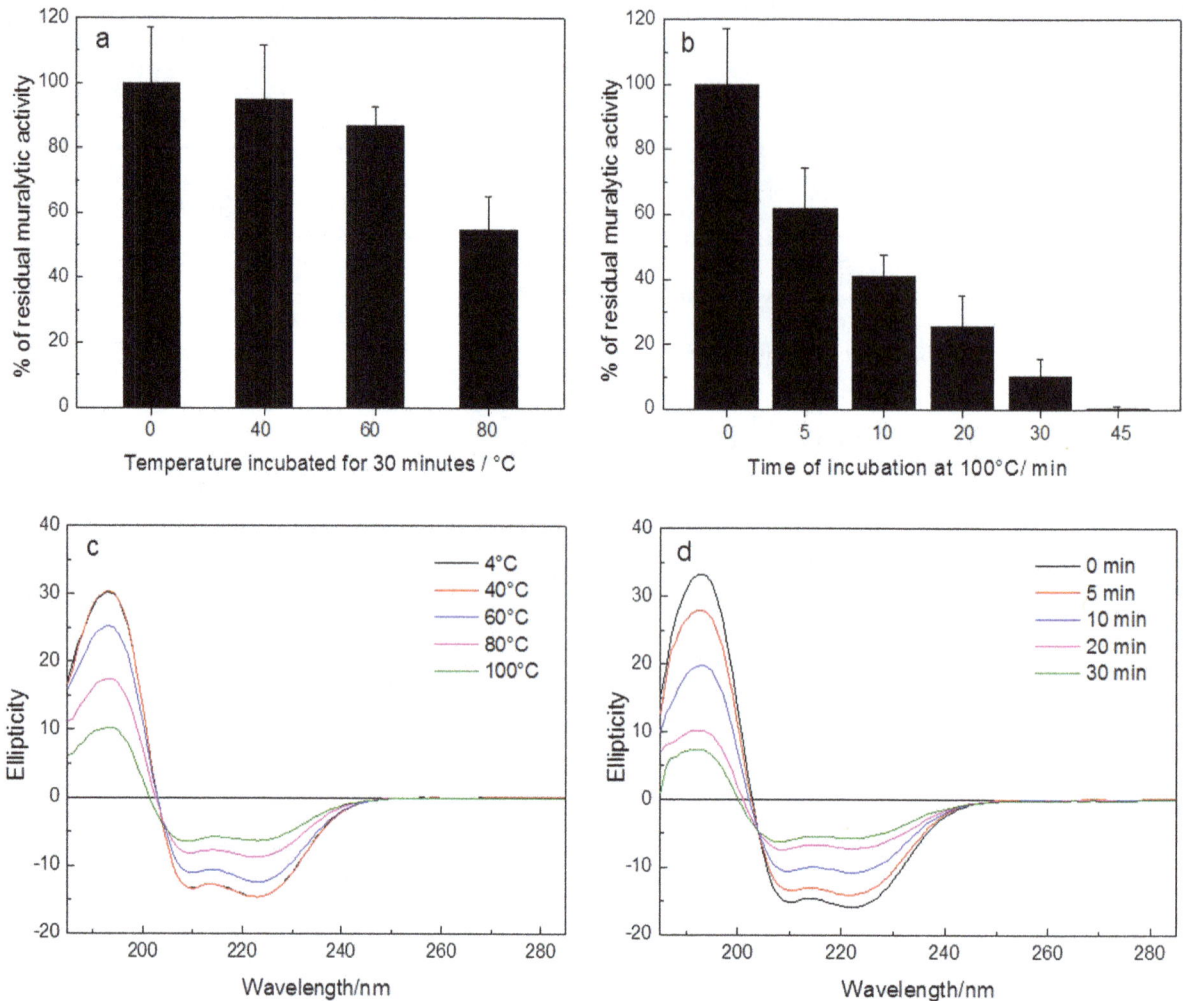

Figure 2. Influence of temperature on the muralytic activity and secondary structure stability of Lys68. a) The residual muralytic activity Lys68 (2 μM) on OM permeabilized *P. aeruginosa* PAO1 cell substrate after incubation at 4, 40, 60 and 80°C for 30 min and b) after heat treatment at 100°C for 0, 5, 10, 20, 30 and 45 min, followed by a 20-min cooling step on ice. Four repeated and independent experiments are shown; c, d) CD spectra of Lys68 incubated under the same conditions.

optimal pH for muralytic activity (pH 7.0) Lys68 demonstrated a sigmoidal thermal denaturation profile, reflecting protein unfolding (**Figure 3**). The T_m was found to be around 44°C according to a two-state model. Interestingly, the renaturation experiments showed a maximum recovery of the native features after the protein was heated up to 75°C under the same experimental conditions, and a loss in ellipticity of only 1.3%. Therefore, a confirmation of protein refolding mechanism is given, explaining how the endolysin can retain activity after exposure of temperatures higher than its T_m of 44°C.

Combinatorial effect of OMPs with Lys68 and HEWL on Gram-negative bacteria

The *in vitro* antibacterial activity of OMP/Lys68 and OMP/HEWL was investigated on several Gram-negative bacteria genera (**Table 1**). In the absence of an OMP agent, the activity of Lys68 or HEWL against exponentially growing cells was insignificant, as expected, due to presence of a protective OM that prevents the endolysin from reaching the PG. In the presence of different OMPs (EDTA, citric and malic acid), a strong antibacterial activity was observed and dependent on the type of OMP used.

EDTA/Lys68 had a more pronounced killing effect on *Pseudomonas* cells than EDTA/HEWL. Both citric and malic acid were significantly better than EDTA in enhancing the access of Lys68 to the PG, but this was not observed for HEWL. Citric acid/Lys68 was not only able to kill *Pseudomonas* cells, but also caused a significant reduction of *S.* Typhimurium LT2 (2.89±0.27 Log reduction) and *A. baumannii* (1.01±0.33 Log reduction) as well of *Shigella sonnei* (1.40±0.33 Log reduction), *E. coli* O157:H7 (1.18±0.12 Log reduction) and *Cronobacter sakazakii* (1.09±0.19 Log reduction). The effect of malic acid/Lys68 was higher in all strains compared to EDTA/Lys68 and citric acid/Lys68. Reductions of 2–3 logs units were observed on *P. aeruginosa* (3.31±0.21), *S.* Typhimurium LT2 (2.62±0.32) and *A. baumannii* (2.75±0.29). Overall, higher log reduction values were obtained with malic acid in combination with Lys68 after 30 min of incubation, killing 11 different Gram-negative bacterial species. In contrast to the other strains, none of the previous OMPs could sensitize *Klebsiella oxytoca* and *Yersinia enterocolitica* strains to Lys68.

Interestingly, extending the reaction period from 30 min to 2 h, the combinatory effect of organic acids and Lys68 against *S.*

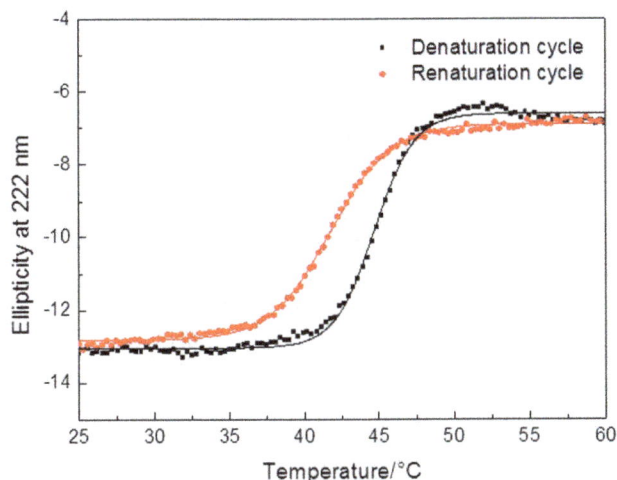

Figure 3. Thermal denaturation and renaturation profiles of Lys68. The melting curve for Lys68 is measured by monitoring the absorbance at 222 nm against increasing (denaturation) and decreasing temperatures (renaturation) at pH 7.0.

Typhimurium LT2 remarkably increased (**Table 2**). Maximum reductions of exponential cells were achieved when citric acid was combined with Lys68 (5.01 ± 0.37 log reduction of viable cells) and this combination was also efficient against stationary cells (1.45 ± 0.15 log units of viable cells) and biofilms (1.26 ± 0.10 log units of viable cells).

OM permeabilization mechanism

To further investigate the permeabilizing effect of citric and malic acid, S. Typhimurium LT2 cells were incubated with Lys68 in the presence of HCl (to drop the pH to 4.2 and 3.8, similar to citric and malic acid, respectively) for a 30 min period. The results summarized in **Table S2** show that the log reduction units of viable cells are mostly due to the acidity. In the presence of HCl, Lys68 is able to kill 2.55 ± 0.37 log of viable cells at pH 4.2 (3.01 ± 0.15 for citric acid) and a 2.31 ± 0.22 log reduction of viable cells was obtained at pH 3.8 (2.60 ± 0.17 for malic acid). In the case of HEWL, no activity towards S. Typhimurium LT2 was found in the presence of both organic acids and HCl.

To assess the effect of organic acids in permeabilizing Gram-negative OMs, an experiment was conducted incubating S. Typhimurium LT2 cells with citric acid in the presence of $MgCl_2$. $MgCl_2$ can contribute to the stability of the OM by providing electrostatic interactions of divalent cations (Mg^{2+}) with proteins and the lipopolysaccharide (LPS) layer, and therefore, decrease the permeability of the OM to external agents. Consequently, the presence of 5 mM of $MgCl_2$ abolished Lys68 antibacterial activity in all tested conditions (**Table S2**).

To visualise the effect of organic acids in permeabilizing Gram-negative OMs, an experiment was conducted incubating S. Typhimurium LT2 cells with citric acid in the presence of a fluorescent probe. This fluorescent probe was constructed earlier and consists of a *Pseudomonas* phage KZ144 endolysin cell wall (peptidoglycan) binding domain fused with an enhanced green fluorescent protein (KZ-EGPF) that has high-affinity to Gram-negative PG [21]. When incubating the cells for a 30-min period, green fluorescent *Salmonella* cells could be observed by epifluorescence microscopy (**Figure S3**), indicative of cell permeabilization.

Lys68 PG digestion mechanism

To assess the Lys68 PG digestion mechanism and explain the activity of the endolysin at low pH, an analysis of its essential catalytic residues was made. Lys68 is classified as a member of the GH24 family [22], which has a glutamine (Glu) residue as the proton donor. Besides Glu, two other catalytic residues (Asp and Thr) may be involved in the catalytic reaction, as previously demonstrated with T4 lysozyme [23]. To identify a putative nucleophile for Lys68, BlastP searches were conducted. Results showed that, besides Glu-18, a second amino acid residue, Thr-35, was indicated as a putative catalytic residue of Lys68, with no Asp identified between both residues. Therefore, mutagenesis of both presumably essential catalytic residues was performed and resulted in one inactive enzyme (Glu18Ala) and one enzyme with similar activity (Thr35Ala) compared to the Lys68 wild-type (**Figure S4a**). Analysis of mutated proteins by CD did not provide any evidence of significant structural alterations due to the amino acid replacements (**Figure S4b**). Therefore the absence of activity was entirely attributed to the Glu18Ala substitution.

Based on the putative catalytic residues identified on other phage endolysins (muramidases) by protein similarity, our BlastP results that identified two catalytic candidates on Lys68 and the experimental data on its deleted mutants , we were not able to identify a second catalytic residue. Recently, Wohlkönig and coworkers, described the existence of two subgroups within the GH24 family; one that has a Glu-Asp-Thr catalytic triad and another one that only has Glu as a catalytic residue [24]. Therefore, we propose that Lys68 operates through a two-step mechanism involving Glu-18 (**Figure 4**). In the first step N-acetylhexosaminidases, which have an acetamido group, are capable of neighboring group participation to form an intermediate oxazolinium ion, with acidic assistance provided by the Glu-18 carboxylate. In the second step the now deprotonated acidic carboxylate acts as a base and assists a nucleophilic water to hydrolyze the oxazolinium ion intermediate, giving the hydrolyzed product (through individual inversions that lead to a net retention of configuration).

Discussion

Endolysins of phages infecting Gram-positive bacteria have been extensively studied [4]. In case of Gram-negative phage endolysins, although their enzymatic mechanisms have been intensively studied, they efficacy in terms of antibacterial (biocontrol) potential have been scarcely explored. This has to do withthe impermeable bacterial OM protects the underlying PG layer from enzymatic hydrolysis by the exogenous endolysin. In this work, a new endolysin from a Gram-negative infecting phage is described. *In silico* characterization of Lys68 from the *Salmonella* phage phi68 shows elevated amino acid sequence similarity (>90% identity) to other *Salmonella* phage endolysins belonging to the same lysozyme-like superfamily. The observed globular structure (catalytic domain alone) is also shared by most other Gram-negative phage endolysins, contrary to modular archetypes (catalytic and a binding domain) typically seen in endolysins from a Gram-positive background [4].

A biochemical characterization showed that Lys68 displays some interesting characteristics. Regarding substrate specificity, Lys68 degrades the PG of all 20 Gram-negative organisms tested, and this broad spectrum is explained by the common PG A1γ chemotype shared by these organisms [25]. The observed muralytic activity of the globular Lys68 (400 Units/µM) is lower compared to other described modular Gram-negative-like endolysins (e.g. KZ144 and PVPSE1gp146 with 2058 and 13613 U/

Table 1. Combinatorial antibacterial activity of outer membrane permeabilizers with (A and B) or without (C) HEWL/Lys68 on Gram-negative bacterial pathogens.

A Bacterial Species	HEWL/Water	HEWL/EDTA	HEWL/Citric	HEWL/Malic
Salmonella Typhimurium *LT2*	0.03±0.07	0.14±0.23	0.08±0.15	0.07±0.09
Acinetobacter baumannii 2	0.01±0.09	0.04±0.06	0.10±0.06	0.17±0.15
Pseudomonas aeruginosa **PAO1**	0.02±0.10	**0.89±0.17**	0.24±0.02	**2.19±0.32**
Pseudomonas fluorescens 7A	0.04±0.04	0.30±0.38	0.28±0.13	**2.88±0.35**
Shigella sonnei **ATCC 25931**	0.22±0.29	0.01±0.11	0.02±0.07	0.06±0.12
E. coli **O157:H7 CECT 4782**	0.04±0.15	0.11±0.12	0.10±0.05	0.09±0.04
Cronobacter sakazakii **CECT 858**	0.11±0.18	0.15±0.20	0.30±0.16	0.20±0.13
Pantoea agglomerans **SA5634**	0.04±0.10	0.08±0.15	0.20±0.19	0.22±0.29
Enterobacter amnigenus **CECT 4878**	0.04±0.07	0.07±0.13	0.18±0.17	0.26±0.04
Proteus mirabilis **SA5445**	0.08±0.03	0.09±0.14	0.03±0.25	0.17±0.14
Salmonella bongori **SGSC 3100**	0.02±0.10	0.10±0.09	0.09±0.13	0.47±0.19
Klebsiella oxytoca **ATCC 13182**	0.01±0.02	0.10±0.05	0.13±0.08	0.26±0.06
Yersinia enterocolitica **SA5429**	0.06±0.05	0.18±0.11	0.11±0.12	0.34±0.11
B	Lys68/Water	Lys68/EDTA	Lys68/Citric	Lys68/Malic
Salmonella Typhimurium *LT2*	0.14±0.16	0.16±0.12	**2.89±0.27**	**2.62±0.32**
Acinetobacter baumannii 2	0.04±0.12	0.14±0.22	**1.01±0.33**	**2.75±0.29**
Pseudomonas aeruginosa **PAO1**	0.04±0.34	**2.41±0.26**	**1.48±0.35**	**3.31±0.21**
Pseudomonas fluorescens 7A	0.02±0.10	**1.55±0.37**	**1.40±0.23**	**5.44±0.24**
Shigella sonnei **ATCC 25931**	0.13±0.28	0.35±0.44	**1.40±0.33**	**2.35±0.31**
E. coli **O157:H7 CECT 4782**	0.02±0.02	0.21±0.14	**1.18±0.12**	**2.37±0.35**
Cronobacter sakazakii **CECT 858**	0.16±0.06	0.30±0.09	**1.09±0.19**	**1.40±0.30**
Pantoea agglomerans **SA5634**	0.03±0.18	0.13±0.12	0.32±0.18	**1.26±0.18**
Enterobacter amnigenus **CECT 4878**	0.04±0.11	0.28±0.18	0.68±0.36	**1.21±0.27**
Proteus mirabilis **SA5445**	0.03±0.12	0.09±0.04	0.41±0.29	**0.97±0.25**
Salmonella bongori **SGSC 3100**	0.06±0.10	0.08±0.16	0.13±0.09	**0.94±0.05**
Klebsiella oxytoca **ATCC 13182**	0.07±0.13	0.12±0.17	0.24±0.14	0.30±0.22
Yersinia enterocolitica **SA5429**	0.11±0.10	0.03±0.06	0.18±0.08	0.22±0.07
C	------	PBS/EDTA	PBS/Citric	PBS/Malic
Salmonella Typhimurium *LT2*		0.09±0.10	0.07±0.05	0.06±0.09
Acinetobacter baumannii 2		0.04±0.03	0.07±0.05	0.07±0.04
Pseudomonas aeruginosa **PAO1**		0.05±0.06	0.60±0.14	**1.83±0.31**
Pseudomonas fluorescens 7A		0.42±0.31	0.70±0.14	**2.70±0.31**
Shigella sonnei **ATCC 25931**		0.01±0.14	0.02±0.14	0.14±0.03
E. coli **O157:H7 CECT 4782**		0.06±0.04	0.01±0.09	0.18±0.18
Cronobacter sakazakii **CECT 858**		0.17±0.12	0.06±0.14	0.28±0.10
Pantoea agglomerans **SA5634**		0.07±0.10	0.03±0.18	0.10±0.18
Enterobacter amnigenus **CECT 4878**		0.08±0.08	0.15±0.15	0.16±0.18
Proteus mirabilis **SA5445**		0.02±0.10	0.01±0.23	0.03±0.12
Salmonella bongori **SGSC 3100**		0.07±0.12	0.05±0.14	0.16±0.03
Klebsiella oxytoca **ATCC 13182**		0.12±0.12	0.01±0.09	0.18±0.09
Yersinia enterocolitica **SA5429**		0.01±0.06	0.30±0.23	0.06±0.07

Exponential growing cells (10^8 CFU/ml) were incubated with 2 µM HEWL/Lys68 in combination with one of the OMPs (0.5 mM EDTA, 2 mM citric acid or 5 mM malic acid). Protein incubation with water or OMPs with PBS served as negative controls. After incubation, the effect of HEWL/Lys68 and OMP mixtures was assessed by quantification of the number of CFUs. Antibacterial activity was quantified as the relative inactivation in logarithmic units ($= \log_{10}(N_0/N_i)$ with N_0 = number of untreated cells and N_i = number of cells after treatment). Averages ± standard deviations are given for n = 4 repeats. Indicated in bold are significant log reduction units observed (≥1 log).

Table 2. Influence of different *Salmonella* Typhimurium LT2 physiological states (planktonic/biofilm) on the combinatorial effect of Lys68 and outer membrane permeabilizers.

S. Tyhpimurium LT2	Planktonic		Biofilm
	Exponential phase	Stationary phase	
Lys68 + Water	0.11±0.14	0.14±0.07	0.15±0.11
PBS + Citric acid	0.10±0.11	0.08±0.11	0.09±0.19
PBS + Malic acid	0.08±0.20	0.19±0.18	0.15±0.25
Lys68 + Citric acid	**5.01±0.37**	**1.45±0.15**	**1.26±0.10**
Lys68 + Malic acid	**3.23±0.33**	0.65±0.28	**1.01±0.13**

In the planktonic and stationary assay, 50 μL cells resuspended in 10 mM HEPES/NaOH (pH 7.2) to a final 10^8 CFU/ml were incubated with 25 μL Lys68 (2 μM final concentration) together with 25 μL of citric and malic acid (2 mM and 5 mM final concentrations, respectively) for 2 h. In biofilms assay, cells were washed twice with 200 μL of 10 mM HEPES/NaOH (pH 7.2). Cells were then incubated with 100 μL of Hepes together with 50 μL Lys68 (2 μM final concentration) and 50 μL of citric and malic acid (2 mM and 5 mM final concentrations, respectively) for 2 h. For both experiments, cells incubated with water (instead of citric/malic acid) or PBS (instead of Lys68) were used as negative controls. After incubation CFUs were counted. Averages ± standard deviations are given for n = 4 repeats. Indicated in bold are significant log reduction units observed (≥1 log).

μM, respectively, calculated using the same method) [8]. Although KZ144 and PVPSE1gp146 have different catalytic activities, belonging to the lytic transglycosylases and glycoside hydrolase family 19 respectively, a more obvious explanation for their higher activity is presence of cell wall binding domains. It has been previously demonstrated that modular Gram-negative endolysins possessing an N-terminal binding domain exponentially increases the muralytic activity [26]. These binding domains have been found in KZ144 and PVPSE1gp146, where in Lys68 is apparently absent. Nevertheless, when compared to the globular homologue HEWL (35 Units/μM, reported elsewhere), Lys68 is 12 times more active [14]. In addition, Lys68 also showed to be highly stable, a feature required for antimicrobial enzymes. From pH values of 4.0 to 10.0, Lys68 does not change its secondary structure, where it has an optimal pH for muralytic activity around 7.0, without significant loss of its activity even when stored for two months at 4°C.

Once subjected to thermostability tests, in general, some endolysins maintain their activity after long heating periods. Within endolysins from a Gram-negative background, *Pseudomonas* phage endolysins KZ144 and EL188 activity still remains high after being exposed to 50°C for 10 min [26]. In contrast, the phage T4 lysozyme only retains a minor fraction of its activity after a 5 min treatment at 65°C [27]. Recently, a highly thermostable

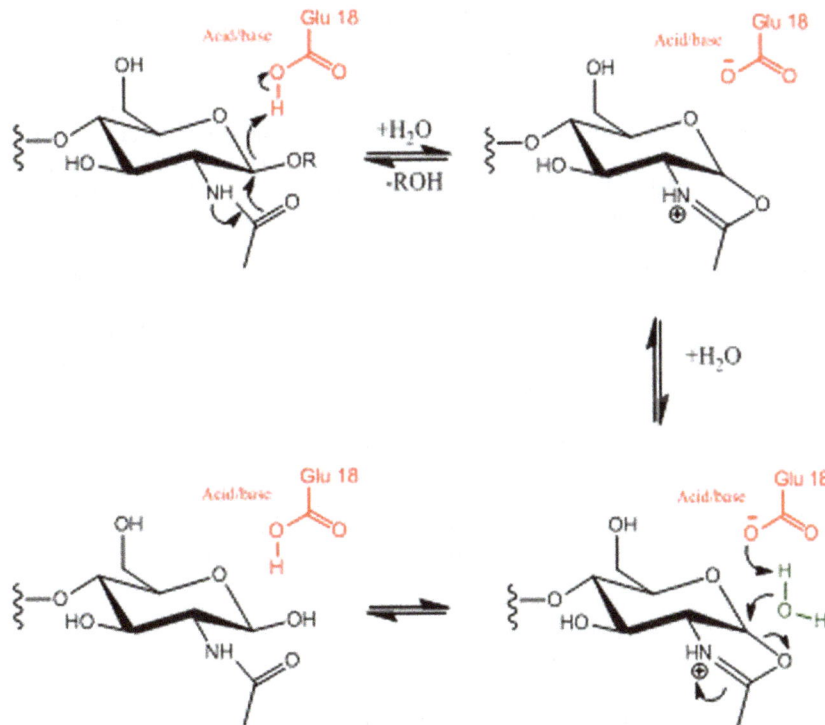

Figure 4. Proposed peptidoglycan hydrolases mechanism of Lys68. Lys68 hydrolyses the 1, 4-linkages between *N*-acetylmuramic acid and *N*-acetyl-D-glucosamine, through a two-step mechanism involving Glu-18 catalytic residue to cleave the peptidoglycan.

modular endolysin (PVP-SE1gp36) that withstands temperatures up to 100°C was described [8]. Lys68 is a globular protein where the muralytic activity is only completely inactivated after 30 min at 100°C. Because the melting temperature of the endolysin at pH 7.0 is 44°C, it seems that under the stress caused by higher temperatures, the endolysin is able to gradually regain its secondary structure to a point where, at high temperatures and prolonged exposures, it cannot refold. Recently, a similar folding/refolding mechanism was suggested by Schmelcher and coworkers http://www.plosone.org/article/info%3Adoi%2F10.1371%2Fjournal.pone.0036991 - pone.0036991-Schmelcher1to explain the high thermostability of the Gram-positive Listeria monocytogenes phage endolysins, HPL118 and HPL511 [28]. In their study, HPL118 and HPL511 retained 35% activity after 30 min at 90°C.

To allow a better diffusion of endolysins through the OM, a combinatorial approach was used on Gram-negative bacteria, using bacterial cell wall-degrading enzyme Lys68 or HEWL with an OMP. Both proteins combined with EDTA could reduce P. aeruginosa to some extent. A similar effect has previously been described with endolysin EL188 from Pseudomonas infecting phage EL of P. aeruginosa strain PAOI [7]. The positive effect of EDTA on the antibacterial activity is caused by the strong OM permeabilization capacity of EDTA through binding and withdrawal of the stabilizing divalent Mg^{2+} and Ca^{2+} cations present in the LPS layer of the Gram-negative OM [29]. As a result, the bacterial PG becomes more prone to the muralytic activity of the externally added endolysins. However, the sensitivity of other bacterial species to the destabilizing effect of EDTA was significantly lower, even when EDTA was increased to 5 mM (data not shown). It seems that the type and structure of the OM plays a key role for the antibacterial efficacy of the EDTA/enzyme combinatorial effect. Possibly, due to the low number of phosphate groups per LPS molecule and the corresponding amount of stabilizing divalent cations in the OM of Enterobacteriaceae compared to the Pseudomonadaceae, EDTA turns out to be ineffective to the former group. The observed synergetic effects of citric acid and malic acid, has never been described for phage endolysins. In this respect, both citric acid and malic acid had a better permeabilizing effect than EDTA, and were able to permeabilize several Gram-negative strains enhancing the activity of Lys68. Exceptions were K. oxytoca and Y. enterocolitica strains that may present some acid tolerance systems described (e.g. consuming or removing excess of protons by decarboxylation reactions and ion transporters, changing membrane composition), protecting them from the organic acids OM-destabilization [30].

In particular against exponential S. Typhimurium LT2 cells, an even more profound effect was observed when extending the reaction from 30 min to 2 h, although proving to be less efficient against stationary cells and biofilms. The lower efficiency in killing stationary cells can be a result of two aspects; i) structural changes in the OM (e.g. LPS biosynthesis), compromising the endolysin entry or LPS permeabilization through the organic acids, or ii) chemical modifications in the PG (e.g. glycosylated or acetylated glycans) known to contribute to high levels of resistance to glycosyl hydrolases [31,32]. As for biofilms, besides the presence of several bacterial physiological states that can give rise to similar problems, additional diffusion limitations due to the high density matrix content can hinder the efficacy of endolysins and antimicrobials in general.

Although the permeabilizing capacity of citric acid by chelating divalent cations from the OM associated with its protonated form has been reported, additional acidity can contribute to OM damage [33]. It has been proposed that the LPS disintegration is accomplished by the ability of organic acids groups to migrate

inside the cells to cause sublethal damages (e.g. enzyme inhibitions, amino acid decarboxylation, membrane disruption) [33,34]. The 2 mM of citric acid (pKa 3.13 at 25°C) and 5 mM of malic acid (pKa 3.4 at 25°C) used, caused a drop of pH to 4.2 and 3.8, respectively. Because similar log reduction units were obtained using HCl instead of citric or malic acid combined with Lys68, the acidity effect (instead of its chelation effect) seems to play a key role in the OM permeabilization. Because millimolar concentrations of these organic acids by themselves did not have an antibacterial effect (with exception of Pseudomonas), they seem to be enough to compromise the cell wall to a level that allows Lys68 to act on the PG, causing bacterial death. Recently, a combined treatment with different organic acids to induce cell permeability in E. coli O157:H7 was reported [35]. Transmission electron microscopy images showed clear membrane disintegration that potentiated the bactericidal action of medium-chain fatty acids. When bacterial suspensions were supplemented with 5 mM of $MgCl_2$, the effect of Lys68 combined with organic acids was completely abolished. $MgCl_2$ has the ability to protect the bacterial OM damaged from the acid challenge, an effect already observed with lactic acid [34].

Based on the carbohydrate-active enzymes classification, several amino acids have been identified within lysozyme-like proteins to play a key role in the glycosidase reaction. HEWL (belonging to the GH22 class), which cleaves polysaccharide chains in bacterial cell walls, contains an active site that is constituted essentially by the two amino acids Glu-35 (pKa ~6.2) and Asp-52 (pKa ~3.6), that are only functional if the former is non-ionized and the latter is ionized [36]. Hence, the absence of HEWL enzymatic activity in the presence of malic and citric acid is related to the deionization of Asp52 at an acidic pH. Lys68 has a different catalytic system. Since the Thr-35 mutation did not affect Lys68 antimicrobial activities, we can conclude that Thr-35 is not involved in the catalysis of PG. Therefore, it is tempting to speculate that Lys68 operates through a two-step mechanism involving Glu-18, capable of digesting the 1, 4-linkages between N-acetylmuramic acid and N-acetyl-D-glucosamine residues of the PG in a low pH environment (pH around 4.0) through a retaining mechanism. There is a lack of studies conducted to confirm the phage endolysins (muramidases) catalytic sites. However, the catalytic activity of amino acid Glu-73 in the goose egg-white lysozyme has been demonstrated by Weaver and co-workers to be sufficient to digest PG bonds [37].

To summarize, Gram-negative bacterial pathogens prevail in various surroundings and their resilience is aided by the presence of an OM that prevents toxic substances from entering the cell. We show that an endolysin from a Gram-negative background can have a broad activity and good (thermo)stability properties. We further demonstrate that Lys68 can have an effective anti-Gram-negative activity when combined not only with EDTA, but specially with citric acid and malic acid, that can now be further explored to other endolysins for several potential applications. EDTA/Lys68 could be used as a therapeutic product to control P. aeruginosa, typically present in burn wound infections or as food preservative against P. fluorescens. The broad antimicrobial effect of citric acid/Lys68 and malic acid/Lys68 acid could be applied in food and clinical settings as an effective sanitizer to prevent food spoilage and nosocomial bacterial infections. Additionally, the powerful effect of Lys68 with organic acids can be used therapeutically to treat for example topical infections as chronic wounds, when applied in a cream or ointment. Altogether, this study demonstrated that bacterial cell wall hydrolase from phage origin (Lys68) combined with citric or malic acid have a powerful killing activity and can be used, as alternatives to chemical antimicrobial agents, to fight Gram-negative pathogens.

Supporting Information

Figure S1 *In silico* **analysis of the Lys68. a**) Amino acid sequence of endolysin Lys68, where a phage-related lysozyme domain, belonging to the Glycoside Hydrolase Family 24 (GH24), was identified using HHpred webserver with the Pfam, Inter-ProScan and COG databases and an E-value of 2.3×10^{-47} and 100% of query coverage (AA 1–151 underlined). Secondary structure analysis based on PSIPRED predicts 9 α-helices and 7 β-sheets. BlastP output indicates conservation in both presumed catalytic residues glutamic acid (Glu-18) and Threonine (Thr-35) for the glycosylase reaction (marked in bold). **b**) Sequence alignment of high Lys68 homologs identified in HHpred output using PBD as database: *Enterobacteria* phages T4 (P00720.2), P1 (Q37875.1) and P21 (P27359.1) with the respective accession numbers given. The N-terminal domains are aligned with the Glu-8aa-Asp-5aa-Thr identified in T4 (highlighted in red), illustrating other putative catalytic residues by homology (highlighted in violet).

Figure S2 **Saturation curves for Lys68 muralytic activity under optimal pH (7.2).** The activity in $OD_{600 \text{ nm}}$/min (Y-axis) for incremental amounts of Lys68 (0 months – squares; 2 months at 4°C – circles; 2 months at −20°C – triangles) is depicted. Muralytic activity was quantified using outer membrane-permeabilized *P. aeruginosa* PAO1 cells as a substrate, resuspended in 80 mM phosphate buffer pH 7.2. Lys68 activity reaches saturation at 2 to 3 μM. A linear regression of the demarcated linear region of the saturation curves gives an activity of 400, 310 (23.3% less) and 100 Units/μM (74.9% less) at 0 months and after 2 months at 4°C and −20°C, respectively. Averages and standard deviations of three repeated experiments are given.

Figure S3 **Epifluorescence microscopy of the *S.* Typhimurium LT2 cells. a**) Cells incubated with 2 mM citric acid and 2 μM of KZ-EGFP protein; **b**) Cells incubated with water and 2 μM of KZ-EGFP; and **c**) Cells incubated with 2 mM citric acid and PBS for 30 min. Cell pellets were then washed twice and visualized using epifluorescence microscopy with a 1500 magni-fication. EGFP proteins are visualized in green targeted to the bacterial cell wall.

Figure S4 **Comparison of muralytic activity and the secondary structure of Lys68 wild-type and its mutants. a**) Muralytic activity on *P. aeruginosa* OM permeabilized cells resuspended in 80 mM phosphate buffer pH 7.2, measured as optical density decrease. **b**) Circular dichroism spectra of Lys68 wild-type and the two mutants (Glu18Ala and Thr35Ala).

Table S1 **Lytic activity of Lys68 against various Gram-negative strains.** Mid-exponential Gram-negative growing cells were plated onto LB agar Petri dishes to form bacterial lawns, after which their OM was permeabilized by chloroform treatments. Then, a 30-μL drop of 2 μM of purified protein was added on top of the lawn and incubated for 30 min, followed by a visualization analysis to spot lysis halos and assess bacterial susceptibility.

Table S2 *In vitro* **antibacterial activity of Lys68 and HEWL in combination with HCl, citric or malic acid against *S.* Typhimurium LT2 cells.** Cell cultures (initial cell density of 10^8 cells/mL) were incubated for 30 min with 2 μM Lys68 or 2 μM of HEWL in presence of either 3.5 mM of HCl (pH 4.2), 4 mM of HCl (pH 3.8), 2 mM of citric acid (pH 4.2) or 5 mM of malic acid (pH 3.8). The use of water instead of acids served as negative control. The antibacterial activity was expressed as the relative inactivation in logarithmic units (= $\log_{10} (N_0/N_i)$ with N_0 = number of untreated cells (negative control) and N_i = number of treated cells counted after incubation). Averages and standard deviations of four repeated and independent experiments are shown. Log reductions considered significant (≥1 log unit) are marked in bold.

Author Contributions

Conceived and designed the experiments: SS RL MNP LK JA. Performed the experiments: HO VT MW. Analyzed the data: SS RL MNP LK JA. Contributed reagents/materials/analysis tools: RL MNP LK JA. Wrote the paper: HO VT.

References

1. Scallan E, Griffin PM, Angulo FJ, Tauxe RV, Hoekstra RM (2011) Foodborne illness acquired in the United States–unspecified agents. Emerg Infect Dis 17: 16–22.
2. Chopra I, Schofield C, Everett M, O'Neill A, Miller K, et al. (2008) Treatment of health-care-associated infections caused by Gram-negative bacteria: a consensus statement. Lancet Infect Dis 8: 133–139.
3. Oliveira H, Melo LD, Santos SB, Nobrega FL, Ferreira EC, et al. (2013) Molecular aspects and comparative genomics of bacteriophage endolysins. J Virol 87: 4558–4570.
4. Fischetti VA (2010) Bacteriophage endolysins: a novel anti-infective to control Gram-positive pathogens. Int J Med Microbiol 300: 357–362.
5. Nikaido H (2003) Molecular basis of bacterial outer membrane permeability revisited. Microbiol Mol Biol Rev 67: 593–656.
6. Mastromatteo M, Lucera A, Sinigaglia M, Corbo MR (2010) Use of lysozyme, nisin, and EDTA combined treatments for maintaining quality of packed ostrich patties. J Food Sci 75: M178–186.
7. Briers Y, Walmagh M, Lavigne R (2011) Use of bacteriophage endolysin EL188 and outer membrane permeabilizers against *Pseudomonas aeruginosa*. J Appl Microbiol 110: 778–785.
8. Walmagh M, Briers Y, dos Santos SB, Azeredo J, Lavigne R (2012) Characterization of modular bacteriophage endolysins from *Myoviridae* phages OBP, 201phi2-1 and PVP-SE1. PLoS One 7: e36991.
9. Lim JA, Shin H, Kang DH, Ryu S (2012) Characterization of endolysin from a *Salmonella* Typhimurium-infecting bacteriophage SPN1S. Res Microbiol 163: 233–241.
10. Doores S (1993) Organic acids. In: Davidson PMB, A.L., editor. Antimicrobials in Food. 2 ed. New York: Marcel Dekker Ltd. pp.95–136.
11. Theron MM IJ (2011) Mechanisms of microbial inhibition. In: Theron MM IJ, editor. Organic Acids and Food Preservation: CRC Press. pp.117–150.
12. Alakomi H-L (2007) Weakening of the Gram-negative bacterial outer membrane: University of Helsinki
13. Sillankorva S, Pleteneva E, Shaburova O, Santos S, Carvalho C, et al. (2010) *Salmonella* Enteritidis bacteriophage candidates for phage therapy of poultry. J Appl Microbiol 108: 1175–1186.
14. Briers Y, Lavigne R, Volckaert G, Hertveldt K (2007) A standardized approach for accurate quantification of murein hydrolase activity in high-throughput assays. J Biochem Biophys Methods 70: 531–533.
15. Raymond Schuch VAF, Daniel C . Nelson (2009) A Genetic Screen to Identify Bacteriophage Lysins. In: Martha R.J. Clokie AMK, editor. Bacteriophages - Methods in Molecular Biology. pp.307–319.
16. Compton LA, Johnson WC Jr (1986) Analysis of protein circular dichroism spectra for secondary structure using a simple matrix multiplication. Anal Biochem 155: 155–167.
17. van Stokkum IH, Spoelder HJ, Bloemendal M, van Grondelle R, Groen FC (1990) Estimation of protein secondary structure and error analysis from circular dichroism spectra. Anal Biochem 191: 110–118.
18. Provencher SW, Glockner J (1981) Estimation of globular protein secondary structure from circular dichroism. Biochemistry 20: 33–37.
19. Whitmore L, Wallace BA (2004) DICHROWEB, an online server for protein secondary structure analyses from circular dichroism spectroscopic data. Nucleic Acids Res 32: W668–673.
20. Whitmore L, Wallace BA (2008) Protein secondary structure analyses from circular dichroism spectroscopy: methods and reference databases. Biopolymers 89: 392–400.

21. Briers Y, Schmelcher M, Loessner MJ, Hendrix J, Engelborghs Y, et al. (2009) The high-affinity peptidoglycan binding domain of *Pseudomonas* phage endolysin KZ144. Biochem Biophys Res Commun 383: 187–191.

22. Carbohydrate-Active enZYmes website. Available: www.cazy.org/Glycoside-Hydrolases. Accessed 2014 May 12

23. Kuroki R, Weaver LH, Matthews BW (1999) Structural basis of the conversion of T4 lysozyme into a transglycosidase by reengineering the active site. Proc Natl Acad Sci U S A 96: 8949–8954.

24. Wohlkonig A, Huet J, Looze Y, Wintjens R (2010) Structural relationships in the lysozyme superfamily: significant evidence for glycoside hydrolase signature motifs. PLoS One 5: e15388.

25. Schleifer KH, Kandler O (1972) Peptidoglycan types of bacterial cell walls and their taxonomic implications. Bacteriol Rev 36: 407–477.

26. Briers Y, Volckaert G, Cornelissen A, Lagaert S, Michiels CW, et al. (2007) Muralytic activity and modular structure of the endolysins of *Pseudomonas aeruginosa* bacteriophages phiKZ and EL. Mol Microbiol 65: 1334–1344.

27. Nakagawa H, Arisaka F, Ishii S (1985) Isolation and characterization of the bacteriophage T4 tail-associated lysozyme. J Virol 54: 460–466.

28. Schmelcher M, Waldherr F, Loessner MJ (2012) *Listeria* bacteriophage peptidoglycan hydrolases feature high thermoresistance and reveal increased activity after divalent metal cation substitution. Appl Microbiol Biotechnol 93: 633–643.

29. Wren BW (1991) A family of clostridial and streptococcal ligand-binding proteins with conserved C-terminal repeat sequences. Mol Microbiol 5: 797–803.

30. Warnecke T, Gill RT (2005) Organic acid toxicity, tolerance, and production in *Escherichia coli* biorefining applications. Microb Cell Fact 4: 25.

31. WVaP B (2009) Bacterial cell envelope peptidoglycan. In: Moran AP, editor. Microbial Glycobiology - Structures, Relevance and Applications: Elsevier pp.15–28.

32. Rolfe MD, Rice CJ, Lucchini S, Pin C, Thompson A, et al. (2012) Lag phase is a distinct growth phase that prepares bacteria for exponential growth and involves transient metal accumulation. J Bacteriol 194: 686–701.

33. Theron MM, Lues JFR (2011) Mechanisms of microbial inhibition. In: Theron MM, Lues JFR, editors. Organic Acids and Food Preservation: CRC Press. pp.117–150.

34. Alakomi HL, Skytta E, Saarela M, Mattila-Sandholm T, Latva-Kala K, et al. (2000) Lactic acid permeabilizes gram-negative bacteria by disrupting the outer membrane. Appl Environ Microbiol 66: 2001–2005.

35. Kim SA, Rhee MS (2013) Marked synergistic bactericidal effects and mode of action of medium-chain fatty acids in combination with organic acids against *Escherichia coli* O157:H7. Appl Environ Microbiol 79: 6552–6560.

36. Webb H, Tynan-Connolly BM, Lee GM, Farrell D, O'Meara F, et al. (2010) Remeasuring HEWL pK(a) values by NMR spectroscopy: methods, analysis, accuracy, and implications for theoretical pK(a) calculations. Proteins 79: 685–702.

37. Weaver LH, Grutter MG, Matthews BW (1995) The refined structures of goose lysozyme and its complex with a bound trisaccharide show that the "goose-type" lysozymes lack a catalytic aspartate residue. J Mol Biol 245: 54–68.

## 14

Glycerol Uptake Is Important for L-Form Formation and Persistence in *Staphylococcus aureus*

Jian Han[1,2], Lili He[1], Wanliang Shi[2], Xiaogang Xu[3], Sen Wang[4], Shuo Zhang[2], Ying Zhang[2,4]*

1 Department of Pathogenic Biology, School of Basic Medical Sciences, Lanzhou University, Lanzhou, China, **2** Department of Molecular Microbiology and Immunology, Bloomberg School of Public Health, Johns Hopkins University, Baltimore, Maryland, United States of America, **3** Institute of Antibiotics, Huashan Hospital, Fudan University, Shanghai, China, **4** Department of Infectious Diseases, Huashan Hospital, Fudan University, Shanghai, China

Abstract

S. aureus is a significant human pathogen and has previously been shown to form cell wall deficient forms or L-forms *in vitro* and *in vivo* during infection. Despite many previous studies on *S. aureus* L-forms, the mechanisms of L-form formation in this organism remain unknown. Here we established the L-form model in *S. aureus* and constructed a transposon mutant library to identify genes involved in L-form formation. Screening of the library for mutants defective in L-form formation identified *glpF* involved in glycerol uptake being important for L-form formation in *S. aureus*. Consistent with this observation, *glpF* was found to be highly expressed in L-form *S. aureus* but hardly expressed in normal walled form. In addition, *glpF* mutant was found to be defective in antibiotic persistence. The defect in L-form formation and antibiotic persistence of the *glpF* mutant could be complemented by the wild type *glpF* gene. These findings provide new insight into the mechanisms of L-form formation and persistence in *S. aureus* and may have implications for development of new drugs targeting persisters for improved treatment.

Editor: Gunnar F. Kaufmann, The Scripps Research Institute and Sorrento Therapeutics, Inc., United States of America

Funding: These authors have no support or funding to report.

Competing Interests: The authors have declared that no competing interests exist.

* Email: yzhang@jhsph.edu

Introduction

L-form bacteria refer to cell wall deficient form of bacteria that were first discovered by Emmy Klienenberger in 1935 [1]. There is a large body of literature on the bacterial L-forms due to their fascinating biology and their potential importance in latent and persistent infections [2,3,4,5]. L-form bacteria do not grow in regular culture medium but require special culture conditions including rich medium, serum, cell wall inhibitors such as penicillin, osmotic protectant such as sucrose or sodium chloride and soft agar. In contrast to normal bacteria with cell wall that divide by binary fission mediated by FtsZ protein, L-form bacteria divide in a FtsZ-independent manner upon cell envelope stress under specialized conditions [6]. Thus L-form bacteria serve as a useful model to study cell division. There is recent interest in the molecular basis of L-form bacteria formation and survival [5,6,7,8,9]. Despite numerous studies, little is known about the mechanisms of L-form formation. Previous studies have mainly identified mutations in genes involved in cell wall synthesis or cell division in stable L-form bacteria being important for L-form formation [6,9,10,11]. Using *E. coli* unstable L-form bacteria as a model, we recently systematically examined the molecular basis of L-form formation by microarray analysis and mutant screens and identified a network of genes and pathways involved in L-form formation or survival [7]. These include DNA repair and protection (SOS response), energy production, efflux/transporters, iron homeostasis, cell envelope stress, protein degradation such as trans-

translation [7]. These pathways share significant similarity to those involved in bacterial persister and biofilm formation [7]. Despite the above progress, the molecular basis of L-form bacteria formation in other bacteria remains largely unknown.

Staphylococcus aureus is the leading cause of wound and nosocomial infections [12]. Methicillin-resistant *S. aureus* (MRSA) poses a significant threat in different parts of the world. *S. aureus* is known to form L-form bacteria *in vitro* [13,14] or *in vivo* during infection [15,16,17] or after antibiotic treatment [18]. Clinical samples from patients suffering from MRSA infection contained L-form bacteria exhibiting typical "fried-egg" morphology [16]. There is the interesting observation that going through L-form stage with initial phenotypic resistance or persistence to beta-lactam cell wall antibiotic led to subsequent stable genetic resistance after reversion to walled normal form in *S. aureus* [13]. In addition, *S. aureus* has been demonstrated to form persisters in different studies [19,20]. However, the molecular mechanisms of L-form bacteria formation and persistence in *S. aureus* are unclear. In this study, we constructed a transposon mutant library of *S. aureus* and performed a preliminary screen to identify mutants defective in L-form formation. We identified *glpF* involved in glycerol uptake being critical for L-form formation in *S. aureus*. In addition, we found that *glpF* is also important for persistence to antibiotics in *S. aureus*.

Table 1. Bacterial strains and plasmids used in this study.

Strain or plasmid	Relevant genotype and property	Source or reference
S. aureus		
Newman	Clinical isolate, ATCC 25904, saeS constitutively active	ATCC
glpF mutant	Derived from strain Newman with transposon insertion in *glpF* gene	This study
glpF mutant	*glpF* mutant transformed with pT181	This study
glpF-pT181-*glpF*	*glpF* mutant complemented with pT181 plus wild type *glpF* gene	This study
RN4220	Restriction-deficient shuttle plasmid host	ATCC
Plasmids		
pBursa	Transposon encoding plasmid	[12]
pFA545	Transposase encoding plasmid	[12]
pT181	Plasmid vector for transformation of *S. aureus*	[26]

Materials and Methods

Antibiotics

Penicillin, ampicillin, chloramphenicol, erythromycin, tetracycline, and norfloxacin were obtained from Sigma-Aldrich Co., and their stock solutions were freshly prepared, filter-sterilized and used at appropriate concentrations as indicated.

Bacterial strains and culture conditions

Bacterial strains and plasmids used in this study are listed in Table 1. All of the *S. aureus* strains except strain RN4220 were derivatives of the Newman strain. *S. aureus* strains were cultivated in tryptic soy broth (TSB) (BBL, 211768) and tryptic soy agar (TSA) (Difco, 236950) at 37°C. Chloramphenicol, tetracycline and erythromycin were used at concentrations 5, 2.5 and 10 μg/ml respectively for generating random transposon mutant library. Tetracycline was used at 5 μg/ml for complementation of the *glpF* mutant.

Construction of *S. aureus* transposon mutant library

S. aureus transposon mutant library was constructed as described [12]. Briefly, *S. aureus* RN4220 and Newman strain electrocompetent cells were prepared from log phase cultures grown in TSB medium. pBursa and pFA545 plasmid DNA were transformed into *S. aureus* competent cells by electroporation (voltage = 2.5 kV, resistance = 100 Ω, capacity = 25 μF) using MicroPulser Electroporation Apparatus (Bio-Rad). pFA545 plasmid DNA was transformed into *S. aureus* strain RN4220 and then was isolated from RN4220 for introduction into *S. aureus* Newman strain. pBursa was transformed into *S. aureus* Newman strain carrying pFA545. The cells were spread on TSA plates containing 2.5 μg/ml tetracycline and 5 μg/ml chloramphenicol followed by incubation at 30°C and then transferred onto a 43°C prewarmed TSA containing 10 μg/ml erythromycin and incubated at 43°C. About 6000 mutant clones were picked and cultured in 96 wells and then stored as transposon mutant library at −80°C until use.

Induction of *S. aureus* L-form colonies

S. aureus was grown in brain heart infusion (BHI) broth (Becton Dickinson, BD) overnight to stationary phase. Undiluted cultures were spotted onto L-form induction media (LIM) which consisted of BHI supplemented with 1% agar (BD), 10% horse serum (Sigma), 3.5% sodium chloride, 20% sucrose, 0.125% magnesium sulfate, and 600 μg (1000 units)/ml of Penicillin G (Sigma). After the inoculum was absorbed into the agar, the plates were inverted and incubated at 33°C for 7~10 days. The bacterial colonies were

Table 2. Oligonucleotide primers used in this study.

Primer name	Sequence	Source or reference
ermF	5′-TTTATGGTACCATTCATTTTCCTGCTTTTTC-3′	[12]
ermR	5′-AAACTGATTTTTAGTAAACAGTTGACGATATTC-3′	
16SF	5′-CGTGCTACAATGGACAATACAAA-3′	[27]
16SR	5′-ATCTACGATTACTAGCGATTCCA-3′	
glpKF	5′-TGGACAAGCTTGCTTCGAAC-3′	This study
glpKR	5′-GATGGAACCTTCAAGCGCAT-3′	This study
glpFF	5′-CTGGCGCGAAATTAGGTGTT -3′	This study
glpFR	5′-CGGACCTAAATCACGTGCTG -3′	This study
glpfF	5′- ATTGAC<u>GGATCC</u>AACGCTTTCATATCG-3′	This study#
glpfR	5′ -CGCTAAC<u>CTGCAG</u>CCATTGTACAAAATC-3′	This study#

#The underlined nucleotide sequences <u>GGATCC</u> and <u>CTGCAG</u> represent *Bam*HI and *Pst*I restriction sites incorporated for cloning the wild type *glpF* gene from *S. aureus* Newman into plasmid pT181 for complementation.

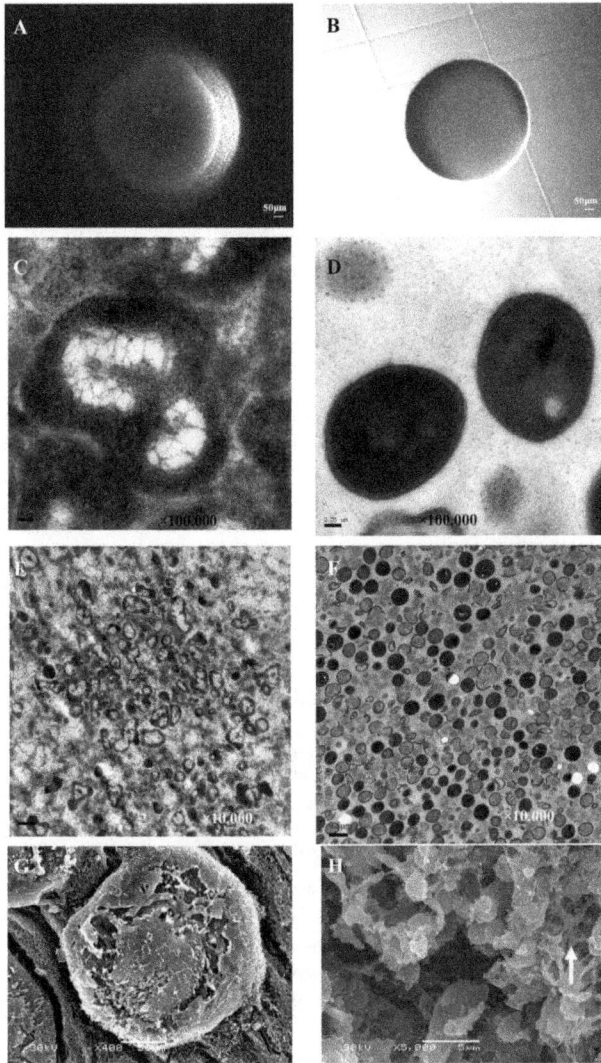

Figure 1. Comparison of L-form and classical form *S. aureus* morphologies. (A). *S. aureus* Newman L-form colony on L-form induction media (LIM) exhibiting typical "fried egg" morphology. (B). Control classical *S. aureus* colony on BHI medium with sucrose control but without penicillin. (C). *S. aureus* L-form shape and structure (TEM, ×100,000). *S. aureus* L-form had irregular morphology with larger size than normal *S. aureus* and contained a large number of vesicles. The L-form bacteria had deficient or fractured cell wall. (D). Normal *S. aureus* with spherical shape and thick cell wall structure (TEM, ×100,000). (E). *S. aureus* L-form (TEM, × 10,000) showing polymorphic L-form bacteria with varying sizes and shapes. (F). *S. aureus* normal classical form (TEM, × 10,000) showing regular spherical morphology with homogeneous size and shapes. (G). *S. aureus* L-form colony (SEM, × 400). *S. aureus* L-form colony exhibited typical "fried egg" morphology. (H). *S. aureus* L-form colony inner structure (SEM, × 5000). In the inside of *S. aureus* L-form colony, the bacteria exhibited polymorphic shapes of varying sizes with the L-form colony structure showing similarity to a multilayered biofilm structure. The polymorphic bacteria are connected with large amounts of extracellular matrix materials (exopolysaccharide (EPS)) and lysed bacteria (arrow).

detected by inverted microscope (Nikon GM3) and the typical L-form colonies appeared as "fried egg".

Microscopy

S. aureus L-form bacteria were examined using a Nikon GM3 inverted microscope for "fried egg" colonies grown on LIM agar along with normal growth on a control medium without penicillin. The typical colonies were fixed by glutaraldehyde before being processed and examined by electron microscopy (EM). Scanning EM and transmission EM were performed with scanning electron microscope (JSM-6380Lv) and transmission electron microscope (JEM-1230), respectively, using procedures as described [7,21].

Library screen to identify mutants with defect in L-form colony formation

The library screening procedure was similar to that as we previously described [7]. Briefly, the mutant library consisting of 6076 transposon mutants of *S. aureus* Newman was grown in 200 μl BHI medium at 37°C overnight in 96-well plates without shaking. Stationary phase culture of the mutant library was transferred onto L-form medium LIM plates (150 mm) by a 96-well replicator. Plates were allowed to dry before being inverted and incubated at 33°C for up to 7 days before mutants were scored for defect in forming L-form colonies on LIM plates.

Inverse PCR and DNA sequencing of PCR products from L-form mutants

Overnight cultures of L-form deficient mutants and the *S. aureus* parent strain Newman were centrifuged and bacterial cells were collected for DNA extraction. The genome DNA was isolated by using lysostaphin (Sigma), glass beads (0.1 mm), RNase solution (4 mg/ml), followed by phenol/chloroform extraction and ethanol DNA precipitation. The purified chromosomal DNA was digested by restriction enzyme *Aci*I (New England Biolabs) and DNA restriction fragments were then circularized using T4 DNA ligase (New England Biolabs). The ligated DNA (5 μl) was used as template for inverse PCR reaction in a 25 μl reaction volume with primers ermF and ermR (Table 2). The PCR cycling parameters were 10 min at 96°C, followed by 40 cycles of 30 s at 94°C, 30 s at 63°C, and 3 min at 72°C. The PCR products were subjected to DNA sequencing with primer ermF. The identity of the DNA sequences was searched in the NCBI database using the BLAST algorithm to identify the gene of interest.

Real-time reverse transcription PCR

Real-time RT-PCR was used to assess the level of expression of *glpF* transcription in *S. aureus* L-form versus classical form bacteria. Cultures of *S. aureus* strain Newman grown overnight in TSB were inoculated onto LIM and BHI media with sucrose control respectively. These plates were incubated at 33°C for 7 days. The colonies grown on LIM and BHI with sucrose were collected for RNA isolation. The culture samples were washed once with DEPC-H₂O and centrifuged at 8,000 g at 4°C for 5 min. RNA was isolated according to manufacturer's instruction (Sangon Biotech Co., Ltd., Shanghai). Primers corresponding to the genes of interest were designed using Primer Express software (Version 2.0, Applied Biosystems) (Table 2). Total RNA was converted to cDNA using Super-Script III First-Strand Synthesis (Takara Bio) as described by the manufacturer. The cDNA was used as template to perform real-time RT-PCR per instruction of the reagent kit SYBR Premix Ex Taq II (Takara Bio). The expression of 16S rRNA was used as the control for estimating the fold changes of genes of interest. Cycling parameters were 95°C for 30 s and followed by 40 cycles of 5 s at 95°C, 30 s at 60°C. Relative expression levels were determined by the comparative threshold cycle (△△Ct) method.

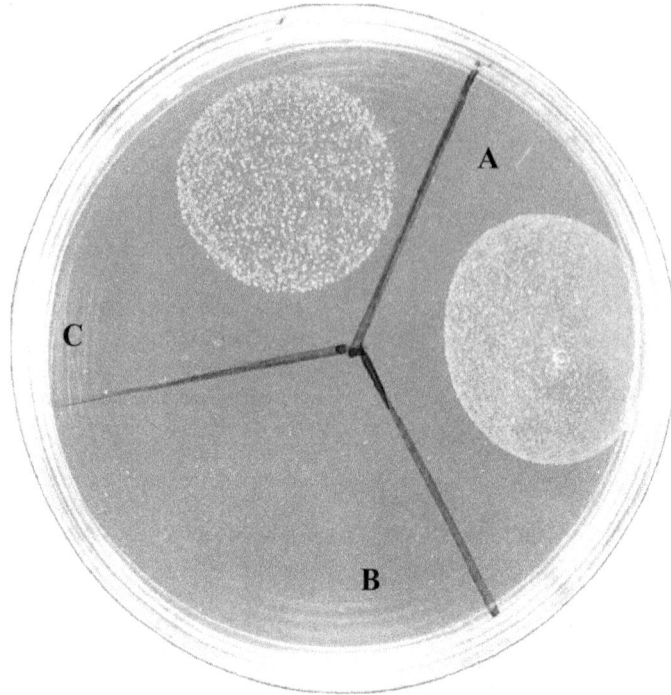

Figure 2. Results of *S. aureus* **Newman,** *glpF* **mutant and the complemented strain in L-form formation.** Stationary phase cultures of (A) *S. aureus* Newman, (B) *glpF* mutant-pT181 vector control, and (C) *glpF*-pT181-*glpF* complemented strain were spotted on L-form induction media (LIM) followed by incubation at 33°C for 5 days when the L-form growth was examined.

Complementation of *S. aureus* mutants

The wild type *glpF* gene from *S. aureus* Newman was amplified by PCR using primers glpfF and glpfR (see Table 2). The PCR primers were taken from −134 bp and 99 bp upstream and downstream of the *glpF* gene and contained restriction sites *Bam*HI and *Pst*I respectively. The PCR parameters were: 94°C 15 min, followed by 35 cycles of 94°C 30 s, 55°C 30 s, and 72°C 2 min, and a final extension at 72°C for 10 min. The PCR products were digested with *Bam*HI and *Pst*I and then cloned into plasmid pT181 cut with the same enzymes. The ligation products were electroporated into *S. aureus* strain RN4220 and the positive clones were identified by restriction digestion and PCR. The

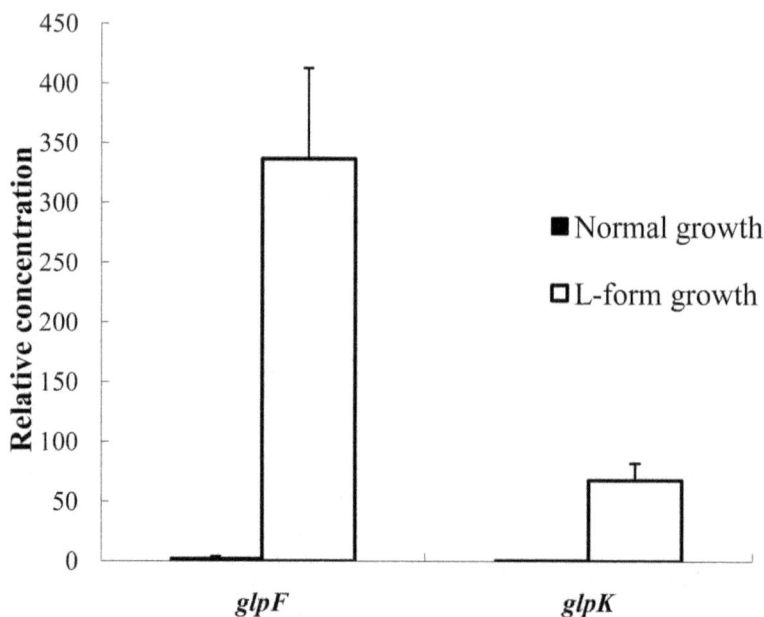

Figure 3. Relative concentrations of *glpF* **and** *glpK* **expression in normal growth and L-form** *S. aureus* **as assessed by RT-PCR.**

Table 3. Effect of glycerol on L-form formation of the *glpF* mutant.

	Wild type	*glpF* mutant	*glpF* complementation
Spotted bacterial number on TSA (cfu/ml)	3.63×10^9	1.73×10^9	5.77×10^9
L-form colony number on LIM (cfu/ml)	3.1×10^5	0	2.15×10^5
L-form colony number on 1% glycerol LIM (cfu/ml)	7.75×10^5	1.03×10^4	2.90×10^5

recombinant plasmid pT181+*glpF* and the pT181 vector alone control were then transformed into *S. aureus glpF* mutant by electroporation as described above under Construction of *S. aureus* transposon mutant library.

For complementation of the *glpF* mutant in L-form formation, *S. aureus* Newman strain was grown in BHI broth, and both *glpF* mutant-pT181 vector control and the *glpF* pT181-glpF complemented strain were grown in BHI broth with 5 µg/ml tetracycline overnight to stationary phase. The strains were spotted onto LIM and incubated aerobically at 33°C for 5 days followed by detection of typical L-form colonies which appeared as "fried egg" by inverted microscope as described above.

Effect of glycerol on L-form formation

The stationary phase cultures of wild type *S. aureus* Newman strain grown in BHI broth and *glpF* mutant and the complemented strain in BHI broth containing 5 µg/ml tetracycline were spotted onto LIM without glycerol and with varying concentrations (0, 0.1%, 1%, and 10%) of glycerol, respectively. After incubation aerobically at 33°C for 5 days, the "fried egg" colonies

were detected to confirm the effects of glycerol on *glpF* mutant L-form formation.

Persister assays

For antibiotic exposure, *S. aureus glpF* mutant, the *glpF* mutant complemented strain, and *S. aureus* parent strain Newman were grown to stationary phase overnight when antibiotics ampicillin (50 µg/ml) and norfloxacin (40 µg/ml) were added to undiluted cultures and incubated without shaking for various times up to 7 or 8 days. Aliquots of bacterial cultures exposed to antibiotics were taken at different time points and washed and then plated onto TSA plates for CFU counting.

Results and Discussion

Generation of antibiotic induced unstable L-forms of *S. aureus*

L-form induction media (LIM) was tested for its ability to induce *S. aureus* (Newman) to grow as L-form colonies. Stationary phase cells of *S. aureus* Newman strain produced L-form colonies when plated directly onto LIM. The minimum bacterial inoculum

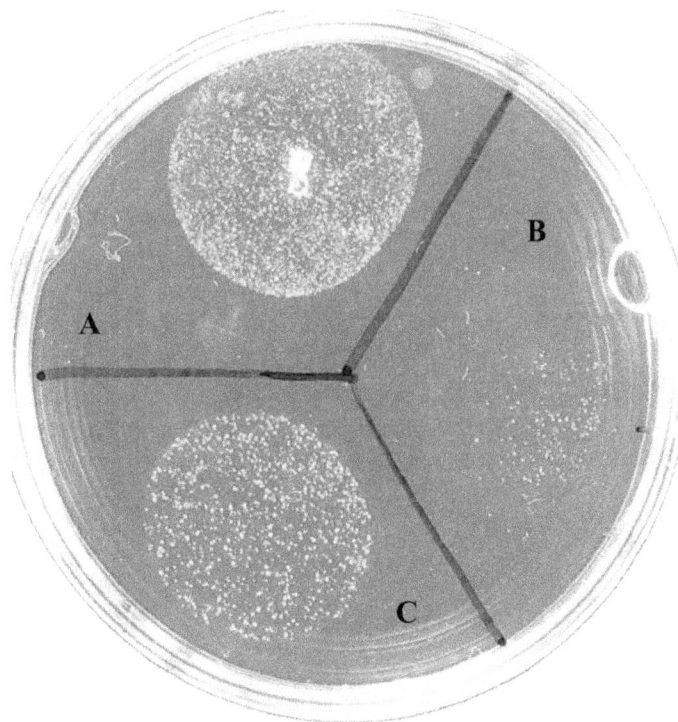

Figure 4. Effect of glycerol on L-form growth for *S. aureus* Newman, *glpF* mutant and the complemented strain. Stationary phase cultures of (A) *S. aureus* Newman, (B) *glpF* mutant-pT181 vector control, and (C) *glpF*-pT181-*glpF* complemented strain were spotted on L-form induction media (LIM) containing 1% glycerol followed by incubation at 33°C for 5 days when the L-form growth was assessed.

Table 4. Survival of stationary phase cultures of the *glpF* mutant, the complemented strain and the parent strain upon antibiotic exposure over time.

Antibiotics	Time	No. of bacteria (log CFU ML⁻¹, mean ±SD)		
		Newman	*glpF* mutant-pT181	*glpF*-pT181-*glpF*
Ampicillin	0	9.88±0.20	9.46±0.11	9.48±0.32
(50 µg/ml)	1d	9.74±0.24	9.26±0.21	9.30±0.17
	2d	8.60±0.09	8.70±0.24	9.28±0.13
	3d	8.37±0.23	7.65±0.07	8.41±0.02
	5d	6.92±0.11	3.84±0.09	7.37±0.10
	7d	5.24±0.34	0	2.65±3.74
Norfloxacin	0	9.88±0.20	9.46±0.11	9.48±0.16
(40 µg/ml)	1d	8.90±0.09	9.2±0.13	9.0±0.17
	2d	9.08±0.22	8.48±0.11	8.90±0.36
	3d	7.91±0.29	7.46±0.04	8.06±0.03
	5d	7.22±0.02	6.04±0.01	7.41±0.03
	8d	5.06±0.16	0	5.22±0.10

required for L-form colony formation was approximately 10^6–10^7 bacteria (the maximum dilution for 10^8–10^9 S. *aureus* bacteria to form L-form colony was 1:100). This frequency is significantly lower than the L-form formation frequency of *E. coli* which is 10^4–10^5 [7]. The typical S. *aureus* L-form colonies had "fried egg" morphology (Fig. 1A) under inverted microscope in contrast to smooth colony of the normal form of S. *aureus* (Fig. 1B). The "fried-egg" L-form colonies had typical embedded growth into the soft agar and could not be scraped off in contrast to the normal classical forms which did not show embedded growth and could be scraped off easily from agar surface. Transmission electron microscopy (TEM) indicated that the S. *aureus* L-form bacteria had complete or partial loss of cell wall and contained a large number of intracellular vesicles (Fig. 1C) in contrast to the normal

forms with cell wall without obvious vesicles (Fig. 1D). S. *aureus* L-form (TEM × 10,000) showed polymorphic sizes and shapes (Fig. 1E) in contrast to normal classical form showing morphology with homogeneous size and round shape with clear and smooth cell boundary (Fig. 1F). Scanning electron microscopy (SEM) indicated that S. *aureus* L-form colony exhibited rough surface morphology (Fig. 1G) while in the inside of the L-form colony the bacterial cells exhibited polymorphic morphologies of varying sizes (Fig. 1H) with the L-form colony structure showing similarity to biofilm structure. This is consistent with previous observation that L-form bacteria secrete exopolysaccharide (EPS) to the surface to prevent desiccation similar to biofilms and that defect in genes involved in EPS synthesis can cause lack of L-form growth [7].

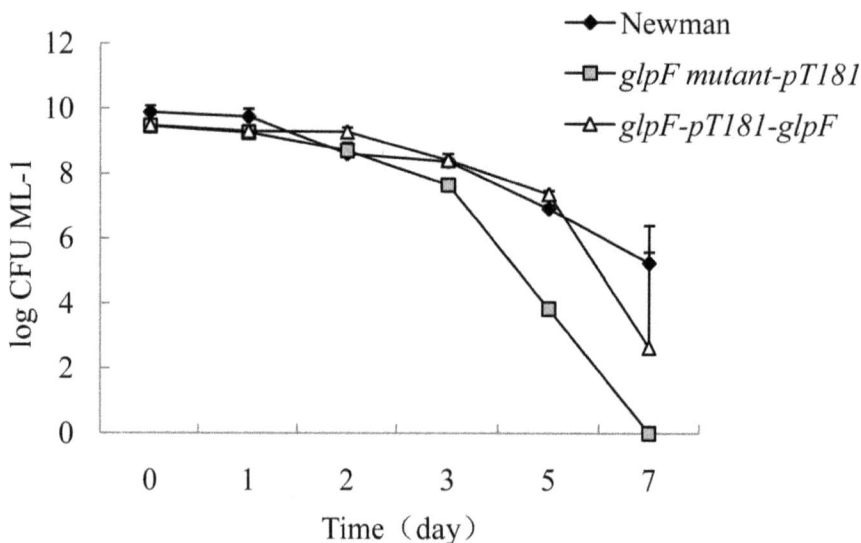

Figure 5. Survival of stationary phase cultures of the *S. aureus glpF* mutant transformed with vector pT181, or pT181+wild type *glpF*, and the parent strain upon ampicillin (50 µg/ml) exposure over time.

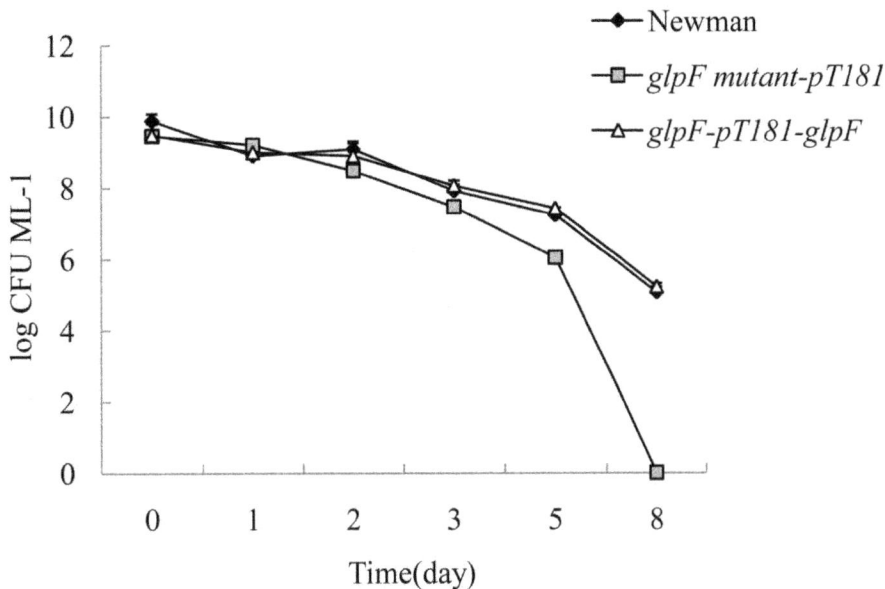

Figure 6. Survival of stationary phase cultures of the *glpF* mutant, the complemented strain, and the parent strain upon norfloxacin (40 μg/ml) exposure over time.

Screening for mutants with defect in L-form formation from *S. aureus* transposon mutant library

Having established the *S. aureus* L-form conditions, we wanted to identify genes that are involved in L-form formation. To do this, we first constructed a *S. aureus* transposon mutant library and then grew the library at 37°C overnight in 96-well plates followed by transfer of the mutants onto LIM plates as described in the Methods. Plates were incubated at 33°C for 7 days when mutants were scored for defect in forming L-form colonies.

To identify the genes whose mutation led to defect in L-form formation, we performed inverse PCR as described in the Methods. Using inverse PCR and DNA sequencing we were able to identify 12 genes from 15 mutants, 3 of which mapped to *glpF* and *NWMN_0623* each, 1 mapped to *glpK*, 1 mapped to gluconate kinase (*gntK*), *NWMN_0623*, *NWMN_0872* (GTP pyrophosphokinase), *NWMN_1269* (sodium:alanine symporter family protein), 2 mapped to hypothetical proteins *NWMN_0333* and *NWMN_0843* of unknown function, 3 mapped to intergenic region. Because we found 4 mutants (3 in *glpF* and 1 in *glpK*) mapped to glycerol metabolism genes which predominate among these identified genes, we therefore focused and further characterized the role of glycerol metabolism genes in this study. To determine if the mutated *glpF* is indeed responsible for defective L-form formation, we attempted to complement the *glpF* mutant with the wild type *glpF* gene using the plasmid vector pT181. However, the initial attempt was unsuccessful when the complemented *glpF* mutant was plated directly on LIM plates. Since our previous work with *E. coli* L-form complementation indicated that inducible expression of the gene involved in L-form formation is critical for successful complementation of L-form defect of the mutants [7], we therefore induced the complemented *glpF S. aureus* strain containing tetracycline inducible vector pT181 with tetracycline in liquid culture prior to plating on LIM plates. This led to successful complementation of the *glpF* mutant with the wild type *glpF* gene. However, the effect of the complementation was partial (Fig. 2C) compared with the parent strain Newman

(Fig. 2A), while the *glpF* mutant transformed with the pT181 vector control did not form any L-form colonies (Fig. 2B).

glpF and *glpK* were overexpressed in *S. aureus* L-form bacteria but not in normal bacteria

Since we identified mutations in *glpF* and *glpK* caused defect in L-form growth, we wanted to know if *glpF* and *glpK* are overexpressed in *S. aureus* L-form bacteria compared with normal classical form *S. aureus*. To confirm this, we prepared *S. aureus* L-form bacteria from L-form media LIM and normal *S. aureus* growth as a control on media without penicillin and isolated RNA from both types of the bacterial cells. The isolated RNA samples were then subjected to RT-PCR. *glpF* and *glpK* were found to be expressed at very low levels in normal growth but were significantly induced to 144-fold and 68-fold higher respectively in L-forms than in the normal control growth ($P < 0.05$) (Fig. 3).

Effect of glycerol on restoring L-form formation in *glpF* mutant

Since GlpF is involved in glycerol transport, we wanted to address the role of exogenously added glycerol in L-form formation in *S. aureus*. To do so, we incorporated varying concentrations of glycerol 0, 0.1%, 1% and 10% glycerol into the LIM. In LIM media with 0 and 0.1% glycerol, only wild type *S. aureus* Newman strain and the *glpF* complemented strain grew but *glpF* mutant failed to grow (Table 3). However, at 1% glycerol, the *glpF* mutant formed 1.03×10^4 L-form colonies but its efficiency was much lower than the wild type (7.75×10^5 L-form colonies) and the *glpF* complemented strain (2.90×10^5 L-form colonies) (Fig. 4, Table 3). The 1% glycerol only marginally increased the number of L-form colonies of the wild type strain by about 2 fold (Table 3). In contrast, at 10% glycerol, none of the strains Newman, the *glpF* mutant, or the *glpF* complemented strain formed L-form colonies.

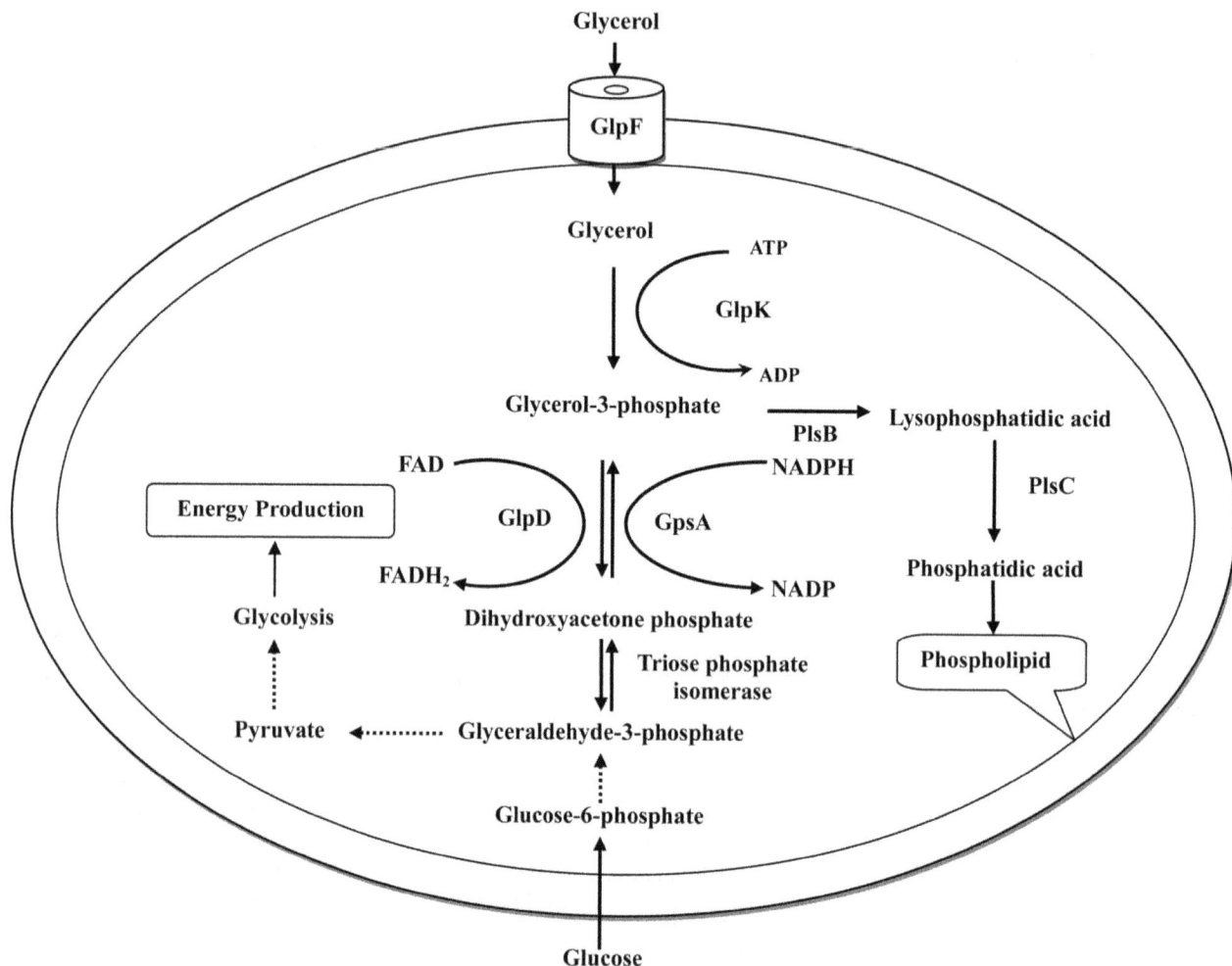

Figure 7. Glycerol uptake and metabolic pathway. GlpF is involved in uptake of glycerol into *S. aureus* while GlpK is glycerol kinase and is involved in converting glycerol to glycerol-3-phosphate which then can be used for synthesis of cell membrane phospholipid and also for energy production.

To address if the role of 1% glycerol is to serve as an osmoprotectant in facilitating L-form formation in the *glpF* mutant, we added known osmoprotectant NaCl at 4.5% to the LIM and assessed if it could allow the *glpF* mutant to grow as L-forms. However, supplementation of NaCl while allowing L-form formation in wild type and complemented *glpF* mutant, failed to facilitate L-form formation in the *glpF* mutant (data not shown).

Defective persistence of *S. aureus glpF* mutant in antibiotic exposure assays

In our previous study with *E. coli* L-form bacteria, we found that genes involved in L-form formation overlapped with those involved in antibiotic persistence [7]. Exposure of the stationary phase culture of *glpF* mutant to ampicillin (50 μg/ml) revealed that the mutant began to show defect in persistence at day 3 but the defect was more obvious after 5 days. After 7 days, no CFU was detectable in the *glpF* mutant transformed with the vector control pT181; in contrast, the complemented strain (transformed with pT181+*glpF*) and the parent strain Newman had 100–1000 and 10,000 CFUs remaining (Fig. 5) (Table 4). For exposure to norfloxacin (40 μg/ml), the *glpF* mutant had a similar trend as ampicillin treatment, where it reached zero CFU by day 8.

Complementation of the *glpF* mutant with the wild type *glpF* gene only partially restored the persister levels compared with parent strain for ampicillin exposure (Fig. 5) but caused full restoration of persister levels for norfloxacin exposure (Fig. 6) (Table 4).

Although many previous studies have demonstrated the formation of L-forms by *S. aureus* [13,14,17], the mechanisms involved in L-form formation has remained unknown in this organism. This study provided the first molecular insight into the mechanism of L-form formation in *S. aureus* by demonstrating the important role of glycerol uptake in the L-form formation. GlpF is involved in glycerol uptake (Fig. 7), and the observation that *glpF* mutation causes defect in L-form growth in *S. aureus* suggests uptake of glycerol is essential for L-form formation. The mechanism by which GlpF is involved in L-form formation in *S. aureus* is likely mediated through its role in production of energy via pyruvate entry into TCA cycle and glycolysis or alternatively through cell membrane synthesis via lysophosphatidic acid (LPA) to phospholipids (Fig. 7) to strengthen membrane integrity required for L-form growth in the absence of cell wall. This study represents the first effort at identifying the mechanisms of L-form formation in *S. aureus,* and further studies are needed to explore other possible mechanisms of L-form formation besides glycerol uptake in future studies.

It is interesting to note that supplementation of appropriate concentration of glycerol (1%) partially compensated for the loss of L-form formation to *glpF* mutant (Fig. 4). GlpF is known to be a member of the aquaporin (AQP) channel family. Some of AQPs channels that conduct water and also conduct glycerol (aqua-glyceroporins) [22]. When the environmental glycerol is increased, the aquaporin (AQP) channel may partially compensate for the loss of GlpF and help the *glpF* mutant to form a small number of L-form colonies (Fig. 4). We believe the role of glycerol in L-form formation is mediated through its uptake and metabolism rather than an osmoprotectant role in L-form formation for the following reasons. First, we demonstrated in the new experiment that adding 4.5% NaCl as an osmoprotectant did not help the *glpF* mutant to form L-forms while it allowed wild type and the complemented *glpF* mutant strains to form L-form colonies. This indicates the defect in *glpF* mutant cannot be complemented by other osmoprotectant like NaCl. Second, if the role of glycerol were to serve as osmoprotectant, we would expect that higher concentration of glycerol would facilitate L-form formation. However, we found higher glycerol content, e.g., 10% glycerol did not allow L-forms to form. Third, our finding that mutation in *glpK* which is glycerol kinase involved in glycerol metabolism caused defect in L-form formation also does not support glycerol serving as an osmoprotectant in facilitating L-form formation. This is because GlpK is glycerol kinase that converts glycerol to glycerol-3-phosphate which can be used for synthesis of cell membrane phospholipid and also for energy production to facilitate L-form formation. However, high glycerol concentration (10%) inhibited the L-form growth for wild type as well as the mutant *S. aureus*, presumably because high concentrations of glycerol produce toxic metabolites thus preventing L-form growth.

It is noteworthy that in addition to its role in L-form formation, *glpF* is also involved in tolerance or persistence to antibiotics in *S. aureus* as demonstrated by a defect in persistence in *glpF* mutant upon exposure to ampicillin or norfloxacin (Fig. 5 and Fig. 6). This finding is consistent with the previous observation that genes involved in glycerol metabolism such as *glpD* encoding sn-glycerol-3-phosphate dehydrogenase and *plsB* encoding sn-glycerol-3-phosphate acyltransferase (Fig. 7), have been found to be involved in persister formation [23]. Our findings that glycerol uptake is important for both L-form and persistence of *S. aureus* and provide further support for the close relationship between the two entities as has been observed for *E. coli* [7]. This finding is consistent with the recent proposal that L-form bacteria are related to persisters and are part of the heterogeneous persister continuum [24]. The only difference is that frequency of L-form forming bacteria is 2–3 orders of magnitude lower than persister frequency [24] and can be considered "deep" persisters [25]. Although previous study with *E. coli* L-form bacteria did not identify *glpF* being critical for L-form formation as in *S. aureus*, microarray analysis indicated that glycerol metabolism gene *glpD* encoding sn-glycerol-3-phosphate dehydrogenase is upregulated in *E. coli* L-form bacteria [7], suggesting glycerol metabolism could be important for L-form bacteria in different bacterial species.

In conclusion, we established an L-form model in *S. aureus*, characterized their morphologies and optima conditions of their formation, and identified genes involved in glycerol uptake and metabolism being important for L-form formation and persistence in *S. aureus*. These findings shed new light on the mechanisms of L-form formation and persister biology in *S. aureus* and may have implications for development of new drugs targeting persisters for improved treatment of persistent bacterial infections.

Acknowledgments

We thank Dominique Missiakas for providing plasmids used in this study. Jian Han was supported by China Scholarship Council.

Author Contributions

Conceived and designed the experiments: YZ. Performed the experiments: JH LLH WLS XGX SW SZ. Analyzed the data: JH YZ. Contributed reagents/materials/analysis tools: JH LLH WLS. Contributed to the writing of the manuscript: YZ JH.

References

1. Klienenberger E (1935) The natural occurrence of pleuropneumonia-like organisms in apparent symbiosis with Streptobacillus moniliforms and other bacteria. J Pathol Bacteriol 40: 93–105.
2. Dienes L, Weinberger HJ (1951) The L forms of bacteria. Bacteriol Rev 15: 245–288.
3. Domingue GJ Sr., Woody HB (1997) Bacterial persistence and expression of disease. Clin Microbiol Rev 10: 320–344.
4. Allan EJ, Hoischen C, Gumpert J (2009) Bacterial L-forms. Adv Appl Microbiol 68: 1–39.
5. Domingue GJ (2010) Demystifying pleomorphic forms in persistence and expression of disease: Are they bacteria, and is peptidoglycan the solution? Discovery medicine 10: 234–246.
6. Leaver M, Dominguez-Cuevas P, Coxhead JM, Daniel RA, Errington J (2009) Life without a wall or division machine in Bacillus subtilis. Nature 457: 849–853.
7. Glover WA, Yang Y, Zhang Y (2009) Insights into the molecular basis of L-form formation and survival in Escherichia coli. PLoS One 4: e7316.
8. Devine KM (2012) Bacterial L-forms on tap: an improved methodology to generate Bacillus subtilis L-forms heralds a new era of research. Molecular microbiology 83: 10–13.
9. Joseleau-Petit D, Liebart JC, Ayala JA, D'Ari R (2007) Unstable Escherichia coli L forms revisited: growth requires peptidoglycan synthesis. J Bacteriol 189: 6512–6520.
10. Siddiqui RA, Hoischen C, Holst O, Heinze I, Schlott B, et al. (2006) The analysis of cell division and cell wall synthesis genes reveals mutationally inactivated ftsQ and mraY in a protoplast-type L-form of Escherichia coli. FEMS Microbiol Lett 258: 305–311.
11. Dominguez-Cuevas P, Mercier R, Leaver M, Kawai Y, Errington J (2012) The rod to L-form transition of Bacillus subtilis is limited by a requirement for the protoplast to escape from the cell wall sacculus. Molecular microbiology 83: 52–66.
12. Bae T, Glass EM, Schneewind O, Missiakas D (2008) Generating a collection of insertion mutations in the Staphylococcus aureus genome using bursa aurealis. Methods Mol Biol 416: 103–116.
13. Fuller E, Elmer C, Nattress F, Ellis R, Horne G, et al. (2005) Beta-lactam resistance in Staphylococcus aureus cells that do not require a cell wall for integrity. Antimicrob Agents Chemother 49: 5075–5080.
14. Banville RR (1964) Factors affecting growth of Staphylococcus aureus L forms on semidefined medium. Journal of bacteriology 87: 1192–1197.
15. Michailova L, Kussovsky V, Radoucheva T, Jordanova M, Markova N (2007) Persistence of Staphylococcus aureus L-form during experimental lung infection in rats. FEMS Microbiol Lett 268: 88–97.
16. Tanimoto A, Kitagaki Y, Hiura M, Fujiwara H, Iijima K, et al. (1995) [Methicillin-resistant Staphylococcus aureus forming the fried egg appearance colonies isolated from a patient with septicemia]. Jap J Clin Pathol 43: 1061–1065.
17. Owens WE (1987) Isolation of Staphylococcus aureus L forms from experimentally induced bovine mastitis. Journal of clinical microbiology 25: 1956–1961.
18. Sears PM, Fettinger M, Marsh-Salin J (1987) Isolation of L-form variants after antibiotic treatment in Staphylococcus aureus bovine mastitis. Journal of the American Veterinary Medical Association 191: 681–684.
19. Lechner S, Lewis K, Bertram R (2012) Staphylococcus aureus persisters tolerant to bactericidal antibiotics. J Mol Microbiol Biotechnol 22: 235–244.
20. Keren I, Kaldalu N, Spoering A, Wang Y, Lewis K (2004) Persister cells and tolerance to antimicrobials. FEMS Microbiol Lett 230: 13–18.
21. Shingaki R, Kasahara Y, Iwano M, Kuwano M, Takatsuka T, et al. (2003) Induction of L-form-like cell shape change of Bacillus subtilis under microculture conditions. Microbiology (Reading, England) 149: 2501–2511.
22. Stroud RM, Miercke LJ, O'Connell J, Khademi S, Lee JK, et al. (2003) Glycerol facilitator GlpF and the associated aquaporin family of channels. Curr Opin Struct Biol 13: 424–431.

23. Spoering AL, Vulic M, Lewis K (2006) GlpD and PlsB participate in persister cell formation in Escherichia coli. J Bacteriol 188: 5136–5144.

24. Zhang Y (2014) Persisters, Persistent Infections and the Yin-Yang Model. Emerging Microbes and Infections (Nature Publishing Group) 3, e3; doi:10.1038/emi.2014.3.

25. Ma C, Sim S, Shi W, Du L, Xing D, et al. (2010) Energy production genes sucB and ubiF are involved in persister survival and tolerance to multiple antibiotics and stresses in Escherichia coli. FEMS Microbiol Lett 303: 33–40.

26. Khan SA, Novick RP (1983) Complete nucleotide sequence of pT181, a tetracycline-resistance plasmid from Staphylococcus aureus. Plasmid 10: 251–259.

27. Luong TT, Dunman PM, Murphy E, Projan SJ, Lee CY (2006) Transcription Profiling of the mgrA Regulon in Staphylococcus aureus. Journal of bacteriology 188: 1899–1910.

## 15

Genome Features of the Endophytic Actinobacterium *Micromonospora lupini* Strain Lupac 08: On the Process of Adaptation to an Endophytic Life Style?

Martha E. Trujillo[1]*, Rodrigo Bacigalupe[1], Petar Pujic[2], Yasuhiro Igarashi[3], Patricia Benito[1], Raúl Riesco[1], Claudine Médigue[4], Philippe Normand[2]

1 Departamento de Microbiología y Genética, Edificio Departamental, Campus Miguel de Unamuno, Universidad de Salamanca, Salamanca, Spain, 2 Université Lyon 1, Université de Lyon, CNRS-UMR5557 Ecologie Microbienne, Villeurbanne, France, 3 Biotechnology Research Center, Toyama Prefectural University, Kurokawa, Imizu, Toyama, Japan, 4 Genoscope, CNRS-UMR 8030, Atelier de Génomique Comparative, Evry, France

Abstract

Endophytic microorganisms live inside plants for at least part of their life cycle. According to their life strategies, bacterial endophytes can be classified as "obligate" or "facultative". Reports that members of the genus *Micromonospora*, Gram-positive Actinobacteria, are normal occupants of nitrogen-fixing nodules has opened up a question as to what is the ecological role of these bacteria in interactions with nitrogen-fixing plants and whether it is in a process of adaptation from a terrestrial to a facultative endophytic life. The aim of this work was to analyse the genome sequence of *Micromonospora lupini* Lupac 08 isolated from a nitrogen fixing nodule of the legume *Lupinus angustifolius* and to identify genomic traits that provide information on this new plant-microbe interaction. The genome of *M. lupini* contains a diverse array of genes that may help its survival in soil or in plant tissues, while the high number of putative plant degrading enzyme genes identified is quite surprising since this bacterium is not considered a plant-pathogen. Functionality of several of these genes was demonstrated *in vitro*, showing that Lupac 08 degraded carboxymethylcellulose, starch and xylan. In addition, the production of chitinases detected *in vitro*, indicates that strain Lupac 08 may also confer protection to the plant. *Micromonospora* species appears as new candidates in plant-microbe interactions with an important potential in agriculture and biotechnology. The current data strongly suggests that a beneficial effect is produced on the host-plant.

Editor: Holger Brüggemann, Aarhus University, Denmark

Funding: MET received financial support from the Spanish Ministerio de Economía y Competitividad under project CGL2009-07287. PN acknowledges financial support from the ANR (Sesam). The funders had no role in study design, data collection and analysis, decisión to publish, or preparation of the manuscript.

Competing Interests: The authors have declared that no competing interests exist.

* Email: mett@usal.es

Background

For a long time, it was considered that a healthy plant was a plant without microbes within its tissues. However, this view has started to change with new approaches to allow strains to grow for a longer time upon isolation as well as the use of NGS, which has permitted the identification of several strains present in the tissues of healthy plants, in particular several actinobacteria [1,2].

Endophytic microorganisms live inside plants for at least part of their life cycle. According to their life strategies, bacterial endophytes can be classified as "obligate" or "facultative". Obligate endophytes are strictly dependent on the host plant for their growth and survival while facultative endophytes have a stage in their life cycle during which they exist outside host plants [3]. These endophytes originate from soil, initially infecting the host plant by colonizing, for instance, the cracks formed at points of emergence of lateral roots from where they quickly spread to the intercellular spaces in the root [4]. Thus, a series of environmental and genetic factors is presumed to have a role in enabling a specific bacterium to become endophytic [5]. Conversely, Marchetti and co-workers [6] recently showed how a pathogen can evolve in a few generations to become a symbiotic endophyte by losing specific transporters and regulators linked to pathogenesis.

Micromonospora is a genus of Gram-positive Actinobacteria that was first isolated from soil [7]. This bacterium has received a lot of attention during natural product screening programs, given its ability to produce a very rich array of secondary metabolites [8,9,10]. The distribution of members of *Micromonospora* is wide-ranging since these bacteria have been isolated from different geographical zones. In addition, its habitats are also diverse and include: soil, freshwater and marine sediments, mangrove soils, rocks, and nitrogen fixing nodules of both leguminous and actinorhizal plants [11,12,13]. The recent report [13] that *Micromonospora* inhabits nitrogen-fixing nodules in a systematic way, has opened up a question as to what is the potential ecological role of this bacterium in the plant and whether this bacterium is in a process of adaptation from a terrestrial to a facultative endophytic life style.

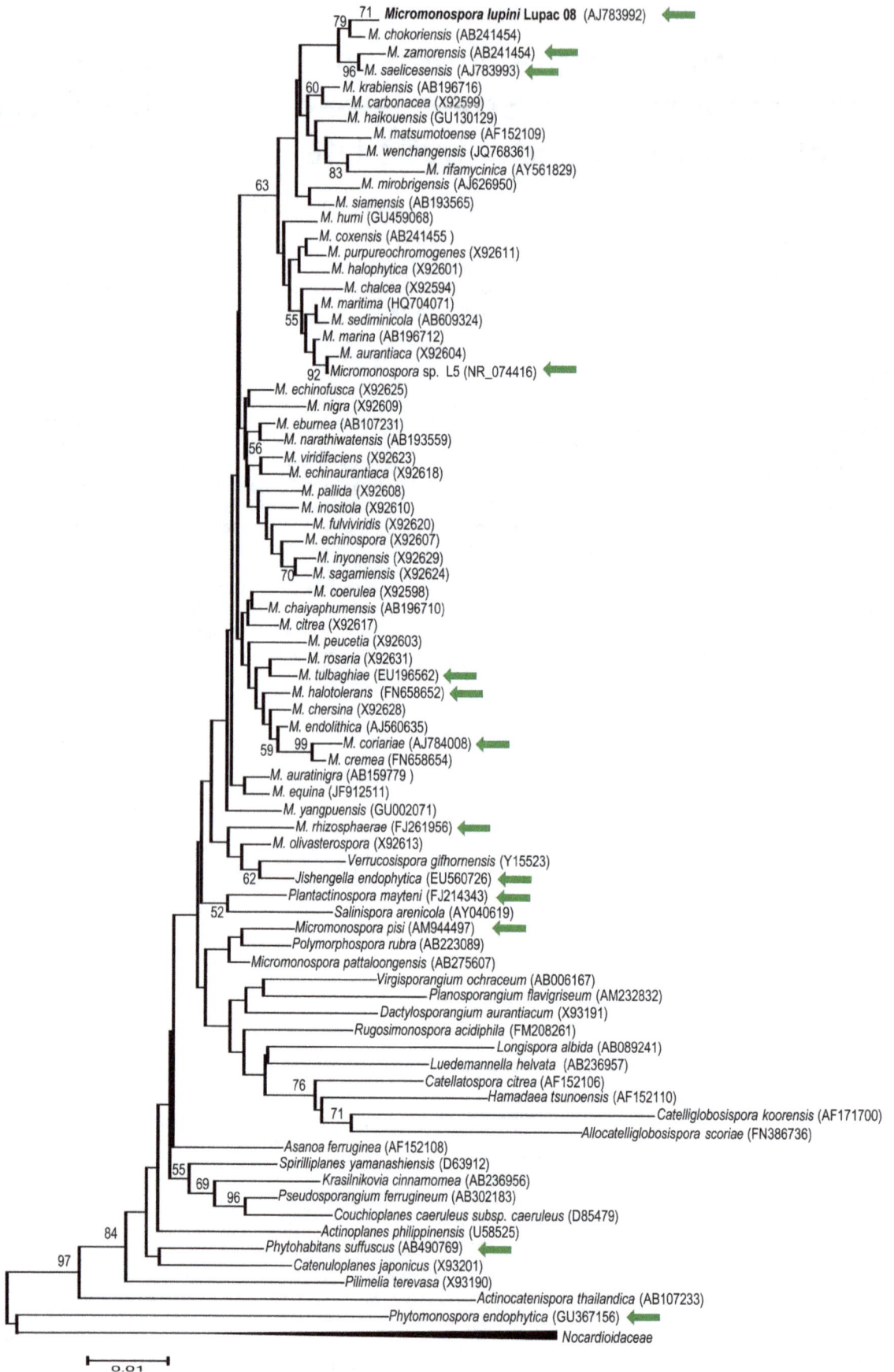

Phylogenetic tree of *Micromonospora* and related genera based on 16S rRNA gene sequences, with scale bar 0.01.

Figure 1. Neighbour-joining tree based on 16S rRNA gene sequences showing the relationship of *Micromonospora* species and other members of the family *Micromonosporaceae*. Strains isolated from plant related sources are indicated by a green arrow.

Taxonomically, *Micromonospora* belongs to the family *Micromonosporaceae* which currently contains 27 genera and includes aerobic, non-acid fast and mesophilic microorganisms. Many strains produce mycelial carotenoid pigments giving the colonies an orange to red appearance, but blue-green, brown or purple pigmented strains have also been isolated. The family *Micromonosporaceae* also harbors the genus *Salinispora*, which is widely distributed in tropical and sub-tropical marine sediments. This taxon was described as the first marine actinomycete given its inability to grow in low salinity medium. Indeed, genomic information obtained from the genomes of *Salinispora tropica* and *Salinispora arenicola* provide evidence of marine adaptation of *Salinispora* species [14]. Thus, it appears that *Salinispora* evolved from a terrestrial environment to a marine habitat. In the case of some *Micromonospora* lineages, the question is whether this bacterium has followed a comparable evolution process, changing from a terrestrial to an endophytic lifestyle.

Further examples of closely related actinobacteria with different lifestyles reflected in their genomes include, among others, the genera *Frankia*, *Mycobacterium* and *Streptomyces*. In the case of *Frankia*, comparative genomic analysis of three representative strains, differing by less than 2% in their 16S rRNA genes revealed significant differences in their genome sizes (5.4–9.0 Mb) suggesting that these differences (e.g. gene deletion, acquisition and duplication, etc.) reflect their rapid adaptation to contrasted host plants and to their environments [15]. Similarly, several mycobacterial genomes were analyzed both at the nucleotide and protein levels. One of the most striking features was lipid metabolism genes with marked expansions of the number of genes related to saturated fatty acid metabolism in the pathogenic mycobacteria compared to the soil-dwelling strains [16].

In an effort to identify the genomic traits which make possible adaptation from a soil dwelling way of life to an endophytic habitat, the aim of this work was to present the genome sequence analysis of a representative strain, *Micromonospora lupini* Lupac 08, isolated from a nitrogen fixing nodule of the legume *Lupinus angustifolius*. This strain is part of a collection of more than 2000 strains isolated from nitrogen fixing root nodules of diverse legume [17,18] and actinorhizal species [13]. Strain Lupac 08 was selected as it showed good plant growth promotion, was used previously for in situ localization studies *in planta* [11] and produced several new secondary metabolites [9,10]. The results presented here show that the genome of *M. lupini* Lupac 08 contains a diverse array of genes that may help its survival in soils or in plant tissues, while the high number of putative plant degrading enzyme genes identified in its genome is quite surprising since this bacterium is not considered a plant-pathogen and may instead reflect their ability to bind to plant structural compounds.

Results

Phylogenetic position of *M. lupini* Lupac 08 and general genome features

The phylogenetic position based on 16S rRNA gene sequence analysis of strain Lupac 08 with respect to currently described *Micromonospora* species and other members of the family *Micromonosporaceae* is presented in Figure 1. Those strains associated with plant/rhizosphere sources are highlighted. Strain Lupac 08 was clearly positioned within the genus *Micromonospora* and forms a subgroup together with the species *Micromonospora*

saelicesensis, *Micromonospora zamorensis* and *Micromonospora chokoriensis*. These strains were isolated from a nitrogen fixing nodule, the rhizosphere of a *Pisum sativum* plant and a sandy soil, respectively. Nevertheless, a clear picture based on the habitat cannot emerge from this analysis.

M. lupini Lupac 08 was shown to have a circular chromosome of 7,327,024 bp with a GC content of 71.96% and no plasmid. A total of 7158 genomic objects were identified: 7,054 protein-coding, 10 rRNAs, 77 tRNAs, and 12 miscRNAs genes. The average gene length was 964 bp with an average intergenic distance of 126 bp. After manual validation of the automatic annotation, 61.5% (4338 CDSs) of the genes were assigned a biological function while 38.5% were registered as open reading frames (ORFs) with an unknown function. Based on the G+C skew analysis and position of *dnaA*, the probable origin of replication (*oriC*), was mapped close to the ribosomal protein *rpmH*. A circular representation of the *M. lupini* chromosome is provided in Figure 2 indicating some of the features described above.

The genomic characteristics of strain Lupac 08 and three additional *Micromonospora* genomes deposited in the public databases including *Micromonospora* sp. strain L5 isolated from root nodules of *Casuarina equisetifolia* [19]; *M. aurantiaca* ATCC 27029[T] and *Micromonospora* sp. ATCC 39149 isolated from soil (Table 1) were compared. An important difference between the four strains was the number of tRNAs identified. *M. lupini* 08 contained by far the highest number with 77 tRNAs while the other strains had between 51 and 53. At present, *M. lupini* Lupac 08 contains one of the largest numbers of tRNAs reported among the actinobacteria sequenced. The number of rRNA and tRNA genes in a genome can be seen as an indication of positive selection. A high number of rRNA genes increases ribosome synthesis, which in turn increases the protein synthesis rate [20] and growth rate [21].

Comparative genome analysis

COG distribution. Nearly 70% of the CDS were classified into clusters of orthologous groups (COGs, Table S1). Thus, 4873 out of 7054 CDS were assigned to 24 different categories, including those for amino acid transport and metabolism (E, 12.7%), transcription (K, 10.8%), carbohydrate transport and metabolism (G, 9.7%), inorganic ion transport and metabolism (P, 8.7%), energy production and conversion (C, 5.5%), and signal transduction mechanisms (5.5%).

The COG distribution of *M. lupini* was similar to that observed in other bacteria in the family *Micromonosporaceae*, however various differences were detected such as the abundance of genes related to carbohydrate transport and metabolism. Among the *Micromonospora* genomes currently available, *M. lupini* Lupac 08 contained the highest percentage of genes (9.7%, 685) related to this category, followed by *Micromonospora* sp. L5 (8.9%, 598) and *M. aurantiaca* ATCC 27029 (8.5%, 576). The gene contents (in the same COG category) of other bacterial genomes classified in the family *Micromonosporaceae* were lower as in the case of *S. tropica* CNS-205 (7.4%, 391) and *S. arenicola* CNH-643 (6.4%, 374) two obligate marine actinomycetes. On the other hand, the overall COG profiles of *Verrucosispora maris* AB-18-032[T] (genome size 6.7 Mb) and *M. lupini* Lupac 08 were very similar and no clear differences were found. Although *V. maris* was isolated from a sea sediment, it does not require sea salts for growth and it is not considered an obligate marine microorganism

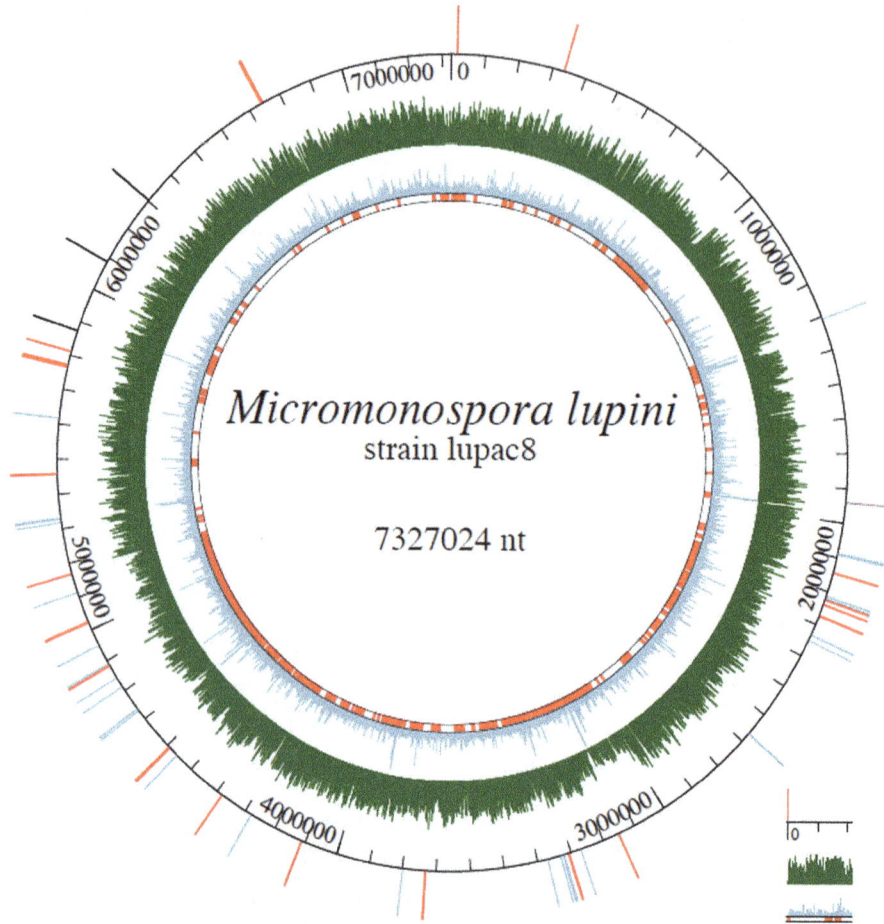

Figure 2. Circular representation of *Micromonospora lupini* **Lupac 08.** Circles displayed from the outside in: 1. Cellulose-binding genes in black, chitin-binding genes in red, lectin genes in lavender blue; 2. Genome coordinates; 3. MW; 4. GC% (linear range between 65 and 80%); 5. Regions of genome plasticity according to the RGP_Finder method (Mage platform) based on synteny breaks between the query genome (Lupac 08) and close genomes (*Micromonospora aurantiaca* ATCC 27029[T], *Micromonospora* sp. L5 and *Verrucosispora maris* AB-18-032[T]) correlated with HGT features (tRNA hotspot, DNA repeats, mobility genes), and compositional bias and GC deviation computation. C1 to C15 indicate the position of the 15 clusters of genes coding for secondary metabolites of Table 4.

unlike *S. tropica* and *S. arenicola*. Thus, its metabolism suggests a terrestrial life style. Micromonosporae are well known for their ability to degrade complex polysaccharides such as cellulose, chitin and lignin [22,23]. In particular, cellulose is frequently utilized as a carbon source [24,25]. Therefore the abundance of these genes in the genome of strain Lupac 08, at first glance may not seem surprising, however, the value of 9.7% is comparable to that of highly active cellulolytic microorganisms such as *Cellulomonas*

Table 1. Comparative genomic characteristics of *M. lupini* Lupac 08 and three *Micromonospora* genomes publicly available.

Feature	*M. lupini* Lupac 08	*M. aurantiaca* ATCC 27029[T]	*Micromonospora* sp. L5	*Micromonospora* sp. ATCC 39149
Size (Mb)	7.3	7.0	6.9	6.8
GC%	72	73	73	72
rRNA Operon	10	9	9	6
tRNA	77	52	53	51
CDS number	7054	6676	6617	5633
Average gene size (kb)	946	964	969	975
Protein-coding density (%)	90.1	90.4	90.4	89.9
Genes in COGs (%)	70.2%	68.3%	69%	nd

nd, not determined.

flavigena 134[T] (9.5%) and *Thermobifida fusca* XY (7.9%), which are abundant in cellulose enriched environments such as soil, or plant tissues.

Synteny. The genome sequence of strain Lupac 08 was aligned with those of *Micromonospora* sp. L5, *M. aurantiaca* ATCC 27029[T] and *Micromonospora* sp. ATCC 39149[T] (Fig. 3). Although the four genomes share a significant amount of genetic characteristics, they have undergone various inversions and translocations and *M. lupini* Lupac 08 contains the highest number of non-conserved regions. In addition, this alignment shows a high homology between strains *Micromonospora* sp. L5 and *M. aurantiaca* ATCC 27029[T] confirming their close phylogenetic relationship as suggested by 16S rRNA gene phylogeny (Fig. 1); nevertheless, strain L5 shows a large inversion event. Thus, although the four *Micromonospora* genomes share many common features, it is also evident that *M. lupini* contains unique genomic regions as compared to *M. aurantiaca* ATCC 27029[T] or *Micromonospora* sp. L5.

Diversity of *Micromonosporae*: core vs. flexible gene pool

Using the *Micromonospora* genomes of strains M. *lupini* Lupac 08, *M. aurantiaca* ATCC 27029[T] and that of *Micromonospora* sp. L5 available in the NCBI [19], the core genome was calculated using the SiLix software [26]. The core genome was composed of 2294 CDSs, which correspond to approximately 32% of the predicted proteome. In addition, *M. lupini* Lupac 08 contained the highest number of strain specific CDSs, 4702 (66.6%), which is a very high value when compared to *Micromonospora* sp. L5 and *M. aurantiaca* ATCC 27029[T] (13–14%, Figure 4), which both share a high gene similarity (85–86%).

Horizontal gene transfer is universally recognized as an efficient mechanism for microorganisms to acquire functions that enable them to adapt to environments with different selective pressures. Therefore insertion elements, transposases, integrated phages, and plasmids can be related to the plasticity of a genome. Strain Lupac 08 contained 49 CDSs (0.7%, of total CDSs) related to gene exchange including eight integrases and eleven recombinases. Except for seven CDSs, most of these genes were grouped into 20 clusters. Interestingly, eight of these mobile element clusters were found near genes related to carbohydrate transport and metabolism.

Metabolic Features. A metabolic pathway reconstruction was performed between the genome of strain Lupac 08 and 20 additional strains among which plant pathogens, symbiotic and saprophytic bacteria were included. The distribution and grouping of the microorganisms analyzed using 798 metabolic routes are presented in Figure 5. A good correlation was obtained between the microorganisms, their life style and phylogeny. Two main groups were obtained, the proteobacteria and actinobacteria. Within the actinobacteria, three clusters were clearly identified: the first one contained strains that belonged to the family *Micromonosporaceae*, the second cluster corresponded to various streptomycetes and the third cluster included the three *Frankia* genomes. Surprisingly, *Micromonospora lupini* Lupac 08 showed a closer metabolic relationship with the three *Frankia* strains (ACN14a, CcI3 and EAN1pec) than with the other two *Micromonospora* genomes.

Plant/Soil-associated life style

Transport systems. Organisms living in endophytic associations need to share resources with their host. Membrane transport systems play essential roles in cellular metabolism and activities. Current data suggest a correlation of transporter profiles to both evolutionary history and the overall physiology and lifestyles of organisms [27].

A total of 631 CDSs were located in the genome of *M. lupini* coding for a large diversity of transporters, representing approximately 8.9% of the genome. The majority of CDSs were related

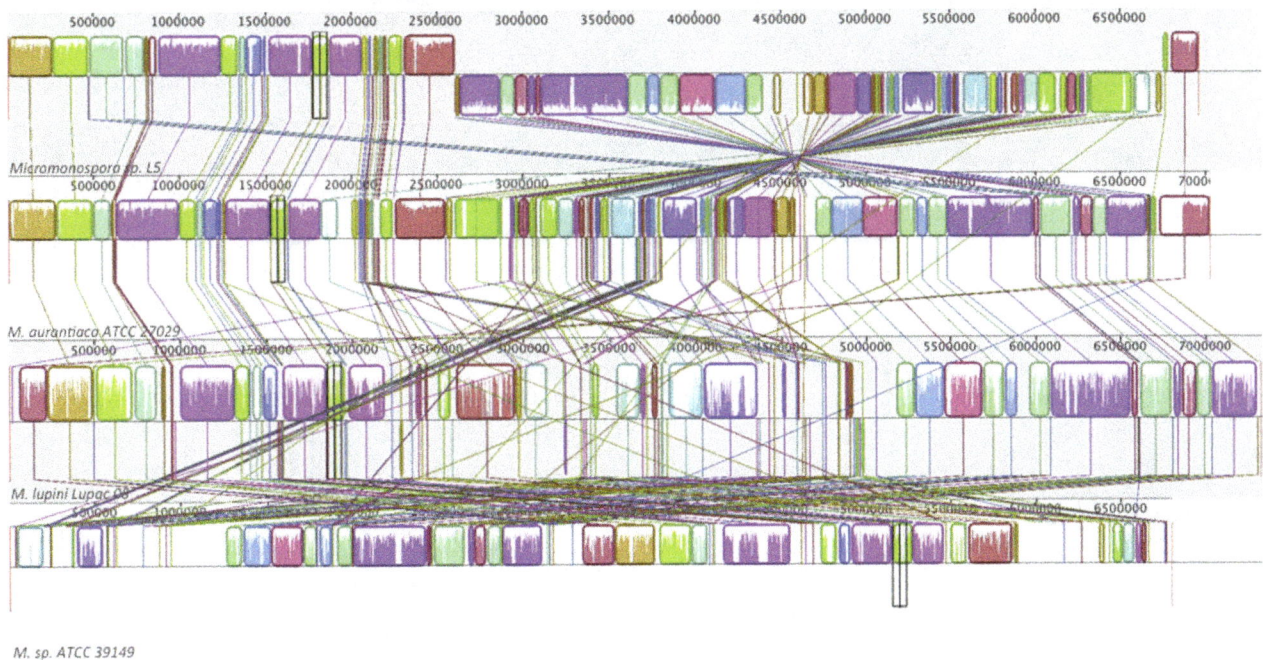

Figure 3. MAUVE alignment of the genome sequences of *Micromonospora lupini* **Lupac 08,** *Micromonospora* **sp. L5,** *Micromonospora aurantiaca* **ATCC 27029**[T] **and** *Micromonospora* **sp. ATCC 39149.** When boxes have the same colour, this indicates syntenic regions. Boxes below the horizontal line indicate inverted regions. Rearrangements are shown by coloured lines. Scale is in nucleotides.

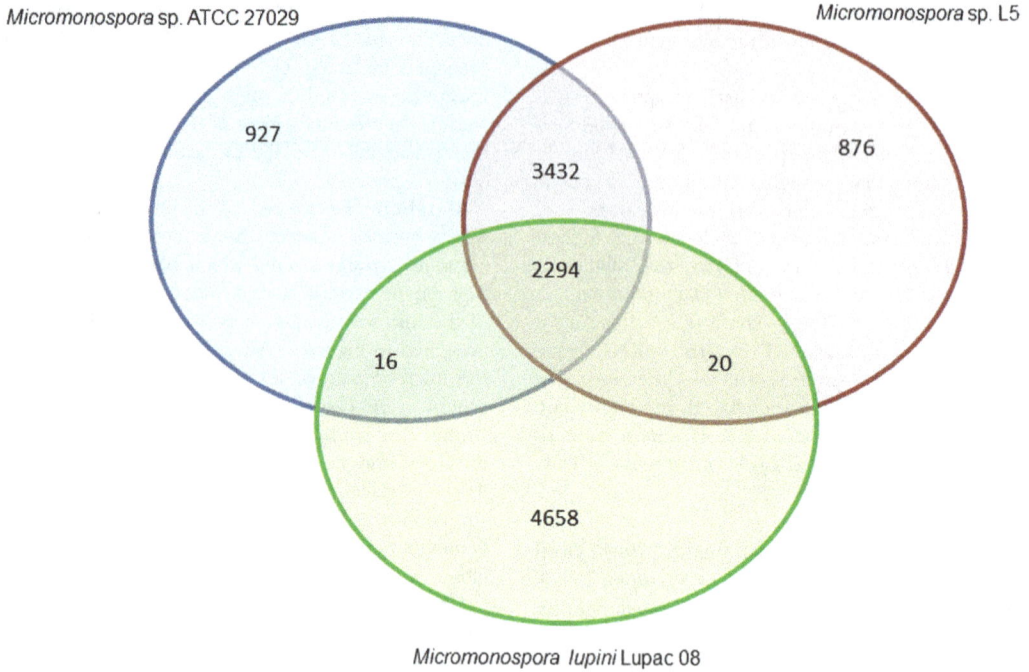

Figure 4. Venn diagram showing the number of clusters of orthologous genes, shared and unique, between *M. lupini* Lupac 08, *Micromonospora* sp. L5 and *M. aurantiaca* ATCC 27029[T].

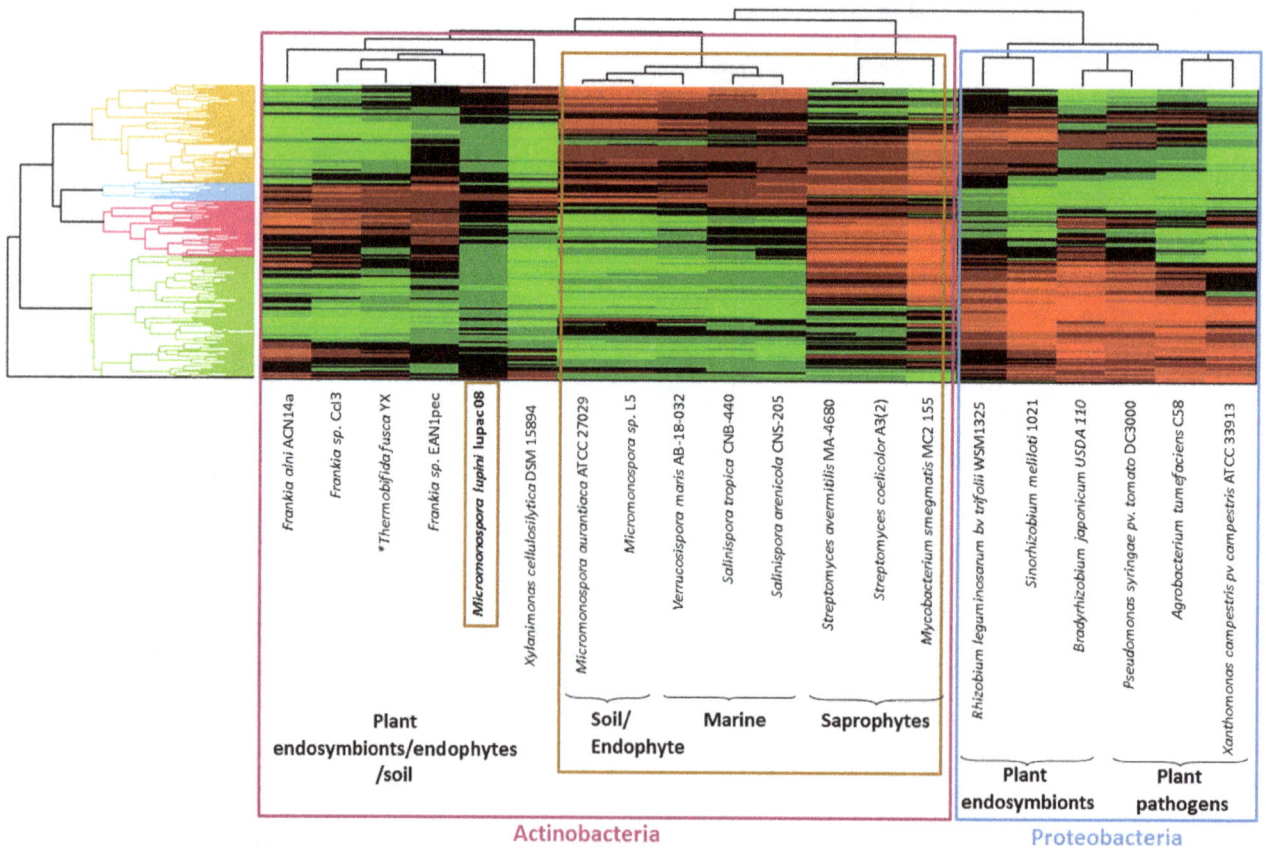

Figure 5. Bicluster plot of the metabolic profiles of *M lupini* Lupac 08 and 20 other bacterial genomes.

to ATP-binding (dependent) transporters of which 362 corresponded to ABC transporters; the next most abundant (215 CDSs) coded for secondary transporters, with 105 classified in the Major Facilitator Superfamily –MFS-); 17 transporters belonged to ion channels and 20 were unclassified. The number of transporters determined in *M. aurantiaca* ATCC 27029[T] and *Micromonospora* sp. L5 were lower with 575 and 587, respectively (Table 2).

The number of transporters identified in the genome of *M. lupini* Lupac 08 is correlated with those in other bacteria with a plant/soil associated life style, which requires an efficient nutrient uptake system to obtain nutrients produced by the host plant, in addition to those found in the rhizosphere and the soil (Table 2) [27,28]. However, the number of transporters identified in strain Lupac 08 was lower than those present in other bacteria such as *Bradyrhizobium japonicum* USDA6[T] (1138, 11.8%), *Mesorhizobium loti* MAF303099 (968, 14.2%), *Sinorhizobium meliloti* Sm1021 (1024, 16.4%) and *Rhizobium leguminosarum* bv. *trifolii* WSM 3125 (1087, 15.5%), which form a very close interaction with legumes. Nevertheless, the overall distribution (types) and the percentages of these values were similar. An additional difference was the absence of phosphotransferase system transporters (PTS) in *M. lupini* as compared to the strains mentioned above and other soil/plant bacteria included in Table 2. On the other hand, the overall profile of *M. lupini* Lupac 08 was very similar to those of *Frankia* sp. ACN14a and *Frankia* sp. CcI3 which also lack a PTS system.

Secretion systems. Secreted proteins play a number of essential roles in bacteria, including the colonization of niches and host–pathogen interactions. In Gram-positive bacteria, the majority of proteins are exported out of the cytosol by the conserved Sec translocase system or, alternately, by the twin-arginine translocation system. In addition, a unique protein export system, the type VII or ESX secretion system also exists in some Gram positive bacteria [29].

The genome of *M. lupini* Lupac 08 encodes for 537 (7.6%) secreted proteins including several protein secretion systems (Table 3). All genes related to the Sec-dependent pathway were located and included the SecY and SecE proteins which form the membrane channel and interact with the cytoplasmic membrane protein SecG; the auxiliary proteins SecD, YajC and the ATPase SecA. In addition, the heterodimer Ffh-FtsY (MiLup08_41486 and Milup08_41460) was also present. As in other Gram-positive bacteria, *M. lupini* Lupac 08 lacks homologs of SecB, the chaperone that targets proteins to the Sec translocon for passage through the cytoplasmic membrane [30].

Genes related to the Sec-independent twin-arginine translocation pathway (TAT), which exports prefolded proteins across the cytoplasmic membrane using the transmembrane proton gradient as the main driving force for translocation were also located in strain Lupac 08 (Table 3). Homologs of TatA and TatC were identified, however no homolog for TatB was found. Similar to other actinobacteria (e.g. *Frankia* sp. ACN14a) the *tatA* gene was found next to *tatC*. Only an ORF encoding TatC was located in the genomes of *Micromonospora* sp. L5 and *M. aurantiaca* ATCC 27029[T] while no copies of *tatA* or *tatB* were found.

A set of fifteen genes identified as part of the type VII secretion system were located in *M. lupini* Lupac 08 (Table 3). These are arranged in three different clusters and included the essential proteins for secretion EccC, EccD, EsxA and EsxB [31]. The first cluster contains eight genes: *eccC, esxA, esxB, eccD, eccB, eccE* and two copies of *mycP*, a subtilisin-like serine protease which also appears essential but the function of which is not yet known [32]. The second cluster includes a copy of *esxA* (MiLup08_40381), *esxB* (MiLup08_40380) and *mycP*, annotated as S8 S53 subtilin

kexin sedolisin (MiLup08_40382). Finally a third cluster contains the genes *eccC* (MiLup08_46744), *eccD* (MiLup08_46743) and *mycP* (MiLup08_46745).

Gram-negative bacteria use the type II secretion system to transport a large number of secreted proteins from the periplasmic space into the extracellular environment. Many of the secreted proteins are major virulence factors in plants and animals [33]. Type II secretion systems have been found in all completely sequenced plant pathogenic bacterial genomes, except in *Agrobacterium tumefaciens*. In addition, other bacteria have been shown to use secretion systems for the delivery of toxins, proteases, cellulases and lipases [34–37]. Genes coding for this system have also been reported for the three symbiotic strains *Frankia* [38].

Fifteen genes in *M. lupini* were annotated as components of the Type II secretion system, grouped into clusters of three to five genes (Table 3). Nine of these genes were annotated as Type II secretion system proteins including protein E and protein F; four were recorded as TadE family proteins and Milup08_40403 was annotated as an uncharacterized protein closest to one found in the *Frankia* symbiont of *Datisca glomerata*.

The secretion systems III and IV which are commonly related to plant-associated bacteria transport a wide variety of effector proteins into the extracellular medium or into the cytoplasm of eukaryotic host cells thus affecting the interaction [39]. In addition, a functional type IV system has been described in the plant symbiont *M. loti* strain R7A [40]. A gene annotated as *virB4* and related to secretion system IV was located in Lupac 08 (MiLup08_42651), this ORF is surrounded by proteins with unknown function related to those present in the genomes of *Micromonospora* sp. L5 and *M. aurantiaca* ATCC 27029[T].

Survival against plant defenses. Reactive oxygen species (ROS) play a major role in plant defense against pathogens. In response to attempted invasion, plants mount a broad range of defense responses, including the synthesis of ROS. *M. lupini* needs to survive under an oxidative environment in the rhizosphere before it can colonize plant roots and its genome revealed several genes encoding proteins to neutralize oxidative stress. The following genes were identified: three *sod* genes (MiLup08_45788, MiLup08_46012 and MiLup08_46604) that code for superoxide dismutases; a catalase HPII *katE* (MiLup08_44247); a catalase-peroxidase (*katG*, MiLup08_44435) and a catalase hydroperoxidase (*katA*, MiLup08_45857); four hydroperoxide reductases (MiLup08_40110, MiLup08_40293, MiLup08_41393, MiLup08_45407); a chloroperoxidase (MiLup08_44157) and a thiol peroxidase (MiLup08_43629).

In addition, a putative organic hydroperoxide resistance protein (Ohr, MiLup08_45098); a 4-hydroxyphenylpyruvate dioxygenase (Hpd, MiLup08_46664) and a homogentisate 1,2-dioxygenase (MiLup08_46677) were identified. Other enzymes include a glutathione peroxidase (MiLup08_45173); two glutathione transferases (MiLup08_46358 and MiLup08_41529) and four glutathione-S-transferases (*fdh*, MiLup08_42270, MiLup08_42834, MiLup08_44416 and 45648). Experimental data indicated that *M. lupini* indeed yields a catalase positive reaction [17] confirming the functionality of some of these genes. Therefore, to successfully reach the internal plant tissues, these genes may defend the bacterium against a ROS release by the plant.

Regulation as a means of adaptation

Lifestyle can be viewed as the set of biotopes an organism can thrive into and the relationships that it establishes with other species and its abiotic components. It is one of the driving forces that contribute to the overall characteristics of bacterial genomes [41].

Table 2. Transporters identified in the genome of *M. lupini* Lupac 08 and comparison with other bacteria with a plant/soil associated life styles.

	M. lupini Lupac 08	M. aurantiaca ATCC 27029[T]	Micromonospora sp. L5	S. arenicola CNS 205	S. tropica CNB 440	F. alni ACN14	Frankia sp. CcI3	F. symbiont Dastia glomerata	B. japonicum USDA6[T]	R. leguminosarum Bv trifolii WSM 3125	Enterobacter sp. 368	S. coelicolor A32	S. scabiei 8722	Pseudomonas syringae pv phaseolicola 1448A
Genome size (Mb)	7.3	7.0	6.9	5.7	5.2	7.5	5.4	5.3	9.6	7.4	4.6	9.1	10	5.9
Total transport proteins	631	575	587	405	413	433	253	300	1138	1087	662	798	775	670
Transporters (%)	8.9	8.5	8.7	7.1	7.8	6.4	4.5	7.1	11.8	15.5	15.6	9.7	8.2	12.5
No. Transporters/Mb genome	0.08	0.08	0.08	0.07	0.08	0.06	0.05	0.06	0.12	0.15	0.14	0.09	0.08	0.11
ATP dependent (% of total)	379 (60.1%)	362 (63.0%)	366 (62.4%)	244 (60.2%)	247 (59.8%)	281 (64.9%)	146 (57.7%)	210 (70%)	684 (60.1%)	800 (73.6%)	317 (49.7%)	461 (57.8%)	480 (61.9%)	392 (58.5%)
ABC family*	362 (95.5%)	341 (94%)	342 (93.4%)	225 (92.2%)	228 (92.3%)	262 (93.2%)	127 (87%)	189 (90%)	645 (94.3%)	769 (96.1%)	287 (90.5%)	433 (93.9%)	455 (94.8%)	346 (88.3%)
Ion channels (% of total)	17 (2.7%)	14 (2.4%)	15 (2.6%)	9 (2.2%)	11 (2.7%)	12 (2.8%)	7 (2.8%)	7 (2.3%)	24 (2.1%)	26 (2.4%)	23 (3.5%)	19 (2.4%)	22 (2.8%)	32 (4.8%)
Phosphotransferase system (PTS)	–	–	–	4 (1%)	–	–	–	–	4 (0.4%)	6 (0.6%)	47 (7.1%)	10 (1.3%)	6 (0.8%)	5 (0.7%)
Secondary transporter	215 (34.1%)	185 (32.3%)	192 (32.7%)	9 (2.2%)	145 (35.1%)	130 (30%)	89 (35.2%)	75 (25%)	408 (35.9%)	234 (21.5%)	256 (38.7%)	286 (35.8%)	252 (32.5%)	226 (33.7%)
MFS family*	105 (48.8%)	68 (36.8%)	70 (36.5%)	65 (46.8%)	74 (51%)	64 (49.2%)	36 (40.4%)	33 (44%)	114 (27.9%)	69 (29.5%)	84 (32.8%)	120 (42%)	111 (44%)	72 (31.9%)
RND family	–	6 (3.2%)	7 (3.6%)	7 (5%)	8 (5.5%)	9 (6.9%)	7 (7.9%)	7 (9.3%)	31 (7.6%)	13 (5.6%)	18 (7%)	15 (5.2%)	18 (7.1%)	16 (7.1%)
Unclassified	20 (3%)	13 (2.3%)	13 (2.2%)	8 (2%)	9 (2.2%)	10 (2.3%)	11 (4.3%)	8 (2.7%)	11 (1%)	16 (1.5%)	14 (2.1%)	21 (2.6%)	14 (1.8%)	11 (1.6%)

*Number and percentage in relation to the total number of ATP dependent and secondary transporters respectively.

Table 3. Secretion system genes present in the genome of *M. lupini* Lupac 08.

Secretion System	Gene (Milup08_X)	Product
Sec-dependent	*secY (prlA)*(46297)	Preprotein translocase, membrane component
	secE (46336)	Preprotein translocase subunit secE
	secG (44961)	Preprotein translocase SecG subunit
	secD (42464)	Protein-export membrane protein secD
	secF (42465)	Protein-export membrane protein secF
	yajC (42463)	Preprotein translocase, YajC subunit
	secA (41087)	Protein translocase subunit secA
	ffh (41468)	Signal recognition particle protein
	scRNA (misc_RNA-12)	SRP, Ribosome-nascent chain complex (RNC)
	yidC (30220)	Cytoplasmic insertase into membrane protein
	yidC-like (43138)	Membrane protein insertase, YidC/Oxa1 family
	yidC (45964)	Inner membrane protein translocase component YidC
	Milup_08_41485	Signal peptidase I
	Milup_08_41486	Signal peptidase I
	Milup_08_42560	Conserved protein of unknown fuction (probable signal peptidase I)
	lspA (45113)	Lipoprotein signal peptidase
	lgt (45071)	Prolipoprotein diacylglyceryltransferase
TAT-	*tatA* (43424)	Sec-independent protein translocase protein tatA/E homolog
	tatC (43425)	Sec-independent protein translocase protein tatC homolog
Type II- (T2SS)	Milup_08_40403	Similar to uncharacterized protein from *Frankia* symbiont of *Diastica glomerata*
	Milup_08_40405	Putative helicase/secretion neighbourhood TadE-like protein
	tadE (40223)	TadE Family protein
	tadE (40224)	TadE Family protein
	tadE (42690)	Similar to TadE family protein
	tadE (42691)	Similar to TadE family protein
	Milup_08_40226	Type II secretion system protein
	Milup_08_40227	Type II secretion system protein
	Milup_08_40228	Type II secretion system protein E
	Milup_08_40398	Type II secretion system protein E
	Milup_08_40399	Similar to Type II secretion system protein E
	Milup_08_40401	Similar to Type II secretion system protein
	Milup_08_42693	Type II secretion system protein F
	Milup_08_42694	Type II secretion system protein F
	Milup_08_42695	Type II secretion system protein
Type IV- (T4SS)	Milup_08_42651	VirB4 protein-like protein
Type VII/WXG100-	*eccB* (40554)	ESX-4 secretion system protein eccB4
	eccC (40438)	FtsK/SpoIIIE family protein
	eccC (40557)	ESX-4 secretion system protein/cell division protein ftsK/spoIIIE
	eccC (46744)	FtsK/SpoIIIE-like transmembrane protein
	eccD (40556)	ESX-4 secretion system protein eccD4/Putative secretion protein snm4
	eccD (46743)	FtsK/SpoIIIE family protein
	eccE (40555)	Putative uncharacterized protein
	esxA (40381)	Putative uncharacterized protein
	esxA (40559)	Putative uncharacterized protein
	esxB (40380)	Putative uncharacterized protein
	esxB (40558)	Putative uncharacterized protein
	mycP (40382)	Peptidase S8 and S53 subtilisin kexin sedolisin
	mycP (40560)	Peptidase S8 and S53 subtilisin kexin sedolisin
	mycP (40564)	Peptidase S8 and S53 subtilisin kexin sedolisin
	mycP (46745)	Peptidase S8 and S53 subtilisin kexin sedolisin

TAT, twin-arginine translocation; X, corresponds to the annotation gene numbers given in parenthesis.

The *M. lupini* genome shows a strong emphasis on regulation, with 643 proteins (~10%) predicted to have a regulatory function. This value is lower than that reported for the saprophytic strain *Streptomyces coelicolor* A3(2) with an exclusively terrestrial lifestyle (965 proteins; 12.3%) [42], but higher than the endosymbiotic strains *M. loti* MAFF303099 (542 proteins, 7.7%) [43]; *Frankia alni* ACN14a (515 proteins, 7.6%); *Frankia* sp. EAN1pec (555, 6.1%) and *Frankia* sp. CcI3 (244 proteins, 4.3%).

The genome codes for various regulator families such as TetR, AraC, LacI, ArsR, MerR, AsnC, MarR, DeoR, GntR and Crp. In addition, thirty-three ECF (extra-cytoplasmic function) sigma factors were located. Furthermore, 147 genes were related to two-component regulatory systems of which 34 were LuxR proteins. These two-component systems appear to play a crucial role in quorum sensing of Gram-positive bacteria and a positive correlation between plant-microbe interactions and the number of LuxR proteins has been suggested [44,45].

Many regulatory genes (~18%) were located near polysaccharide related loci including those involved in plant cell wall degradation. Specifically, 63% of cellulose degradation or cellulose binding genes had a nearby regulator (proximity ranged from 2–4 genes up or downstream). In the case of xylan metabolism, regulators were identified for 50% of the genes, while 43% of pectin metabolism genes also had a regulator nearby. An extended overview of the regulators and their associated carbohydrate genes is presented in Table S2.

An endophytic bacterium highly equipped with an array of plant cell wall degrading enzymes

The ability of *M. lupini* lupac 08 to assimilate a wide range of sugars was previously reported [19] and this is clearly reflected in its genome. The range of simple and complex saccharides assimilated by this strain include cellobiose, cellulose, glucose, mannitol, starch, sucrose, trehalose, xylan and xylose among others. Genomic analyses confirmed the presence of a large number of genes devoted to the metabolism of carbohydrates, including many compounds of plant origin. Plant-polymer degrading enzymes such as cellulases, xylanases and pectinases have been suspected to play a role in internal plant colonization [46]. In the case of plant pathogenic bacteria and fungi, these gain access by actively degrading plant cell wall compounds using glycoside hydrolases including cellulases and endoglucanases. However, genomic analyses show that non-pathogenic endophytic microorganisms such as *Enterobacter* sp. 638 [47], *Azoarcus* BH72 [39] or the symbiotic actinobacterium *Frankia sp.* [48] have only a reduced set of cell-wall degrading enzymes.

The genome of *M. lupini* Lupac 08 revealed a significant number of genes encoding enzymes potentially involved in plant-polymer degradation but also an important number of cellulose-binding related genes. Overall, about 10% of the genome coded for genes related to carbohydrate metabolism of which 192 had a hydrolytic function. At least 79 genes putatively involved in interactions with plants and with the potential to hydrolyze plant polymers were identified (Table S1). These genes were placed into the glycosyl hydrolase families GH5, GH6, GH9, GH10, GH11, GH18, GH20, GH43, GH44 and GH62, or into the carbohydrate binding modules CBM2, CBM13, CBM33, CBM3, CBM46, CBM42, CBM5, CBM4, CBM6 and CBM32. The CBM2 family was the most abundant appearing in 46 of the 79 genes identified.

Fourteen genes were further identified as lectins or proteins with lectin binding domains, which presumably bind to and interact with carbohydrates. Some of these loci (e.g. Milup_42969, Milup_42975, Milup_44484, and Milup_44962) appear to be related to cellulases and xylanases, respectively. These proteins are important as they serve as a means of attachment between a bacterium and its host (animal or plant) and are produced by either of the two interacting organisms [48].

Compared to the 45 enzymes predicted to act on oligo- and/or polysaccharides reported for *T. fusca* XY [49], the number of these enzymes present in the genome of *M. lupini* is significantly higher.

Cellulose metabolism. Aerobic cellulolytic actinobacteria have been shown to use a system for cellulose degradation consisting of sets of soluble cellulases and hemicellulases. Most of these independent cellulolytic enzymes contain one or more carbohydrate binding domains [50].

A total of 46 genes were found to present a hydrolytic or binding fuction towards cellulose (Table S1). Several endoglucanases were detected in strain Lupac 08 (e.g. C1, C2 C10 and C14), these enzymes hydrolyze internal bonds at random positions of amorphous regions of cellulose and generate chain ends for the processive action of cellobiohydrolases (exoglucanases). A copy of the exoglucanase gene *cbhA* (C16) was also located in the genome. Exoglucanases act on the ends of cellulose polysaccharide chains, liberating cellobiose as the major product. β-D-glucosidases such as M108 and M109 which would further hydrolyze cellobiose were also identified. In addition, several extracellular cellulase coding genes were identified including *celA* (C3 and C6), *celB* (C5) and *celD* (C13). These results strongly suggest that strain Lupac 08 is potentially capable of completely degrading cellulose.

Strain Lupac 08 was tested for *in vitro* production of cellulases. Very high cellulase activity was detected in minimal agar supplemented with carboxymethylcellulose (CMC, 0.5%) (Fig. 6A). When the culture medium was supplemented with glucose (1%) similar results were obtained indicating that this sugar did not repress nor derepress the expression of the genes responsible for the production of cellulases.

Hemicellulosic substrates. Genome analysis also revealed the ability of *M. lupini* to convert various hemicellulosic substrates to sugars. Twelve putative genes related to the metabolism of xylan included several copies of extracellular xylanases (X1, X3, X4, X5, X6, X7, X9, X10 and X12; see Table S1); an extracellular bifunctional xylanase/deacetylase (X8); and an arabinofuranosidase (X2) which work synergistically with xylanases to degrade xylan to its component sugars. Genes for several α-arabinofuranosidases were also identified (C17, M33, and M39); these are exo-acting enzymes which hydrolyze nonreducing arabinofuranose residues from arabinoxylan, pectins, and shorter oligosaccharides.

In vitro xylanase activity was detected in strain Lupac 08 when tested in a minimal medium supplemented with xylan (1%). Production of xylanases was detected after incubation for 4 days increasing significantly after 14 days (Fig. 6D). The substrate was assayed with and without glucose with similar results.

Starch degradation. Starch is a ubiquitous and easily accessible source of energy. In plant cells it is usually deposited as large granules in the cytoplasm. Several genes coding for amylo-α-1,6-glucosidases (e.g. M26, M32, M44, M63, M111 and M121; Table S1) were located in addition to two *amyE* homologs that code for an extracellular α-amylase. Furthermore, strain Lupac 08 was able to degrade this polymer under laboratory conditions (Fig. 6C) and it was previously shown that Lupac 08 can utilize this substrate as a carbon source [17].

Pectin degradation. Pectinolytic enzymes can degrade pectic substances either through hydrolysis (hydrolases) or trans-elimination (lyases) [51] and are important virulence mechanisms in many soft-rotting and macerating pathogens [52]. Six pectate lyases (P1, P3, P4, P5 and P6; Table S1) were located in the

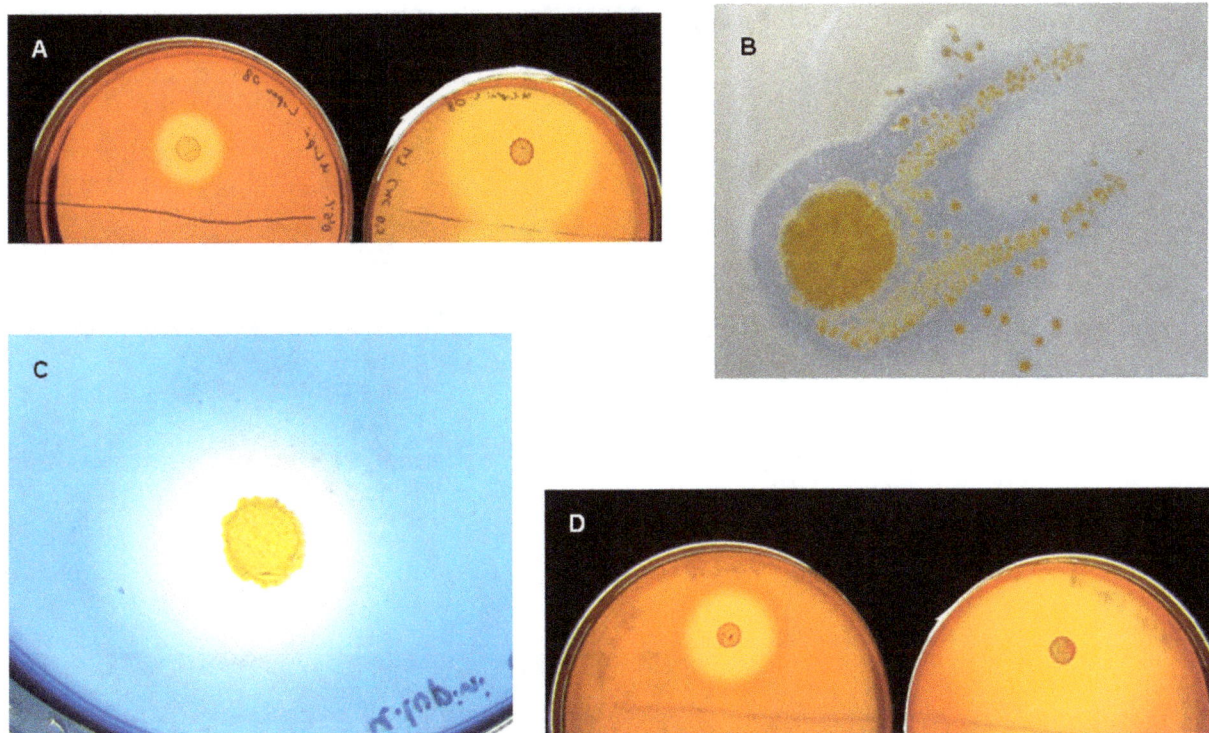

Figure 6. Expression of cellulose, starch, xylan and chitin degrading genes in *Micromonospora lupini* **Lupac 08.** (A) carboxymetheylcellulose hydrolysis at 4 (left) and 14 (right) days after inoculation. (B), starch hydrolysis at 4 days after inoculation. (C), chitin degradation at 7 days after inoculation. (D), xylan degradation at 4 (left) and 14 (right) days after inoculation.

genome of *M. lupini*, two of which were annotated as virulence factors (P5 and P6). In addition, an extracellular pectin methylesterase gene, *pmeA*, and a gene coding for a pectate lyase involved in D-galacturonic acid hydrolysis (P7) were also identified. Interestingly, *T. fusca* XY contains two pectin lyase homologs but does not appear to possess a pectin methylesterase or a pectin acetylesterase gene. Pectinase-encoding genes are reported to be absent in other endophytic microorganisms such as *Azoarcus* sp. BH72 or *Enterobacter* sp. 638 [39,47]. Production of pectinases was observed under laboratory conditions and activity was visualized after 8 days of incubation (Fig. 7D).

Expansin-like proteins. Expansins are proteins that were first described from plants [53]. These molecules function as cell wall loosening proteins by disrupting the noncovalent binding of matrix polysaccharides to cellulose [54], resulting in physical effects, such as polymer creep and stress relaxation of extended cell walls [55,56]. Many plant-associated microorganisms including several pathogenic actinobacterial species have been shown to contain proteins with expansin-like domains [57].

Two genes (MiLup_41274 and MiLup_45306) were identified in the genome of strain Lupac 08 that encode for a secreted protein showing 42% and 48% sequence similarity to the corresponding *celA* genes of *Clavibacter michiganensis* subsp. *michiganensis* and *Clavibacter michiganensis* subsp. *sepedonicus*, respectively. This gene corresponds to a secreted β-1,4-endoglu-canase (CelA) that is required for virulence and contains a C-terminal α-expansin like domain [58,59]. In the case of *C. michiganesis* subsp. *michiganensis* CelA, this expansin-like domain is essential for development of wilting symptoms [58]. It is suggested that microbial expansins function to promote microbe-plant interactions, both harmful and beneficial ones [60].

Plant growth promotion traits of *Micromonospora* lupini Lupac 08

Our current knowledge of plant-microbe interactions indicates that populations inhabiting a host plant are not restricted to a single microbial species but comprise several genera and species. Few reports are available regarding the presence of other microorganisms (associated or endophytic) in nitrogen fixing nodules, in spite of the fact that nodules are much richer in nutrients as compared to roots [61]. The recent reports on the isolation of large *Micromonospora* populations from nitrogen fixing nodules clearly suggest that this bacterium plays an important role which has yet to be defined.

Effect of *M. lupini* **Lupac 08 on** *Trifolium*. *Micromonospora lupini* Lupac 08 clearly produced a plant growth enhancing effect when it was co-inoculated with *Rhizobium* sp. E11 under laboratory conditions on clover plantlets. In general, the number of nodules was higher in those plants co-inoculated (18–24 nodules) with both bacteria as compared to the plants inoculated only with *Rhizobium* sp. E11 (11–15 nodules). Overall, the co-inoculated plants showed better growth and were larger in size as compared to the other two treatments (Fig. 7C). Similar results were previously observed when strain Lupac 08 was inoculated in its original host, *Lupinus* [62].

Nitrogen fixing capacity. Indirect evidence of nitrogen fixing genes was obtained by partial amplification of *nifH*-like gene fragments in strains *Micromonospora* sp. L5 [12] and *M. lupini* Lupac 08 [11]. In the present work the genomes of *Micromonospora lupini* Lupac 08 and *Micromonospora sp.* L5 were screened for the presence of nitrogen fixing genes to confirm this earlier finding. After thorough analysis of the complete genome, no sequences related to this biological process were detected, supporting the results reported for strain

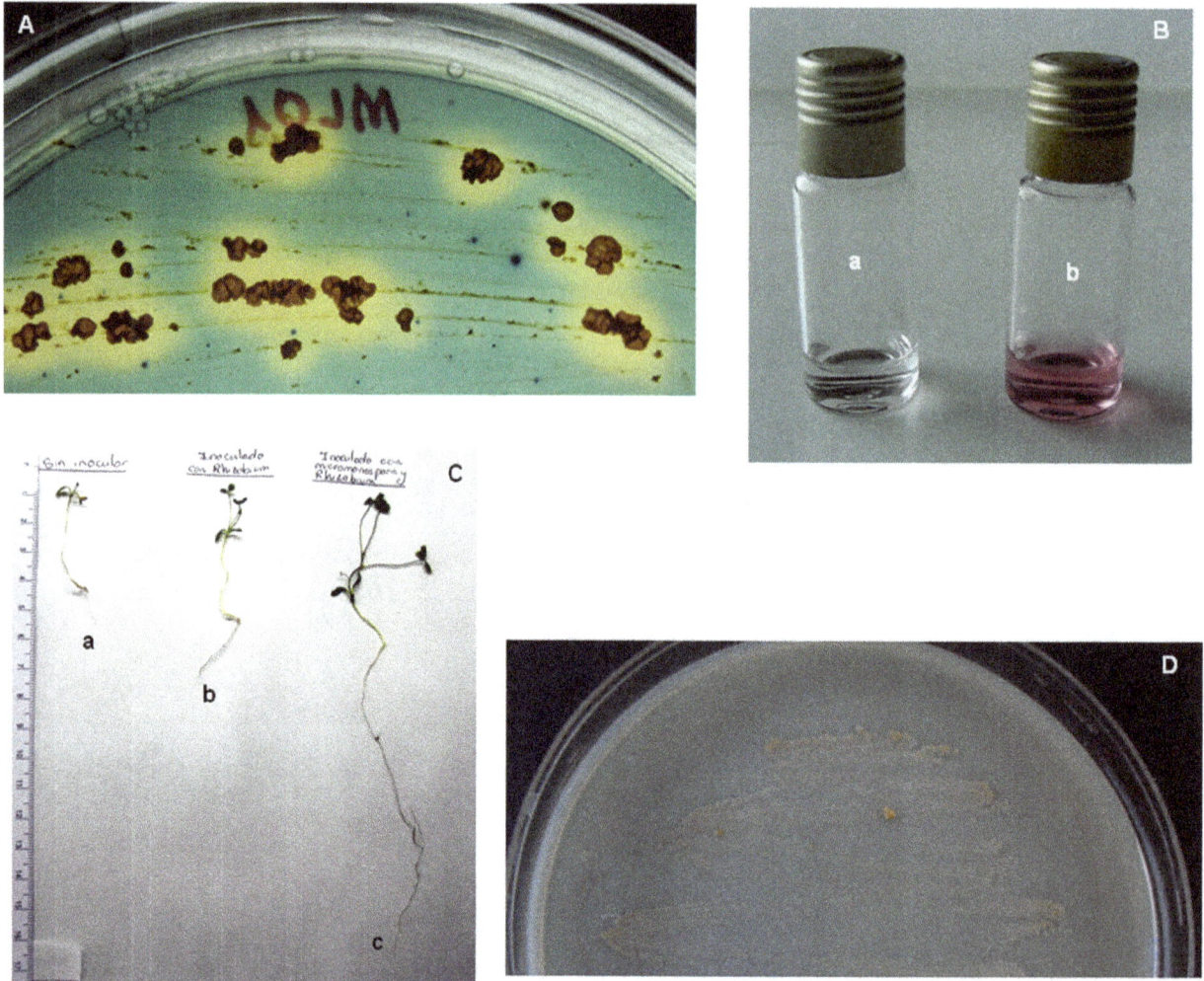

Figure 7. Plant growth promotion and biological control features of *M. lupini* Lupac 08. (A) Siderophore, (B) indole-3-acetic acid [a, negative control *E. coli* DH5α; b, Lupac 08] and pectinase production (D) by *M. lupini* strain Lupac 08;. (C) Plant growth promoting effect of *M. lupini* Lupac 08 on clover plantlets. a) control; b) inoculated with *Rhizobium* sp. E11; c) co-inoculated with *Rhizobium* sp. E11 and *M. lupini* Lupac 08.

Micromonospora sp. L5 [19]. Nitrogenase activity detection by acetylene reduction assays carried out with strain Lupac 08 over a period of two weeks were negative. A positive result was reported for strain L5 [12.].

Trehalose and its role in nodulation and bacteroid survival. Trehalose is a common reserve disaccharide in the root nodules of legumes, present at high concentrations in bacteroids at the onset of nitrogen fixation [63]. It has been reported that in the interaction between *Phaseolus vulgaris* and *Rhizobium,* enhanced germination, quality and grain yield have been correlated with trehalose content, and a higher tolerance to abiotic stress [64,65]. On the other hand, the trehalose content appears to be regulated by trehalase, a nodule stimulated plant enzyme [66,67]. Although trehalose metabolism in leguminous plants is still poorly understood, it has been shown that in senescent nodules, trehalose becomes the most abundant non-structural carbohydrate [68] and it is proposed that trehalose, a stress protectant accumulated in bacteria, could offset membrane injuries and/or serve as an intermediate energy reserve. Indeed, Müller *et al.* [68] showed that during terminal senescence of nodules an appreciable part of the bacteria maintained their trehalose pools and survived.

Eight genes related to the metabolism of trehalose were detected in the genome of Lupac 08; seven genes were related with trehalose synthesis (Mlup08_40949, Mlup08_43225, Mlup08_43226, Mlup_45189, Mlup_45758, Mlup08_45759 and Mlup08_45961) and one (*treA*, Mlup08_45961) with the enzyme trehalase. Barraza *et al.* [67] proposed that modification of the trehalose content in the nodules could trigger physiological alterations that would enhace carbon and nitrogen metabolism, as well as bacteroid fitness (greater survival) and nitrogen fixation, which in turn would positively impact on symbiotic interactions. *Micromonospora* may contribute to the survival of rhizobia by helping to maintain high levels of trehalose.

Chitin degradation and protection against pathogens. Plant β-1,3-glucanases are directly involved in defense by hydrolyzing the cell walls of fungal pathogens most commonly in combination with chitinases. Nine chitin-related ORFs were identified in *M. lupini*. Specifically, six code for a chitooligosaccharide deacetylase, several extracellular endo- and exo-chitinases and a β-N-acetyl-hexosaminidase (MiLup08_41789, MiLup08_41912, MiLup08_43481, MiLup08_44343, MiLup08_45172, MiLup08_45568), while three CDS code for putative chitin-binding domain proteins (MiLup08_41110, MiLup08_41724, MiLup08_41729).

Chitinases often work synergistically with chitin-binding proteins (CBPs). The biological roles of bacterial chitinases and carbon binding proteins are easily understood in an environmental context, especially in soil (that harbour fungi and insects) and marine (shellfish) habitats and their impact on chitin cycling. However, there is an increasing amount of direct or indirect evidence suggesting that some chitinases and CBPs additionally serve as virulence factors for bacterial pathogens during infection of non-chitinous substrates [69].

Experimental data confirmed *in vitro* chitin degradation of strain Lupac 08 (Figure 6B). As with other endophytic bacteria *Micromonospora* may produce chitinases to inhibit fungal pathogens, or may produce these molecules to elicit the plant defense mechanism. Either way, it seems that *Micromonospora* would provide a benefit to its host.

Siderophores (Iron-transport) and other secondary metabolites

Iron is an element essential for every living organism, as a cofactor of numerous proteins. Siderophores produced by plant growth promoting bacteria may reduce the growth of phytophathogens by depriving them of iron. Thus, an efficient iron uptake system can contribute to protect the host plant against pathogens. Interestingly, siderophores can also act as important virulence determinants for both plant and animal pathogens [70].

The genome of strain Lupac 08 revealed several siderophore related genes including specific iron uptake transporters, secretion of different siderophores and synthesis of siderophore receptors. Namely, a zinc/iron permease (MiLup_40258), a ferrous iron permease FTR1 (efeU, MiLup_41076) and eight iron ABC transporters (MiLup_42281-MiLup_42285). The number of the latter transporters is similar to the number of those found in the genome of the endophytic bacterium *Enterobacter* sp. 638 [47] while the plant pathogen *Erwinia amylovora* CFBP 1430 presents only three such transporters [71].

Several gene clusters related to the biosynthesis and transport of the siderophores enterobactin (MiLup_44069-MiUp_44071), aerobactin (iucA/iucC family protein, MiLup_44063 and MiLup_44064; MiLup_40326) and alcaligin (MiLup_44065) were also located. In the case of aerobactin, the gene *iucA* is highly correlated with virulence in avian pathogenic *E. coli* strains [72].

Two siderophore-interacting proteins (MiLup_40648 and MiLup_45559) were also found. One of these genes (MiLup_40648) was located next to a siderophore transporter of the RhtX/FptX family; RhtX from *S. meliloti* 2011 and FptX from *Pseudomonas aeruginosa* appear to be single polypeptide transporters from the major facilitator family for import of siderophores as a means to import iron [73]. In addition, a thiazolinyl imide reductase involved in siderophore biosynthesis was also identified (MiLup_43551).

The genome of Lupac 08 also contained several regulators including an iron-dependent repressor (IdeR, MiLup_41668), two ferric uptake regulation proteins (MiLup_40794 and MiLup_44436) and a putative iron-regulated membrane protein which suggests that these systems are highly regulated. Production of siderophores was detected experimentally (Fig. 7A).

Actinobacteria are well known to be capable of producing a vast diversity of natural secondary metabolite compounds with applications in medicine, agriculture, and other biotechnological areas [74]. Endophytic bacteria are currently of significant interest as an untapped resource of novel bioactive small molecules because their metabolites are speculated to affect the physiological conditions of host plants including growth and disease resistance. *Micromonospora* strains are well known for their capacity to produce many secondary metabolites and *M. lupini* Lupac 08 was previously screened for the production of novel compounds with antitumoral activity and the results obtained confirmed the production of a new family of molecules named Lupinacidins A, B and C [9,10].

Fifteen clusters involved in the biosynthesis of secondary metabolites were identified in the genome of *M. lupini* Lupac 08. These included siderophores (see above), terpenes, butyrolactones, polyketides (PKS), nonribosomal peptides (NRPS), chalcone synthases and bacteriocins (Table 4). A DNA stretch of 544 kb was estimated to code predominantly for secondary metabolites, accounting for about 7.4% of the genome. This percentage is lower than that reported for the marine actinobacterium *S. tropica* (9.9%, [75]) but it is within the range of other actinobacteria e.g. *S. coelicolor* (8.2%, [43]). Interestingly, *Frankia* strains ACN14a and EAN1pec dedicate about 5% of their genomes to natural product assembly while the potential of CcI3, which has the smallest genome of the three *Frankia* strains, has a much reduced host range and is absent from most soils is significantly smaller (~3%) [76].

Several clusters identified in the genome of *M. lupini* were also located in other genomes of phylogenetically related bacteria, especially in *S. tropica* CNB-440, *S. arenicola* CNS-205 and *V. maris* AB-18-032. Nevertheless other clusters were unique to *M lupini* (Table 4). Eight of the 15 clusters identified were located in the region between coordinates 4,000 kb and 5,000 kb of the genome, close to the terminus of replication. This area of the genome also contains a high density of genes coding for the biosynthesis of various plant cell wall degrading enzymes and several transposases.

Terpene related enzymes present in the genome of *M. lupini* are involved in the synthesis of carotenoids, sugar-binding lipids and the production of pentalenolactone type antibiotics. Similar molecules have also been predicted from the genomes of the three sequenced *Frankia* strains [77]. Various polyketide biosynthetic and non-ribosomal peptide synthase pathways were also identified specifically as PKI, PKII (2 clusters), PKIII types, NRPS (2 clusters) and hybrid PKS/NRPS clusters (2 clusters). The presence of these gene clusters suggests that *M. lupini* is capable of producing a vast diversity of secondary metabolites such as the antitumor anthraquinone derivative lupinacidins reported earlier [9,10]. Some of these metabolites may perform specialized functions in ecological niches and recent studies have reported on the importance of PKS and NRPS molecules and their potential role in communication during root colonization [78,79]. In addition, cluster 10 contains genes that putatively code for the production of granaticin. Granaticins are antibiotics of the benzoisochromanequinone class of aromatic polyketides, the best known member of which is actinorhodin produced by *S. coelicolor* A3(2). Production of granaticins has mainly been reported from *Streptomyces* strains [80]. NRPS cluster 11 (see Table 4), appeared to be unique to strain Lupac 08 as this group of genes was not detected in any of the other genomes compared except in *S. tropica* CNB-440 where it seems to be only partially conserved.

The PKS type III cluster corresponds to several genes that code for the production of naringenin, a central precursor of many flavonoids. It has recently been proposed that flavonoids play an important role in the establishment of plant root endosymbioses. In the case of legume-*Rhizobium* interactions, flavonoids released by plant roots induce genes involved in nodulation [81]. In a similar way it has also been suggested that these molecules play an important role during the early stages of the symbiotic association between *Frankia* and actinorhizal plants [82]. *Micromonospora* flavonoids may contribute to support communication between the nitrogen fixing bacteria and their host plants.

Table 4. Comparison of secondary metabolite clusters found in the genome of *M. lupini* Lupac 08 and other related microorganisms.

Cluster	Type	*M. lupini* Lupac 08 (Milup08_X)	*Micromonospora* sp. L5	*M. aurantiaca* ATCC 27029^T	*Verrucosispora maris* AB-18-032	*Salinispora tropica* CNB-440	*Salinispora arenicola* CNS-205	*Streptomyces coelicolor* A3(2)
1	Terpene	40204–40210	Conserved	Conserved	Conserved	Conserved	Conserved	Conserved
2	Terpene	40306–40320	Conserved	Conserved	Conserved	Absent	Absent	Present
3	Butyrolactone	40602–40668	Absent	Absent	Absent	Conserved	Conserved	Conserved
4	Type I PKS	41995–42009	Conserved	Conserved	Conserved	Conserved	Conserved	Absent
5	Terpene	43134–43144	Conserved	Conserved	Conserved	Conserved	Conserved	Conserved
6	NRPS+PKS	43546–43581	Absent	Absent	Conserved	Conserved	Absent	Partially conserved
7	Type II PKS	43804–43844	Conserved	Conserved	Conserved	Conserved	Conserved	Partially conserved
8	Siderophore	44063–44071	Conserved	Conserved	Conserved	Conserved	Conserved	Conserved
9	NRPS-Type I PKS	44386–44405	Conserved	NRPS Absent	NRPS Absent	NRPS Absent	NRPS Absent	NRPS Absent
10	Type II PKS	44613–44624	Partially conserved	Partially conserved	Partially conserved	Partially conserved	Partially conserved	Partially conserved
11	NRPS	44684–44691	Absent	Absent	Absent	Partially conserved	Absent	Absent
12	Bacteriocin	44929–44933	Conserved	Conserved	Conserved	Conserved	Conserved	Partially conserved
13	Terpene	45087–45093	Conserved	Conserved	Conserved	Absent	Absent	Conserved
14	NRPS	45439–45446	Conserved	Conserved	Conserved	Conserved	Conserved	Conserved
15	Type III PKS	46684–46700	Conserved	Conserved	Conserved	Conserved	Conserved	Absent

PKS, polyketide synthases; NRPS, non-ribosomal peptide synthases.

Genomic information also revealed that *M. lupini* has the potential to produce bacteriocins (cluster 12) as suggested by the presence of a putative short-chain dehydrogenase/reductase.

Phytohormones

Indole-3 acetic acid. Diverse bacterial species have the ability to produce auxinic phytohormones such as indole-3-acetic acid (IAA) and a few can also produce phenyl-acetate (PAA) such as *Frankia alni* [83,84]. Different biosynthesis pathways have been identified and redundancy for IAA biosynthesis is widespread among plant-associated bacteria [83]. Interactions between IAA-producing bacteria and plants may lead to several outcomes, from pathogenesis to phytostimulation [85]. The genome of *M. lupini* Lupac 08 contains a gene (Milup_45687) potentially involved in the biosynthesis of IAA via the indole-3-acetonitrile pathway. This gene corresponds to the conversion of indole-3-acetonitrile to indole-3-acetic acid. Nitrilases with specificity for indole-3-acetonitrile have been reported in *Alcaligenes faecalis* [86]. In *A. tumefaciens* and *Rhizobium* spp. nitrile hydratase and amidase activity could be identified, indicating the conversion of indole-3-acetonitrile to indole-3-acetic acid via indole-3-acetamide [87]. Analysis of IAA production by strain Lupac 08 was carried out, yielding a positive result (Fig. 7B).

Acetoin and 2,3-butanediol. The volatile compounds acetoin and 2,3-butanediol produced by bacteria such as *Bacillus subtilis* GB03 and *Bacillus amyloliquefaciens* IN937a have been reported as plant growth promoting hormones [88]. Several genes were located in the genome of strain Lupac 08 which could be involved in the production of these compounds. Two copies of the gene *pdhB* (MiLup_40114 and MiLup_43782) that encode the enzyme pyruvate dehydrogenase were identified. This enzyme transforms pyruvate to acetaldehyde and in this process a small fraction of pyruvate is converted to acetoin as a by-product. In addition, three acetolactate synthases involved in the synthesis of acetolactate from pyruvate are present (MiLup_41336, MiLup_41383 and MiLup_41384). Under aerobic conditions, acetolactate is converted to acetoin by the enzyme acetoin dehydrogenase (MiLup_41670).

Discussion

The most extensively studied bacteria interacting with plants are Gram-negative proteobacteria because they are readily isolated from plant tissues and genetically handled for interaction studies. However, the impact of Gram-positive bacteria on plants should not be underestimated as has been done for many years mainly due to their slow growth. *M. lupini* Lupac 08, a Gram-positive actinobacterium was isolated from the internal root nodule tissues of *Lupinus angustifolius* but it is only a representative of a large collection of more than 2000 *Micromonospora* strains isolated from diverse legumes and actinorhizals from different geographical locations. So far, several genomes of root symbionts and soil saprophytes have been studied; therefore we decided to focus on an intermediary category, that of facultative endophytes.

Lupac 08 was isolated from lupine nodules, and shown to produce the anticancer agents Lupinacidin A, B and C. The genome of strain Lupac 08 was sequenced to obtain information about the potential ecological role of *Micromonospora* in interaction with legumes and actinorhizal plants. Genomic analysis revealed several strategies which are probably necessary to lead a successful lifestyle as a saprophyte in the rhizosphere, a competitive and harsh environment, and as an endophyte capable of colonizing the internal plant tissues. *Micromonospora* species have less than 3% distance in their 16S rRNA genes, which can be

roughly translated to a time of 150MY according to the equivalence proposed by Ochman and Wilson [89]. In the current phylogeny (Fig. 1), *M. lupini* has as closest neighbours *M. chokoriensis M. saelicesensis* and *M. zamorensis*, isolated from sandy soil, root nodules of *L. angustifolius* and the rhizosphere of *P. sativum* respectively. *Micromonospora* sp. L5 and *M. aurantiaca* are located further away, with a distance of 1.2% that would translate to 60 MY for the emergence of a group of species that interact with plants, a date that would be close to the postulated time of emergence of *Fabaceae* and that of many actinorhizal plant families [90]. The separation from the *Salinispora* and *Verrucosispora* lineages would constitute two independent events that would have occurred slightly earlier at 160MY and 170MY, while the emergence of the *Actinoplanes* and that of the *Dactylosporangium* would have occurred 250MY ago, a time when dicotyledons had not yet appeared but when continents and thus soils had appeared that did permit the growth of primitive plants such as the gymnosperms.

The size of the *Micromonospora* genomes analyzed is quite uniform, with that of Lupac 08 slightly larger. The chromosome size of *M. lupini* Lupac 08 appears to reflect a wealth of genes allowing for adaptation to a complex saprophytic/endophytic lifestyle, which means adapting to a wider range of environmental conditions with the ability to metabolize a large variety of nutrient sources. Considering that *Micromonospora* sp. L5 and *M. lupini* Lupac 08 were both isolated from nitrogen fixing nodules (actinorhizal and legume plants, respectively), it would be expected that the genomes of these strains be more similar to each other than to *M. aurantiaca* ATCC 27029[T] which was originally isolated from soil. Surprisingly this was not the case as confirmed by the high number of strain specific genes identified in the genome of Lupac 08, suggesting a high capacity of adaptation to a fluctuating environment by this microorganism. On the other hand, *Micromonospora* sp. L5 and *M. aurantiaca* ATCC 27029[T] share a high number of orthologous genes (86%) suggesting that the niche of origin is not crucial.

An interesting result was the distribution of the metabolic profiles of 20 bacteria representing different living environments (Fig. 5). There was a clear proximity between *M. lupini* Lupac 08 and the three *Frankia* genomes. This result suggests that strain Lupac 08 contains metabolic functions similar to those found in *Frankia* strains that are probably useful for its interaction with plants. This metabolic versatility combined with a diverse transport system make Lupac 08 an organism fit to adapt to a soil/plant environment.

The emergence and evolution of nitrogen fixation ability among the domains *Bacteria* and *Archaea* is complex and has not yet been fully elucidated. Although it was previously reported that *Micromonospora* strains isolated from legume and actinorhizal root nodules contained *nifH*-like gene fragments [11,12], we could not confirm these results. In a similar approach based on PCR-amplification, other authors reported the presence of *nif*-H like sequences for bacterial isolates obtained from legumes collected in arid zones including *Microbacterium*, *Agromyces*, *Mycobacterium* and *Ornithinicoccus* [91]. One recurrent problem with the use of a PCR-based approach is that it is limited to a single gene amplified billions of times, which may provide false-positive results [92] and for this reason must always be confirmed by an independent approach.

Plant-polymer degrading enzymes such as cellulases and pectinases have been suspected to play a role in internal colonization. Most plant pathogens secrete cellulases, pectinases, xylanases, or other enzymes to hydrolyze plant cell wall polymers, while a lack of secreted hydrolases has been proposed to be favourable for microorganisms that form beneficial association with plants. Examples of endophytic plant growth promoting bacteria that lack large amounts of cell wall degrading enzymes include *Frankia* [38], *Enterobacter* sp. 638 [47], *Azoarcus* sp. BH72 [60] and *Herbaspirillum seropedicae* [93]. An *Azospirillum* sp. that does not colonize root tissues proper, but only the rhizosphere, has a genome containing a large number of putative cellulases similar to soil cellulolytic bacteria with 26–34 glycosyl hydrolases [94], as compared to the 37 present in *T. fusca*, a highly cellulolytic actinobacterium isolated from soil.

The genome of *M. lupini* revealed a high number of putative genes that encode for hydrolytic enzymes and specifically cellulolytic, xylanolytic, chitinolytic and pectinolytic activities were confirmed in the laboratory, indicating the capacity of *Micromonospora* to degrade plant polymers in a way similar to that of plant pathogen microorganisms. In the case of *Micromonospora*, there seems to be a paradox since strain *M. lupini* Lupac 08 shows a very high *in vitro* activity for cellulases and xylanases, however, preliminary inoculation experiments in our laboratory indicate that the microorganism does not behave as a pathogen, on the contrary, *Micromonospora* appears to interact in a tripartite relationship stimulating nodulation and plant growth (Fig. 7c). Therefore the question arises as to what is the function of these enzymes when *Micromonospora* interacts with the host plant. Alternatively some of the genes present, especially those related to the metabolism of cellulose may not necessarily imply that the bacterium is involved in plant cell wall degradation but have a different role, yet to be defined [95]. In addition many of the cellulose-related genes contain binding-domains suggesting that these may be related to the adhesion of the bacterium to the plant. These genes could also help *Micromonospora* digest plant cell walls upon senescence of the nodules.

M. lupini Lupac 08 contains several secondary metabolite gene clusters, many of which appear to be involved in the synthesis of siderophores and also of antibiotics. These would also in all likelihood be involved in the synthesis of the antitumor anthraquinone molecules described previously [9,10]. *Micromonospora*, like many other endophytic bacteria is a facultative plant colonizer that must compete with other microorganisms in the rhizosphere before entering the plant. In this sense, the NRPS and PKS gene clusters identified in the genome of *M. lupini* Lupac 08 may be involved in defense as well as in interaction and communication with its host plant. Thus, it will be necessary to identify these compounds and their functional attributes to further expand our knowledge of this plant-microbe interaction.

Conclusions

We have provided experimental data which supports the hypothesis that *M. lupini* Lupac 08 is a plant growth promoting bacterium. *Micromonospora lupini* Lupac 08 clearly produces a plant growth enhancing effect as observed in laboratory experiments. The localization of several genes involved in plant growth promotion traits such as the production of siderophores, phytohormones, the degradation of chitin (biocontrol) and the biosynthesis of trehalose may all contribute to the welfare of the host plant. *Micromonospora* appears as a new candidate in plant-microbe interactions with an important potential in agriculture and other biotechnological applications. The current data is promising but it is still too early to determine which specific roles are played by this microorganism in interactions with nitrogen fixing plants.

Methods

Genome sequencing, annotation and analysis

The genome sequence of *M. lupini* Lupac 08 was determined using the 454 FLX system and Titanium platform (454 Life Sciences) as previously reported [96]. Sequences were assembled into 50 contigs and four scaffolds ranging from 583 to 7,083,659 nucelotides using the MaGe (Magnifying Genomes) interface [97]. This Whole Genome Shotgun project has been deposited at European Nucleotide Archive under accession number NZ_CAIE00000000.01.

16S rRNA gene phylogeny

Sequences obtained from public databases (Genbank/EMBL) were manually aligned using clustal X software [98]. Phylogenetic distances were calculated with the Kimura 2-parameter model [99] and the tree topologies were inferred using the maximum-likelihood method [100]. All analyses were carried out using the MEGA5 program [101].

Comparative genome analysis

Genome rearrangement of the *Micromonospora* strains *M. lupini* Lupac 08, *Micromonospora* sp. L5, *M. aurantiaca* ATCC 27029[T] and *Micromonospora* sp. ATCC 39149 were carried out using MAUVE software [102]. The number of shared and unique genes present in the respective genomes were calculated and represented by a Venn-diagram using the EDGAR software [103]. Potential horizontally transferred genes were predicted using the "Region of Genomic Plasticity Finder" method implemented on the MicroScope platform. First we selected genomes included in the PkGDB and NCBI RefSeq databases that presented high synteny percentages with the Lupac 08 strain. Automatic results were manually curated according to several features such as base composition, DNA repetitions, presence of near mobile elements and information provided by SIGI and IVON programs [104].

Comparative analysis of metabolic profiles

A bicluster plot of the metabolic profiles for *M lupini* Lupac 08 and 20 other bacterial genomes was performed with Multibiplot [105]. A comparison of 798 MicroCyc metabolic pathways was made using MaGe. This comparison is based on the calculation of 'pathway completion' values, scaled from 0 to 1, where 0 means that a particular organism does not contain any enzyme for a given pathway and 1 that it contains all the reactions of the pathway. These values were transformed applying row standardization and a JK-Biplot was constructed after performing a PCA (Singular Value Decomposition estimation method). The heatmap was then obtained with the expected values computing the Euclidian distance and average linkage for rows and columns.

Transport proteins identification and classification

Information about transport proteins of genomes was obtained from the TransportDB relational database when available (http://www.membranetransport.org/). The identification and classification of the transporters of strain Lupac 08 was performed using the TransAAP tool based on TransportDB [106] followed by manual validation.

Cellulose, starch and xylan degradation

Strain Lupac 08 was cultivated on yeast-malt agar for 5 days and subsequently transferred to M3 agar [107] with and without glucose and supplemented with one of the substrates in the following way: carboxymethylcellulose (CMC, 0.5%), starch (1%)

and xylan (1%). A bacterial suspension of 10^6 per ml was prepared in saline solution (0.85%) and 200 µl were inoculated on the different plates which were then incubated at 28°C and results were recorded at 4, 7 and 14 days after inoculation. Xylan and CMC plates were stained with Congo red [108] while starch plates were flooded with iodine solution [109].

Pectin degradation

Pectinolytic acitivity was determined as described in Williams et al. [109]. Agar plates supplemented with pectin (0.5%) were streaked with strain Lupac 08 and incubated at 28°C. Hydrolysis zones were detected after 14 days incubation by flooding plates with an aqueous solution of cetyltrimethyl ammonium bromide (CTAB, 1%) and examining them after 30 min.

Chitin degradation

Chitinolytic acitivity was determined as described in Murthy et al. [110]. Agar plates supplemented with colloidal chitin at 0.5% (Gift of France-Chitine, Orange, http://france-chitine.com/), partly hydrolysed by stirring in 0.5 M HCl for 2h, were inoculated with strain Lupac 08 and incubated at 28°C. Hydrolysis zones were detected as cleared zones after 14 days incubation.

Siderophore production

Siderophore production was assessed using a modified chrome azurol S (CAS) assay [111]. Strain Lupac 08 was cultured on yeast-malt agar and incubated for 7 days, subsequently it was streaked onto CAS agar plates and incubated at 28°C for 7–10 days. A positive result was indicated by an orange halo around bacterial colonies.

Indole-3-acetic acid production

Production of indole acetic acid was assayed following the method of Glickmann and Dessaux [112]. Strain Lupac 08 was inoculated in 5 ml of yeast-malt medium supplemented with L-tryptophan (0.2%) and incubated at 28°C at 150 rpm during 7 days. The culture was then centrifuged at 12,000 x g for 10 min and 1 ml of the supernatant was mixed with 2 ml of Salkowski's reagent [113] and incubated at room temperature for 30 min. IAA

production was measured spectrophotometrically at 530 nm to assess the development of a pink colour.

Plant growth

Surface-sterilized seeds of *Trifolium* sp. were germinated axenically in Petri dishes on 1.4% w/v agar. Seedlings were transferred to sterile square plastic plates that contained a nitrogen-free nutrient solution [114]. Fifteen plantlets were inoculated in the following manner (5 per treatment): 500 µl (each) of bacterial suspensions (10^6 cells per ml) of *M. lupini* Lupac 08 and *Rhizobium* sp. E11 for coinoculation treatment; inoculation with *Rhizobium* sp. E11; and uninoculated plants as negative controls.

Acetylene reduction activity

Nitrogenase activity was measured using acetylene reduction [115] in sterile 150 ml plasma flasks with a rubber stopper. Cells of Lupac 08 were cultured in liquid minimal glucose medium without nitrogen at 28°C with shaking. The air in flasks was replaced with mixture of air and acetylene (ration 90:10 v/v). One mililiter of mixture was sampled for each measure using gas chromatography with a flame ionization detector (Girdel 30, France). *Mesorhizobium melitolti* Sm1021 was used as positive control.

Supporting Information

Table S1 ***M. lupini* Lupac 08 genome distribution of 4873 CDS (70.2%) based on COG categories.**

Table S2 **Carbohydrate related loci including cell-wall degrading enzymes and their potential regulators located on the genome of *M. lupini* Lupac 08.**

Author Contributions

Conceived and designed the experiments: MET PN. Performed the experiments: MET PP PB RR. Analyzed the data: MET RB PP PN CM. Contributed reagents/materials/analysis tools: MET PN PP YI CM. Contributed to the writing of the manuscript: MET PN RB PP.

References

1. Conn VM, Franco CM (2004) Analysis of the endophytic actinobacterial population in the roots of wheat (*Triticum aestivum* L.) by terminal restriction fragment legth polymorphism and sequencing of 16S rRNA clones. Appl Environ Microbiol 70: 1787–94.
2. Kaewkla O, Franco CM (2013) Rational approaches to improving the isolation of endophytic actinobacteria from Australian native trees. Microb Ecol 65: 384–393.
3. Hardoim PR, van Overbeek LS, Elsas DJ (2008) Properties of bacterial endophytes and their proposed role in plant growth. Trends Microbiol 16: 463–471.
4. Chi F, Shen SH, Cheng HP, Jing YX, Yanni YG, et al. (2005) Ascending migration of endophytic rhizobia, from roots to leaves, inside rice plants and assessment of benefits to rice growth physiology. Appl Environ Microbiol 71: 7271–7278.
5. Reinhold-Hurek B, Hurek T (1998) Life in grasses: diazotrophic endophytes. Trends Microbiol 6: 139–144.
6. Marchetti M, Capela D, Glew M, Cruvellier S, Chane-Woon-Ming B, et al. (2010) Experimental evolution of a plant pathogen into a legume symbiont Plos Biol 8: e1000280. doi: 10.1371/journal.pbio.1000280.
7. Ørskov J (1923) Investigations into the morphology of the ray fungi. Copenhagen.
8. Igarashi Y, Ogura H, Furihata K, Oku N, Indananda C, et al. (2011a) Maklamicin, an antibacterial polyketide from an endophytic *Micromonospora* sp. J Nat Prod 74: 670–674.
9. Igarashi Y, Trujillo ME, Martinez-Molina E, Yanase S, Miyanaga S, et al. (2007) Antitumor anthraquinones from an endophytic actinomycete *Micromonospora lupini* sp. nov. Bioorg Med Chem Lett 17: 3702–3705.
10. Igarashi Y, Yanase S, Sugimoto K, Enomoto M, Miyanaga S, et al. (2011b) Lupinacidin C, an inhibitor of tumor cell invasion from *Micromonospora lupini*. J Nat Prod 74: 862–865.
11. Trujillo ME, Alonso-Vega P, Rodriguez R, Carro L, Cerda E, et al. (2010) The genus *Micromonospora* is widespread in legume root nodules: the example of *Lupinus angustifolius*. ISME J 4: 1265–1281.
12. Valdés M, Perez NO, Estrada-de Los Santos P, Caballero-Mellado J, Pena-Cabriales JJ, et al. (2005) Non-*Frankia* actinomycetes isolated from surface-sterilized roots of *Casuarina equisetifolia* fix nitrogen. Appl Environ Microbiol 71: 460–466.
13. Carro L, Pujic P, Trujillo ME, Normand P (2013) *Micromonospora* is a normal inhabitant of actinorhizal nodules. J Biosci 38: 685–693.
14. Penn K, Jensen PR (2012) Comparative genomics reveals evidence of marine adaptation in *Salinispora* species. BMC Genomics 13: 86.
15. Normand P, Lapierre P, Tisa LS, Gogarten JP, Alloisio N, et al. (2007) Genome characteristics of facultatively symbiotic *Frankia* sp. strains reflect host range and host plant biogeography. Genome Res 17: 7–15.
16. Smith SE, Corneli-Showers P, Dardenne CN, Harpending HH, Martin DP, et al. (2012) Comparative genomic and phylogenetic approaches to characterize the role of genetic recombination in mycobacterial evolution. PLoS ONE 7: e50070.
17. Trujillo ME, Kroppenstedt RM, Fernandez-Molinero C, Schumann P, Martinez-Molina E (2007) *Micromonospora lupini* sp. nov. and *Micromonospora saelicesensis* sp. nov., isolated from root nodules of *Lupinus angustifolius*. Int J Syst Evol Microbiol 57: 2799–2804.
18. Carro L, Spröer C, Alonso P, Trujillo ME (2012) Diversity of *Micromonospora* strains isolated from nitrogen fixing nodules and rhizosphere of *Pisum sativum* analyzed by multilocus sequence analysis. Syst Appl Microbiol 35: 73–80.

19. Hirsch AM, Alvarado J, Bruce D, Chertkov O, De Hoff PL, et al. (2013) Complete genome sequence of *Micromonospora* strain L5, a potential plant-growth regulating actinomycete, originally isolated from *Casuarina equisetifolia* root nodules. Genome Announc 1: 2–00759–13.

20. Lethlefsen L, Schmidt TM (2007) Performance of the translational apparatus varies with the ecological strategies of bacteria J Bacteriol 189: 3237–3245.

21. Yano K, Wada T, Suzuki S, Tagami K, Matsumoto T, et al. (2013) Multiple rRNA operons are essential for efficient cell growth and sporulation as well as outgrowth in *Bacillus subtilis*. Microbiol 159: 2225–2236.

22. McCarthy AJ, Broda P (1984) Screening for lignin degrading actinomycetes and characterization of their activity against ¹⁴C-lignin labelled wheat lignocellulose. J Gen Microbiol 130: 905–2913.

23. Jendrossek D, Tomasi G, Kroppenstedt R (1997) Bacterial degradation of natural rubber: a privilege of actinomycetes? FEMS Microbiol Lett 150: 179–188.

24. Sandrak NA (1977) Degradation of cellulose by micromonospores. Mikrobiologiia 46 478–481.

25. de Menezes AB, Lockhart RJ, Cox MJ, Allison HE, McCarthy AJ (2008) Cellulose degradation by micromonosporas recovered from freshwater lakes and classification of these actinomycetes by DNA gyrase B gene sequencing. Appl Environ Microbiol 74: 7080–7084.

26. Miele V, Penel S, Duret L (2011) Ultra-fast sequence clustering from similarity networks with SiLiX. BMC Bioinformatics 12: 116 doi: 10.1186/1471-2105-12-116.

27. Ren Q, Paulsen IT (2007) Large-scale comparative genomic analyses of cytoplasmic membrane transport systems in prokaryotes. J Mol Microbiol Biotechnol 12: 165–179.

28. Ren Q, Paulsen IT (2005) Comparative analyses of fundamental differences in membrane transport capabilities in prokaryotes and eukaryotes. PLoS Comp Biol 1: e27.

29. Sutcliffe I (2011) New insights into the distribution of WXG100 protein secretion systems. Antonie van Leeuwenhoek 99: 127–131.

30. Scott JR, Barnett TC (2006) Surface proteins of Gram-positive bacteria and how they get there. Annu Rev Microbiol 60: 397–423.

31. Fyans JK, Bignell D, Loria R, Toth IT, Palmer T (2013) The ESX7type VII secretion system modulates development, but not virulence, of the plant pathogen *Streptomyces scabies* Mol Plant Pathol 14: 119–130.

32. Abdallah AM, Gey van Pittius NC, DiGiuseppe Champion PA, Cox J, Luirink J, et al. (2007) Type VII secretion – mycobacteria show the way. Nat Rev 5: 883–891.

33. Johnson TL, Abendroth J, Hol WGJ, Sandkvist M (2006) Type II secretion: from structure to function. FEMS Microbiol Lett 255: 175–186.

34. Dow JM, Daniels MJ, Dums F, Turner PC, Gough C (1989) Genetic and biochemical analysis of protein export from *Xanthomonas campestris*. J Cell Sci Suppl 11: 59–72.

35. Filloux A, Bally M, Ball G, Akrim M, Tommassen J, et al. (1990) Protein secretion in gram-negative bacteria: transport across the outer membrane involves common mechanisms in different bacteria. EMBO J 9: 4323–4329.

36. Reeves PJ, Whitcombe D, Wharam S, Gibson M, Allison G, et al. (1993) Molecular cloning and characterization of 13 out genes from *Erwinia carotovora* subspecies *carotovora*: genes encoding members of a general secretion pathway (GSP) widespread in Gram-negative bacteria. Mol Microbiol 8: 443–456.

37. DeShazer D, Brett PJ, Burtnick MN, Woods DE (1999) Molecular characterization of genetic loci required for secretion of exoproducts in *Burkholderia pseudomallei*. J Bacteriol 181: 4661–4664.

38. Mastronunzio JE, Tisa LS, Normand P, Benson DR (2008) Comparative secretome analysis suggests low plant cell wall degrading capacity in *Frankia* symbionts. BMC Genomics 9: 47 doi: 10.1186/1471-2164-9-47.

39. Krause A, Ramakumar A, Bartels D, Battistoni F, Bekel T, et al. (2007) Complete genome of the mutualistic, N₂-fixing grass endophyte *Azoarcus* sp. strain BH72. Nat Biotechnol 24: 1385–1391.

40. Hubber A, Vergunst AC, Sullivan JT, Hooykaas PJJ, Ronson CW (2004) Symbiotic phenotypes and translocated effector proteins of the *Mesorhizobium loti* strain R7A VirB/D4 type IV secretion system. Mol Microbiol 54: 561–574.

41. Cases I, de Lorenzo V, Ouzounis CA (2003) Transcription regulation and environmental adaptation in bacteria. Trends Microbiol 11: 248–253.

42. Bentley SD, Chater KF, Cerdeño-Tárraga AM, Challis GL, Thomson NR, et al. (2002) Complete genome sequence of the model actinomycete *Streptomyces coelicolor* A3(2). Nature 417: 141–147.

43. Kaneko T, Nakamura Y, Sato S, Asamizu E, Kato T et al. (2000) Complete genome structure of the nitrogen-fixing symbiotic bacterium *Mesorhizobium loti*. DNA Res 7: 331–338.

44. Lopes-Santos C, Correia-Neves M, Moradas-Ferreira P, Vaz-Mendes M (2009) A walk into the LuxR regulators of actinobacteria: phylogenomic distribution and functional diversity. Plos One 7: e46768.

45. Patankar AV, Gonzalez JE (2009) Orphan LuxR regulators of quorum sensing. FEMS Microbiol Rev 33: 739–756.

46. Compant S, Duffy B, Nowak J, Clément C, Barka EA (2005) Use of plant-growth-promoting bacteria for biocontrol of plant diseases: principles, mechanisms of action, and future prospects. Appl Environ Microbiol 71: 4951–4959.

47. Taghavi S, van der Lelie D, Hoffman A, Zhang YB, Walla MD, et al. (2010) Genome sequence of the plant growth promoting endophytic bacterium

Enterobacter sp. 638. Plos Genet 6 e1000943. doi:10.1371/jounal.pgen.1000943.

48. Pujic P, Fournier P, Alloisio N, Hay AE, Maréchal J, et al. (2012) Lectin genes in the *Frankia alni* genome. Arch Microbiol 194: 47–56.

49. Lykidis A, Mavromatis K, Ivanova N, Anderson I, Land M, et al. (2007) Genome sequence and analysis of the soil cellulolytic actinomycete *Thermobifida fusca* YX. J Bacteriol 189: 2477–2486.

50. Anderson I, Abt B, Lykidis A, Klenk HP, Kyrpides N, et al. (2012) Genomics of aerobic cellulose utilization sytems in actinobacteria. PlosOne 7 e39331.

51. Jayani RS, Saxena S, Gupta R (2005) Microbial pectinolytic enzymes: a review. Process Biochem 40: 2931–2944.

52. Jakob K, Kniskern J, Bergelson MJ (2007) The role of pectate lyase and the jasmonic acid defense response in *Pseudomonas viridiflava* virulence. Mol Plant Microb Interact 20: 146–158.

53. McQueen-Mason S, Durachko DM, Cosgrove DJ (1992) Two endogenous proteins that induce cell wall extension in plants. Plant Cell 4: 1425–1433.

54. Georgelis N, Tabuchi A, Nikolaidis N, Cosgrove DJ (2011) Structure-function analysis of the bacterial expansin EXLX1. J Biol Chem 286: 16814–16823.

55. McQueen-Mason SJ, Cosgrove DJ (1995) Expansin Mode of Action on Cell Walls - Analysis of wall hydrolysis, stress relaxation, and binding. Plant Physiol 107: 87–100.

56. Cosgrove DJ (2000) Loosening of plant cell walls by expansins. Nature 407: 321–326.

57. Bignell DR, Huguet-Tapia JC, Joshi MV, Pettis GS, Loria R (2010) What does it take to be a pathogen: genomic insights from *Streptomyces* species. Antonie van Leeuwenhoek 98: 179–194.

58. Jahr H, Dreider J, Meletzus D, Bahro R, Eichenbalub R (2000) The endo-beta-1,4-glucanase CelA of *Clavibacter michiganensis* subsp. *michiganensis* is a pathogenicity determinant required for induction of bacterial wilt of tomato. Mol Plant Microb Interact 13: 703–14.

59. Laine MJ, Haapalainen M, Wahlroos T, Kankare K, Nissinen R, et al. (2000) The cellulase encoded by the native plasmid of *Clavibacter michiganensis* ssp. *sepedonicus* plays a role in virulence and contains an expansin-like domain. Physiol Mol Plant Pathol 57: 221–233.

60. Kerff F, Amoroso A, Herman R, Sauvage E, Petrella S, et al. (2008) Crystal structure and activity of *Bacillus subtilis* YoaJ (EXLX1), a bacterial expansin that promotes root colonization. Proc natl Acad Sci USA 105: 16876–16881.

61. Dudeja SS, Giri R, Saini R, Suneja-Madan P, Kothe E (2012) Interaction of endophytic microbes with legumes. J Basic Microbiol 52: 248–60.

62. Cerda ME (2008) Aislamiento de *Micromonospora* de nódulos de leguminosas tropicales y análisis de su interés como promotor del crecimiento vegetal. Ph.D. Thesis. Universidad de Salamanca, Spain.

63. Streeter JG (1985) Accumulation of alpha, alpha-trehalose by *Rhizobium* bacteria and bacteroids. J Bacteriol 164: 78–84.

64. Farías-Rodriguez R, Mellor RB, Arias C, Peña-Cabriales JJ (1998) The accumulation of trehalose in nodules of several cultivars of common bean (*Phaseolus vulgaris*) and its correlation with resistance to drought stress. Physiol Plant 102 353–359.

65. Altamirano-Hernández J, López MG, Acosta-Gallegos JA, Farías-Rodríguez R, Peña-Cabriales JJ (2007) Influence of soluble sugars on seed quality in nodulated common bean (*Phaseolus vulgaris* L.): the case of trehalose. Crop Sci 47: 1193–1205.

66. Aeschbacher RA, Müller J, Boller T, Wiemken A (1999) Purification of the trehalase GMTRE1 from soybean nodules and cloning of its cDNA. GMTRE1 is expressed at a low level in multiple tissues. Plant Physiol 119: 489–495.

67. Barraza AG, Estrada-Navarrete G, Rodriguez-Alegria ME, Lopez-Munguia A, Merino E, et al. (2013) Down-regulation of PvTRE1 enhances nodule biomass and bacteroid number in the common bean. New Phytol 197: 194–206.

68. Müller J, Boller T, Wiemken A (2001) Trehalose becomes the most abundant non-structural carbohydrate during senescence of soybean nodules. J Experimental Bot 52 943–947.

69. Frederiksen RF, Paspaliari DK, Larsen T, Storgaard BG, Larsen MH, et al. (2013) Bacterial chitinases and chitin-binding proteins as virulence factors. Microbiology 159: 833–847.

70. Taguchi F, Suzuki T, Inagaki Y, Toyoda K, Shiraishi T, et al. (2010) The siderophore pyoverdine of *Pseudomonas syringae* pv. tabaci 6605 is an intrinsic virulence factor in host tobacco infection. J Bacteriol 192: 117–126.

71. Smits THM, Jaenicke S, Rezzonico F, Kamber T, Blom J, et al. (2010) Complete genome sequence of the fire blight pathogen *Erwinia amylovora* CFBP 1430 and comparsion to other *Erwinia* spp. Mol Plant Microbe Interact 23: 384–393.

72. Tivendale KA, Allen JL, Ginns CA, Crabb BS, Browning GF (2004) Association of *iss* and *iucA*, but not *tsh*, with plasmid-mediated virulence of avian pathogenic *Escherichia coli*. Infect Immun 72: 6554–6560.

73. Cuív PO, Clarke P, Lynch D, O'Connell M (2004) Identification of *rhtX* and *fptX*, novel genes encoding proteins that show homology and function in the utilization of the siderophores rhizobactin 1021 by *Sinorhizobium meliloti* and pyochelin by *Pseudomonas aeruginosa*, respectively. J Bacteriol 186: 2996–3005.

74. Genilloud O (2014) The re-emerging role of microbial natural products in antibiotic discovery. Antonie van Leeuwenhoek 106: 173–178.

75. Udwary DW, Zeigler L, Asolkar RN, Vasanth S, Alla L, et al. (2007) Genome sequencing reveals complex secondary metabolome in the marine actinomycete *Salinispora tropica*. Pro Natl Acad Sci 104: 10376–10381.

76. Nett M, Ikeda H, Moore BS (2009) Genomic basis for natural product biosynthetic diversity in the actinomycetes. Nat Prod Rep 26: 1362–1384.

77. Udwary DW, Gontang EA, Jones AC, Jones CS, Schultz AW, et al. (2011) Significant natural product biosynthetic potential of actinorhizal symbionts of the genus *Frankia*, as revealed by comparative genomic and proteomic analyses. Appl Environ Microbiol 77: 3617–3625.

78. Velázquez-Robeldo R, Contreras-Cornejo H, Macías-Rodriguez LI, Hernández-Morales A, Aguirre J, et al. (2011) Role of the 4-phosphopantetheinyl transferase of *Trichoderma virens* in secondary metabolism, and induction of plant defense responses. Mol Plant Microb Interact 24: 1459–1471.

79. Mukherjee PK, Buensanteai N, Moran-Diez ME, Druzhinina IS, Kenerley CM (2012) Functional analysis of non-ribosomal peptide synthetases (NRPs) in *Trichoderma virens* reveals a polyketide synthase (PKS)/NRPS hybrid enzyme involved in the induced systemic resistance response in maize. Microbiol 158: 155–165.

80. Tornus D, Floss HG (2001) Identification of four genes from the granaticin biosynthetic gene cluster of Streptomyces violaceoruber Tü22 involved in the biosynthesis of L-rhodisone. J Antibiot 54: 91–101.

81. Tadra-Sfeir MZ, Souza EM, Faoro H, Müller-Santos M, Baura VA, et al. (2011) Naringenin regulates expression of genes involved in cell wall synthesis in *Herbaspirillum seropedicae* Appl Environ Microbiol 77: 2180–2183.

82. Abdel-Lateif K, Boguz D, Hocher V (2012) The role of flavonoids in the establishment of plant roots endosymbioses with arbuscular mycorrhiza fungi, rhizobia and *Frankia* bacteria. Plant Signal Behav 7: 636–641.

83. Hammad Y, Nalin R, Marechal J, Fiasson K, Pepin R, et al. (2003) A possible role for phenyl acetic acid (PAA) on *Alnus glutinosa* nodulation by *Frankia* bacteria. Plant and Soil 254: 193–206.

84. Duca D, Lorv J, Patten CL, Rose D, Glick BR (2014) Indole-3-acetic acid in plant-microbe interactions. Antonie van Leeuwenhoek 106: 85–125.

85. Spaepen S, Vanderleyden J, Remans R (2007) Indole-3-acetic acid in microbial and microorganism-plant signaling. FEMS Microbiol Rev 31: 425–448.

86. Kobayashi M, Izui H, Nagasawa T, Yamada H (1993) Nitrilase in biosynthesis of the plant hormone indole-3-acetic acid from indole-3-acetonitrile: cloning of the *Alcaligenes* gene and site-directed mutagenesis of cysteine residues. Proc Natl Acad Sci 90: 247–251.

87. Kobayashi M, Suzuki T, Fujita T, Masuda M, Shimizu S (1995) Occurrence of enzymes involved in biosynthesis of indole-3-acetic acid from indole-3-acetonitrile in plant-associated bacteria, *Agrobacterium* and *Rhizobium*. Proc Natl Acad Sci 92: 714–718.

88. Ryu CM, Farag MA, Hu CH, Reddy MS, Wei HX, et al. (2003) Bacterial volatiles promote growth in Arabidopsis. Proc Natl Acad Sci 100: 4927–4933.

89. Ochman H, Wilson AC (1987) Evolution in bacteria: evidence for a universal substitution rate in cellular genomes. J Mol Evol 26: 74–86.

90. Bell CD, Soltis DE, Soltis PS (2010) The age and diversification of the angiosperms re-revisited. Amer J Bot 97: 1–8.

91. Zakhia F, Jeder H, Willems A, Gillis M, Dreyfus B, et al. (2006) Diverse bacteria associated with root nodules of spontaneous legumes in Tunisia and first report for nifH-like gene within the genera Microbacterium and Starkeya. Microb Ecol 51: 375–393.

92. Gtari M, Ghodhbane-Gtari F, Nouioui I, Beauchemin N, Tisa LS (2012) Phylogenetic perspectives of nitrogen-fixing actinobacteria. Arch Microbiol 194: 3–11.

93. Pedrosa FO, Monteiro RA, Wassem R, Cruz LM, Ayub RA, et al. (2011) Genome of *Herbaspirillum seropedicae* strain SmR1, a specialized diazotrophic endophyte of tropical grasses. PLoS Genet 7: e1002064. doi:10.1371/journal.pgen.1002064.

94. Wisniewski-Dyé F, Borziak K, Khalsa-Moyers G, Alexandre G, Sukharnikov LO, et al. (2011) *Azospirillum* genomes reveal transition of bacteria from aquatic to terrestrial environments. Plos Genetics 7: e1002430. doi: 10.1371/journal.pgen.1002430.

95. Medie FM, Davies GJ, Drancourt M, Henrissat B (2012) Genome analyses highlight the different biological roles of cellulases. Nat Rev Microbiol 10: 227–234.

96. Alonso-Vega P, Normand P, Bacigalupe R, Pujic P, Lajus A, et al. (2012) Genome sequence of *Micromonospora lupini* Lupac 08, isolated from root nodules of *Lupinus angustifolius*. J Bacteriol 194: 4135.

97. Vallenet D, Labarre L, Rouy Z, Barbe C, Bocs S, et al. (2006) MaGe–a microbial genome annotation system supported by synteny results. Nucleic Acids Res 34: 53–65.

98. Thompson JD, Gibson TJ, Plewniak F, Jeanmougin F, Higgins DG (1997) The CLUSTAL_X windows interface: flexible strategies for multiple sequence alignment aided by quality analysis tools. Nucleic Acids Res 25: 4876–4882.

99. Kimura M (1980) A simple method for estimating evolutionary rates of base substitutions through comparative studies of nucleotide sequences. J Mol Evol 16: 111–120.

100. Felsenstein J (1981) Evolutionary trees from DNA sequences: a maximum likelihood approach. J Mol Evol 17: 368–376.

101. Tamura K, Peterson D, Peterson N, Stecher G, Nei M, et al. (2011) MEGA5: molecular evolutionary genetics analysis using maximum likelihood, evolutionary distance, and maximum parsimony methods. Mol Biol Evol 10: 2731–2739.

102. Darling AE, Mau B, Perna NT (2010) progressiveMauve: multiple genome alignment with gene gain, loss and rearrangement. PLoS ONE 5: e11147. doi: 10.1371/journal.pone.0011147.

103. Blom J, Albaum S, Doppmeier D, Pühler A, Vorhölter FJ, et al. (2009) EDGAR: a software framework for the comparative analysis of prokaryotic genomes. BMC Bioinformatics 10: 154. doi: 10.1186/1471–2105–10–154.

104. Vernikos GS, Parkhill J (2006) Interpolated variable order motifs for identification of horizontally acquired DNA: revisiting the *Salmonella* pathogenicity islands. Bioinformatics 22: 2196–2203.

105. Vicente-Villardón JL (2010) MULTBIPLOT: A package for Multivariate Analysis using Biplots. Departamento de Estadística. Universidad de Salamanca. Available: http://biplot.usal.es/ClassicalBiplot/index.html.

106. Ren Q, Chen K, Paulsen IT (2007) TransportDB: a comprehensive database resource for cytoplasmic membrane transport systems and outer membrane channels. Nucleic Acids Res35: 274–279.

107. Rowbotham TJ, Cross T (1977) Ecology of *Rhodococcus coprophilus* and associated actnomycetes in fresh water and agricultural habitats. J Gen Microbiol 100: 231–240.

108. Mateos PF, Jimenez-Zurdo JI, Chen J, Squartini AS, Haack SK, et al. (1992) Cell-associated pectinolytic and cellulolytic enzymes in *Rhizobium leguminosarum* biovar trifolii. Appl Environ Microbiol 58: 1816–1822.

109. Williams ST, Goodfellow M, Alderson G, Wellington EMH, Sneath PHA, et al. (1983) Numerical classification of *Streptomyces* and related genera J Gen Microbiol 129: 1743–1813.

110. Murthy N, Bleakley B (2012) *Simplified method of preparing colloidal chitin used for screening of chitinase-producing microorganisms*. Internet J Microbiol 10: DOI:10.5580/2bc3.

111. Milagres AMF, Machuca A, Napoleão D (1999) Detection of siderophore production from several fungi and bacteria by a modification of chrome azurol S (CAS) agar plate assay. J Microbiol Meth 37: 1–6.

112. Glickmann E, Dessaux Y (1995) A critical examination of the specificity of the salkowski reagent for indolic compounds produced by phytopathogenic bacteria. Appl Environ Microbiol 61: 793–796.

113. Gordon SA, Weber RP (1951) Colorimetric estimation of indole acetic acid. Plant Physiol 26: 192–195.

114. Rigaud J, Puppo A (1975) Indole-3-acetic acid catabolism by soybean bacteroids J Gen Microbiol 88: 223–228.

115. Hardy RWF, Hoilsten RD, Jackson EK, Burns RC (1968) The acetylene-ethylene assay for N₂ fixation: laboratory and field evaluation. Plant Physiol 43: 1185–1207.

## 16

Development and Validation of an *Haemophilus influenzae* Supragenome Hybridization (SGH) Array for Transcriptomic Analyses

Benjamin A. Janto[1,2], N. Luisa Hiller[1,3], Rory A. Eutsey[1], Margaret E. Dahlgren[1], Joshua P. Earl[1], Evan Powell[1], Azad Ahmed[1], Fen Z. Hu[1,2,4]*, Garth D. Ehrlich[1,2,4]*

1 Center for Genomic Sciences, Allegheny-Singer Research Institute, Pittsburgh, Pennsylvania, United States of America, **2** Department of Microbiology and Immunology, Drexel University College of Medicine, Allegheny Campus, Pittsburgh, Pennsylvania, United States of America, **3** Department of Biological Sciences, Carnegie Mellon University, Pittsburgh, Pennsylvania, United States of America, **4** Department of Otolaryngology Head and Neck Surgery, Drexel University College of Medicine, Allegheny Campus, Pittsburgh, Pennsylvania, United States of America

Abstract

We previously carried out the design and testing of a custom-built *Haemophilus influenzae* supragenome hybridization (SGH) array that contains probe sequences to 2,890 gene clusters identified by whole genome sequencing of 24 strains of *H. influenzae*. The array was originally designed as a tool to interrogate the gene content of large numbers of clinical isolates without the need for sequencing, however, the data obtained is quantitative and is thus suitable for transcriptomic analyses. In the current study RNA was extracted from *H. influenzae* strain CZ4126/02 (which was not included in the design of the array) converted to cDNA, and labelled and hybridized to the SGH arrays to assess the quality and reproducibility of data obtained from these custom-designed chips to serve as a tool for transcriptomics. Three types of experimental replicates were analyzed with all showing very high degrees of correlation, thus validating both the array and the methods used for RNA profiling. A custom filtering pipeline for two-condition unpaired data using five metrics was developed to minimize variability within replicates and to maximize the identification of the most significant true transcriptional differences between two samples. These methods can be extended to transcriptional analysis of other bacterial species utilizing supragenome-based arrays.

Editor: Holger Fröhlich, University of Bonn, Bonn-Aachen International Center for IT, Germany

Funding: This work was supported by Allegheny General Hospital, Allegheny-Singer Research Institute and National Institutes of Health grant numbers DC002148, DC02148 – 16S1, and AI080935 to GDE. The funders had no role in study design, data collection and analysis, decision to publish, or preparation of the manuscript.

Competing Interests: The authors have declared that no competing interests exist.

* Email: garth.ehrlich@drexelmed.edu (GDE); fhu168168@gmail.com (FZH)

Introduction

The tremendous advancement of sequencing technologies combined with rapid reductions in associated costs has allowed researchers not only to sequence more diverse organisms but also to sample individual bacterial species with much greater resolution. Extensive strain sequencing has led to the realization and appreciation that most bacterial species harbour enormous genomic diversity among strains [1–10] This diversity is manifested in small scale as single nucleotide polymorphisms (SNPs), and also the more dramatic swapping in and out of entire genes and/or operons through the process of horizontal gene transfer as predicted by the Distributed Genome Hypothesis [11–12]. Analyses of multiple genomes from individual bacterial species have led to the recognition that there exists a supragenome [11,13] or pan-genome [1] at the species-level that is far larger than the genome of any single strain. The supragenome is composed of the core genome (those genes shared among all strains, and the distributed/accessory genome (those genes that are present in a subset of strains). The ability to take up and incorporate DNA from the distributed genome by sampling from other strains' DNA during polyclonal infections has been hypothesized to give these

organisms an important mechanism for rapid diversity generation [11,14–17]. This genetic diversity manifests as phenotypic diversity as different strains within the same species have been found to possess enormous differences in complex processes such as quorum sensing, biofilm formation and pathogenesis [12,17–21] (Janto et al., Kress-Bennett et al. unpublished observations).

Haemophilus influenzae (Hi) is one such species of bacteria that has been demonstrated to possess enormous genomic variability [2,5,22]. These bacteria are commensals of the human respiratory tract but some have pathogenic potential. Un-encapsulated non-typeable Hi (NTHi) are most often associated with localized disease such as chronic obstructive pulmonary disease (COPD) [23–27], otorrhea [21], chronic otitis media with effusion (COME) and acute otitis media (AOM) [27–33], however, they are increasingly being found as the major source of invasive disease [34–38]. Individual NTHi strains share only ∼80% of their ∼1,800 genes with all other strains (core genes) with the rest being distributed (or accessory) genes [2]. The Finite Supragenome Model [2,5] predicts the Hi supragenome to contain 4547 genes of which only ∼33% represent the core genome while the rest are present at various other frequencies among strains within the species [22]. The tremendous genic variability among NTHi strains presents a significant challenge when studying whole genome transcriptional patterns among many different strains. Traditionally, genic content must be known *a priori* in order to target genes with sequence-specific probes for measurement. Since two different strains might at a minimum share only the core genes, an array of probes designed for genes found in any single strain does not appropriately represent the species and therefore a significant amount of information will be lost (hundreds of genes) when using an array developed from a single strain. Thus, a more robust strategy for the design of bacterial microarrays is to use probes based on defined supragenomic sequences.

We previously designed and tested an *H. influenzae* supragenome hybridization (SGH) array in order to perform DNA-DNA hybridizations for the purpose of determining gene content in unsequenced strains [22]. This array was designed based on 3,100 genes that were identified in whole genome sequencing (WGS) of 24 geographically and clinically diverse NTHi strains and which includes >98% of all non-rare ($v>0.1$) genes. Since genes are either present or absent from genomic DNA (gDNA) of any given strain, the signal obtained for each probe is essentially binary and a signal threshold cut-off was used to determine whether a gene was present or not. Nevertheless the data collected is quantitative and these arrays can also be used to hybridize labelled RNA instead of DNA thereby acting as a transcriptomic tool. Here we report the testing and validation of these custom SGH arrays for this application, as well as the design of an analysis pipeline for suggested use.

Materials and Methods

Design of the *H. influenzae* supragenome hybridization (SGH) array

Design and testing of the *H. influenzae* supragenome hybridization (SGH) array is described by [22]. Briefly, annotations from 24 sequenced *H. influenzae* strains were clustered using a custom supragenome pipeline to obtain unique clusters of genes [14]. NimbleGen probe design software was used to design between three and thirteen, 60 mer probe sequences to the longest sequence in each gene subcluster. Probes were tested and graded *in silico* based on uniqueness, distribution and probe manufacturing parameters. In all, 31,307 *H. influenzae* specific probes were synthesized by Roche/NimbleGen. Each array contained duplicates of each probe and each slide contained 12 arrays. An additional 185 negative control probes based on *Streptococcus pneumoniae* chromosomal sequences were also attached to the slides in duplicate and a further 9,053 random sequence probes were included. These arrays containing a total of 72,037 probes (72 K) are referred to as the SGH arrays.

Genomic Hybridization

Genomic DNA (gDNA) was isolated from strain CZ4126/02 and Cy3-labeled using a NimbleGen One-Color DNA Labeling Kit. NimbleGen Hybridization Kits and Sample Tracking Control Kits were used to hybridize this labeled DNA to the custom-designed *H. influenzae* SGH arrays as well as for array washing. Images were acquired on an Axon Instruments GenePix 4200AL array scanner.

Genomic Hybridization data processing

Images were processed and data were normalized within chips using a Robust Multichip Average (RMA) algorithm and quantile normalization via the NimbleScan software v2.5 [30,40]. Raw data was converted into gene possession or absence by applying a

combination of an expression threshold (1.5X the median background value in log_2 scale) and a measure of probe variance [22]. Subclusters producing a signal above this value were set to a value of 1 (present) and subclusters with values below this value were set to a value of 0 (absent). The list of present subclusters was then used as a reference list for filtering transcription-based microarray data.

Experimental design and sample collection for two-condition transcriptional microarray analyses

Parallel work has focused on the role of AI-2 signalling in Hi by comparative studies between CZ4126/02 and an AI-2 sensing mutant (CZ4126/02ΔLsr::Cm^r [KO]) (Janto et al. unpublished observations). These strains were used to test and validate the use of the SGH Array for unpaired two-condition transcriptional microarray analysis. The CZ4126/02 WT and its cognate KO strain were grown in two different media (BHI and CDM) and sampled at multiple time points (the combinations of which are referred to here as "conditions", **Table S3**) for RNA extraction. The RNAs were converted to cDNA, Cy-3 labelled and hybridized to the *H. influenzae* SGH arrays as described below. For each condition, RNA samples were collected twice which are referred to as replicates A and B. Each condition/replicate was hybridized on two different SGH arrays referred to as chip 1 and chip 2. Finally each array outputs separate information for two duplicate probe-sets referred to as probe-set 1 and probe-set 2. All transcriptional data in this study has been deposited in NCBI's Gene Expression Omnibus (GEO) [41] and are accessible through GEO Series accession number GSE41690 [42].

H. influenzae culture media

Brain-Heart Infusion broth (BHI - Oxoid) was made using 37 g of powdered media/L and supplemented with hemin (Sigma-Aldrich) to a final concentration of 10 μg/mL and β-nicotinamide adenine dinucleotide (β-NAD) to a final concentration of 2 μg/mL. Chemically Defined Media (CDM) was made exactly as described [43] with the following minor catalog change (Dr. Arnold Smith, personal communication): 1X RPMI 1640 with glutamine and 25 mM HEPES (Gibco, catalog #22400-089).

Bacterial growth for RNA extraction

For microarray experiments frozen stocks were used to inoculate BHI plates that were incubated overnight at 37°C with 5% CO_2. Isolated colonies from these plates were used to inoculate 5 mL BHI cultures that were incubated overnight at 37°C with shaking at 200 rpm. The cultures were diluted to an OD_{A600} of 0.02 in 40 mL BHI or CDM. At selected time-points (**Table S3**), 1 mL culture samples were collected and transferred immediately into 2 mL RNAProtect (Qiagen). Samples were incubated for 10 minutes at room temperature and then stored overnight at 4°C.

RNA extraction and quality check

Samples stored in RNAProtect were spun for 10 minutes at 2,500×g (Sorvall RT-7), the supernatant removed and the cell pellets resuspended in 100 μL of 1X Tris-EDTA (TE) +1 mg/mL lysozyme (Worthington Biochemical) and 1 mg/mL proteinase K (Qiagen). RNA was extracted using a Qiagen RNeasy Mini Plus kit with the standard protocol including genomic DNA (gDNA) eliminator columns. The eluted RNA (~85 μL) was DNased by adding 10 μL 10X TurboDNase buffer and 5 μL TurboDNase (2 units/μL) (Ambion) and incubating at 37°C for 1.5 hours. 2 μL more TurboDNase was added and incubation continued for an additional 1.5 hours. The DNased RNA samples were cleaned by passing the samples through the RNeasy protocol a second time (including the gDNA eliminator column steps). Samples were eluted in nuclease free (n.f.) water, quantitated on a Nanodrop 1000 spectrophotometer and stored at −80°C. 200 ng of each RNA sample was run on an Agilent 2100 Bioanalyzer using RNA Nano6000 chips to check for RNA degradation. We performed paired reverse transcription reactions on every RNA sample where one reaction received reverse transcriptase (+RT, M-MLV, Promega) and the other did not (−RT). Both reactions were PCR amplified using primers directed against a housekeeping gene (GAPDH) and observation of amplification in the +RT reaction as well as lack of amplification in the −RT reaction verified removal of gDNA from each RNA sample.

qRT-PCR

Single-stranded cDNA was synthesized from the extracted RNA samples using the Roche Transcriptor First Strand Synthesis kit. Specific primers for the housekeeping and experimental genes were designed using Roche Probe Finder online software in order to design ~75 bp amplicons. qRT-PCR was performed on the Roche Light Cycler 480 using a SYBR green master mix. Reactions were performed in a 20 µl volume containing 2 µl cDNA (1:5 dilution) and primers at 0.5 µm each. Primer efficiency was determined by testing all primer pairs ahead of time with gDNA template. All reactions were measured in triplicate. The experimental data were normalized using the hpr and ldhA genes as internal standards. Independent data analysis was carried out using both the Pfaffl-$\Delta\Delta C_T$ method with the Roche Light Cycler software as well as a linear regression method using the LinRegPCR [44] software package. Fold changes presented are the mean results from both methods of analysis and from normalization against both of the housekeeping genes.

Generation of labelled double-stranded cDNA for SGH array hybridization

First and second-strand cDNA synthesis was performed using a SuperScript One-Cycle cDNA Kit (Invitrogen) as outlined in the NimbleGen Microarray Experienced User's Guide including RNaseA and cDNA precipitation steps. 1 µg of cDNA was Cy3-labeled using a NimbleGen One-Color DNA Labeling Kit. NimbleGen Hybridization Kits and Sample Tracking Control Kits were used to hybridize the labelled cDNA to the custom-designed H. influenzae SGH arrays as well as for array washing. Images were acquired on an Axon Instruments GenePix 4200AL array scanner.

Analysis of microarrays

Images were processed to.pair files containing expression values for both sets of duplicate probes representing all the subclusters on the H. influenzae SGH array using the NimbleScan software. These.pair tables were merged with a reference list of subclusters that had been determined to be present in the CZ4126/02 genome (see above) in order to remove non-relevant probe/subcluster data. These parsed.pair files were then normalized within and across chips using a Robust Multichip Average (RMA) algorithm and quantile normalization using the NimbleScan v2.5 software followed by a median polish whereby the 3 to 13 probe values/subcluster were condensed to a single value (in duplicate) [39,40]. Duplicate probe-set values were treated as independent replicates. For comparison of technical and biological replicate data, CyberT was used to obtain Bayesian corrected p-values, Bonferroni corrected p-values and Benjamini-Hochberg values

[45]. Significance Analysis of Microarrays (SAM v3.0) was used to obtain lists of genes with associated permutation-based false discovery rates (FDR) [46]. These data were combined and filtered in the following order: 1) SAM FDR <10%, Bayesian p-values< .05, Benjamini-Hochberg FDR<10%, Bonferroni corrected p-value<.05, raw values in at least one of the two conditions being compared >256 normalized intensity.

Results

Removal of non-relevant subclusters for microarray analysis

The custom-designed H. influenzae SGH array contains 31,307 unique probes that target 2,890 of the 3,100 gene clusters identified in 24 geographically diverse clinical strains. Gene clusters were further subdivided into "subclusters" with more stringent alignment parameters in order to capture allelic differences within more variable genes. The power of this array is its ability to capture information for any strain of H. influenzae since the probes represent the majority of the predicted supragenome (>85% of all "non-rare" genes). Non-rare genes are defined as those that are present in more than 10% of Hi strains) [22]. However, since a large proportion of the gene probes present in the SGH array do not correspond with any gene for any given single strain, once a strain is selected for study, an in silico analysis should be performed to ensure that only the relevant subset of probes is included in the final analysis. This is important for the purposes of obtaining a Gaussian distribution of data needed for both normalization and statistical testing. Therefore, it is necessary to remove all data from so-called "non-relevant" gene clusters, defined as those present in one of the 24 strains used to design the array but not present in the strain being interrogated.

For testing purposes we used a strain (CZ4126/02) that was not included in the design of the H. influenzae SGH array. Our first task was to determine the gene content of this strain for the purposes of removing non-relevant data later. This we accomplished by two methods 1) WGS of CZ4126/02 and mapping of identified genes back to the SGH array clusters (Janto et al. unpublished observations) and 2) hybridization of genomic DNA (gDNA) to the SGH array and application of a signal threshold to determine whether a gene was present or not [22]. A comparison of the WGS and SGH data sets from strain CZ4126/02 revealed that 2805/2890 (97%) of the identified gene clusters were in agreement between the two methods. Using WGS as the gold standard we identified 39 false positives (some of which could potentially be true positives present in WGS contig gaps) and 46 false negatives. In addition, we found only four genes in the WGS that were not represented on the SGH array [22]. Because of this accuracy we used SGH data for the purposes of removing data not relevant to strain CZ4126/02.

From the SGH gene possession experiment, hybridized CZ4126/02 gDNA gave a positive signal for 1702 of the 2890 total gene clusters and 2194 of the 4052 total gene subclusters represented on the array. This list of 2194 gene subclusters was then merged with the raw output from transcriptional experiments to isolate data only associated with those CZ4126/02 strain-specific subclusters. All transcriptional data in this study has been deposited in NCBI's Gene Expression Omnibus (GEO) [41] and are accessible through GEO Series accession number GSE41690 [42]. A representative histogram of the distribution of $\log_2$ transformed raw intensity values obtained before and after removal of non-relevant subclusters is shown in **Figure 1**. A summary of the distributions of each data set (relevant and non-

Table 1. Comparison of the distributions of raw data and filtered sub sets of probe-level data.

	mean	median	max	min	N
ALL probe-set 1	6.85	5.49	15.99	3.45	31,307
ALL probe-set 2	6.85	5.49	15.99	3.43	31,307
CZ specific probe-set 1	10.22	10.42	15.99	3.79	10,161
CZ specific probe-set 2	10.23	10.44	15.99	3.86	10,161
Not CZ probe-set 1	5.31	5.16	14.62	3.69	21,146
Not CZ probe-set 2	5.31	5.16	14.65	3.68	21,146
Negative probe-set 1	5.15	5.09	6.65	3.71	185
Negative probe-set 2	5.27	5.17	7.46	3.78	185
Random control probes	5.12	5.01	10.83	3.53	9,053

Expression data in $\log_2$ from condition 3, replicate A, chip 1. Raw output (ALL) was filtered using SGH data to produce subsets of data containing subclusters (and associated probes) present in CZ4126/02 (CZ specific) and subclusters not present in CZ4126/02 (Not CZ). Normalization was performed after filtering. Negative: probes synthesized from *Streptococcus pneumoniae* genes expected to be absent in *H. influenzae*. N: number of probes in each set. Probes are synthesized in duplicate on each chip (probe-set 1, probe-set 2).

relevant) compared to random and control data at the probe-level is presented in **Table 1**.

After removal of non-relevant subcluster data, the remaining data was normalized between and across chips where necessary using the NimbleGen NimbleScan software. In the same representative sample we found only fourteen false positives (14/2194 or 0.64%) (gene subclusters that gave a transcriptional signal significantly above background that did not give a SGH signal over threshold) illustrating the consistency across methods of analysis.

Step-wise filtering with statistical tests and testing on technical and biological replicates

In analyzing the data here, a Bayesian-corrected variance was applied and t-tests were performed using the web-based microarray analysis tool, CyberT [45]. We found that using a Bayesian p-value cut-off of 0.05 alone is not stringent enough as it results in reporting of a large number of false positives. This is illustrated by comparing the technical replicates for condition 4 (the same RNA sample run on two different chips), in this case biological replicate B, chip 1 vs. chip 2, which produced nearly identical results, with a R^2 of .9941 (**Figure 2**). Submitting this comparison to CyberT and obtaining Bayesian corrected p-values results in a list of 120 subclusters with $p < 0.05$, each of them a false positive (**Table S2**). In this same dataset applying a fold change cut-off alone is similarly inappropriate. In this comparison of a technical replicate, 30 subclusters are found with a fold change > 1.5, all false positives (**Table S3**). Twenty-three (23)/30 of these subclusters have expression values below 2^8 (256) raw intensity on both chips. These two parameters applied together (fold changes and t-tests) give some measure of biological and statistical significance that compensate for each others' weaknesses as far fewer clusters meet both cut-offs of fold > 1.5 and $p < 0.05$ than either alone, however, there are still several that slip through as false positives.

Therefore, a permutation-based false discovery rate (FDR) was calculated with the Significance Analysis of Microarrays (SAM) excel plug-in [46]. An FDR (q-value) of 10% calculated by SAM was used for this filtering step. A non-permutation-based estimate of the FDR was also used as an additional filter again with a cut-off of 10% (0.1) (Benjamini-Hochberg [BH]). The extremely stringent Bonferroni-corrected p-value was used to identify only the most significant findings. In the technical replicate comparison discussed above (Condition 4, replicate B, chip 1 vs. chip 2), the application of any one of these three additional statistical filters (SAM FDR, BH FDR, or Bonferroni-corrected p-value) results in no significant findings between the technical replicates, which is the reality (**Table S1**).

Thus, a custom step-wise filtering process was developed and implemented roughly in order of stringency. This involved obtaining the set of genes associated with the SAM permutation-

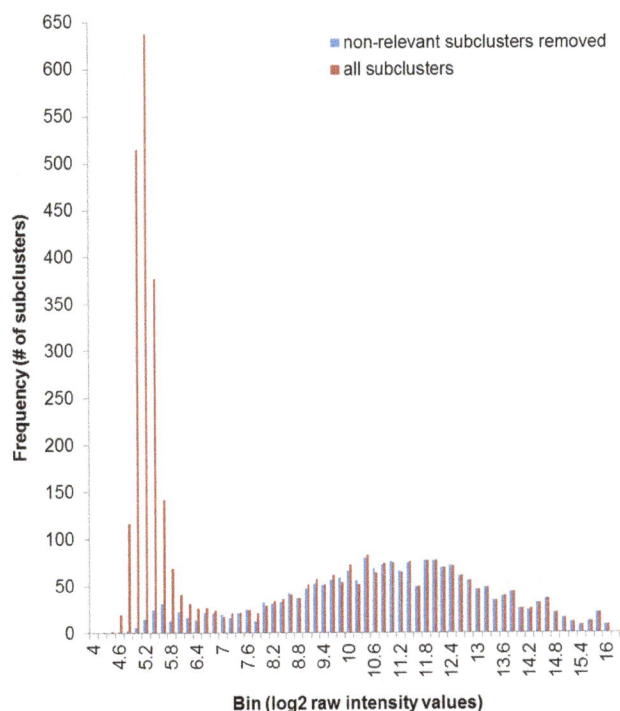

Figure 1. Data distribution with and without removal of non-relevant subclusters. Data from condition 3 (WT CZ4126/02 grown in in CDM media to OD_{A600} 1.0), replicate A, chip 1. Intensity values are binned by 0.2 in $\log_2$ scale. Non-relevant subclusters are defined as genes that are not present in the strain being interrogated and thus are uninformative (and detrimental) to the analysis.

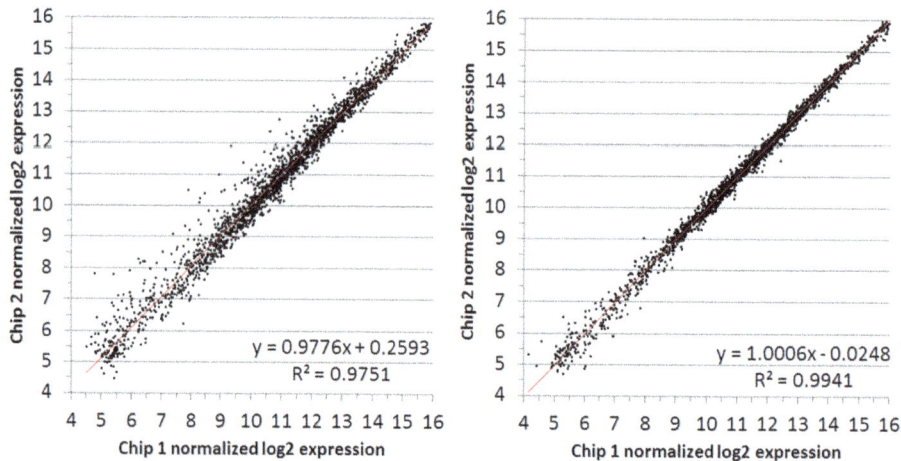

Figure 2. Comparison of chip replicate values for RNA from condition 4. Log$_2$ expression values (average of two probe-sets) from the same RNA samples run on two different chips: chip1 (x-axes) and chip2 (y-axes). Condition 4 (CZ4126/02ΔLsr::Cm^r grown in CDM media to OD$_{A600}$ 1.0), replicate A (left) and replicate B (right).

based FDR<10% and then removing genes with a Bayesian-corrected p-value>.05. In some samples the SAM FDR filter was more stringent than the Bayesian-corrected p-value, therefore, the order of filtering was flipped in these cases. After this, genes were removed that had a Benjamini-Hochberg FDR>10%. The remaining genes with Bonferroni corrected p-values<.05 were selected for the next step. Any findings with raw intensity values below 2^8 (256 = ~1.5X background in log$_2$) in both conditions were not considered sufficiently above background levels and also removed. This step-wise application of filters complemented each test's weaknesses and allowed us to observe lists of differentially regulated genes at varying levels of stringency. Fold change was not considered until after the final filter and is presented as is with no cut-off. In most cases this step-wise filtering resulted in final lists of genes up or downregulated by more than 1.5 fold which we consider to be reasonably biologically relevant.

Technical replication for microarray studies

An exhaustive analysis of technical replicates was performed to assess the reproducibility of the *H. influenzae* SGH chips for transcriptomic analyses. Two levels of technical replication were applied. The first level is contained within the array design wherein each probe is represented in duplicate on each chip. Since each subcluster is represented by between 3 and 13 probes on each chip, two sets of 3–13 values are obtained per subcluster. The NimbleScan normalization process includes a median polish whereby the 3–13 probe values are condensed to a single value, which again is calculated in duplicate. We describe this type of technical replication as probe-replication. Therefore the first validation test was to evaluate these duplicate normalized probe values as shown in **Figures 3**, and **4** for two separate RNA samples from a single condition. Each figure displays two plots which represent the same RNA sample run on two separate chips. There is no major skew in any of the data as evidenced by the best-fit lines associated with very high correlation coefficients (R^2) > 0.98. Most variation occurs at expression values at or below background levels (log$_2$ value of ~5.5). Comparison of probe-sets was performed for an additional 9 conditions (18 RNA samples; 36 chips) the results of which are displayed in figures **S1–S18**.

A second level of technical replication was performed in microarray studies by hybridizing each RNA sample to two separate chips. Therefore this level of replication evaluated the

reproducibility of the labeling, hybridization, scanning and normalization processes. Similar expression plots were generated by averaging the two duplicate-probe intensity values for chip 1 and plotting against the average for chip 2 (**Figures 1** and **5**). We observe that the inter-chip variability is just as low or lower as the probe variability with R^2 values ranging from 0.9751 to as high as 0.9951. These extremely high correlation coefficients were found consistently throughout all of the transcriptomic studies performed and are presented in full in figures **S19–S26**.

Biological Replication

To establish whether the results are reproducible across different RNA samples, two separate cultures of CZ4126/02 were grown to obtain duplicate RNA samples under the same conditions (**Table S3**). In chemically defined media (CDM), R^2-values ranged between 0.9751 and 0.9945 (**Figure 6**, Figures **S27 and S28**). When grown in the complex media, brain-heart infusion (BHI), the R^2-values ranged between .9385 and .9696 (Figures **S29** and **S30**). Furthermore, we compared the results of our microarray analysis pipeline with qRT-PCR analysis on pairs of biological replicates (**Figure 7**). As these were biological replicates we did not expect to see any differences. However, our microarray analysis pipeline indicated that two genes (cluster2554 and cluster2443aa) were significantly differentially regulated between these particular biological replicates and this was confirmed in the qRT-PCR analysis. This demonstrates that comparable results are obtained between the two methods of analysis.

Discussion

In this paper we repurpose an established CGH array for microarray-based transcriptional analysis of virtually any *H. influenzae* strain without the need for sequencing of the genome. The current format of the *H. influenzae* SGH array allows for 12 independent hybridizations (samples) per chip and here we describe a single-color fluorescence analysis pipeline. Strain-specific genome variability in terms of gene possession is captured by the SGH array [22] via-hybridization with genomic DNA. This provides all the information required for removal of non-relevant gene clusters during subsequent transcriptional analysis. While WGS provides more information than SGH it is not necessary for

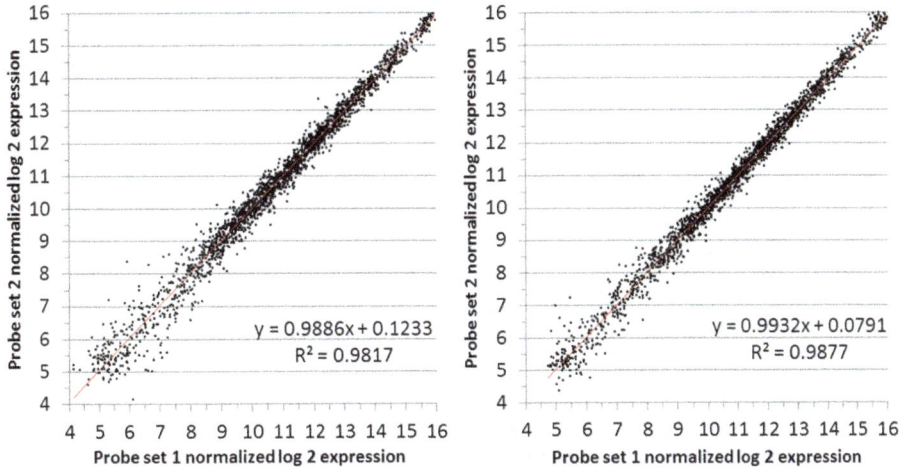

Figure 3. Comparison of probe replicate values within RNA from condition 3, replicate A. Log_2 expression values from the two probe sets in condition 3 (WT CZ4126/02 grown in in CDM media to OD_{A600} 1.0), replicate A on two separate chips (left/right).

the purposes of transcriptional experiments. SGH produces excellent results more rapidly, at a lower cost and is more easily integrated because the reagents, the array itself and the data output are exactly the same in SGH and transcriptional experiments. For transcriptional experiments we estimate the costs per sample to be approximately $110 with a workflow of four days from sample collection to data output (including analyses). Compared with RNAseq this is a cheaper alternative at this time which can run several hundred dollars more for a small bacterial genome. Organisms with larger genomes require more sequencing and thus the cost per sample increases. Significantly adding to the cost and workflow of RNAseq is the requirement for pre-processing of RNA to remove ribosomal RNA before being sequenced. However, besides cost there are other considerations to be taken into account if deciding between these technologies. Although microarrays remain the cheaper option RNAseq has been demonstrated to have higher sensitivity and a higher dynamic range than microarray analysis [47]. If the objective is to detect genes expressed at very low levels RNAseq may be a better choice. Furthermore, because RNAseq is non-probe-based

it allows measurement of all expressed transcripts without *a priori* knowledge of the genome content thereby bypassing any artifacts of gene annotation and potentially identifying novel gene transcripts. In contrast, the described SGH microarray is targeted and those targets are fixed, thus a small amount of data will inevitably be missing for a few genes in any given strain. However, we have demonstrated here that we are able to capture the vast majority of annotated genes in any *H. influenzae* strain by incorporating the supragenome into the design of the array [22]. Another major consideration is that microarray analysis methods and software are well-established and user-friendly while RNAseq data analysis has not reached the same level of development, requiring knowledge and expertise in handling large datasets as well as the command line.

Because of the multiple strain design of the SGH array some extra steps in data analysis are required that are not normally encountered in single organism microarrays. Primarily this involves removing hybridization data that is not relevant to the strain under investigation (ie probes that are designed to genes that are not present in that particular strain). A comparison of data

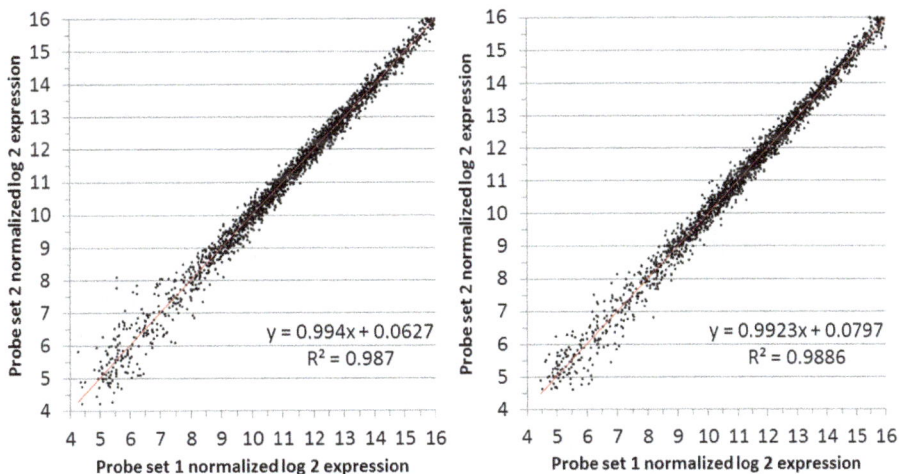

Figure 4. Comparison of probe replicate values within RNA from condition 3, replicate B. Log_2 expression values from the two probe sets in condition 3 (WT CZ4126/02 grown in in CDM media to OD_{A600} 1.0), replicate B on two separate chips (left/right).

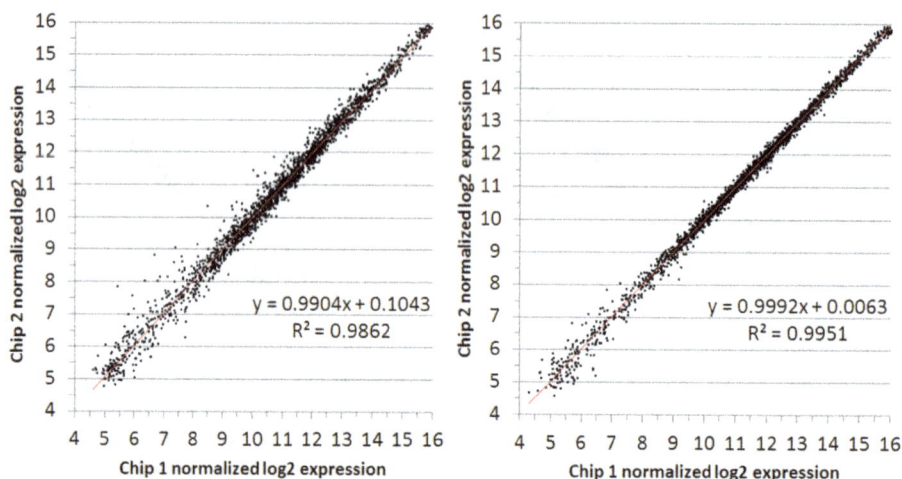

Figure 5. Comparison of chip replicate values for RNA from condition 3. Log_2 expression values (average of the two probe-sets) from the same RNA samples run on two different chips: chip1 (x-axes) and chip2 (y-axes). Condition 3 (WT CZ4126/02 grown in in CDM media to OD_{A600} 1.0), replicate A (left) and replicate B (right).

distributions before and after the removal of non-relevant gene clusters confirmed that the transcriptional datasets are converted from non-Gaussian to Gaussian distributions (**Figure 1, Table 1**).

For sample comparison we developed an analysis pipeline to be used with the normalized strain-relevant data that incorporates two freely available microarray analysis tools (SAM and CyberT) that together incorporate four diverse methods of statistical analysis. The most logical measure of difference used to analyze transcriptional data is the fold change, obtained simply by dividing the sample means. The weakness in drawing conclusions using fold changes alone lies in the blindness to expression levels illustrated in **Figure S3**. To compensate, a statistical comparison of mean values using T-tests is commonly used. However, this test is influenced greatly by the variance of the samples, something that is determined by both the biology of the gene and also by the equipment and methods used to measure the sample. The Bayesian-corrected variance is a method for reducing noise in microarrays which can overwhelm biological differences with

limited sampling. This method infers single gene variance based on the variance of other similarly expressed genes. We found in developing our transcriptional comparison pipeline that relying on calculated Bayesian-corrected p-values in conjunction with fold change cut-offs was not stringent enough as false positives were still being detected between replicate samples (**Tables S2, S3**).

To further control for the discovery of false positives (FDR) we used both permutation and non-permutation based methods using SAM and CyberT. One weakness of SAM is that it is inappropriate for small numbers of replicates. This is due to the fact that the modified p-values (q-values) are calculated based on the number of times the software can randomly rearrange the replicate values (permutations). Thus SAM's power, accuracy and significance increase with increased replication.

As a final filtering step we used Bonferroni-corrected p-values calculated by CyberT. In many microarray analyses applying the Bonferroni correction is not performed because it is too stringent and often removes data that is known to differ between two samples (false negatives). In these *H. influenzae* microarray

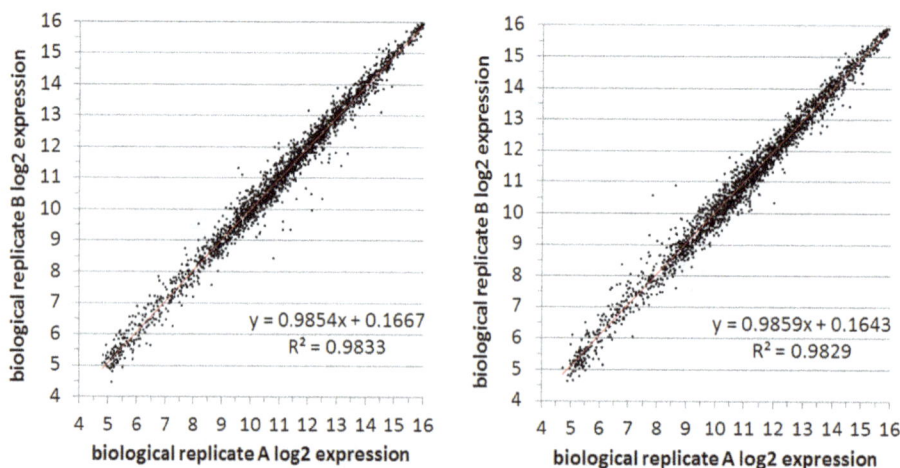

Figure 6. Representative biological replicate comparisons. Log_2 expression values (averages of both probe-sets from both chips) from the two biological replicate samples for both condition 3 (left, WT CZ4126/02 grown in in CDM media to OD_{A600} 1.0) and condition 4 (right, CZ4126/02ΔLsr::Cm^r grown in CDM media to OD_{A600} 1.0).

	Pair 1 qRT-PCR	Pair 1 microarray	Pair 2 qRT-PCR	Pair 2 microarray
cluster1730	-1.15	1.04	-1.11	-1.02
cluster2051e	1.70	1.11	1.83	1.19
cluster748	1.30	1.02	1.77	1.04
cluster2554	-27.94	-5.15*	-2.70	-1.28
cluster2164	-1.57	1.19	-1.17	-1.14
cluster2001	1.18	-1.47	2.82	-1.05
cluster2739	1.38	-1.22	1.45	-1.23
cluster1002	-1.80	-1.10	1.70	-1.32
cluster2443qq	-1.16	-1.10	1.56	1.31
cluster2443gg	1.25	1.12	1.92	1.61
cluster2443s	-1.62	-1.06	1.18	1.36
cluster2443aa	-2.46	1.00	4.83	2.04*

Figure 7. Comparison of SGH microarray analysis vs qRT-PCR in two biological replicate pairs. All values represent fold changes between biological replicates. Pair 1 = Condition 3 (WT CZ4126/02 grown in in CDM media to OD_{A600} 1.0) replicate A vs. replicate B. Pair 2 = Condition 4 (CZ4126/02ΔLsr::Cm^r grown in CDM media to OD_{A600} 1.0) replicate A vs. replicate B. qRT-PCR data are the average from normalization against two different house-keeping genes and two different methods of analysis (Pfaffl-$\Delta\Delta C_T$ and linear regression) *a statistically significant difference from SGH microarray analysis (passed all five filtering steps). Colors highlight fold changes above 2 (red) and below −2 (blue).

studies, however, we found that it was appropriate and useful for removing many genes that had fold changes under 1.5 but were still found significant by the other tests as well as for removing genes for which there was high variance between biological samples. These four filtering steps (T-tests, permutation-based FDR, BH FDR and Bonferroni-corrected T-tests) used together in conjunction with a raw hybridization value cut-off provide a robust, flexible method for two-condition microarray analyses.

Replicate samples were examined in detail and conclusively demonstrated that duplicate probe sets provide consistent repeatable quantitative data (**Figure 3–4, S1–S18**). We also found that technical replication by hybridization of the same cDNA samples on two different chips was just as robust as probe replication (**Figures 1, 5, S19–S26**). The correlation between biological replicate experiments is also extremely high, especially when bacteria were grown in the defined media, CDM, where the growth conditions are more tightly controlled (**Figures 6, S27–S28**). Not surprisingly, the variation observed is somewhat higher when the bacteria were grown in the complex media, BHI, likely due to variation in the growth conditions and not the array (**Figures S29–S30**).

These results establish that the methods developed to obtain transcriptional data from the *H. influenzae* SGH arrays are consistent and robust within and across chips as well as among biological replicates. Used in conjunction with the filtering pipeline described above we have described the establishment of a unique, robust microarray analysis pipeline for *multi-strain* comparisons specific to *Haemophilus influenzae*. We believe this tool could be of great use to others studying this organism and to those studying other organisms that make use of similar supragenome based arrays who may not have the resources to perform RNAseq or the expertise to deal with the associated downstream bioinformatic data analysis required.

Supporting Information

Figure S1   Comparison of probe replicate values within RNA from condition 1, replicate A. Log$_2$ expression values from the two probe sets in condition 1, replicate A on two separate chips (left/right).

Figure S2   Comparison of probe replicate values within RNA from condition 1, replicate B. Log$_2$ expression values from the two probe sets in condition 1, replicate B on two separate chips (left/right).

Figure S3   Comparison of probe replicate values within RNA from condition 2, replicate A. Log$_2$ expression values from the two probe sets in condition 2, replicate A on two separate chips (left/right).

Figure S4   Comparison of probe replicate values within RNA from condition 2, replicate B. Log$_2$ expression values from the two probe sets in condition 2, replicate B on two separate chips (left/right).

Figure S5   Comparison of probe replicate values within RNA from condition 4, replicate A. Log$_2$ expression values from the two probe sets in condition 4, replicate A on two separate chips (left/right).

Figure S6   Comparison of probe replicate values within RNA from condition 4, replicate B. Log$_2$ expression values from the two probe sets in condition 4, replicate B on two separate chips (left/right).

Figure S7   Comparison of probe replicate values within RNA from condition 5, replicate A. Log$_2$ expression values from the two probe sets in condition 5, replicate A on two separate chips (left/right).

Figure S8   Comparison of probe replicate values within RNA from condition 5, replicate B. Log$_2$ expression values from the two probe sets in condition 5, replicate B on two separate chips (left/right).

Figure S9   Comparison of probe replicate values within RNA from condition 6, replicate A. Log$_2$ expression values from the two probe sets in condition 6, replicate A on two separate chips (left/right).

Figure S10   Comparison of probe replicate values within RNA from condition 6, replicate B. Log$_2$ expression values from the two probe sets in condition 6, replicate B on two separate chips (left/right).

Figure S11   Comparison of probe replicate values within RNA from condition 7, replicate A. Log$_2$ expression values from the two probe sets in condition 7, replicate A on two separate chips (left/right).

Figure S12   Comparison of probe replicate values within RNA from condition 7, replicate B. Log$_2$ expression values from the two probe sets in condition 7, replicate B on two separate chips (left/right).

Figure S13   Comparison of probe replicate values within RNA from condition 8, replicate A. Log$_2$ expression values from the two probe sets in condition 8, replicate A on two separate chips (left/right).

Figure S14   Comparison of probe replicate values within RNA from condition 8, replicate B. Log$_2$ expression values from the two probe sets in condition 8, replicate B on two separate chips (left/right).

Figure S15   Comparison of probe replicate values within RNA from condition 9, replicate A. Log$_2$ expression values from the two probe sets in condition 9, replicate A on two separate chips (left/right).

Figure S16   Comparison of probe replicate values within RNA from condition 9, replicate B. Log$_2$ expression values from the two probe sets in condition 9, replicate B on two separate chips (left/right).

Figure S17   Comparison of probe replicate values within RNA from condition 10, replicate A. Log$_2$ expression values from the two probe sets in condition 10, replicate A on two separate chips (left/right).

Figure S18   Comparison of probe replicate values within RNA from condition 10, replicate B. Log$_2$ expression values from the two probe sets in condition 10, replicate B on two separate chips (left/right).

Figure S19   Comparison of chip replicate values for RNA from condition 1. Log$_2$ expression values (average of two probe-sets) from the same RNA samples run on two different chips: chip1 (x-axes) and chip2 (y-axes). Condition 1, replicate A (left) and replicate B (right).

Figure S20   Comparison of chip replicate values for RNA from condition 2. Log$_2$ expression values (average of two probe-sets) from the same RNA samples run on two different chips: chip1 (x-axes) and chip2 (y-axes). Condition 2, replicate A (left) and replicate B (right).

Figure S21   Comparison of chip replicate values for RNA from condition 5. Log$_2$ expression values (average of two probe-sets) from the same RNA samples run on two different chips: chip1 (x-axes) and chip2 (y-axes). Condition 5, replicate A (left) and replicate B (right).

Figure S22   Comparison of chip replicate values for RNA from condition 6. Log$_2$ expression values (average of two probe-sets) from the same RNA samples run on two different chips: chip1 (x-axes) and chip2 (y-axes). Condition 6, replicate A (left) and replicate B (right).

Figure S23   Comparison of chip replicate values for RNA from condition 7. Log$_2$ expression values (average of two probe-sets) from the same RNA samples run on two different chips: chip1 (x-axes) and chip2 (y-axes). Condition 7, replicate A (left) and replicate B (right).

Figure S24   Comparison of chip replicate values for RNA from condition 8. Log$_2$ expression values (average of two probe-sets) from the same RNA samples run on two different chips: chip1 (x-axes) and chip2 (y-axes). Condition 8, replicate A (left) and replicate B (right).

Figure S25   Comparison of chip replicate values for RNA from condition 9. Log$_2$ expression values (average of two probe-sets) from the same RNA samples run on two different chips: chip1 (x-axes) and chip2 (y-axes). Condition 9, replicate A (left) and replicate B (right).

Figure S26   Comparison of chip replicate values for RNA from condition 10. Log$_2$ expression values (average of two probe-sets) from the same RNA samples run on two different chips: chip1 (x-axes) and chip2 (y-axes). Condition 10, replicate A (left) and replicate B (right).

Figure S27   Biological replicate comparison from Condition 1 and 2. Log$_2$ expression values (averages of both probe-sets from both chips) from the two biological replicate samples for both condition 1 (left) and condition 2 (right).

Figure S28   Biological replicate comparison from Condition 5 and 6. Log$_2$ expression values (averages of both probe-sets from both chips) from the two biological replicate samples for both condition 5 (left) and condition 6 (right).

Figure S29 Biological replicate comparison from Condition 7 and 8. Log_2 expression values (averages of both probesets from both chips) from the two biological replicate samples for both condition 7 (left) and condition 8 (right).

Figure S30 Biological replicate comparison from Condition 9 and 10. Log_2 expression values (averages of both probesets from both chips) from the two biological replicate samples for both condition 9 (left) and condition 10 (right).

Table S1 Source of RNA for each "condition".

Table S2 False positives found using a p-value<0.05 cutoff. Condition 4, replicate B, chip 1 and chip 2 were compared. Raw expression values are shown. FDR: False discovery rate, BH: Benjamini-Hochberg, Bon. pVal: Bonferroni-corrected p-value.

Table S3 False positives using a fold change cutoff of >1.5. Condition 4, replicate B, chip 1 and chip 2 were compared. Raw expression values are shown. FDR: False discovery rate, BH: Benjamini-Hochberg, Bon. pVal: Bonferroni-corrected p-value.

Acknowledgments

We thank Dr. Helena Zemlickova for providing strain CZ4126/02. We thank Mary O'Toole and Carol Hope for help in the preparation of the manuscript.

Author Contributions

Conceived and designed the experiments: BAJ NLH FZH GDE. Performed the experiments: BAJ RAE EP AA. Analyzed the data: BAJ NLH MED JPE FZH GDE. Wrote the paper: BAJ NLH FZH GDE.

References

1. Tettelin H, Masignani V, Cieslewicz MJ, Donati C, Medini D, et al. (2005) Genome analysis of multiple pathogenic isolates of Streptococcus agalactiae: implications for the microbial "pan-genome". Proc Natl Acad Sci USA 102: 13950–13955.
2. Hogg J, Hu F, Janto B, Boissy R, Hayes J, et al. (2007) Characterization and modeling of the Haemophilus influenzae core and supragenomes based on the complete genomic sequences of Rd and 12 clinical nontypeable strains. Genome Biol 8: R103.
3. Hiller N, Janto B, Hogg J, Boissy R, Yu S, et al. (2007) Comparative genomic analyses of seventeen Streptococcus pneumoniae strains: insights into the pneumococcal supragenome. J Bacteriol 189: 8186–8195.
4. Donati C, Hiller NL, Tettelin H, Muzzi A, Croucher NJ, et al. (2010) Structure and dynamics of the pan-genome of Streptococcus pneumoniae and closely related species. Genome Biol 11: R107.
5. Boissy R, Ahmed A, Janto B, Earl J, Hall BG, et al. (2011) Comparative supragenomic analyses among the pathogens Staphylococcus aureus, Streptococcus pneumoniae, and Haemophilus influenzae using a modification of the finite supragenome model. BMC Genomics 12: 187.
6. Davie JJ, Earl J, de Vries SP, Ahmed A, Hu FZ, et al. (2011) Comparative analysis and supragenome modeling of twelve Moraxella catarrhalis clinical isolates. BMC Genomics 12: 70.
7. Ahmed A, Earl J, Retchless A, Hillier SL, Rabe LK, et al. (2012) Comparative genomic analyses of 17 clinical isolates of Gardnerella vaginalis provide evidence of multiple genetically isolated clades consistent with subspeciation into genovars. J Bacteriol 194; 3922–3937.
8. Borneman AR, McCarthy JM, Chambers PJ, Bartowsky EJ (2012) Comparative analysis of the Oenococcus oeni pan genome reveals genetic diversity in industrially-relevant pathways. BMC Genomics 13: 373.
9. He M, Sebaihia M, Lawley TD, Stabler RA, Dawson LF, et al. (2010) Evolutionary dynamics of Clostridium difficile over short and long time scales. Proc Natl Acad Sci USA 107: 7527–7532.
10. Conlan S, Mijares LA, NISC Comparative Sequencing Program, Becker J, Blakesley RR, et al. (2012) Staphylococcus epidermidis pan-genome sequence analysis reveals diversity of skin commensal and hospital infection-associated isolates. Genome Biol 13: R64.
11. Ehrlich GD, Hu FZ, Shen K, Stoodley P, Post JC (2005) Bacterial plurality as a general mechanism driving persistence in chronic infections. Clin Orthop Relat Res 437: 20–24.
12. Ehrlich GD, Ahmed A, Earl J, Hiller NL, Costerton JW, et al. (2010) The distributed genome hypothesis as a rubric for understanding evolution in situ during chronic bacterial biofilm infectious processes. FEMS Immunol. Med. Microbiol 59: 269–279.
13. Shen K, Antalis P, Gladitz J, Sayeed S, Ahmed A, et al. (2005) Identification, distribution, and expression of novel genes in 10 clinical isolates of nontypeable Haemophilus influenzae. Infect Immun 73: 3479–3491.
14. Hogg JS, Hu FZ, Janto B, Boissy R, Hayes J, et al. (2007) Characterization and modeling of the Haemophilus influenzae core and supragenomes based on the complete genomic sequences of Rd and 12 clinical nontypeable strains. Genome Biol 8: R103.
15. Hu FZ, Ehrlich GD (2008) Population-level virulence factors amongst pathogenic bacteria: relation to infection outcome. Future Microbiol 3: 31–42.
16. Shen K, Antalis P, Gladitz J, Sayeed S, Ahmed A, et al. (2005) Identification, distribution, and expression of novel genes in 10 clinical isolates of nontypeable Haemophilus influenzae. Infect Immun 73: 3479–3491.
17. Ehrlich GD, Hiller NL, Hu FZ (2008) What makes pathogens pathogenic. Genome Biol 9: 225.
18. Buchinsky FJ, Forbes ML, Hayes JD, Shen K, Ezzo S, et al. (2007) Virulence phenotypes of low-passage clinical isolates of nontypeable Haemophilus influenzae assessed using the chinchilla laniger model of otitis media. BMC Microbiol 7; 56.
19. Forbes ML, Horsey E, Hiller NL, Buchinsky FJ, Hayes JD, et al. (2008) Strain-specific virulence phenotypes of Streptococcus pneumoniae assessed using the Chinchilla laniger model of otitis media. PloS One 3: e1969.
20. Hall-Stoodley L, Nistico L, Sambanthamoorthy K, Dice B, Nguyen D, et al. (2008) Characterization of biofilm matrix, degradation by DNase treatment and evidence of capsule downregulation in Streptococcus pneumoniae clinical isolates. BMC Microbiol 8: 173.
21. Dohar JE, Hebda PA, Veeh R, Awad M, Costerton JW, et al. (2005) Mucosal biofilm formation on middle-ear mucosa in a nonhuman primate model of chronic suppurative otitis media. The Laryngoscope 115: 1469–1472.
22. Eutsey RA, Hiller NL, Earl JP, Janto BA, Dahlgren ME, et al. (2013) Design and validation of a supragenome array for determination of the genomic content of Haemophilus influenzae isolates. BMC Gemonics 14: 484. doi:10.1186/1471-2164-14-484.
23. Erwin AL, Nelson KL, Mhlanga-Mutangadura T, Bonthuis PJ, Geelhood JL, et al. (2005) Characterization of genetic and phenotypic diversity of invasive nontypeable Haemophilus influenzae. Infect Immun 73: 5853–5863.
24. Gilsdorf JR, Marrs CF, Foxman B (2004) Haemophilus influenzae: genetic variability and natural selection to identify virulence factors. Infect Immun 72: 2457–2461.
25. Murphy TF, Sethi S, Klingman KL, Brueggemann AB, Doern GV (1999) Simultaneous respiratory tract colonization by multiple strains of nontypeable haemophilus influenzae in chronic obstructive pulmonary disease: implications for antibiotic therapy. J Infect Dis 180: 404–409.
26. Sethi S, Murphy TF (2008) Infection in the pathogenesis and course of chronic obstructive pulmonary disease. N Engl J Med 359: 2355–2365.
27. Murphy TF, Faden H, Bakaletz LO, Kyd JM, Forsgren A, et al. (2009) Nontypeable Haemophilus influenzae as a pathogen in children. Pediatr Infect Dis J 28: 43–48.
28. Pereira MB, Pereira MR, Cantarelli V, Costa SS (2004) Prevalence of bacteria in children with otitis media with effusion. J Pediatr (Rio J) 80: 41–48.
29. Bluestone CD, Stephenson JS, Martin LM (1992) Ten-year review of otitis media pathogens. Pediatr Infect Dis J 11: S7–11.
30. Post JC, Preston RA, Aul JJ, Larkins-Pettigrew M, Rydquist-White J, et al. (1995) Molecular analysis of bacterial pathogens in otitis media with effusion. JAMA 273: 1598–1604.
31. Dohar JE, Roland P, Wall GM, McLean C, Stroman DW (2009) Differences in bacteriologic treatment failures in acute otitis externa between ciprofloxacin/dexamethasone and neomycin/polymyxin B/hydrocortisone: results of a combined analysis. Curr Med Res Opin 25: 287–291.
32. Ehrlich GD, Veeh R, Wang X, Costerton JW, Hayes JD, et al. (2002) Mucosal biofilm formation on middle-ear mucosa in the chinchilla model of otitis media. JAMA 287: 1710–1715.
33. Hall-Stoodley L, Hu FZ, Gieseke A, Nistico L, Nguyen D, et al. (2006) Direct detection of bacterial biofilms on the middle-ear mucosa of children with chronic otitis media. JAMA 296: 202–211.
34. Ladhani S, Slack MP, Heath PT, von Gottberg A, Chandra M, et al. (2010) Invasive Haemophilus influenzae Disease, Europe, 1996–2006. Emerg Infect Dis 16: 455–463.
35. Agrawal A, Murphy TF (2011) Haemophilus influenzae infections in the H. influenzae type b conjugate vaccine era. J Clin Microbiol 49: 3728–3732.

36. Tsang RS, Sill ML, Skinner SJ, Law DK, Zhou J, et al. (2007) Characterization of invasive Haemophilus influenzae disease in Manitoba, Canada, 2000–2006: invasive disease due to non-type b strains. Clin Infect Dis 44: 1611–1614.

37. Kelly L, Tsang RS, Morgan A, Jamieson FB, Ulanova M (2011) Invasive disease caused by Haemophilus influenzae type a in Northern Ontario First Nations communities. J Med Microbiol 60: 384–390.

38. Shuel M, Law D, Skinner S, Wylie J, Karlowsky J, et al. (2010) Characterization of nontypeable Haemophilus influenzae collected from respiratory infections and invasive disease cases in Manitoba, Canada. FEMS Immunol Med Microbiol 58: 277–284.

39. Irizarry RA, Hobbs B, Collin F, Beazer-Barclay YD, Antonellis KJ, et al. (2003) Exploration, normalization, and summaries of high density oligonucleotide array probe level data. Biostatistics 4: 249–264.

40. Bolstad BM, Irizarry RA, Astrand M, Speed TP (2003) A comparison of normalization methods for high density oligonucleotide array data based on variance and bias. Bioinformatics 19: 185–193.

41. Edgar R1, Domrachev M, Lash AE (2002) Gene Expression Omnibus: NCBI gene expression and hybridization array data repository. Nucleic Acids Res. 30: 207–10.

42. National Center for Biotechnology Information website. Available: http://www.ncbi.nlm.nih.gov/geo/query/acc.cgi?acc=GPL16197. Accessed 2014 Sep 3.

43. Coleman HN, Daines DA, Jarisch J, Smith AL (2003) Chemically defined media for growth of Haemophilus influenzae strains. J Clin Microbiol 41: 4408–4410.

44. Ramakers C, Ruijter JM, Deprez RH, Moorman AF (2003) Assumption-free analysis of quantitative real-time polymerase chain reaction (PCR) data. Neurosci Lett, 339(1), 62–6.

45. Baldi P, Long AD (2001) A Bayesian framework for the analysis of microarray expression data: regularized t -test and statistical inferences of gene changes. Bioinformatics 17: 509–519.

46. Tusher VG, Tibshirani R, Chu G (2001) Significance analysis of microarrays applied to the ionizing radiation response. Proc Natl Acad Sci USA 98: 5116–5121.

47. Zhao S, Fung-Leung WP, Bittner A, Ngo K, Liu X (2014) Comparison of RNA-Seq and microarray in transcriptome profiling of activated T cells. PLoS One 16; 9(1).

## 17

Age and Microenvironment Outweigh Genetic Influence on the Zucker Rat Microbiome

Hannah Lees[1], Jonathan Swann[2], Simon M. Poucher[3¤], Jeremy K. Nicholson[1], Elaine Holmes[1], Ian D. Wilson[1], Julian R. Marchesi[4,5]*

1 Section of Computational and Systems Medicine, Department of Surgery and Cancer, Faculty of Medicine, Imperial College London, London, United Kingdom, 2 Department of Food and Nutritional Sciences, School of Chemistry, Food and Pharmacy, University of Reading, Reading, United Kingdom, 3 Cardiovascular and Gastro-Intestinal Disorders Innovative Medicines, AstraZeneca Pharmaceuticals, Alderley Park, Cheshire, United Kingdom, 4 School of Biosciences, Cardiff University, Cardiff, United Kingdom, 5 Centre for Digestive and Gut Health, Imperial College London, London, United Kingdom

Abstract

Animal models are invaluable tools which allow us to investigate the microbiome-host dialogue. However, experimental design introduces biases in the data that we collect, also potentially leading to biased conclusions. With obesity at pandemic levels animal models of this disease have been developed; we investigated the role of experimental design on one such rodent model. We used 454 pyrosequencing to profile the faecal bacteria of obese (n = 6) and lean (homozygous n = 6; heterozygous n = 6) Zucker rats over a 10 week period, maintained in mixed-genotype cages, to further understand the relationships between the composition of the intestinal bacteria and age, obesity progression, genetic background and cage environment. Phylogenetic and taxon-based univariate and multivariate analyses (non-metric multidimensional scaling, principal component analysis) showed that age was the most significant source of variation in the composition of the faecal microbiota. Second to this, cage environment was found to clearly impact the composition of the faecal microbiota, with samples from animals from within the same cage showing high community structure concordance, but large differences seen between cages. Importantly, the genetically induced obese phenotype was not found to impact the faecal bacterial profiles. These findings demonstrate that the age and local environmental cage variables were driving the composition of the faecal bacteria and were more deterministically important than the host genotype. These findings have major implications for understanding the significance of functional metagenomic data in experimental studies and beg the question; what is being measured in animal experiments in which different strains are housed separately, nature or nurture?

Editor: Yolanda Sanz, Instutite of Agrochemistry and Food Technology, Spain

Funding: The funder provided support in the form of salaries for authors [IDW, SMP], but did not have any additional role in the study design, data collection and analysis, decision to publish, or preparation of the manuscript. The specific roles of these authors are articulated in the 'author contributions' section.

Competing Interests: The authors have read the journal's policy and have the following conflicts: SMP and IDW were employees of AstraZeneca, the funder of this study, at the time of this study.

* Email: MarchesiJR@cardiff.ac.uk

¤ Current address: CrystecPharma, Crystec Ltd, Bradford, United Kingdom

Introduction

Emerging evidence of an obesity-associated altered microbiome with the potential to influence caloric extraction from the diet and host energy metabolism [1–3] has fuelled a surge in both scientific and public interest in the role of the microbiome in the etiopathogenesis of obesity, with particular interest in the functional properties of the gut microbiota, microbe-host signaling and the possibility of using the microbiome as a therapeutic target. However, evidence also suggests that the relationship between the microbiota and obesity is complex, with contradictory findings relating to the nature of the shift in the relative contributions of phyla to the microbiota composition in obesity, and the question of whether the observed shift in the microbiome is more associated with a high-fat diet than genetically induced obesity *per se* [4–9]. Given the complexity of the host-microbiome relationship, it is vital that experimental studies on microbiota composition are well-founded at the most basic level as well as at the high end levels of analytical phenotyping, genotyping and functional ecological analysis.

Several rodent models have been developed to investigate the role of the host's genotype on the development of obesity. One such model is the homozygous Zucker (fa/fa) obese rat, which is characterised by an autosomal recessive mutation of the *fa*-gene, encoding for the leptin receptor. This results in reduced sensitivity to leptin, leading to hyperphagia, obesity and hyperinsulinaemia. In contrast, the heterozygous (fa/+) and homozygous (+/+) Zucker genotypes remain lean as they age and do not develop insulin resistance. Previous analyses of the intestinal microbiota of the Zucker rat found differences between obese and lean strains when the animals were housed according to strain [10]. Therefore, we have designed an experiment to explore the effect of age, genotype, obese/lean phenotype, and cage environment on the evolution and development of the faecal microbiota of the male Zucker rat. We aimed to test the hypothesis that the obese phenotype will result in the evolution of a faecal microbiome and host metabotype distinct from the lean Zucker rats, independent of cage or age. We evaluated this by including each of the three different genotypes in each cage.

Methods

Ethics statement

All animal work was carried out in accordance with the U.K. Home Office Animals (Scientific Procedures) Act 1986 under a Project Licence which was approved by the AstraZeneca Ethical Review Committee. The specific protocols described in this paper were also reviewed and approved by the local Departmental Review to ensure that they adhered to the principals of minimising animal suffering. The hypothesis/ethical review study code for the animal study conducted at AstraZeneca was HETP24. The protocol review document was ETP40.

Animals and sample collection

Three strains of male rat were used in this study, Zucker (fa/fa) obese, heterozygous Zucker lean (fa/+), and Zucker lean (+/+) (n = 6 per strain). The animals were bred on site, (Alderley-Park, AstraZeneca) and housed in a conventional animal room in Techniplast P2000 cages at standard room temperature and humidity on a 12 h:12 h light:dark cycle. The pups were reared with their mothers until separated at weaning; they were housed as littermates in six cages, each containing one rat from each genotype (n = 3 per cage). The rats in all six cages had different mothers and fathers, and the three rats inside each single cage were littermates. Food (SDS breeding diet RM-3) and water were available *ad libitum* throughout the study. At weekly intervals, from 5 to 14 weeks of age, the animals were transferred to a procedures room, weighed, and placed individually in metabolism cages, for no more than 2 hours, for urine and faeces collection. Samples were collected at the same time of day to remove diurnal effects on profiles. The rats had access to food and water whilst in the metabolism cages. At 14 weeks of age, following urine and faeces collection, animals were rendered insentient by inhalation of a 5:1 mixture of $CO_2:O_2$, and a blood sample taken by cardiac puncture into lithium heparin blood syringes. Urine was also collected for metabolite analysis (data not shown, Lees *et al.*, in preparation) together with a terminal blood sample. Euthanasia was confirmed by cervical dislocation. Faeces were stored at $-40°C$ prior to 16S rRNA gene profiling analysis.

Sample preparation

For 16S rRNA gene profiling, four faeces collection time points were selected from the ten time points of the study, when the animals were: five, seven, ten and fourteen weeks of age. The faecal DNA was extracted from at least two different pellets, with a total weight of approximately 200 mg. The Qiagen QIAamp DNA stool kit was used for DNA extraction, as per the manufacturer's instructions, with an additional bead-beating step for homogenisation of sample and lysis of bacterial cells (0.1 g 0.1 mm sterile glass beads, FastPrep bead-beater (Q-BIOgene), setting six (6 metres per second) for 20 seconds, repeated a further two times with 5 minutes on ice between cycles). Following DNA extraction, DNA concentration and purity was determined using a NanoDrop Spectrophotometer (Thermo Scientific, Wilmington, DE, USA), and diluted to a working concentration of 10 ng/μl. The polymerase chain reaction (PCR) was used to amplify the V1-V3 regions of the 16S rRNA gene from each DNA sample using the primers shown in Table S1 and was performed in triplicate on all samples using a C1000 Thermal Cycler (Bio-Rad, USA). PCR mixtures (50 μl) contained *Taq* polymerase (0.25 μl, 5 U/μl solution), buffer (10 μl), $MgCl_2$ (3 μl, 1.5 mM), deoxynucleoside triphosphates (dNTPs, 0.4 μl, 0.2 mM of each dNTP), 1 μl of each barcoded primer, 1 μl of each sample DNA (10 ng), and 34.35 μl H_2O. The PCR cycle conditions were: 95°C for 5 min initial

denaturation, 25 cycles of amplification at 95°C denaturation for 30 s, annealing at 55°C for 40 s, and extension of 72°C for 1 min, with a final extension of 72°C for 5 min. PCR products (created in triplicate) were pooled for each sample, and purified using a Qiagen QIAquick PCR purification kit, quantified, again using a NanoDrop Spectrophotometer. The samples were normalised to 5 ng/μl, and 4 μl was transferred to a new micro-centrifuge tube for pooling of samples. The samples were run on three PTPs (Pico Titre Plates), and so were pooled in to three 1.5 ml micro-centrifuge tubes. Samples were sent to the University of Liverpool to be sequenced on a Roche 454 GS FLX sequencer. All sequences are deposited in the European Nucleotide Archive under accession number PRJEB5969.

Data processing

Samples were processed using the Ribosomal Database Project (RDP) pyropipeline [11] to remove any reads that were less than 250 base pairs, <Q20 and contained any ambiguities (Ns). The filtered sequences were classified using the RDP classifier [12] and the relative proportions of phyla and families calculated. To account for variation in sequence reads per sample, the samples were normalised to the lowest sequence count per animal [13] (Table S2). The resultant relative abundance values were used for multivariate (PCA) and univariate (one-way ANOVA) statistical analysis. UniFrac distances (both unweighted and weighted [14]) were calculated using Mothur v 1.28.1 [13].

Statistical analysis

UniFrac unweighted distances were analysed by non-metric multidimensional scaling (NMDS) in R [15]. The UniFrac unweighted distances were analysed at each time point using an unpaired Student's t test after normality of data had been ensured. Univariate statistical analysis of relative abundance values was performed using GraphPad Prism version 6 software (GraphPad Software, San Diego, CA). To meet the assumptions of the one-way analysis of variance (ANOVA), the data were assessed for normality prior to analysis using the D'Agostino-Pearson test, and the Bartlett's test for equality of variance. The differences between samples from differing time points were assessed using one-way ANOVA and Tukey-Kramer multiple comparisons test. Analysis of the samples at the individual operational taxonomic unit (OTU) level was undertaken in STAMP [16] using genotype, cage and week as the three main discriminators. The means for each OTU were tested using an ANOVA and corrected for multiple testing using the Bonferroni correction. In addition, the data were divided into four time points and tested independently of each other to remove the time factor from the analysis and to allow for the effect of cage and phenotype to be measured at the OTU level.

Multivariate analysis of relative abundance values

To aid interpretation of the data and quickly visualise trends associated with age, genotype and cage environment, principal component analysis (PCA) was applied to the relative abundance data [17]. The relative abundance values were filtered so that only bacteria detected in at least 75% of animals per group were included in models. PCA was performed on mean-centred, Pareto-scaled [18] data for phylum-level data, using SIMCA 12.0 (Umetrics 2009). For PCA modelling of family-level profiles, data were again mean-centred and a log_{10} transformation was required due to the distribution of the data [19].

Results

Metataxonomic characterisation of the faecal microbiota

Data generated from the 16S rRNA gene profiling of faeces from rats aged five, seven, ten and fourteen weeks of age were examined with respect to age- and phenotype-related variation, and also the effects of housing (cage effect) were considered.

Age-related development of the gut microbiota

Based on UniFrac distances (Figure 1) and the 16S rRNA gene profiling of the faecal samples, the intestinal microbiota showed clear age-related trends at the phylum, family and OTU level. At the phylum level there was a decrease in the *Firmicutes:Bacteroidetes* ratio (from an average ratio of 5.38 at week five, to 1.05 at week fourteen), with both phyla varying with increasing age (Figure 2A). At the family level, aging in the Zucker rat was associated with a reduction in *Bacteroidaceae* and *Peptostreptococcaceae*, and an increase in *Ruminococcaceae* and *Bifidobacteriaceae* (Figure 2B). Statistical analysis using one-way ANOVA was not appropriate due to the heteroscedasticity of the relative abundance data at both the phylum and family level (when comparing values from differing time points, the variance of the groups differed significantly), as judged by Bartlett's test for equal variances. Transformation of the data failed to resolve this issue. When each dataset was tested across the four time points, 24 OTUs were found to vary significantly due to age (Table S3 and Figure S2). The differences ranged from 15-25% enrichment for OTU001 (*Clostridium* XI (family *Peptostreptococcaceae*)) in week 5 compared to weeks 7, 10 and 14. While OTUs 035 and 051 changed between 0.4 and 0.5% and were enriched in week 14 compared to the other weeks for both OTUs. Seventeen OTUs varied when each time point was analysed independently of each other time point (Table S4 and Figure S3). For week five, 3 OTUs varied between the cages; at week seven, 5 OTUs; at week ten, 3 OTUs; and at week fourteen, 8 OTUs varied. There were no consistent changes in the OTUs between cages. For example, cage 3 at week 5 showed enrichment of OTU017 (genus *Bacteroides* enriched between 10-15% over all other cages) and OTU032 (genus *Subdoligranulum* enriched between 5–6% over all other cages) and for cage 1 at week 5 OTU001 (genus Clostridium XI enriched between 34–52% over all cages) was enriched. Only OTU002 and OTU019 showed any changes from week to week and only OTU019, changed from one to another i.e. week 10 to week 14; however, only some of the cages showed the same change between the two time points. In addition, the age of the animals was the largest source of systematic variation in the PCA models of the phylum and family level data (Figures S4A and S5A).

Impact of the cage environment

The intestinal bacteria profiles of animals from within the same cage exhibited similarities at the phylum and family level, in spite of the differing obese and lean phenotypes present within each cage. In the taxon-based analysis, cage environment-associated trends in the phylum and family-level datasets were not obvious when all time points were considered together (Figures S4C and S5C), as age at sample collection was the dominant source of systematic variation, and obscured any cage-associated trends. However, there was evidence of cage-environment associated trends, at both the phylum and family-level, when each timepoint was considered independently (Figure 3, Figure S6 and S7). Cage-associated clustering of samples was also evident in the NMDS plot based on the unweighted UniFrac distances between faecal samples (Figure 1). The mean unweighted UniFrac distances of animals from within the same cage were significantly lower (P<

0.0001) than animals from differing cages at each time point (Figure 4), and significant differences between cohoused and non-cohoused animals were also observed in the weighted UniFrac distances at week 5 (P<0.001), week 7 (P<0.0001) and week 14 (P<0.01) (Figure S8). The effect of animal housing was most prominent at the beginning of the study in samples from animals at five and seven weeks of age, but differences persisted until the end of the study (Figures S9 and S10). Significant differences were found in the relative abundances of *Bacteroidetes* and *Firmicutes* at the phylum level, and *Bacteroidaceae*, *Lachnospiraceae*, *Peptostreptococcaceae*, *Porphyromonadaceae*, *Prevotellaceae* and *Ruminococcaceae*, at the family level, between the cages at weeks 5, 7 and 14 (P<0.05) (Table S5 and Table S6), with cages three and four showing significantly higher *Bacteroidetes* at week 5; cages one and two showing significantly higher *Firmicutes* at week 7; and cage four showing significantly higher *Firmicutes* at week 14, compared to all other cages. At the OTU level, only OTU061 was different between cages (corrected P-value = 0.036) across all time points. This OTU was found to be enriched in cage 3 when compared to cages 2, 4, 5 and 6 and clusters in the genus *Bifidobacterium* (Figure S11).

Phenotypic variation in the faecal microbiota

Food was available *ad libitum* and, despite exhibiting the normal weight-gain-associated-phenotypes expected for these animals (Figure S12 and S13), both multivariate and univariate statistical analyses of the relative abundance values at the phylum, family and OTU levels for samples across all time points, and each timepoint separately, found no differences between the lean and obese phenotypes (Figure 5, Figures S4B and S5B). No statistically significant differences (P<0.05) were found in the relative abundance values of bacterial phyla and families between the three genotypes, except in the relative abundance of *Proteobacteria*, which was higher in samples from homozygous lean animals at week 5 (Figure S14). In the phylogenetic analysis, the NMDS plot based on the unweighted UniFrac distances failed to show any clear genotype-based clustering of samples at any of the time points (Figure S1). No differences were found when comparing the mean unweighted (Figure 4) or weighted (Figure S8) UniFrac distances from animals of the same and different genotypes.

Discussion

In this study, the age of the rats was found to be the most significant source of systematic variation in the faecal bacterial profile analyses at the phylum, family and OTU levels. Cohabitation had a significant impact on the intestinal microbiota, with more similar communities derived from co-housed animals. The impact of differences in host genotype and phenotype were largely undetected.

The predominant phyla detected in the faecal samples of the Zucker rats in this study were *Firmicutes* and *Bacteroidetes*, with significantly lower detection of *Actinobacteria* and *Tenericutes*; this is consistent with previous analyses of faecal bacterial profiles from rats [20,21], mice [22–24], and humans [1,3,25,26]; although certain studies have seen much greater representation of bacteria from the *Actinobacteria* phylum in humans [27,28], mice [8] and rats [29] and the *Proteobacteria* phylum in rats [29]. Interestingly, the average relative abundance of *Tenericutes* exceeded that of *Proteobacteria* in samples from animals at five weeks old, in contrast to other analyses of rat faecal microbiota [30,31]. The observed actinobacterial variability may be due to the primers used for the PCR [32] or the DNA extraction kit used [33], and it is important to note that the hypervariable region of the 16S

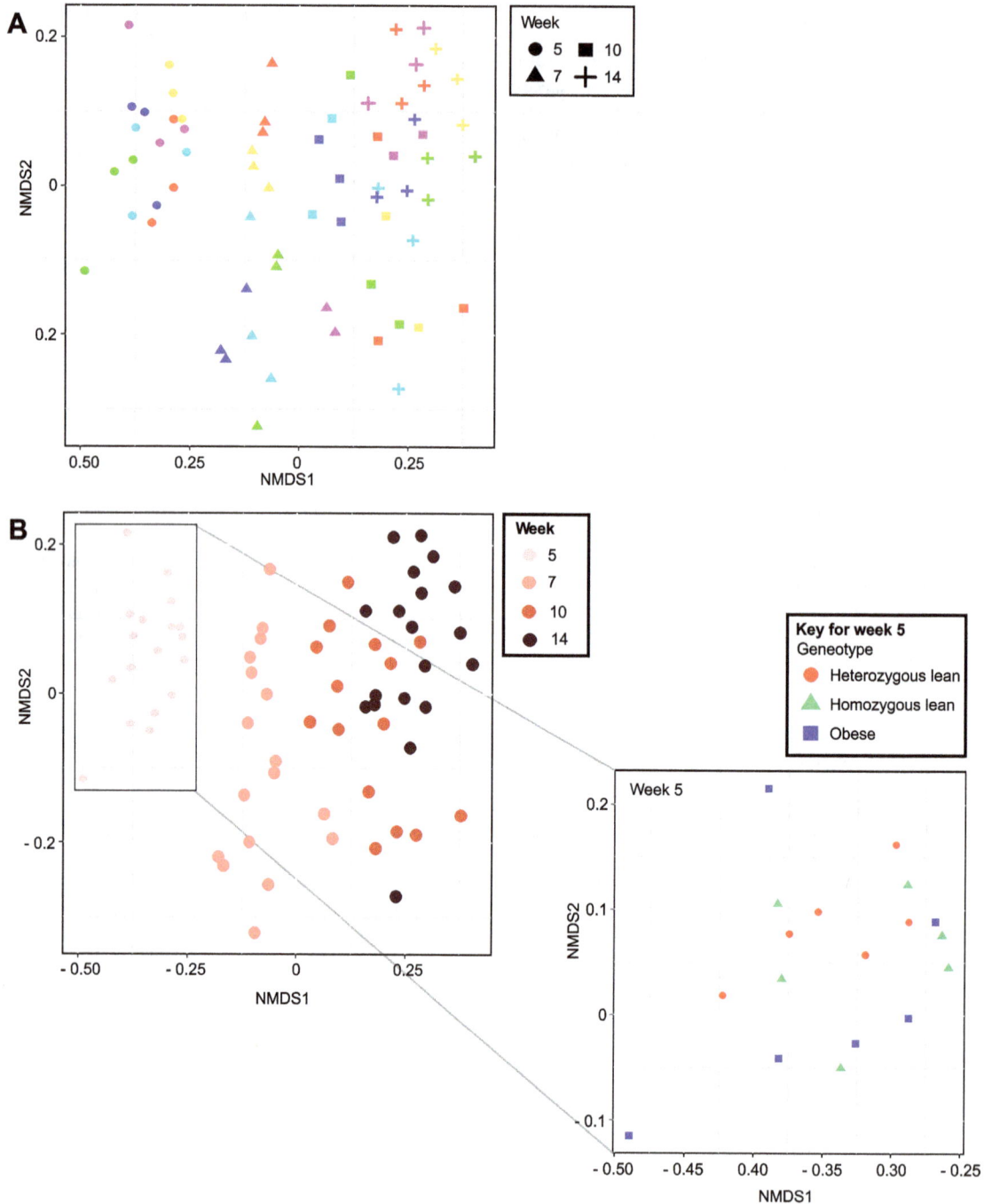

Figure 1. Non-Metric Multidimensional Scaling (NMDS) based on the unweighted UniFrac distances between the faecal samples. A: Samples are coloured by cage (1, red; 2, yellow; 3, green; 4, cyan; 5, dark blue; 6, purple). B: Samples are coloured by the age of the animals at sample collection; the genotype of the animals is shown for week 5. All time points coloured according to genotype are shown in supplementary information (Figure S1).

rRNA gene we selected to amplify (V1-V3) may underestimate the contribution of *Bifidobacteria* to the faecal bacterial profile [34].

At the phylum level, the most significant age-related trend was a decrease in the *Firmicutes:Bacteroidetes* ratio with increasing age, in contrast to the findings of previous investigators [8,35]. Given that the ages of the rats, 5–14 weeks, is more representative of maturation than aging *per se*, it is likely that the age-related trends observed here in the Zucker rat reflect normal development of the

microbiota towards a stable climax community. The composition of the intestinal microbiota is known to vary throughout infancy to adulthood, with further variation described in the elderly [36–38]. The increasing use of culture-independent direct sequencing techniques will facilitate our understanding of precisely how the intestinal microbiota varies with age, but these results demonstrate the significance of age on the composition of the intestinal microbiota and the importance of the consideration of this

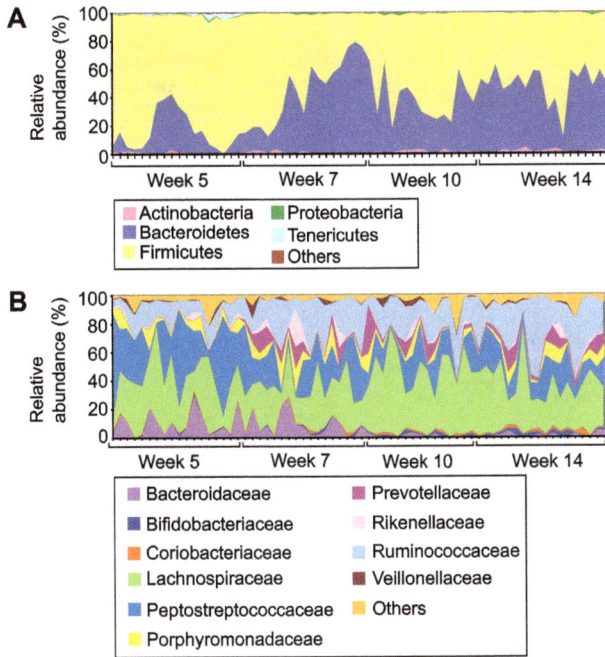

Figure 2. Relative abundances of bacteria across all 68 animal samples ordered by time point. A: Phylum-level; key: 'Others' composed of TM7 and *Verrucomicrobia*. B: Family-level; key: 'Others' composed of the families: *Alcaligenaceae, Anaeroplasmataceae, Bacillaceae, Clostridiaceae, Enterobacteriaceae, Erysipelotrichaceae, Eubacteriaceae, Halomonadaceae, IncertaeSedis XIII, IncertaeSedis XIV, Lactobacillaceae, Peptococcaceae, Pseudomonadaceae* and *Sphingomonadaceae*. Plot labels: O = obese, L = homozygous lean, H = heterozygous lean; number indicates cage number 1–6.

variable in the context of designing and interpreting animal studies.

No significant differences were found between the intestinal bacteria profiles of the three Zucker rat genotypes at either the phylum or the family level in the taxon-based analyses, and bacterial communities from the same genotype were not found to be more similar than communities from animals of differing genotypes when the UniFrac distance measures were compared. This result is interesting in light of the attention given to the possibility of an obesity-associated altered microbiome, with an increased potential for energy harvest [1–3], and also considering the clear phenotype-based differentiation observed in the ¹H NMR spectroscopy-based metabolite profiles of the urine, plasma and tissues of these animals (Lees *et al.*, in preparation).

In a previous study of the faecal bacterial profiles of the Zucker rat, employing DGGE and fluorescence *in situ* hybridization, differences between all three strains of the Zucker rat were observed, in spite of no phenotypic difference between the two lean strains. It was proposed that the microbiotal differences between the two lean strains were due to host genotype influence on the composition of the faecal microbiota [10]. However, in contrast to the present study, the animals were housed according to genotype, thus the cage environment (and coprophagic activity of the animals) is likely to have been influential in the experimental outcomes and may have reinforced or potentially enhanced any differences.

Certain studies have alluded to a more complex involvement of the microbiota in obesity than perhaps first indicated [4] and the nature of the shift in the relative contributions of phyla to the

microbiota composition in obesity has also been contested [5]. Additionally, there is gathering support for the role of diet, rather than obesity itself, in altering bacterial profiles, with shifts in the intestinal microbiome found to be associated with a high-fat diet rather than genetically induced obesity [4,6–8,39,40]. With these studies in mind, it is perhaps unsurprising that a quantitative difference in chow consumption, as would be expected between the obese and lean phenotypes analysed here [41–45], did not result in a difference in bacterial profiles between the obese and two lean phenotypes. Nevertheless, a more recent analysis of the leptin-resistant *db/db* mouse model identified compositional differences in the gut microbiota between the genetically obese and lean mice [46]; although, again it is unclear to what extent the arrangement of animal housing contributed to these results.

Several studies have explored the regulation of the intestinal microbiota by both host genes and the microenvironment in rodents [7,47–50]. In a quantitative PCR-based analysis of several germfree inbred strains of mice colonised with altered Schaedler flora (ASF), the microenvironment was found to influence the intestinal microbiota, with animals in differing cages showing divergence in ASF profiles. However, cohabitation of differing inbred strains of mice preserved most of the interstrain variation, with species variation in coprophagic behaviour suggested as a potential cause [49]. Further to this, Dimitriu and colleagues found that the response of faecal bacteria profiles to cohousing was strongly dependent on mouse genotype, with immunodeficient mice being more resistant to bacterial colonisation than wild type mice [51]. Similarly, Campbell and colleagues found host genetics to significantly correlate with bacterial phylotypes. Cohabitation of different strains revealed an interaction between host genetic and environmental factors, with bacterial communities more similar between co-housed animals, but with strain specificity maintained [50]. However, in a study of five common laboratory mouse strains, caging was found to contribute more variance to the murine microbiota composition than variation in genetics (31.7% compared to 19%, respectively), but inter-individual variance was the largest contribution (45.5%) [7]. Here, the intestinal bacteria profiles of animals from within the same cage showed clear similarities at the phylum and family level in the taxon-based analysis, in spite of the differing genotypes/phenotypes present. Additionally, comparison of UniFrac distances demonstrated that rats co-housed had significantly more similar bacterial communities than animals from different cages.

The obese and lean Zucker rats from within the same cage shared the same mother and the same cage environment from an early age and throughout the study. The maternal microbiota has been shown to be a significant indicator of offspring microbiota composition, irrespective of genetic background, resulting in similarities between progeny despite strain differences [52]. Furthermore, a study comparing knock-out mice, deficient in Toll-like receptors, with wild type animals, found that this genetic difference had a minimal impact on the composition of the microbiota, and that familial transmission of the maternal microbiota was the dominant source of variation in progeny microbiota composition [53]. The inheritance of the microbiota was also shown by Ley and colleagues in lean and *ob/ob* mice at the genus level; however, phylum-level distinctions between the two phenotypes were also observed [22], indicating that phenotypic differences may dominate in certain circumstances.

In addition to the influence of the maternal microbiota on the intestinal bacteria of offspring, the immediate cage environment has been shown to be a highly influential factor in microbiota development [52,54] and cohousing of litters will likely have reinforced inter-cage differences in the bacterial profiles of the

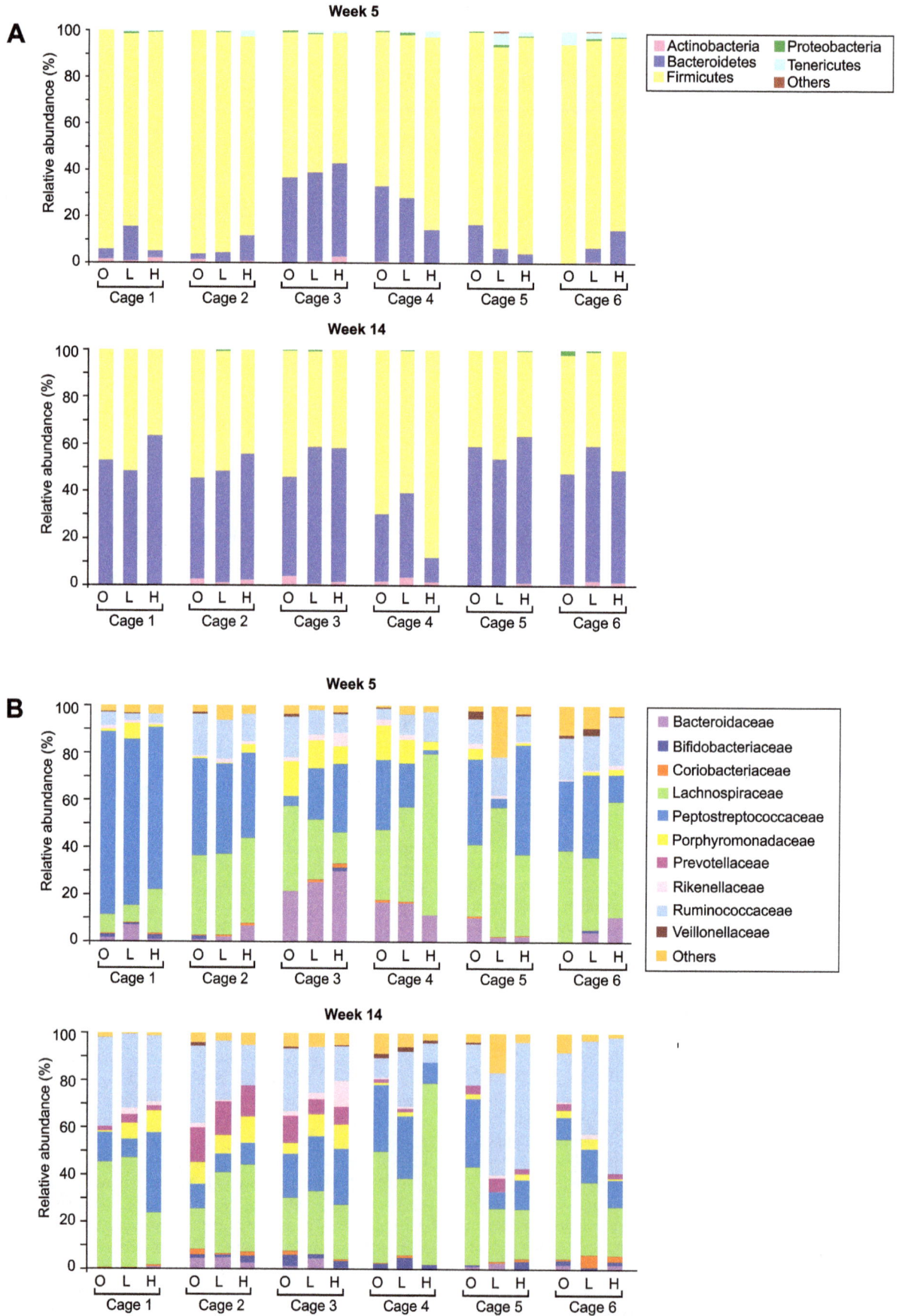

Figure 3. Relative abundances of bacteria for all animals grouped according to cage, at weeks 5 and 14. A: Phylum-level; key: see Figure 2 legend. B: Family-level; key: see Figure 2 legend. Data for weeks 7 and 10 are shown in Figure S9 (phylum) and S10 (family). Key: O = obese, L = homozygous lean, H = heterozygous lean.

Zucker rats. Rodents are coprophagic and ingestion of phenotypically differing littermates' faeces will have occurred from an early age, contributing towards the development of a common microbiome in animals occupying the same cage [55]. The influence of the cage environment on the developing intestinal microbiome was clearly demonstrated by Friswell and colleagues; marked changes were observed in the gut microbiota of mice re-located to new housing at four weeks of age, but not when mice were re-located at eight weeks of age [52]. Additionally, Ma and co-workers found that relocation of mice to new cages in a different intracampus facility was associated with transient variation in the composition of the faecal microbiota [53]. Furthermore, the effect of cage-environment has proved significant in a previous analysis of bacterial recolonisation profiles in rats following antibiotic exposure [56].

Germ free animal models have also been utilised to understand the contributions of various factors to the development of the microbiome; in a comparison of germ free mice either gavaged

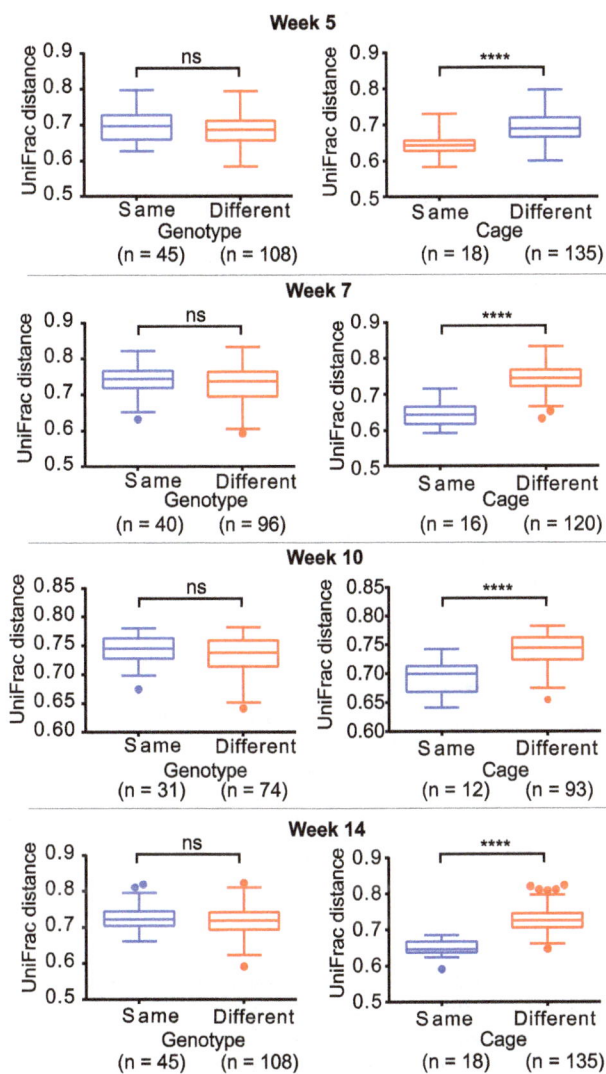

Figure 4. Box plots of the unweighted UniFrac distances. Box plots showing the median, lower and upper quartiles of the unweighted UniFrac distances at each time point comparing the effect of genotype and cage on the community structure. Whiskers were calculated using the Tukey method; filled circles represent outliers. A lower UniFrac distance indicates greater similarity between two microbial communities (Student's *t* test: ns = not significant; asterisks indicate significant differences: ****P<0.0001).

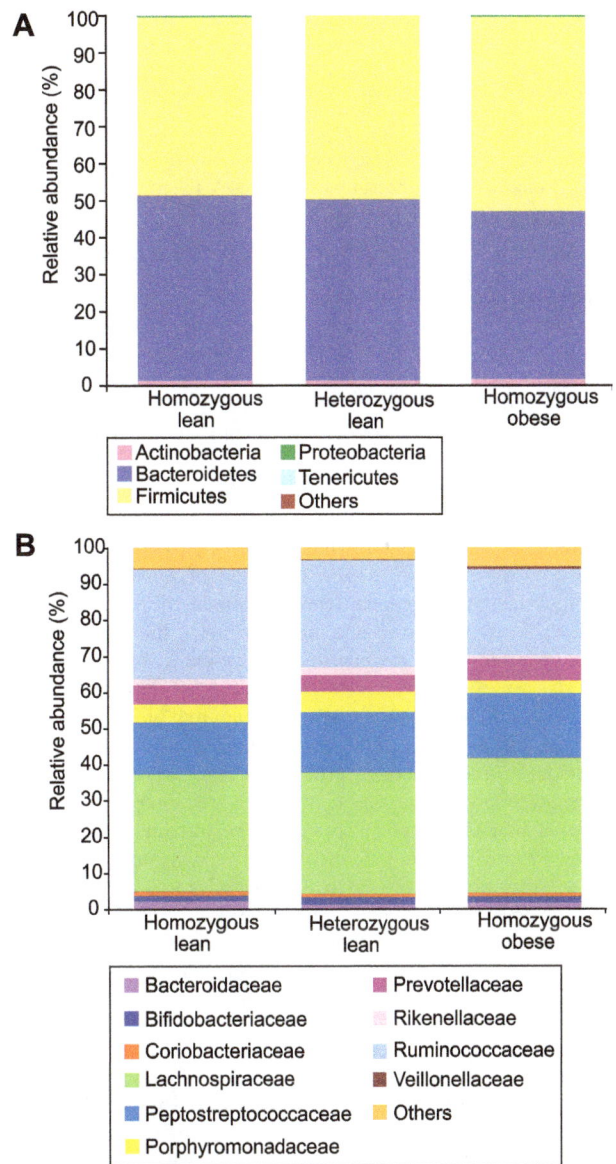

Figure 5. Mean relative abundances of bacteria for each genotype at week 14 (n = 6 per genotype). A: Phylum level; key: see Figure 2 legend. B: Family level; key: see Figure 2 legend. Mean relative abundances of each phylum and family for each genotype at each time point (weeks 5, 7, 10 and 14) are shown in Figure S15 (phylum) and S16 (family).

with a microbiota harvested from adult wild type mice, or allowed to acquire an intestinal microbiome from the cage microenvironment, authors found that the cage microenvironment mitigated the effects of the founding community [54]. More recently, a study of germ-free mice gavaged with the cultured microbiota of a twin pair discordant for obesity, demonstrated the significant impact of within-cage coprophagy on host metabolism. Recipients of the obese and lean microbiotas were co-housed, leading to certain bacterial species successfully invading the microbiome of co-housed animals, an effect that was diet dependent [57].

A potential limitation of our study is the lack of accurate measurement of food intake, prohibited by the complex nature of the animal housing design, which might have further strengthened our conclusions. However, we are satisfied our assumptions are reasonable, due to previous studies in our facility and a number of publications detailing the relative food intake of obese and lean Zucker rats of the same approximate age and bodyweight. Thus, obese Zucker rats, fed *ad libitum*, were found to have an increased food intake of between 30–60%, compared to the lean animals [58–60]. Additionally, we acknowledge that the use of 454 technology, and level of sequencing employed here, will have broadly characterized the samples in terms of the major patterns of variation, and that less abundant species of the populations sampled may not have been represented.

Conclusions

This study presents novel findings relating to how the faecal microbiota in the Zucker rat develops with age through juvenile, pubertal and post-pubertal stages. In addition, these results clearly demonstrate the significance of both age and cage environment on the composition of the faecal microbiota, in the context of an obese animal model, with both variables exerting a greater pressure on intestinal microbiota community structure than obese or lean phenotype and chow consumption.

In the context of the recent explosion of research into the compositional and functional aspects of the intestinal microbiota, these data emphasise the need to control for the effect of the microenvironment on the intestinal microbiome. As a minimum requirement, researchers need to be transparent regarding the specific animal housing arrangements when publishing studies, to allow for informed interpretation of data. This may be particularly important in studies whereby group-housing of animals according to genotype/phenotype acts to positively reinforce a particular compositional or functional aspect of the intestinal microbiota, effectively amplifying any differences between groups in differing cages. The profound effects of the housing of experimental animals on outcomes demonstrated here have clear implications for investigations relating to the development of the intestinal microbiota, and to microbiome-host co-metabolism, and should be given greater attention when designing studies.

Supporting Information

Figure S1 Non-Metric Multidimensional Scaling (NMDS) based on the unweighted UniFrac distances between the faecal samples. Central plot shows samples coloured according to animal cage (1–6), with sample marker shape representing time of sample collection. Enlarged plots show the genotype of the animals at each sample collection time point.

Figure S2 ANOVA of the means of OTUs, demonstrating that several OTUs varied between different time points across all the animals tested.

Figure S3 ANOVA of the means of OTUs, demonstrating that several OTUs varied between cages at each time point.

Figure S4 PCA scores plots generated using relative abundance values of the three most abundant phyla: *Bacteroidetes*, *Firmicutes* and *Actinobacteria*, in samples collected from all animals at all time points (mean centred, Pareto-scaled data; $R^2 = 0.99$, $Q^2 = 0.96$). Principal components 1 and 2 (PC1 and PC2) are shown with the percentage of explained variance described by each component. A: Samples are coloured according to the age (in weeks) at which the sample was collected. B: Samples are coloured according to the genotype of the animal. C: Samples are coloured according to the cage (1–6) of each animal. The scores plot in (A) can be used as a reference for the sample time points; the time points are not shown in (B) and (C) to aid visualisation of potential trends.

Figure S5 PCA scores plots generated using relative abundance values of the six most abundant families: *Bacteroidaceae*, *Porphyromonadaceae*, *Rikenellaceae*, *Lachnospiraceae*, *Ruminococcaceae* and *Peptostreptococcaceae*, in samples collected from all animals at all time points (Log$_{10}$ transformed, mean centred data; $R^2 = 0.83$, $Q^2 = 0.01$). Principal components 1 and 3 (PC1 and PC3) are shown with the percentage of explained variance described by each component. A: Samples are coloured according to the age (in weeks) at which the sample was collected. B: Samples are coloured according to the genotype of the animal. C: Samples are coloured according to the cage (1–6) of each animal. The scores plot in (A) can be used as a reference for the sample time points; the time points are not shown in (B) and (C) to aid visualisation of potential trends.

Figure S6 PCA scores plots generated using relative abundance values of the three most abundant phyla: *Bacteroidetes*, *Firmicutes* and *Actinobacteria*. Plots are shown for samples collected from all animals at weeks 5, 7, 10 and 14 (mean centred, Pareto-scaled data; Week 5: $R^2 = 1.00$ $Q^2 = 0.92$; Week 7: $R^2 = 1.00$ $Q^2 = 0.98$; Week 10: $R^2 = 1.00$ $Q^2 = 0.97$; Week 14: $R^2 = 1.00$ $Q^2 = 0.95$). In each plot principal components 1 and 2 (PC1 and PC2) are shown with the percentage of explained variance described by each component. Samples are coloured according to the cage (1–6) of each animal.

Figure S7 PCA scores plots generated using relative abundance values of the six most abundant families: *Bacteroidaceae*, *Porphyromonadaceae*, *Rikenellaceae*, *Lachnospiraceae*, *Ruminococcaceae* and *Peptostreptococcaceae*. Plots are shown for samples collected from all animals at weeks 5, 7, and 10 (Log$_{10}$ transformed, mean centred data; Week 5: $R^2 = 0.87$ $Q^2 = 0.53$; Week 7: $R^2 = 0.82$ $Q^2 = 0.06$; Week 10: $R^2 = 0.78$ $Q^2 = 0.29$). In each plot principal components 1 and 2 (PC1 and PC2) are shown with the percentage of explained variance described by each component. Samples are coloured according to the cage (1–6) of each animal. Week 14 is not shown here, as the Q^2 was negative with the first component, and was thus not considered a valid model.

Figure S8 Box plots showing the median, lower and upper quartiles of the weighted UniFrac distances at each time point comparing the effect of genotype and cage on the community structure. Whiskers were calculated using the Tukey method; filled circles represent outliers. A lower UniFrac distance indicates greater similarity between two microbial communities (Student's t test: ns = not significant; asterisks indicate significant differences: ** P<0.01; *** P<0.001; ****P<0.0001).

Figure S9 Relative abundances of bacteria at the phylum-level for all animals grouped according to cage, at each time point separately. Key: O = obese, L = homozygous lean, H = heterozygous lean. Phylum key: 'Others' composed of TM7 and *Verrucomicrobia*.

Figure S10 Relative abundances of bacteria at the family-level for all animals grouped according to cage, at each time point separately. Key: O = obese, L = homozygous lean, H = heterozygous lean. Family key: 'Others' composed of the families: *Alcaligenaceae, Anaeroplasmataceae, Bacillaceae, Clostridiaceae, Enterobacteriaceae, Erysipelotrichaceae, Eubacteriaceae, Halomonadaceae, Incertae Sedis XIII, Incertae Sedis XIV, Lactobacillaceae, Peptococcaceae, Pseudomonadaceae* and *Sphingomonadaceae*.

Figure S11 ANOVA of the means of the OTU061 shows that this OTU was the only one to vary at any significant levels between cages across the 4 time points.

Figure S12 Body weights for each animal at 4 weeks (pre-study) and at every urine sample collection point (weeks 5 to 14). (A) obese (fa/fa) animals, (B) lean (+/+) animals and (C) lean (fa/+) animals. Colour of data points indicates cage number of the animal.

Figure S13 Body weights for each strain at each week including pre-study (at four weeks for age), data expressed as mean ± standard error of the mean. Asterisks indicate significant differences (one-way ANOVA, followed by Tukey-Kramer multiple comparisons test, * P<0.05; ** P<0.01; *** P<0.001; **** P<0.0001). Green asterisks relate to the comparison of (fa/fa) and (+/+); red asterisks relate to the comparison of (fa/fa) and (fa/+).

Figure S14 Box plots of the relative abundance of *Proteobacteria* for each genotype at each time point. The median, lower and upper quartiles are shown. Whiskers were calculated using the Tukey method; filled circles represent outliers. Asterisks indicate significant differences (one-way ANOVA, followed by Tukey-Kramer multiple comparisons test, * P<0.05; ** P<0.01; *** P<0.001).

Figure S15 A: mean relative abundances of each phylum for each genotype (all time points included). B: mean relative abundances of each phylum for each genotype at each time point separately. Phylum key: 'Others' composed of TM7 and *Verrucomicrobia*.

Figure S16 A: mean relative abundances of each family for each genotype (all time points included). B: mean relative abundances of each family for each genotype at each time point separately. Family key: 'Others' composed of the families: *Alcaligenaceae, Anaeroplasmataceae, Bacillaceae, Clostridiaceae, Enterobacteriaceae, Erysipelotrichaceae, Eubacteriaceae, Halomonadaceae, Incertae Sedis XIII, Incertae Sedis XIV, Lactobacillaceae, Peptococcaceae, Pseudomonadaceae* and *Sphingomonadaceae*.

Table S1 Primers used to amplify the V1-V3 regions of the 16S rRNA gene. The unique barcode for each sample is shown in red, and allowed for multiplexing of the samples on the 454 sequencer on three different PTPs (Pico Titre Plate 8ths, 1 2 or 3).

Table S2 Sequence counts per sample.

Table S3 The OTUs identified by STAMP to be significantly altered in the faecal samples when grouped by week. All the means for each group were compared using an ANOVA and multiple testing using the Bonferroni correction (see Figure S2 for more detail).

Table S4 OTUs which were significantly changed at each time point between cages (P<0.05, corrected for multiple testing), see Figure S3 for more detail.

Table S5 Significant differences in the relative abundances of *Bacteroidetes* and *Firmicutes* between cages (no other phyla were found to be significantly different). Level of significance: * P<0.05; ** P<0.01; *** P<0.001. Difference between means of cages assessed using one-way ANOVA, followed by Tukey-Kramer multiple comparisons test.

Table S6 Significant differences in the relative abundances of families between cages (no other families were found to be significantly different). Level of significance: * P<0.05; ** P<0.01; *** P<0.001. Difference between means of cages assessed using one-way ANOVA, followed by Tukey-Kramer multiple comparisons test. Univariate statistical comparison of all cages was not possible at week 10 due to the small sample numbers per cage (n<3).

Author Contributions

Conceived and designed the experiments: IDW. Performed the experiments: HL. Analyzed the data: HL JRM. Contributed reagents/materials/analysis tools: IDW SMP. Wrote the paper: HL JS SMP JKN EH IDW JRM.

References

1. Ley RE, Turnbaugh PJ, Klein S, Gordon JI (2006) Microbial ecology: human gut microbes associated with obesity. Nature 444: 1022–1023.

2. Turnbaugh PJ, Ley RE, Mahowald MA, Magrini V, Mardis ER, et al. (2006) An obesity-associated gut microbiome with increased capacity for energy harvest. Nature 444: 1027–1131.

3. Turnbaugh PJ, Hamady M, Yatsunenko T, Cantarel BL, Duncan A, et al. (2009) A core gut microbiome in obese and lean twins. Nature 457: 480–484.

4. Fleissner CK, Huebel N, Abd El-Bary MM, Loh G, Klaus S, et al. (2010) Absence of intestinal microbiota does not protect mice from diet-induced obesity. Br J Nutr 104: 919–929.

5. Schwiertz A, Taras D, Schafer K, Beijer S, Bos NA, et al. (2009) Microbiota and SCFA in Lean and Overweight Healthy Subjects. Obesity 18: 190–195.

6. Duncan SH, Lobley GE, Holtrop G, Ince J, Johnstone AM, et al. (2008) Human colonic microbiota associated with diet, obesity and weight loss. Int J Obes 32: 1720–1724.

7. Hildebrandt MA, Hoffmann C, Sherrill-Mix SA, Keilbaugh SA, Hamady M, et al. (2009) High-Fat Diet Determines the Composition of the Murine Gut Microbiome Independently of Obesity. Gastroenterology 137: 1716–1724.e1712.

8. Murphy EF, Cotter PD, Healy S, Marques TM, O'Sullivan O, et al. (2010) Composition and energy harvesting capacity of the gut microbiota: relationship to diet, obesity and time in mouse models. Gut 59: 1635–1642.

9. Jumpertz R, Le DS, Turnbaugh PJ, Trinidad C, Bogardus C, et al. (2011) Energy-balance studies reveal associations between gut microbes, caloric load, and nutrient absorption in humans. The American Journal of Clinical Nutrition 94: 58–65.

10. Waldram A, Holmes E, Wang Y, Rantalainen M, Wilson ID, et al. (2009) Top-Down Systems Biology Modeling of Host Metabotype−Microbiome Associations in Obese Rodents. Journal of Proteome Research 8: 2361–2375.

11. Cole JR, Wang Q, Cardenas E, Fish J, Chai B, et al. (2009) The Ribosomal Database Project: improved alignments and new tools for rRNA analysis. Nucleic Acids Res 37: D141–145.

12. Wang Q, Garrity GM, Tiedje JM, Cole JR (2007) Naive Bayesian classifier for rapid assignment of rRNA sequences into the new bacterial taxonomy. Appl Environ Microbiol 73: 5261–5267.

13. Schloss PD, Westcott SL, Ryabin T, Hall JR, Hartmann M, et al. (2009) Introducing mothur: Open-Source, Platform-Independent, Community-Supported Software for Describing and Comparing Microbial Communities. Appl Environ Microbiol 75: 7537–7541.

14. Lozupone C, Knight R (2005) UniFrac: a New Phylogenetic Method for Comparing Microbial Communities. Applied and Environmental Microbiology 71: 8228–8235.

15. Team RDC (2008) R: A Language and Environment for Statistical Computing. In: Computing RFfS, editor. Vienna.

16. Parks DH, Beiko RG (2010) Identifying biologically relevant differences between metagenomic communities. Bioinformatics 26: 715–721.

17. Wold S, Esbensen K, Geladi P (1987) Principal component analysis. Chemometrics and Intelligent Laboratory Systems 2: 37–52.

18. Wold S, Johansson E, Cocchi M (1993) In: Kubinyi H, editor. 3D-QSAR in Drug Design: Theory, Methods and Applications: ESCOM Science.

19. Benson AK, Kelly SA, Legge R, Ma F, Low SJ, et al. (2010) Individuality in gut microbiota composition is a complex polygenic trait shaped by multiple environmental and host genetic factors. Proceedings of the National Academy of Sciences 107: 18933–18938.

20. Manichanh C, Reeder J, Gibert P, Varela E, Llopis M, et al. (2010) Reshaping the gut microbiome with bacterial transplantation and antibiotic intake. Genome Research 20: 1411–1419.

21. Li JV, Ashrafian H, Bueter M, Kinross J, Sands C, et al. (2011) Metabolic surgery profoundly influences gut microbial-host metabolic cross-talk. Gut 60: 1214–1223.

22. Ley RE, Bäckhed F, Turnbaugh P, Lozupone CA, Knight RD, et al. (2005) Obesity alters gut microbial ecology. Proceedings of the National Academy of Sciences of the United States of America 102: 11070–11075.

23. Claus SP, Ellero SL, Berger B, Krause L, Bruttin A, et al. (2011) Colonization-induced host-gut microbial metabolic interaction. MBio 2: e00271–00210.

24. Turnbaugh PJ, Backhed F, Fulton L, Gordon JI (2008) Diet-induced obesity is linked to marked but reversible alterations in the mouse distal gut microbiome. Cell Host Microbe 3: 213–223.

25. Eckburg PB, Bik EM, Bernstein CN, Purdom E, Dethlefsen L, et al. (2005) Diversity of the Human Intestinal Microbial Flora. Science 308: 1635–1638.

26. Larsen N, Vogensen FK, van den Berg FWJ, Nielsen DS, Andreasen AS, et al. (2010) Gut Microbiota in Human Adults with Type 2 Diabetes Differs from Non-Diabetic Adults. PLoS One 5: e9085.

27. Gill SR, Pop M, DeBoy RT, Eckburg PB, Turnbaugh PJ, et al. (2006) Metagenomic Analysis of the Human Distal Gut Microbiome. Science 312: 1355–1359.

28. Andersson AF, Lindberg M, Jakobsson H, Bäckhed F, Nyrén P, et al. (2008) Comparative Analysis of Human Gut Microbiota by Barcoded Pyrosequencing. PLoS One 3: e2836.

29. Nelson TA, Holmes S, Alekseyenko AV, Shenoy M, Desantis T, et al. (2011) PhyloChip microarray analysis reveals altered gastrointestinal microbial communities in a rat model of colonic hypersensitivity. Neurogastroenterology & Motility 23: 169–e142.

30. Duca FA, Sakar Y, Lepage P, Devime F, Langelier B, et al. (2014) Replication of obesity and associated signaling pathways through transfer of microbiota from obese prone rat. Diabetes 63(5): 1624–1636.

31. Zhu Q, Jin Z, Wu W, Gao R, Guo B, et al. (2014) Analysis of the Intestinal Lumen Microbiota in an Animal Model of Colorectal Cancer. PLoS ONE 9: e90849.

32. Turroni F, Marchesi JR, Foroni E, Gueimonde M, Shanahan F, et al. (2009) Microbiomic analysis of the bifidobacterial population in the human distal gut. ISME Journal 3: 745–751.

33. Maukonen J, Simões C, Saarela M (2012) The currently used commercial DNA-extraction methods give different results of clostridial and actinobacterial populations derived from human fecal samples. FEMS Microbiology Ecology 79: 697–708.

34. Chen HM, Yu YN, Wang JL, Lin YW, Kong X, et al. (2013) Decreased dietary fiber intake and structural alteration of gut microbiota in patients with advanced colorectal adenoma. Am J Clin Nutr 97: 1044–1052.

35. Mariat D, Firmesse O, Levenez F, Guimaraes V, Sokol H, et al. (2009) The Firmicutes/Bacteroidetes ratio of the human microbiota changes with age. BMC Microbiol 9: 123.

36. O'Toole PW, Claesson MJ (2010) Gut microbiota: Changes throughout the lifespan from infancy to elderly. International Dairy Journal 20: 281–291.

37. Hopkins MJ, Sharp R, Macfarlane GT (2002) Variation in human intestinal microbiota with age. Digestive and Liver Disease 34, Supplement 2: S12–S18.

38. Makivuokko H, Tiihonen K, Tynkkynen S, Paulin L, Rautonen N (2010) The effect of age and non-steroidal anti-inflammatory drugs on human intestinal microbiota composition. Br J Nutr 103: 227–234.

39. Cani PD, Neyrinck AM, Fava F, Knauf C, Burcelin RG, et al. (2007) Selective increases of bifidobacteria in gut microflora improve high-fat-diet-induced diabetes in mice through a mechanism associated with endotoxaemia. Diabetologia 50: 2374–2383.

40. Zhang C, Zhang M, Wang S, Han R, Cao Y, et al. (2010) Interactions between gut microbiota, host genetics and diet relevant to development of metabolic syndromes in mice. ISME J 4: 232–241.

41. Zucker TF, Zucker LM (1962) Hereditary Obesity in the Rat Associated with High Serum Fat and Cholesterol. Proceedings of the Society for Experimental Biology and Medicine Society for Experimental Biology and Medicine (New York, NY) 110: 165–171.

42. Barry WS, Bray GA (1969) Plasma triglycerides in genetically obese rats. Metabolism 18: 833–839.

43. Jenkins TC, Hershberger TV (1978) Effect of Diet, Body Type and Sex on Voluntary Intake, Energy Balance and Body Composition of Zucker Rats. The Journal of Nutrition 108: 124–136.

44. Harris RB, Tobin G, Hervey GR (1988) Voluntary food intake of lean and obese Zucker rats in relation to dietary energy and nitrogen content. The Journal of Nutrition 118: 503–514.

45. Lindborg KA, Jacob S, Henriksen EJ (2011) Effects of Chronic Antagonism of Endocannabinoid-1 Receptors on Glucose Tolerance and Insulin Action in Skeletal Muscles of Lean and Obese Zucker Rats. Cardiorenal Med 1: 31–44.

46. Geurts L, Lazarevic V, Derrien M, Everard A, Van Roye M, et al. (2011) Altered gut microbiota and endocannabinoid system tone in obese and diabetic leptin-resistant mice: impact on apelin regulation in adipose tissue. Front Microbiol 2: 149.

47. McKnite AM, Perez-Munoz ME, Lu L, Williams EG, Brewer S, et al. (2012) Murine gut microbiota is defined by host genetics and modulates variation of metabolic traits. PLoS One 7: e39191.

48. Hufeldt MR, Nielsen DS, Vogensen FK, Midtvedt T, Hansen AK (2010) Variation in the gut microbiota of laboratory mice is related to both genetic and environmental factors. Comp Med 60: 336–347.

49. Deloris Alexander A, Orcutt RP, Henry JC, Baker J, Jr., Bissahoyo AC, et al. (2006) Quantitative PCR assays for mouse enteric flora reveal strain-dependent differences in composition that are influenced by the microenvironment. Mamm Genome 17: 1093–1104.

50. Campbell JH, Foster CM, Vishnivetskaya T, Campbell AG, Yang ZK, et al. (2012) Host genetic and environmental effects on mouse intestinal microbiota. Isme j 6: 2033–2044.

51. Dimitriu PA, Boyce G, Samarakoon A, Hartmann M, Johnson P, et al. (2013) Temporal stability of the mouse gut microbiota in relation to innate and adaptive immunity. Environ Microbiol Rep 5: 200–210.

52. Friswell MK, Gika H, Stratford IJ, Theodoridis G, Telfer B, et al. (2010) Site and strain-specific variation in gut microbiota profiles and metabolism in experimental mice. PLoS One 5: e8584.

53. Ubeda C, Lipuma L, Gobourne A, Viale A, Leiner I, et al. (2012) Familial transmission rather than defective innate immunity shapes the distinct intestinal microbiota of TLR-deficient mice. J Exp Med 209: 1445–1456.

54. McCafferty J, Muhlbauer M, Gharaibeh RZ, Arthur JC, Perez-Chanona E, et al. (2013) Stochastic changes over time and not founder effects drive cage effects in microbial community assembly in a mouse model. Isme j 7: 2116–2125.

55. Barnes RH, Fiala G, McGehee B, Brown A (1957) Prevention of Coprophagy in the Rat. The Journal of Nutrition 63: 489–498.

56. Swann JR, Tuohy KM, Lindfors P, Brown DT, Gibson GR, et al. (2011) Variation in Antibiotic-Induced Microbial Recolonization Impacts on the Host Metabolic Phenotypes of Rats. Journal of Proteome Research 10: 3590–3603.

57. Ridaura VK, Faith JJ, Rey FE, Cheng J, Duncan AE, et al. (2013) Gut microbiota from twins discordant for obesity modulate metabolism in mice. Science 341: 1241214.

58. Vasselli JR, Cleary MP, Jen KL, Greenwood MR (1980) Development of food motivated behavior in free feeding and food restricted Zucker fatty (fa/fa) rats. Physiol Behav 25: 565–573.

59. Alonso-Galicia M, Brands MW, Zappe DH, Hall JE (1996) Hypertension in obese Zucker rats. Role of angiotensin II and adrenergic activity. Hypertension 28: 1047–1054.

60. Becker EE, Grinker JA (1977) Meal patterns in the genetically obese Zucker rat. Physiol Behav 18: 685–692.

## 18

Nanoparticle-Encapsulated Chlorhexidine against Oral Bacterial Biofilms

Chaminda Jayampath Seneviratne[1,2,9], Ken Cham-Fai Leung[3,9], Chi-Hin Wong[3], Siu-Fung Lee[4], Xuan Li[1], Ping Chung Leung[5], Clara Bik San Lau[5], Elaine Wat[5], Lijian Jin[1]*

1 Faculty of Dentistry, The University of Hong Kong, Hong Kong SAR, China, 2 Faculty of Dentistry, National University of Singapore, Singapore, Singapore, 3 Department of Chemistry, Institute of Creativity, and Partner State Key Laboratory of Environmental & Biological Analysis, The Hong Kong Baptist University, Hong Kong SAR, China, 4 Department of Chemistry, The Chinese University of Hong Kong, Hong Kong SAR, China, 5 Institute of Chinese Medicine and Partner State Key Laboratory of Phytochemistry & Plant Resources in West China, The Chinese University of Hong Kong, Hong Kong SAR, China

Abstract

Background: Chlorhexidine (CHX) is a widely used antimicrobial agent in dentistry. Herein, we report the synthesis of a novel mesoporous silica nanoparticle-encapsulated pure CHX (Nano-CHX), and its mechanical profile and antimicrobial properties against oral biofilms.

Methodology/Principal Findings: The release of CHX from the Nano-CHX was characterized by UV/visible absorption spectroscopy. The antimicrobial properties of Nano-CHX were evaluated in both planktonic and biofilm modes of representative oral pathogenic bacteria. The Nano-CHX demonstrated potent antibacterial effects on planktonic bacteria and mono-species biofilms at the concentrations of 50–200 µg/mL against *Streptococcus mutans, Streptococcus sobrinus, Fusobacterium nucleatum, Aggregatibacter actinomycetemcomitans* and *Enterococcus faecalis*. Moreover, Nano-CHX effectively suppressed multi-species biofilms such as *S. mutans, F. nucleatum, A. actinomycetemcomitans* and *Porphyromonas gingivalis* up to 72 h.

Conclusions/Significance: This pioneering study demonstrates the potent antibacterial effects of the Nano-CHX on oral biofilms, and it may be developed as a novel and promising anti-biofilm agent for clinical use.

Editor: Zezhang Wen, LSU Health Sciences Center School of Dentistry, United States of America

Funding: This study has been supported by the General Research Fund (GRF) from the Hong Kong Research Grants Council (HKU767512M and 768713M) and the Modern Dental Laboratory/HKU Endowment Fund to LJJ. The funders had no role in study design, data collection and analysis, decision to publish, or preparation of the manuscript.

Competing Interests: The authors have declared that no competing interests exist.

* Email: ljjin@hku.hk

9 These authors contributed equally to this work.

Introduction

Dental plaque biofilm is the aetiological agent for common oral diseases such as dental caries, periodontal disease and emerging peri-implant infections [1,2]. The pathogenic microorganisms in the plaque biofilm critically contribute to the aforementioned oral diseases.

Chlorhexidine (CHX) is a widely used antimicrobial agent in various formulations in dentistry [3,4]. Recently, nano-encapsulation has emerged as a novel approach to delivering biologically potent compounds more effectively to specific-targets, while maintaining their original capacity in the development of new health products and drugs [5,6]. It has been shown that nano-encapsulated drugs can enhance the overall biological effectiveness through fast penetration and bioavailability while reducing potential cytotoxicity, drug dosage and the production costs, with reference to the controls [7].

Herein, we report a novel synthesis of mesoporous silica nanoparticle encapsulated with pure (non-salt form) CHX, namely Nano-CHX; and present its morphological profile and mechanical properties. Its antimicrobial properties were comprehensively characterized by using planktonic bacteria, mono-species and mixed-species models of oral biofilms.

Materials and Methods

Synthesis of the Nano-CHX

The mesoporous silica nanoparticles were used and prepared according to the previously validated protocol, with an average particle diameter of around 140 nm and pore size of approximately 2.5 nm, the Brunauer, Emmett and Teller (BET) surface area of $\sim$1,000 m^2/g, and pore volume $\sim$1.0 cm^3/g [8]. In brief, 50 mg of pure CHX (non-salt form, Sigma-Aldrich) was dissolved in 5 mL of ethanol. Fifty milligram of mesoporous silica nanoparticles were added into 4 mL of the CHX solution and then incubated for 24 h at room temperature. This process allowed the CHX molecules to swell and assemble into the inner pores of the nanoparticles. The mixture was then centrifuged and the Nano-CHX particles were collected by membrane filtration (0.2 micron, Macherey-Nagel, Germany).

Morphology, thermostability and release profile

The morphology of the Nano-CHX particles was analyzed by using transmission electron microscopy (TEM, Tecnai G2 20 S-TWIN). Prior to the analysis, an ethanol solution of Nano-CHX particles was drop-casted, followed by solvent evaporation at room temperature on a carbon-coated copper grid. The thermostability of the nanomaterials was analyzed by the Thermogravimetric Analyzer (Perkin Elmer TGA-6) [9]. This method determined the change of a sample weight with slowly increasing temperature, and weight loss was usually observed with a specific range of temperature. Weight percentages of CHX loaded to the blank nanoparticles were then determined by observing the weight loss at a range of 100–900°C. For detecting the release profile, the Nano-CHX (30 mg) was dispersed in 10 mL of deionized water (Barnstead RO Pure System) at 37°C for 72 h. The release profile of CHX from Nano-CHX in water over time was observed at the local maximum absorption of CHX (254 nm) by UV/visible absorption spectroscopy (Cary UV-100).

Anti-biofilm properties in mono-species and mixed-species oral biofilms

Oral pathogenic bacteria, viz. Streptococcus mutans (ATCC 35668), Streptococcus sabrinus (ATCC 33478), Fusobacterium nucleatum (ATCC 25586), Aggregatibacter actinomycetemcomitans (ATCC 43718), Enterococccus faecalis (ATCC 29212) and P. gingivalis (ATCC 33277) were obtained from the archival collection at the Oral Biosciences, Faculty of Dentistry, The University of Hong Kong, and used for the experiments. These species were inoculated on horse blood agar plates and incubated in an anaerobic chamber with 5% CO_2, 10% H_2 and 85% of N_2 at 37°C for two days. The bacterial cultures were harvested and suspended in phosphate-buffered saline (PBS) for subsequent microbiological experiments.

Planktonic mode. Broth microdilution assay was performed to determine the minimum inhibitory concentration (MIC) and minimum bactericidal concentration (MBC), according to the standard NCCLS criteria with slight modifications as previously described [2]. S. mutans and E. faecalis were used for this assay. Briefly, bacteria cultures were harvested and suspended with Brain Heart Infusion Broth (BHI) at a concentration of 10^6 cells/mL using a calibrated spectrophotometer. Bacterial cultures were incubated with serially double diluted Nano-CHX particles for 48 h at 37°C under anaerobic conditions. The lowest concentration without visual bacterial growth was recorded as the MIC. Afterward, 20 μL of bacterial suspensions was inoculated in horse blood agar and kept for 48 h for further observation. The lowest drug concentration that yielded no bacterial growth was documented as MBC.

Mono-species biofilms. Mono-species biofilms of oral pathogens were formed according to the previous protocol [10,11]. S. mutans, S. sabrinus, F. nucleatum, A. actinomycetemcomitans and E. faecalis were resuspended in BHI at a concentration of 10^8 cells/mL to develop a biofilm, respectively. 100 μL of standard cell suspension was pipetted to polystyrene 96-well plates and incubated under anaerobic conditions for 24 h. The biofilms were washed with PBS and treated with a diluted series of Nano-CHX for 24 h before taking MIC readings. In parallel, planktonic bacterial cell suspension at 10^8 cells/mL was used to obtain the MIC readings.

Mixed-species biofilms. Three selected combinations of mixed-species were made, viz. i) S. mutans with F. nucleatum and P. gingivalis; ii) S. sobrinus with F. nucleatum and P. gingivalis; and iii) S. mutans with F. nucleatum, A. actinomycetemcomitans and P. gingivalis. These mixed-species biofilms were formed by mixing equal amount of bacterial suspensions as an initial inoculum and they were grown under anaerobic conditions for 24, 48 and 72 h as described above. Following each time point, the biofilms were treated with Nano-CHX for 24 h using blank nanoparticles as the negative controls. The MIC of Nano-CHX against the mixed-species biofilms was determined by treating the biofilms with Nano-CHX as described. During this experiment, it was observed that Nano-CHX was unable to eradicate the mature biofilm at 72 h. Therefore, another assay was performed to examine the preventive potential of Nano-CHX against the mixed-species biofilms. In this assay, Nano-CHX was introduced to the wells of the plate together with mixed-species bacterial inoculum. Biofilm growth was monitored for 24, 48 and 72 h.

Microscopic observations of the efficacy of Nano-CHX against oral biofilms

In a separate set of experiments, the mixed-species biofilms of S. sobrinus, F. nucleatum and P. gingivalis were developed in 1×1 cm sterilized coupons made of polystyrene material (IWAKI, Tokyo, Japan). The biofilms were formed on coupons in a 12-well plate (IWAKI, Tokyo, Japan) under anaerobic conditions as described above. After 24 h, the biofilms were treated with Nano-CHX particles for 24 h. Blank nanoparticles were used as the controls. These specimens were subjected to Scanning Electron Microscopy and Confocal Scanning Laser Microscopy as described previously [11]. Experiments were repeated in three separate occasions.

Scanning Electron Microscopy (SEM) analysis. Each coupon with the mixed-species biofilms was post-fixed in a 5% gluteraldehyde and 3% formaldehyde solution (5% v/v 0.1 M phosphate buffer, pH 7.4) for 4 h, air dried, and then placed in 1% osmium tetroxide for 1 h. Coupons were subsequently washed in distilled water, dehydrated in a series of ethanol washes (70% for 10 mins, 95% for 10 mins and 100% for 20 mins), and air dried in a desiccator. Afterward, the specimens were mounted on aluminum stabs with copper tape, coated with gold in a low-pressure atmosphere with an ion sputter coater (JEOL JFC1 100: JEOL, Tokyo, Japan). The topographic features of each biofilm were visualized with a SEM (Hitachi S-3400N SEM at EM Unit, HKU) in high vacuum mode at 10 kV.

Confocal Scanning Laser Microscopy (CSLM). Molecular Probes' Live/Dead BacLight Viability kit comprising SYTO-9 and propidium iodine (PI) (Molecular Probes, Eugene, OR) was used for the CSLM analysis of the mixed-species biofilms and selected mono-species biofilms of S. mutants, S. sobrinus and E. faecalis. The Control and Nano-CHX treated biofilms were washed once with 1.5 mL of PBS and stained with Live/Dead BacLight Viability kit using previously established protocol by our group [11]. In brief, coupons were incubated with SYTO-9 and PI for 30 min in the dark at 37°C. Subsequently, coupons were mounted on microscope slides and the images of stained biofilms were captured using a CSLM system (FLUOVIEW FV 1000, Olympus, Tokyo, Japan). All tests were repeated three times.

Statistical analysis

The statistical significance of MIC and MBC values of Nano-CHX against both planktonic and biofilm modes as well as

A

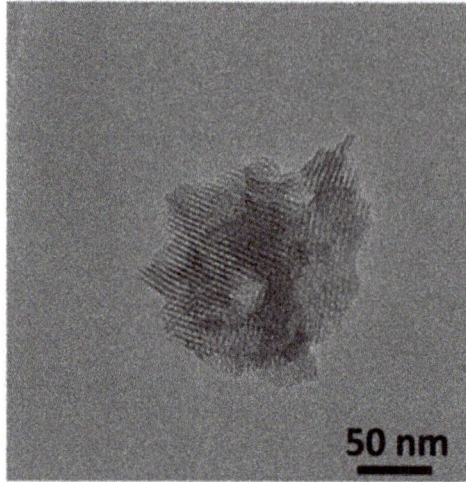

B

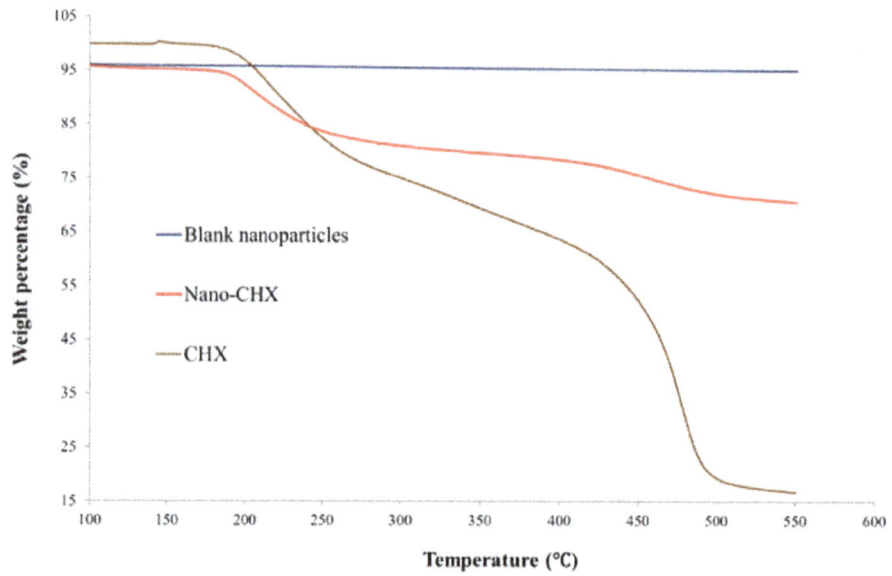

C

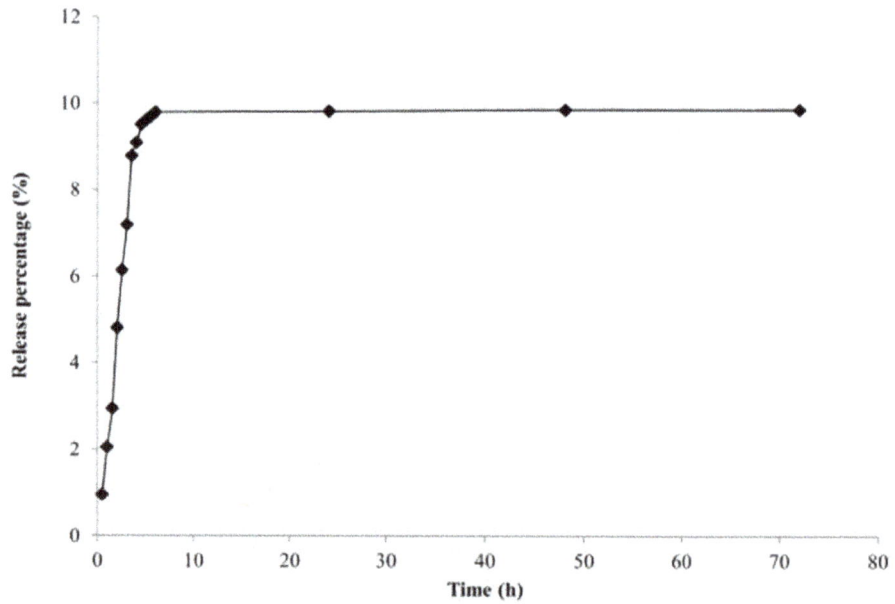

Figure 1. The characteristics of the Nano-CHX particles. A transmission electron microscopic image of single nanoparticle (A). The thermogravimetric analysis on the weight losses of i) CHX between 150–500°C (80.6%); ii) blank nanoparticles between 150–500°C (1.3%); and iii) Nano-CHX between 150–500°C (23.1%) (B). The release profile of CHX (%) from the Nano-CHX assessed by UV/visible absorption spectroscopy at 254 nm (C).

different species were analyzed by Student's t-test or ANOVA as appropriate by using IBM SPSS Statistics for Windows, Version 21.0, IBM Corp. Armonk, NY, USA.

Results

Characterization of the Nano-CHX particles

The transmission electron microscopic image showed that a single blank nanoparticle exhibited ordered and structured inner porous channels with around 2.5 nm diameter (Figure 1A). As shown in Figure 1B, the thermogravimetric analyses revealed that a significant weight loss (80.6%) of CHX was observed from 150 to 500°C, and the weight loss of blank nanoparticles was 1.3% between 150–500°C. As the weight loss of nanoparticles Nano-CHX was 23.1% between 150–500°C, the CHX loading efficiency was calculated to be 21.8% after a background subtraction with the blank nanoparticles. This finding demonstrated that a significant amount of CHX was capable to be loaded onto the nanoparticles by swelling and assembling processes. Furthermore, the release experiment showed a quick and marked release of CHX from Nano-CHX that reached a maximal saturation of around 10% at 6 h, and this saturated level of released CHX persisted till the end of the observation period of time at 72 h (Figure 1C).

Antimicrobial activity against planktonic and biofilm modes of oral pathogens

The MICs of Nano-CHX against *S. mutans* and *E. faecalis* were 19.5 µg/mL and 156 µg/mL, respectively; while the respective MBC values were 312.5 µg/mL and 1250 µg/mL. Notably, the blank nanoparticles did not demonstrate any antimicrobial activity (5000 µg/mL was ineffective). Hence, the antimicrobial activity of Nano-CHX was due to the action of CHX on the microorganisms, instead of the residual effect of nanoparticles *per se*. Comparative analysis of the mono-species biofilm and planktonic modes (10^8 cells/mL inoculum) showed a promising antibacterial activity of Nano-CHX treatment for 24 h (Table 1). Interestingly, the Nano-CHX treatment (100 or 200 µg/mL) for 24 h was highly effective against the mono-species biofilms, including *S. mutans*, *S. sobrinus*, *F. nucleaturm*, *A. actinomycetemcomitans* and *E. faecalis*. The CLSM images of Nano-CHX-treated (24 h) mono-species biofilms of *S. mutants*, *S. sobrinus* and *E. faecalis* are presented in Figure S1.

Furthermore, in the mixed-species biofilm assay it was observed that the Nano-CHX was unable to eradicate the 72 h mixed-species biofilm. Hence, we proceed to observe the preventive action of Nano-CHX against mixed-species biofilms. The MIC data showed a potent antibacterial activity of Nano-CHX against all three groups of mixed-species biofilms at 24, 48 and 72 h (Table 2). Microscopic observations using SEM and CLSM further confirmed the above observation on the mixed-species biofilms of *S. sobrinus*, *F. nucleatum* and *P. gingivalis*. As shown in the SEM and CLSM images, only a few scattered bacterial cells (Figure 2B), or a few viable cells existed (Figure 2D) existed, with reference to the control treated with the blank nanoparticles (Figures 2A and 2C).

Discussion

CHX is well known to be effective against both Gram-positive and Gram-negative bacteria [4,12]. It is also less likely to develop antimicrobial resistance as compared to antibiotics, and hence serves as a good candidate for further developing new antimicrobial products [13]. In the present study, we attempted to develop novel Nano-CHX particles and examine their antibacterial effects on both planktonic and biofilm modes of representative oral pathogenic bacteria, such as *S. mutans*, *S. sobrinus*, *E. faecalis*, *F. nucleatum*, *A. actinomycetemcomitans* and *P. gingivalis*. These bacteria are significantly involved in common oral diseases, such as dental caries, pulpal infections and periodontal disease.

In the present study, the mesoporous silica nanoparticles with inner pore channels of approximately 2.5 nm were used [9,14,15] for loading of pure CHX. By swelling the blank mesoporous nanoparticles in an ethanolic solution of pure CHX, a significant amount of over 20 weight percent of CHX was encapsulated in the nanoparticles. Theoretically, the salt forms of CHX such as CHX gluconate, chloride and acetate may not achieve the loading efficacy as high as the pure CHX (21.8%), mainly due to their relatively large molecular size. Moreover, the assembly process of CHX toward the nanoparticles proceeded smoothly. Notably, the Nano-CHX particles were capable of being dried and well re-dispersed in water without jeopardizing the structure of these particles. The silica material employed in this study has recently been shown to be relatively non-cytotoxic [7,16–18]. This material has also been clinically approved to be used as an implant material [19]. Therefore, we assume that the novel Nano-CHX could have a promising clinical usage, although more *in vitro* and *in vivo* studies are highly warranted for affirmative evidence.

One of the major concerns with antimicrobial agents in clinical setting is the growth mode of microbial biofilms [20]. Numerous studies have shown that the biofilm mode of microbial growth can be highly resistant to antimicrobial agents with reference to their planktonic counterparts [21,22]. Pathogenic plaque biofilms consisting of mixed pathogens are the etiological agent for major oral diseases like dental caries and periodontal diseases [1,2]. Over the years, CHX containing oral healthcare products such as mouth rinses have been frequently used in the clinical practice, due to its board-spectrum antimicrobial properties [3,23,24]. On the other hand, studies have also shown the limitation of CHX formulations owing to its inability to effectively penetrate the mature dental plaque biofilm and low substantivity [25,26]. A recent study demonstrates that CHX is more effective against *de novo* biofilms and its effects may reduce with time [27]. Hence, the substantivity of CHX only persisted for 6–10 h after mouth rinse, and the salivary and plaque bacteria gradually obtained a higher viability with time [28,29]. The foregoing fact highlights the need to develop novel delivery vehicles for slow-release and relatively high penetration of CHX with increased efficiency and effectiveness against oral biofilm for better clinical usage. Therefore, great attention has been increasingly paid to develop improved application form of CHX by incorporating nanotechnology. However, there are only a very few studies assessing the efficacy and effectiveness of Nano-CHX against oral pathogens. For instance, a recent study has shown that Nano-CHX particles could be effective against *S. mutans* biofilms *in vitro* [30].

Table 1. The minimal inhibitory concentration (MIC, µg/mL) of Nano-CHX against the planktonic mode and mono-species biofilms of pathogenic oral bacteria at 24 h.

Bacterial species	Planktonic mode	Mono-species biofilms
S. mutans	50	100**
S. sobrinus	100	200*
F. nucleatum	50	100**
A. actinomycetemcomitans	50	100**
E. faecalis	100	200*

Significant difference from the planktonic mode,
*$p<0.05$,
**$p<0.01$.

Interestingly, the novel Nano-CHX particles developed in the present study demonstrate promising antimicrobial effects against both planktonic and biofilm bacteria. Hence, Nano-CHX is effective against major oral pathogens such as *S. mutans*, *S. sobrinus*, *F. nucleatum*, *A. actinomycetemcomitans* and *E. faecalis* at 50–100 µg/mL for their planktonic modes and at 100–200 µg/mL for their mono-species biofilms, respectively. Moreover, our study further shows that the Nano-CHX at 12.5–100 µg/mL can inhibit the development of multi-species oral biofilm up to 72 h, and this finding has been supported by SEM and CLSM observations on the mixed-species biofilms of *S. sobrinus*, *F. nucleatum* and *P. gingivalis*. Technically, the finding from the present release experiment also shows a maximal saturated release of CHX from Nano-CHX at 6 h without consumption, and the anticipated further release of CHX could persist against micro-organisms. Further study is required to confirm this point. It is

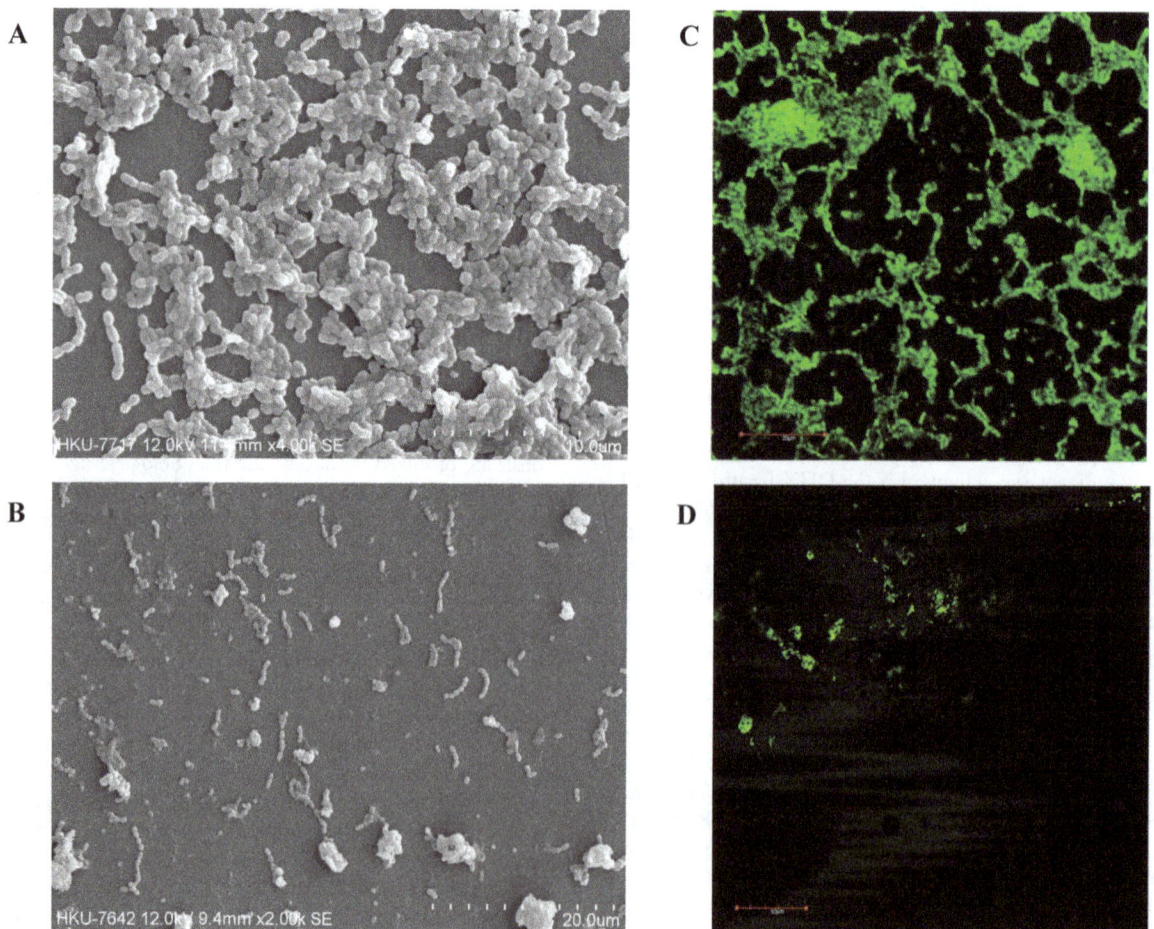

Figure 2. The antibacterial effects of Nano-CHX treatment for 24 h on the mixed-species biofilms of *S. sobrinus*, *F. nucleatum* and *P. gingivalis*. Representative scanning electron microscopy images: B vs. A (blank nanoparticles); and confocal laser scanning microscopy images: D vs. C (blank nanoparticles).

Table 2. The minimal inhibitory concentration (MIC, μg/mL) of the preventive effect of Nano-CHX on the mixed-species oral biofilms in a time-dependent assay from 24 to 72 h.

Mixed-species biofilms	24 h	48 h	72 h
S. mutans, F. nucleatum & P. gingivalis	12.5	50	50
S. sobrinus, F. nucleatum & P. gingivalis	25	50	100
S. mutans, F. nucleatum, A. actinomycetemcomitans & P. gingivalis	25	50	50

worthy to note that in the present study *P. gingivalis* was selected as a pathogen in the three groups of multi-species oral biofilms for the experiments. It is worth noting that *P. gingivalis* has been implicated as a keystone bacteria, through facilitating a critical change in the composition of normal oral microbiota and modulating immuno-inflammatory response, thereby contributing to the development of periodontal disease and tissue destruction [31–33]. Based on the positive results observed, it could be assumed that the Nano-CHX may be able to penetrate effectively the oral biofilms and control the bacteria. Further intensive study is certainly required to determine the exact effects of Nano-CHX on different forms of multi-species biofilms and the underlying mechanisms of antimicrobial actions, with reference to CHX alone. The present 'proof of concept' study on the Nano-CHX may lay a foundation for further *in vitro* and *in vivo* studies on this promising novel antimicrobial agent.

In summary, the mesoporous silica nanoparticle has been effectively loaded with CHX, and its effective release from the generated Nano-CHX has been confirmed. The MIC and MBC of the Nano-CHX have been evaluated on the planktonic mode of pathogenic oral bacteria, mono-species bacterial biofilms and multi-species oral biofilms, respectively. Its anti-biofilm effects have been characterized by SEM and CLSM. In conclusion, this pioneering study demonstrates the potent antibacterial effects of

the Nano-CHX on oral biofilms, and it may be developed as a novel and promising anti-biofilm agent for clinical use.

Supporting Information

Figure S1 The antibacterial effects of Nano-CHX treatment for 24 h on the selected mono-species biofilms. Representative confocal laser scanning microscopy images of blank nanoparticles- and Nano-CHX-treated mono-species biofilms of *Streptococcus mutants* (A vs. D), *Streptococcus sobrinus* (B vs. E) and *Enterococcus faecalis* (C vs. F), respectively.

Acknowledgments

The authors are grateful to Ms. Joanne Yip, Dr. Sarah Wong and Ms. Suhasini Rajan for their technical assistance.

Author Contributions

Conceived and designed the experiments: CJS KCFL PCL LJJ. Performed the experiments: CJS KCFL CHW XL. Analyzed the data: CJS KCFL SFL PCL CBSL EW LJJ. Contributed reagents/materials/analysis tools: CJS KCFL SFL PCL LJJ. Contributed to the writing of the manuscript: CJS KCFL LJJ.

References

1. Marsh PD, Moter A, Devine DA (2011) Dental plaque biofilms: communities, conflict and control. Periodontol 2000 55:16–35.
2. Seneviratne CJ, Zhang CF, Samaranayake LP (2011) Dental plaque biofilm in oral health and disease. Chin J Dent Res 14:87–94.
3. Papas AS, Vollmer WM, Gullion CM, Bader J, Laws R, et al. (2012) Efficacy of chlorhexidine varnish for the prevention of adult caries: a randomized trial. J Dent Res 91:150–5.
4. Varoni E, Tarce M, Lodi G, Carrassi A (2012) Chlorhexidine (CHX) in dentistry: state of the art. Minerva Stomatol 61:399–419.
5. Li L, Pan H, Tao J, Xu X, Mao C, et al. (2008) Repair of enamel by using hydroxyapatite nanoparticles as the building blocks. J Mater Chem 18:4079–84.
6. Barbour ME, Maddocks SE, Wood NJ, Collins AM (2013) Synthesis, characterization, and efficacy of antimicrobial chlorhexidine hexametaphosphate nanoparticles for applications in biomedical materials and consumer products. Int J Nanomed 8:3507–19.
7. Zhu XM, Wang YXJ, Leung KCF, Lee SF, Zhao F, et al. (2012) Enhanced cellular uptake of aminosilane-coated superparamagnetic iron oxide nanoparticles in mammalian cell lines. Int J Nanomed 7:953–64.
8. Nguyen TD, Liu Y, Saha S, Leung KCF, Stoddart JF, et al. (2007) Design and optimization of molecular nanovalves based on redox-switchable bistable rotaxanes. J Am Chem Soc 129:626–34.
9. Lee SF, Zhu XM, Wang YXJ, Xuan S, You Q, et al. (2013) Ultrasound, pH, and magnetically responsive crown-ether-coated core/shell nanoparticles as drug encapsulation and release systems. ACS Appl Mater Interfaces 5:1566–74.
10. Chu CH, Mei L, Seneviratne CJ, Lo EC (2012) Effects of silver diamine fluoride on dentine carious lesions induced by Streptococcus mutans and Actinomyces naeslundii biofilms. Int J Paediatr Dent 22:2–10.
11. Seneviratne CJ, Yip JW, Chang JW, Zhang CF, Samaranayake LP (2013) Effect of culture media and nutrients on biofilm growth kinetics of laboratory and clinical strains of Enterococcus faecalis. Arch Oral Biol 58:1327–34.
12. Milstone AM, Passaretti CL, Perl TM (2008) Chlorhexidine: expanding the armamentarium for infection control and prevention. Clin Infect Dis 46:274–81.
13. Sreenivasan P, Gaffar A (2002) Antiplaque biocides and bacterial resistance: a review. J Clin Periodontol 29:965–74.
14. Xuan S, Lee SF, Lau JTF, Zhu XM, Wang YXJ, et al. (2012) Photocytotoxicity and magnetic relaxivity responses of dual-porous γ-Fe$_2$O$_3$@meso-SiO$_2$ microspheres. ACS Appl Mater Interfaces 4:2033–40.
15. Wang DW, Zhu XM, Lee SF, Chan HM, Li HW, et al. (2013) Folate-conjugated Fe$_3$O$_4$@SiO$_2$@gold nanorods@mesoporous SiO$_2$ hybrid nanomaterial. J Mater Chem B 1:2934–42.
16. Wang HH, Wang YXJ, Leung KCF, Au DWT, Xuan S, et al. (2009) Durable mesenchymal stem cell labeling by using polyhedral superparamagnetic iron oxide nanoparticles. Chem Eur J 15:12417–25.
17. Wang YXJ, Quercy-Jouvet T, Wang HH, Li AW, Chak CP, et al. (2011) Efficacy and durability in direct labeling of mesenchymal stem cells using ultrasmall superparamagnetic iron oxide nanoparticles with organisilica, dextran, and PEG coatings. Materials 4:703–15.
18. Xuan S, Wang F, Gong X, Kong SK, Yu JC, et al. (2011) Hierarchical core/shell Fe$_3$O$_4$@SiO$_2$@γ-AlOOH@Au micro/nanoflowers for protein immobilization. Chem Commun 47:2514–6.
19. Hench LL, Wilson J (1999) Surface-active processes in materials. In: Clarke ED, Simmons J, Folz D eds. Ceramic Transactions, American Ceramic Society, OH, v101.
20. Costerton JW, Stewart PS, Greenberg EP (1999) Bacterial biofilms: a common cause of persistent infections. Science 284:1318–22.
21. Wilson M (1996) Susceptibility of oral bacterial biofilms to antimicrobial agents. J Med Microbiol 44:79–87.
22. Seneviratne CJ, Wang Y, Jin LJ, Wong SS, Herath TD, et al. (2012) Unraveling the resistance of microbial biofilms: has proteomics been helpful? Proteomics 12:651–65.
23. Sandham HJ, Nadeau L, Phillips HI (1992) The effect of chlorhexidine varnish treatment on salivary mutans streptococcal levels in child orthodontic patients. J Dent Res 71:32–5.
24. Goutham BS, Manchanda K, Sarkar AD, Prakash R, Jha K, et al. (2013) Efficacy of two commercially available Oral Rinses - Chlorhexidine and

Listrine on Plaque and Gingivitis - A Comparative Study. J Int Oral Health 5:56–61.

25. Shen Y, Stojicic S, Haapasalo M (2011) Antimicrobial efficacy of chlorhexidine against bacteria in biofilms at different stages of development. J Endod 37:657–61.

26. Yamaguchi M, Noiri Y, Kuboniwa M, Yamamoto R, Asahi Y, et al. (2013) Porphyromonas gingivalis biofilms persist after chlorhexidine treatment. Eur J Oral Sci 121:162–8.

27. García-Caballero L, Quintas V, Prada-López I, Seoane J, Donos N, et al. (2013) Chlorhexidine substantivity on salivary flora and plaque-like biofilm: an in situ model. PLoS One 8:e83522.

28. Cousido MC, Tomás Carmona I, García-Caballero L, Limeres J, Alvarez M, et al. (2010) In vivo substantivity of 0.12% and 0.2% chlorhexidine mouthrinses on salivary bacteria. Clin Oral Investig 14:397–402.

29. Tomás I, García-Caballero L, López-Alvar E, Suárez-Cunqueiro M, Diz P, et al. (2013) In situ chlorhexidine substantivity on saliva and plaque-like biofilm: influence of circadian rhythm. J Periodontol 84:1662–72.

30. Cheng L, Weir MD, Xu HHK, Kraigsley AM, Lin NJ, et al. (2012) Antibacterial and physical properties of calcium-phosphate and calcium-fluoride nanocomposites with chlorhexidine. Dent Mater 28:573–83.

31. Hajishengallis G, Liang S, Payne MA, Hashim A, Jotwani R, et al. (2011) Low-abundance biofilm species orchestrates inflammatory periodontal disease through the commensal microbiota and complement. Cell Host Microbe 10:497–506.

32. Darveau RP, Hajishengallis G, Curtis MA (2012) Porphyromonas gingivalis as a potential community activist for disease. J Dent Res 91:816–20.

33. Herath TD, Darveau RP, Seneviratne CJ, Wang CY, Wang Y, et al. (2013) Tetra- and penta-acylated lipid A structures of Porphyromonas gingivalis LPS differentially activate TLR4-mediated NF-κB signal transduction cascade and immuno-inflammatory response in human gingival fibroblasts. PLoS One 8:e58496.

PERMISSIONS

LIST OF CONTRIBUTORS

Alka Gupta, Murali Gopal and George V. Thomas
Central Plantation Crops Research Institute, Kasaragod, Kerala, India

Vinu Manikandan, Preeti Rajesh and Ravi Gupta
SciGenom Labs Pvt. Ltd., Plot 43A, SDF 3rd Floor CSEZ, Kakkanad, Cochin, Kerala, India

John Gajewski
Center for Comparative Genomics and Bioinformatics, Pennsylvania State University, 310 Wartik Lab, University Park, Pennsylvania, United States of America

George Thomas
SciGenom Research Foundation, Cochin, Kerala, India

Somasekar Seshagiri
Department of Molecular Biology, Genentech Inc., South San Francisco, California, United States of America

Stephan C. Schuster
Center for Comparative Genomics and Bioinformatics, Pennsylvania State University, 310 Wartik Lab, University Park, Pennsylvania, United States of America
Singapore Centre on Environmental Life Sciences Engineering, Nanyang Technical University, Singapore, Singapore

Michael V. Airola, Justin Snider and Yusuf A. Hannun
Department of Medicine and the Stony Brook University Cancer Center, Stony Brook University, Stony Brook, New York, United States of America

Jessica M. Tumolo
Department of Biochemistry and Molecular Biology, Medical University of South Carolina, Charleston, South Carolina, United States of America

Sarah L. Svensson, Mark Pryjma and Erin C. Gaynor
Department of Microbiology and Immunology, University of British Columbia, Vancouver, British Columbia, Canada

Rebekah F. Hare
Department of Biology and Wildlife, Institute of Arctic Biology, University of Alaska Fairbanks, Fairbanks, Alaska, United States of America

Karsten Hueffer
Department of Veterinary Medicine, University of Alaska Fairbanks, Fairbanks, Alaska, United States of America

Florence Reibel, Coralie Pintard, Cecile Gateau, Adrien Launay, Alice Ledda and Olivier Tenaillon
Institut National de la Santéet de la Recherche Médicale (INSERM), Unité Mixte de Recherche (UMR) 1137, Paris, France

Mathilde Lescat
Institut National de la Santéet de la Recherche Médicale (INSERM), UnitéMixte de Recherche (UMR) 1137, Paris, France
Laboratoire de Microbiologie, Hôpital Jean Verdier, Assistance Publique-Hôpitaux de Paris, Bondy, France et Universite´ Paris Nord, Sorbonne Paris Cité, Paris, France

Sara Dion and Jérémy Glodt
Institut National de la Santéet de la Recherche Médicale (INSERM), Unité Mixte de Recherche (UMR) 1137, Paris, France
UMR 1137, Universite´ Paris Diderot, Sorbonne Paris Cité, Paris, France

Jérôme Tourret
Institut National de la Santéet de la Recherche Médicale (INSERM), Unité Mixte de Recherche (UMR) 1137, Paris, France
Département d'Urologie, Néphrologie et Transplantation, Hôpital Pitié-Salpêtriére
Assistance Publique-Hôpitaux de Paris et Université Pierre et Marie Curie, Paris, France

Stephane Cruvellier
Laboratoire de Génomique Comparative, Centre national de la Recherche Scientifique (CNRS) UMR 8030, Institut de Génomique, Commissariat á l'énergie atomique et aux énergies alternatives (CEA), Genoscope, Evry, France

Karola Prior and Dag Harmsen
Department for Periodontology, University of Münster, Münster, Germany

Sebastian Jünemann
Department for Periodontology, University of Münster, Münster, Germany

Andreas Albersmeier and Jörn Kalinowski
Technology Platform Genomics, Center for Biotechnology, Bielefeld University, Bielefeld, Germany

Stefan Albaum
Bioinformatics Resource Facility, Center for Biotechnology, Bielefeld University, Bielefeld, Germany

Alexander Goesmann
Bioinformatics and Systems Biology, Justus-Liebig-Univeristy Gießen, Gießen, Germany

Jens Stoye
Institute for Bioinformatics, Center for Biotechnology, Bielefeld University, Bielefeld, Germany
Genome Informatics Group, Faculty of Technology, Bielefeld University, Bielefeld, Germany

Justin L. Eddy, Lindsay M. Gielda, Adam J. Caulfield, Stephanie M. Rangel and Wyndham W. Lathem
Department of Microbiology-Immunology, Northwestern University Feinberg School of Medicine, Chicago, Illinois, United States of America

Sayanti Brahmachari, Subhra Kanti Mandal and Prasanta Kumar Das
Department of Biological Chemistry, Indian Association for the Cultivation of Science, Kolkata, India

Richard W. Hammack
National Energy Technology Laboratory, Pittsburgh, Pennsylvania, United States of America

Arvind Murali Mohan and Kelvin B. Gregory
National Energy Technology Laboratory, Pittsburgh, Pennsylvania, United States of America
Department of Civil and Environmental Engineering, Carnegie Mellon
University, Pittsburgh, Pennsylvania, United States of America

Kyle J. Bibby
National Energy Technology Laboratory, Pittsburgh, Pennsylvania, United States of America
Department of Civil and Environmental Engineering, University of Pittsburgh, Pittsburgh, Pennsylvania, United States of America
Department of Computational and Systems Biology, University of Pittsburgh Medical School, Pittsburgh, Pennsylvania, United States of America

Daniel Lipus
National Energy Technology Laboratory, Pittsburgh, Pennsylvania, United States of America
Department of Civil and Environmental Engineering, University of Pittsburgh, Pittsburgh, Pennsylvania, United States of America

Raymond N. Allan
Academic Unit of Clinical and Experimental Sciences, Faculty of Medicine and Institute for Life Sciences, University of Southampton, Southampton, United Kingdom
Southampton NIHR Wellcome Trust Clinical Research Facility, University Hospital Southampton NHS Foundation Trust, Southampton, United Kingdom

Johanna Jefferies
Academic Unit of Clinical and Experimental Sciences, Faculty of Medicine and Institute for Life Sciences, University of Southampton, Southampton, United Kingdom,
Public Health England, Southampton, United Kingdom

Saul N. Faust
Academic Unit of Clinical and Experimental Sciences, Faculty of Medicine and Institute for Life Sciences, University of Southampton, Southampton, United Kingdom
Southampton NIHR Wellcome Trust Clinical Research Facility, University Hospital Southampton NHS Foundation Trust, Southampton, United Kingdom
Southampton NIHR Respiratory Biomedical Research Unit, University Hospital Southampton NHS Foundation Trust, Southampton, United Kingdom

Luanne Hall-Stoodley
Academic Unit of Clinical and Experimental Sciences, Faculty of Medicine and Institute for Life Sciences, University of Southampton, Southampton, United Kingdom
Southampton NIHR Wellcome Trust Clinical Research Facility, University Hospital Southampton NHS Foundation Trust, Southampton, United Kingdom

Microbial Infection and Immunity, Centre for Microbial Interface Biology, The Ohio State University, Columbus, Ohio, United States of America

Stuart C. Clarke
Academic Unit of Clinical and Experimental Sciences, Faculty of Medicine and Institute for Life Sciences, University of Southampton, Southampton, United Kingdom,
Public Health England, Southampton, United Kingdom
Southampton NIHR Respiratory Biomedical Research Unit, University Hospital Southampton NHS Foundation Trust, Southampton, United Kingdom

Paul Skipp
Centre for Biological Sciences, University of Southampton, Southampton, United Kingdom
Centre for Proteomic Research, Institute for Life Sciences, University of Southampton, Southampton, United Kingdom

Jeremy Webb
Centre for Biological Sciences, University of Southampton, Southampton, United Kingdom
Southampton NIHR Respiratory Biomedical Research Unit, University Hospital Southampton NHS Foundation Trust, Southampton, United Kingdom

Francisco Pérez-Montaño, Irene Jiménez-Guerrero, Pablo Del Cerro, Francisco Javier López-Baena, Francisco Javier Ollero, Ramón Bellogín, Javier Lloret and Rosario Espuny
Departamento de Microbiología, Facultad de Biología, Universidad de Sevilla, Sevilla, Spain

Irene Baena-Ropero
Departamento de Biología, Facultad de Ciencias, Universidad Autónoma
de Madrid, Madrid, Spain

Marco Plomp and Alexander J. Malkin
Biosciences and Biotechnology Division, Physical and Life Sciences Directorate, Lawrence Livermore National Laboratory, Livermore, California, United States of America

Alicia Monroe Carroll and Peter Setlow
Department of Molecular Biology and Biophysics, University of Connecticut Health Center, Farmington, Connecticut, United States of America

Hugo Oliveira, Sanna Sillankorva, Leon D. Kluskens and Joana Azeredo
Centre of Biological Engineering, University of Minho, Braga, Portugal

Viruthachalam Thiagarajan
School of Chemistry, Bharathidasan University, Tiruchirappalli, India

Maarten Walmagh and Rob Lavigne
Laboratory of Gene Technology, Katholieke Universiteit Leuven, Leuven, Belgium

Maria Teresa Neves-Petersen
Nanomedicine Department, International Iberian Nanotechnology Laboratory, Braga, Portugal
Medical Faculty, Aalborg University, Aalborg, Denmark

Lili He
Department of Pathogenic Biology, School of Basic Medical Sciences, Lanzhou University, Lanzhou, China

Jian Han
Department of Pathogenic Biology, School of Basic Medical Sciences, Lanzhou University, Lanzhou, China
Department of Molecular Microbiology and Immunology, Bloomberg School of Public Health, Johns Hopkins University, Baltimore, Maryland, United States of America

Wanliang Shi and Shuo Zhang
Department of Molecular Microbiology and Immunology, Bloomberg School of Public Health, Johns Hopkins University, Baltimore, Maryland, United States of America

Ying Zhang
Department of Molecular Microbiology and Immunology, Bloomberg School of Public Health, Johns Hopkins University, Baltimore,Maryland, United States of America
Department of Infectious Diseases, Huashan Hospital, Fudan University, Shanghai, China

Xiaogang Xu
Institute of Antibiotics, Huashan Hospital, Fudan University, Shanghai, China

Sen Wang
Department of Infectious Diseases, Huashan Hospital, Fudan University, Shanghai, China

Martha E. Trujillo, Rodrigo Bacigalupe, Patricia Benito and Raú l Riesco
Departamento de Microbiología y Genética, Edificio Departamental, Campus Miguel de Unamuno, Universidad de Salamanca, Salamanca, Spain

Petar Pujic and Philippe Normand
Université Lyon 1, Université de Lyon, CNRS-UMR5557 Ecologie Microbienne, Villeurbanne, France

Yasuhiro Igarashi
Biotechnology Research Center, Toyama Prefectural University, Kurokawa, Imizu, Toyama, Japan

Claudine Médigue
Genoscope, CNRS-UMR 8030, Atelier de Génomique Comparative, Evry, France

Rory A. Eutsey, Margaret E. Dahlgren, Joshua P. Earl, Evan Powell and Azad Ahmed
Center for Genomic Sciences, Allegheny-Singer Research Institute, Pittsburgh, Pennsylvania, United States of America

Benjamin A. Janto
Center for Genomic Sciences, Allegheny-Singer Research Institute,
Pittsburgh, Pennsylvania, United States of America
Department of Microbiology and Immunology, Drexel University College of Medicine, Allegheny Campus, Pittsburgh, Pennsylvania, United States of America

Fen Z. Hu and Garth D. Ehrlich
Center for Genomic Sciences, Allegheny-Singer Research Institute, Pittsburgh, Pennsylvania, United States of America
Department of Microbiology and Immunology, Drexel University College of Medicine, Allegheny Campus, Pittsburgh, Pennsylvania, United States of America
Department of Otolaryngology Head and Neck Surgery, Drexel University College of Medicine, Allegheny Campus, Pittsburgh, Pennsylvania, United States of America

N. Luisa Hiller
Center for Genomic Sciences, Allegheny-Singer Research Institute, Pittsburgh, Pennsylvania, United States of America
Department of Biological Sciences, Carnegie Mellon University, Pittsburgh, Pennsylvania, United States of America

Hannah Lees, Jeremy K. Nicholson, Elaine Holmes and Ian D. Wilson
Section of Computational and Systems Medicine, Department of Surgery and Cancer, Faculty of Medicine, Imperial College London, London, United Kingdom

Jonathan Swann
Department of Food and Nutritional Sciences, School of Chemistry, Food and Pharmacy, University of Reading, Reading, United Kingdom

Simon M. Poucher
Cardiovascular and Gastro- Intestinal Disorders Innovative Medicines, AstraZeneca Pharmaceuticals, Alderley Park, Cheshire, United Kingdom

Julian R. Marchesi
School of Biosciences, Cardiff University, Cardiff, United Kingdom, 5 Centre for Digestive and Gut Health, Imperial College London, London, United Kingdom

Xuan Li and Lijian Jin
Faculty of Dentistry, The University of Hong Kong, Hong Kong SAR, China

Chaminda Jayampath Seneviratne
Faculty of Dentistry, The University of Hong Kong, Hong Kong SAR, China
Faculty of Dentistry, National University of Singapore, Singapore, Singapore

Ken Cham-Fai Leung and Chi-Hin Wong
Department of Chemistry, Institute of Creativity, and Partner State Key Laboratory of Environmental & Biological Analysis, The Hong Kong Baptist University, Hong Kong SAR, China

Siu-Fung Lee
Department of Chemistry, The Chinese University of Hong Kong, Hong Kong SAR, China

Ping Chung Leung, Clara Bik San Lau and Elaine Wat
Institute of Chinese Medicine and Partner State Key Laboratory of Phytochemistry & Plant Resources in West China, The Chinese University of Hong Kong, Hong Kong SAR, China

Index